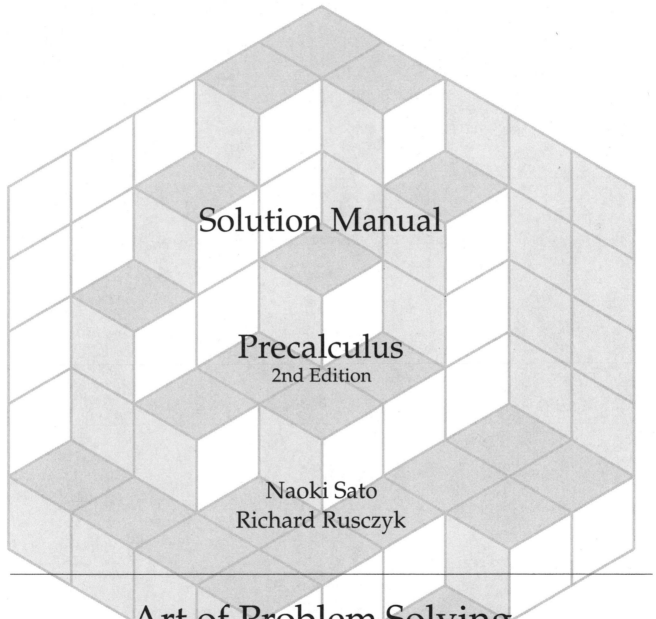

Solution Manual

Precalculus
2nd Edition

Naoki Sato
Richard Rusczyk

Art of Problem Solving

Books • Online Classes • Videos • Interactive Resources

www.artofproblemsolving.com

Published by: AoPS Incorporated
                 10865 Rancho Bernardo Rd Ste 100
                 San Diego, CA 92127-2102
                 books@artofproblemsolving.com

ISBN #: 978-1-934124-27-7

Visit the Art of Problem Solving website at http://www.artofproblemsolving.com

Cover image designed by Vanessa Rusczyk using KaleidoTile software.

Printed in the United States of America

Second Edition 2019

Contents

CHAPTER 1

Functions Review

Exercises for Section 1.1

1.1.1

(a) $f(2) = 2^2 - 2 - 6 = \boxed{-4}$.

(b) Since $f(x) = x^2 - x - 6 = (x-3)(x+2)$, we have $f(x) = 0$ when $\boxed{x = 3 \text{ and } x = -2}$.

(c) $f(x-1) = (x-1)^2 - (x-1) - 6 = x^2 - 2x + 1 - x + 1 - 6 = \boxed{x^2 - 3x - 4}$.

(d) If $f(x) = 6$, then $x^2 - x - 6 = 6$, so $x^2 - x - 12 = 0$. Factoring gives $(x-4)(x+3) = 0$, which means $x = 4$ or $x = -3$. Hence, $f(4) = f(-3) = 6$, so $\boxed{\text{yes}}$, we can have $f(x) = 6$.

(e) Completing the square, we get $f(x) = x^2 - x - 6 = \left(x - \dfrac{1}{2}\right)^2 - \dfrac{25}{4}$. Hence, the range of f is $\boxed{[-25/4, +\infty)}$.

1.1.2

(a) We can evaluate the absolute value of $1 - x$ for any real number x, so the domain of f is $\boxed{\mathbb{R}}$. The absolute value of a real number is always nonnegative, and $1 - x$ can equal any nonnegative real number, so the range of f is $\boxed{[0, +\infty)}$.

(b) We can only take the square root of a nonnegative real number, so $2 - t$ must be nonnegative. Hence, the domain of f is $\boxed{(-\infty, 2]}$. Also, the square root of a nonnegative real number is always nonnegative, so the range of f is $\boxed{[0, +\infty)}$.

(c) We can evaluate $1 - x^2$ for any real number x, so the domain of h is $\boxed{\mathbb{R}}$. Since x^2 is always nonnegative, $1 - x^2$ is always less than or equal to 1, so the range of h is $\boxed{(-\infty, 1]}$.

(d) First, we note that we cannot have $u = 0$, since this makes the denominator of $1/u$ equal to 0. We also cannot have the denominator of the whole function, $1 + 1/u$, equal to 0. If $1 + 1/u = 0$, then $u = -1$, so the domain of g is $\boxed{\text{all real numbers except } 0 \text{ and } -1}$.

To find the range of g, let $v = g(u) = \frac{1}{1+1/u}$. Taking the reciprocal of both sides and rearranging gives

$$1 + \frac{1}{u} = \frac{1}{v} \quad \Rightarrow \quad \frac{1}{u} = \frac{1}{v} - 1 = \frac{1-v}{v} \quad \Rightarrow \quad u = \frac{v}{1-v}.$$

Hence, for every value of v, there is a corresponding value of u, except when $v = 1$. Furthermore, as determined above, u cannot be equal to 0 or -1. There is no real number v that corresponds to $u = -1$, but the real number $v = 0$ corresponds to $u = 0$, so v cannot be equal to 0 either. Therefore, the range of g is all real numbers except 0 and 1 .

1.1.3 The function $(f \cdot g)(x) = \sqrt{4-x} \cdot \sqrt{2x-6}$ is defined only where both the functions $f(x) = \sqrt{4-x}$ and $g(x) = \sqrt{2x-6}$ are defined. The function $\sqrt{4-x}$ is defined only for $x \le 4$, and the function $\sqrt{2x-6}$ is defined only for $x \ge 3$. Hence, the domain of the function $f \cdot g$ is $[3, 4]$.

As with $f \cdot g$, the function

$$\left(\frac{f}{g}\right)(x) = \frac{\sqrt{4-x}}{\sqrt{2x-6}}$$

is defined only where both the functions $f(x) = \sqrt{4-x}$ and $g(x) = \sqrt{2x-6}$ are defined. However, we must also exclude points where the denominator is 0. In this case, the denominator is 0 when $x = 3$. Hence, the domain of the function f/g is $(3, 4]$.

The domains of $f \cdot g$ and f/g are different because we must exclude those values where $g(x) = 0$ from the domain of f/g, but we include these values in the domain of $f \cdot g$ if they are also in the domain of f.

1.1.4 The function

$$f(x) = \sqrt{\frac{2x-5}{x-8}}$$

is defined only where $(2x-5)/(x-8) \ge 0$. For this inequality to hold, either both $2x-5$ and $x-8$ must be positive, both of them must be negative, or $2x-5 = 0$.

If both $2x-5$ and $x-8$ are positive, then $x > 5/2$ and $x > 8$, so in this case, we get $x > 8$. If both $2x-5$ and $x-8$ are negative, then $x < 5/2$ and $x < 8$, so in this case, we get $x < 5/2$. Finally, if $2x-5 = 0$, then $x = 5/2$. Therefore, the set of x such that $(2x-5)/(x-8) \ge 0$ is $(-\infty, 5/2] \cup (8, +\infty)$, which is the domain of f.

On the other hand, the function

$$g(x) = \frac{\sqrt{2x-5}}{\sqrt{x-8}}$$

is defined only where both the functions $\sqrt{2x-5}$ and $\sqrt{x-8}$ are defined, and the denominator is nonzero. The function $\sqrt{2x-5}$ is defined for $x > 5/2$, and the function $\sqrt{x-8}$ is defined for $x > 8$. The denominator is zero when $x = 8$, so the domain of g is $(8, +\infty)$.

Thus, the functions f and g have different domains , because those values of x for which $2x-5$ and $x-8$ are negative are in the domain of f, but not in the domain of g.

1.1.5 $f(x) \cdot f(-x) = \dfrac{x+1}{x-1} \cdot \dfrac{-x+1}{-x-1} = \dfrac{x+1}{x-1} \cdot \dfrac{x-1}{x+1} = \boxed{1}$, for $|x| \ne 1$.

1.1.6 We break up the interval $[-3, 4]$ into the intervals $[-3, -2]$ and $[-2, 4]$. If $-3 \le x \le -2$, then $-1 \le x+2 \le 0$, so $0 \le (x+2)^2 \le 1$. If $-2 \le x \le 4$, then $0 \le x+2 \le 6$, so $0 \le (x+2)^2 \le 36$. Hence, every value in the range of g is in the interval $[0, 36]$.

Intuitively, it seems clear that every value in this interval is in the range. To prove it we let $y = (x+2)^2$ and solve for x. Taking the square root of both sides gives $x+2 = \pm\sqrt{y}$, so $x = \pm\sqrt{y} - 2$. Consider the solution $x = \sqrt{y} - 2$. For any y such that $0 \le y \le 36$, we have $0 \le \sqrt{y} \le 6$, so $-2 \le \sqrt{y} - 2 \le 4$. Therefore, the number $\sqrt{y} - 2$ is in the domain of g if $0 \le y \le 36$. So, if we let $x = \sqrt{y} - 2$ for any y such that $0 \le y \le 36$, we have $f(x) = y$. Therefore, every value in $[0, 36]$ is in the range, and our range is $[0, 36]$.

1.1.7

(a) We have $T(2, 3, -5) = 3 \cdot 2^3 - (-5) = 3 \cdot 8 + 5 = \boxed{29}$.

(b) Since $T(x, 2, 6) = 3x^2 - 6$, we have $3x^2 - 6 = 21$. Isolating x^2 gives $x^2 = 9$, from which we find $x = \boxed{\pm 3}$.

Exercises for Section 1.2

1.2.1

(a) If $3x - 7 = 0$, then $x = 7/3$, so the x-intercept is $\boxed{(7/3, 0)}$. Since $f(0) = -7$, the y-intercept is $\boxed{(0, -7)}$.

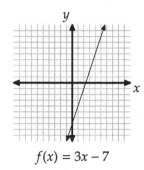

$f(x) = 3x - 7$

(b) If $2 - |x| = 0$, then $|x| = 2$, so $x = \pm 2$. Hence, the x-intercepts are $\boxed{(2, 0) \text{ and } (-2, 0)}$. Since $f(0) = 2$, the y-intercept is $\boxed{(0, 2)}$.

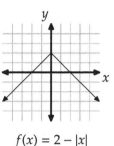

$f(x) = 2 - |x|$

(c) Note that $f(x) = x^2 - 5x + 6$ factors as $(x - 2)(x - 3)$. If $(x - 2)(x - 3) = 0$, then $x = 2$ or $x = 3$, so the x-intercepts are $\boxed{(2, 0) \text{ and } (3, 0)}$. Since $f(0) = 6$, the y-intercept is $\boxed{(0, 6)}$.

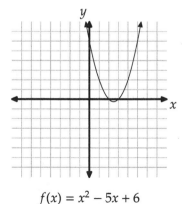

$f(x) = x^2 - 5x + 6$

1.2.2

(a) Every vertical line intersects the graph at most once, so the graph represents a function.

(b) Every vertical line intersects the graph at most once, so the graph represents a function. Note that every vertical line does not need to intersect the graph, such as the line $x = 0$.

(c) There are many vertical lines that intersect the graph more than once, such as the line $x = 1$, so the graph does not represent a function.

1.2.3

(a) All we know about the function f is that $f(9) = 2$. Hence, if we set $x - 3 = 9$, then we have $x = 12$, and $y = f(12 - 3) + 5 = f(9) + 5 = 7$. Therefore, a point on the graph of $y = f(x - 3) + 5$ is $\boxed{(12, 7)}$. We can also find this point by noting that the graph of $y = f(x - 3) + 5$ results from shifting the graph of $y = f(x)$ to the right by 3 and up by 5. So, because $(9, 2)$ is on the graph of $y = f(x)$, the point $(9 + 3, 2 + 5) = (12, 7)$ is on the graph of $y = f(x - 3) + 5$.

(b) If we set $x/4 = 9$, then $x = 36$, and $y = 2f(9) = 4$. Therefore, a point on the graph of $y = 2f(x/4)$ is $\boxed{(36, 4)}$. We can also see this by noting that the graph of $y = 2f(x/4)$ results from scaling the graph of $y = f(x)$ vertically by a factor of 2 and horizontally by a factor of 4. So, because $(9, 2)$ is on the graph of $y = f(x)$, the point $(9 \cdot 4, 2 \cdot 2) = (36, 4)$ on the graph of $y = 2f(x/4)$.

(c) If we set $3x - 1 = 9$, then $x = 10/3$, and $y = 2f(9) + 7 = 11$. Therefore, a point on the graph of $y = 2f(3x - 1) + 7$ is $\boxed{(10/3, 11)}$.

1.2.4 The graph of $y = f(2x)$ is the same as the graph of $y = f(x)$ compressed by a factor of 2 horizontally, and the graph of $y = f(x - 6)$ is the same as the graph of $y = f(x)$ shifted 6 units to the right.

Knowing these results, we may think that the graph of $y = f(2x - 6)$ results from compressing the graph of $y = f(x)$ by a factor of 2 horizontally to give the graph of $y = f(2x)$, and then shifting the result 6 units to the right. However, this is not true. The first step is correct; compressing the graph of $y = f(x)$ by a factor of two horizontally gives the graph of $y = f(2x)$. To see how we produce the graph of $y = f(2x - 6)$, we let $h(x) = f(2x)$. Then, the graph of $y = h(x)$ is the same as the graph of $y = f(2x)$. To get the graph of $y = f(2x - 6)$, we note that $h(x - 3) = f(2(x - 3)) = f(2x - 6)$. The graph of $y = h(x - 3)$, which is the same as the graph of $y = f(2x - 6)$, is the result of shifting the graph of $y = h(x)$, which is the same as the graph of $y = f(2x)$, by 3 units to the right.

Therefore, the answer is (c). The graph of $y = f(2x - 6)$ is the same as the graph of $y = f(2x)$ shifted 3 units to the right.

Exercises for Section 1.3

1.3.1 If $g(x) = 3x + 7$, then $g(g(x)) = g(3x + 7) = 3(3x + 7) + 7 = \boxed{9x + 28}$. If $g(x) = -3x - 14$, then $g(g(x)) = g(-3x - 14) = -3(-3x - 14) - 14 = \boxed{9x + 28}$. Hence, both solutions work.

1.3.2

(a) First, we note that the domain of g is $(-\infty, 1]$. Since the domain of f is all reals, all outputs of g are in the domain of f. For $x \le 1$, we therefore have $f(g(x)) = f(\sqrt{1 - x}) = \boxed{1 - x - 2\sqrt{1 - x}, \text{ where } x \le 1}$.

(b) The domain of $f(g(x))$ is the set of values of x such that both $g(x)$ is defined, and $g(x)$ lies in the domain of $f(x)$. From $g(x) = \sqrt{1 - x}$, we see that $g(x)$ is defined only for $x \le 1$. The domain of $f(x)$ consists of all real numbers, so the domain of $f(g(x))$ is $\boxed{(-\infty, 1]}$.

(c) First, $f(-2) = (-2)^2 - 2 \cdot (-2) = 8$. Since $g(x)$ is only defined for $x \le 1$, we see that $g(f(-2))$ is $\boxed{\text{not defined}}$.

1.3.3 We show that $f^3(x)$ is the same as $f^2(f(x))$ by expanding both expressions. By definition, $f^3(x) = f(f(f(x)))$. Since $f^2(x) = f(f(x))$, we may substitute to get $f^2(f(x)) = f(f(f(x)))$, so $f^2(f(x)) = f^3(x)$. Similarly, $f(f^2(x)) = f(f(f(x)))$, so $f^3(x)$ and $f(f^2(x))$ are the same, as well.

1.3.4 To get a feel for the problem, we compute $f^n(x)$ for the first few values of n:

$$f^1(x) = f(x) = ax,$$
$$f^2(x) = f(f(x)) = f(ax) = a^2 x,$$
$$f^3(x) = f(f^2(x)) = f(a^2 x) = a^3 x,$$
$$f^4(x) = f(f^3(x)) = f(a^3 x) = a^4 x,$$

and so on.

Applying the function f to x multiplies it by a. Therefore, applying the function f to x exactly n times multiplies it by a exactly n times; in other words, it multiplies it by a^n. Hence, $f^n(x) = \boxed{a^n x}$.

1.3.5 All the statement "(3, 0) is the only x-intercept of the graph of g" tells us is that $g(3) = 0$. So, we have $h(3) = f(g(3)) = f(0)$. This tells us that $(3, f(0))$ is on the graph of $y = h(x)$, but this doesn't tell us anything about the intercepts of h. We cannot conclude anything about the intercepts of h from the given information.

Exercises for Section 1.4

1.4.1 Let $y = f^{-1}(3)$. So, we have $f(y) = f(f^{-1}(3)) = 3$. This means we have

$$\frac{y-1}{y-2} = 3 \quad \Rightarrow \quad y - 1 = 3(y-2) = 3y - 6 \quad \Rightarrow \quad y = \boxed{\frac{5}{2}}.$$

1.4.2

(a) Let $x = f(y) = 2y - 7$. Then $y = (x+7)/2$, so $f^{-1}(x) = \boxed{(x+7)/2}$.

(b) Let $x = f(y) = \frac{1}{2y+3}$. Then $2y + 3 = \frac{1}{x}$, so $y = (\frac{1}{x} - 3)/2 = (1 - 3x)/(2x)$. Therefore, $f^{-1}(x) = \boxed{\dfrac{1-3x}{2x}}$.

(c) Let $x = f(y) = \sqrt{2-y}$. Then $x^2 = 2 - y$, so $y = 2 - x^2$. Therefore, $f^{-1}(x) = 2 - x^2$. Note that since the range of $f(x)$ is the set of all nonnegative real numbers, the domain of f^{-1} is also all nonnegative real numbers. So, we have $\boxed{f^{-1}(x) = 2 - x^2, \text{ where the domain of } f^{-1} \text{ is } [0, +\infty)}$.

(d) Since $f(4) = |5 - 4| = 1$ and $f(6) = |5 - 6| = 1$, there are two different values of x that give the same value of $f(x)$. Therefore, the function $f(x)$ $\boxed{\text{does not have an inverse}}$.

(e) Since $f(2) = \sqrt{4-2} + \sqrt{2-2} = \sqrt{2}$ and $f(4) = \sqrt{4-4} + \sqrt{4-2} = \sqrt{2}$, there are two different values of x that give the same value of $f(x)$. Therefore, the function $f(x)$ $\boxed{\text{does not have an inverse}}$.

(f) Completing the square gives $f(x) = (x - 3)^2 - 6$. We might at first think that this function does not have an inverse, since $f(1) = f(5)$. However, $f(x)$ is only defined for $x \geq 4$. By considering the graph of $f(x)$, or by noting that the function $f(x)$ strictly increases as x increases for $x \geq 4$, we see that no two values of x in the domain of f give the same output from f. In particular, $f(4) = -5$, so the range of f is $[-5, +\infty)$. So, f does have an inverse. But what is it?

To find the inverse of f, we solve $x = f(y) = (y - 3)^2 - 6$ for y in terms of x. Adding 6 to both sides gives $(y - 3)^2 = x + 6$. Since x is in the range of f, we have $x \geq -5$, so $x + 6 \geq 1$. Therefore, we take the positive square root to find $y = \sqrt{x+6} + 3$. Hence, $\boxed{f^{-1}(x) = \sqrt{x+6} + 3, \text{ where the domain of } f^{-1} \text{ is } [-5, +\infty)}$. Note that $f^{-1}(x)$ is defined only for $x \geq -5$, because this is the range of f.

1.4.3 To find $f^{-1}(x)$, let $x = f(y) = ay + b$, so $y = (x - b)/a$. Then the equation $f(x) = f^{-1}(x)$ becomes

$$ax + b = \frac{x-b}{a} = \frac{x}{a} - \frac{b}{a}.$$

This holds for all x if and only if $a = 1/a$ and $b = -b/a$. From the first equation, we have $a^2 = 1$, so $a = 1$ or $a = -1$.

If $a = 1$, then the second equation becomes $b = -b$, so $b = 0$. If $a = -1$, then the second equation becomes $b = b$, which is true for any real number b. Therefore, the ordered pairs (a, b) that satisfy the problem are $\boxed{(1, 0) \text{ and } (-1, b) \text{ for any any real number } b}$.

We can check these solutions quickly. For $(a,b) = (1,0)$, we have $f(x) = x$, so $f(f(x)) = f(x) = x$, which means $f^{-1}(x) = f(x)$. If $(a,b) = (-1,b)$, then $f(x) = -x + b$, so $f(f(x)) = f(-x + b) = -(-x + b) + b = x$, and again we have $f^{-1}(x) = f(x)$.

1.4.4 First, we find $g^{-1}(x)$. Let $x = g(y) = (ay + b)/(cy + d)$. Then

$$x(cy + d) = ay + b \quad \Rightarrow \quad cxy + dx = ay + b \quad \Rightarrow \quad cxy - ay = b - dx \quad \Rightarrow \quad y(cx - a) = b - dx$$

$$\Rightarrow \quad y = \frac{b - dx}{cx - a}.$$

Therefore, we have $g^{-1}(x) = \dfrac{b - dx}{cx - a}$. We see that $g^{-1}(x)$ is defined for all x except when $cx - a = 0$, or $x = \boxed{a/c}$. (Note that we must have $ad \neq bc$ because if $ad = bc$, then $g(x)$ is a constant for all x such that $cx + d \neq 0$, which means that g does not have an inverse.)

Review Problems

1.23

(a) Since $A(x) = 4x^2 + 1$ is defined for all real numbers, the domain of A is $\boxed{\mathbb{R}}$. Since $4x^2$ can equal any nonnegative real number, but no negative number, the range of $A(x) = 4x^2 + 1$ is $\boxed{[1, +\infty)}$.

(b) The function $o(x) = 3 + \sqrt{16 - (x - 3)^2}$ is defined only when $(x - 3)^2 \leq 16$, so $x - 3$ must be between -4 and 4. In other words, $-4 \leq x - 3 \leq 4$, so $-1 \leq x \leq 7$. Therefore, the domain of o is $\boxed{[-1, 7]}$.

As x varies from -1 to 3, $\sqrt{16 - (x - 3)^2}$ varies from 0 to 4, so $o(x) = 3 + \sqrt{16 - (x - 3)^2}$ varies from 3 to 7. As x varies from 3 to 7, $\sqrt{16 - (x - 3)^2}$ varies from 4 to 0, so $o(x) = 3 + \sqrt{16 - (x - 3)^2}$ varies from 7 to 3. Therefore, the range of o is $\boxed{[3, 7]}$.

(c) The function $P(x) = 1/(3 + \sqrt{x + 1})$ is defined if and only if $x + 1 \geq 0$, or $x \geq -1$. Therefore, the domain of P is $\boxed{[-1, +\infty)}$.

Since $\sqrt{x + 1}$ can take on any nonnegative value, we see that $P(x) = 1/(3 + \sqrt{x + 1})$ varies from $1/3$ to 0 (without ever reaching 0). Therefore, the range of P is $\boxed{(0, 1/3]}$.

(d) The function $S(x) = (12x - 9)/(6 - 9x) = (4x - 3)/(2 - 3x)$ is defined for all x except when the denominator $2 - 3x$ is zero. We have $2 - 3x = 0$ when $x = 2/3$, so the domain of S is $\boxed{\text{all reals except } 2/3}$.

To find the range of S, we let $y = S(x) = (4x - 3)/(2 - 3x)$. Then we have

$$y(2 - 3x) = 4x - 3 \quad \Rightarrow \quad 2y - 3xy = 4x - 3 \quad \Rightarrow \quad 3xy + 4x = 2y + 3 \quad \Rightarrow \quad x = \frac{2y + 3}{3y + 4}.$$

Hence, for every value of y except $y = -4/3$, there is a corresponding value of x such that $S(x) = y$. Therefore, the range of S is $\boxed{\text{all reals except } -4/3}$.

1.24

(a) The function $\frac{1}{\sqrt{2x-5}}$ is defined only when $2x - 5 > 0$, or $x > 5/2$, and the function $\sqrt{9 - 3x}$ is defined only when $9 - 3x \geq 0$, or $x \leq 3$. Combining these, the domain of f is $\boxed{(5/2, 3]}$.

(b) The function $|\sqrt{x} - 2|$ is defined only when $x \geq 0$ and the function $|\sqrt{x - 2}|$ is defined only when $x - 2 \geq 0$, or $x \geq 2$. Therefore, the domain of f is $\boxed{[2, +\infty)}$.

(c)   The function $\sqrt{|x| - 2}$ is defined only when $|x| \geq 2$. This is equivalent to $x \leq -2$ or $x \geq 2$. The function $\sqrt{|x - 3|}$ is defined for all x, since $|x - 3|$ is always nonnegative. Therefore, the domain of g is $\boxed{(-\infty, -2] \cup [2, +\infty)}$.

1.25 $f(g(x)) = f\left(\dfrac{2x}{x+4}\right) = \dfrac{4 \cdot \frac{2x}{x+4}}{\frac{2x}{x+4} + 2} = \dfrac{8x}{2x + 2(x+4)} = \dfrac{8x}{4x + 8} = \boxed{\dfrac{2x}{x+2}}$, where $x \neq 4$, because -4 is not in the domain of g.

1.26   The equation $f(1) = g(1) + 2$ gives us $a + b + c = a - b + c + 2$, so $b = 1$. Then the equation $f(2) = 2$ gives $4a + 2b + c = 2$, so $4a + c = 2 - 2b = 0$. Therefore, $g(2) = 4a - 2b + c = (4a + c) - 2b = 0 - 2 = \boxed{-2}$.

1.27   For part (a), we shift the graph 3 units to the right. For part (b), we scale the graph vertically away from the x-axis by a factor of 2, then shift the graph 1 unit upwards. For part (c), we reflect the graph over the y-axis.

For part (d), we reflect the graph over the y-axis, then shift the graph 2 units to the right. To see why we shift to the right rather than the left, let $h(x) = f(-x)$. Then, the graph of $y = f(2 - x)$ is the same as the graph of $y = h(x - 2)$. The graph of $y = h(x - 2)$ is the result of shifting the graph of $y = h(x)$ to the right 2 units. So, the graph of $y = f(2 - x)$ is the result of shifting the graph of $y = f(-x)$ to the right 2 units.

For part (e), we compress the graph horizontally by a factor of $1/2$, then shift the graph $1/2$ unit to the right. As with part (d), we can see that this shift is $1/2$ unit by letting $h(x) = f(2x)$; therefore, $h(x - \frac{1}{2}) = f(2x - 1)$, so we must shift the graph of $h(x)$, which is the same as the graph of $y = f(2x)$, to the right $1/2$ unit to get the graph of $y = f(2x - 1)$. We then compress the graph of $y = f(2x - 1)$ vertically by a factor of $1/2$, then shift the resulting graph 3 units upwards.

The results for all five parts are shown below.

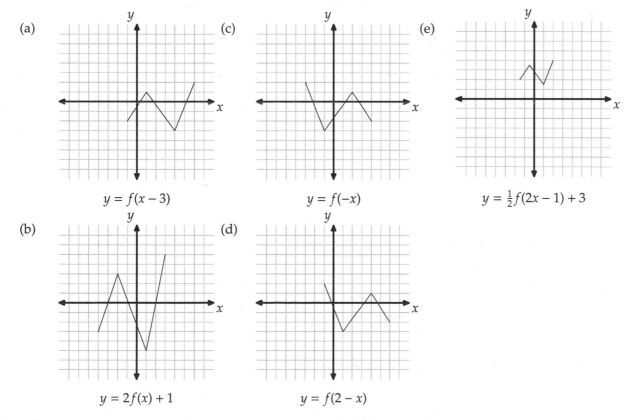

(a) $y = f(x - 3)$

(c) $y = f(-x)$

(e) $y = \frac{1}{2}f(2x - 1) + 3$

(b) $y = 2f(x) + 1$

(d) $y = f(2 - x)$

1.28   The functions f and g need not be the same. For example, let $f(x) = |x|$ and $g(x) = x$. Then $f(f(x)) = |f(x)| = ||x|| = |x|$, and $g(f(x)) = g(|x|) = |x|$. So, we have $f(f(x)) = g(f(x))$, but $f(x) \neq g(x)$.

1.29

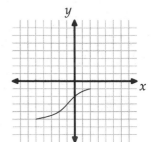

(a) Every vertical line intersects the graph at most once, so the graph can represent a function. We also see that every horizontal line intersects the graph at most once, so the function has an inverse, the graph of which is shown at right. (It is simply the graph in the problem reflected over the line $y = x$.)

(b) There are vertical lines that intersect the graph more than once, so the graph cannot represent a function.

(c) Every vertical line intersects the graph at most once, so the graph can represent a function. However, there are horizontal lines that intersect the graph more than once, so the function does not have an inverse.

1.30 From $f(a, b, c) = 1$, we have $\frac{a-c}{b-c} = 1$, so $a - c = b - c$, which means $a = b$. Then $f(a, c, b) = \frac{a-b}{c-b} = \boxed{0}$. (Note that we know that $b \neq c$ because $f(a, b, c)$ is defined, so we know that we don't have division by 0 in evaluating $f(a, c, b)$.)

1.31 If $f(x) = -x^2 + bx + c$ and $g(x) = dx + e$, then

$$f(g(x)) = f(dx + e) = -(dx + e)^2 + b(dx + e) + c = -d^2x^2 - 2dex - e^2 + bdx + be + c = -d^2x^2 + (bd - 2de)x - (e^2 - be - c).$$

For $f(g(x))$ to be equal to x^2 for all x, the coefficient of x^2 must be 1. However, there is no real number d for which $-d^2 = 1$, so it is $\boxed{\text{not possible}}$ for $f(g(x))$ to be x^2.

1.32 To find the inverse of $f(x) = ax + b$, set $x = f(y) = ay + b$. Then $y = \frac{x-b}{a} = \frac{x}{a} - \frac{b}{a}$, so we have $f^{-1}(x) = \frac{1}{a}x - \frac{b}{a}$. This is a linear function. We conclude that the inverse of a linear function is always a linear function.

1.33 Consider a point $(x, |f(x)|)$ on the graph of $y = |f(x)|$. If $f(x) \geq 0$, then $|f(x)| = f(x)$. On the other hand, if $f(x) < 0$, then $|f(x)| = -f(x)$, and the point $(x, |f(x)|) = (x, -f(x))$ is the reflection of the point $(x, f(x))$ over the x-axis.

Thus, the graph of $y = |f(x)|$ can be constructed from the graph of $y = f(x)$ by reflecting over the x-axis the portion of the graph of $y = f(x)$ that lies below the x-axis. An example is shown below. The dashed lines in the diagram on the right indicates portions of the graph of $y = f(x)$ that are reflected over the x-axis to produce the graph of $y = |f(x)|$.

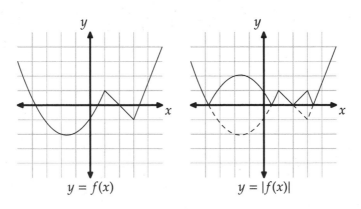

$$y = f(x) \qquad\qquad y = |f(x)|$$

1.34 If $y = f(x) = x/(1 - x)$, then

$$y(1 - x) = x \quad\Rightarrow\quad y - xy = x \quad\Rightarrow\quad xy + x = y \quad\Rightarrow\quad x(y + 1) = y \quad\Rightarrow\quad x = \frac{y}{y + 1}.$$

We compare this to each of the given possibilities:

$$f\left(\frac{1}{y}\right) = \frac{1/y}{1 - 1/y} = \frac{1}{y - 1},$$

$$-f(y) = -\frac{y}{1 - y},$$

$$-f(-y) = -\frac{-y}{1 + y} = \frac{y}{y + 1},$$

$$f(-y) = \frac{-y}{1 + y} = -\frac{y}{y + 1},$$

$$f(y) = \frac{y}{1 - y}.$$

Therefore, we have $x = \dfrac{y}{y + 1} = \boxed{-f(-y)}$.

1.35   We have $f(f(x)) = f\left(\dfrac{cx}{2x + 3}\right) = \dfrac{c \cdot cx/(2x + 3)}{2 \cdot cx/(2x + 3) + 3} = \dfrac{c^2 x}{2cx + 3(2x + 3)} = \dfrac{c^2 x}{(2c + 6)x + 9}.$

If $f(f(x)) = x$ for all x except $-3/2$, then

$$\frac{c^2 x}{(2c + 6)x + 9} = x \quad \Rightarrow \quad c^2 x = x[(2c + 6)x + 9] \quad \Rightarrow \quad c^2 x = (2c + 6)x^2 + 9x.$$

This equation holds for all x except $-3/2$ if and only if $2c + 6 = 0$ and $c^2 = 9$. The only value of c that satisfies both equations is $c = \boxed{-3}$.

1.36   Let $g(x) = f(2x)$. As discussed in the text, we find the inverse of $g(x)$ by solving the equation $x = g(y)$ for y in terms of x. From $x = g(y)$, we have $x = f(2y)$. Because f is invertible, $x = f(2y)$ tells us that $2y = f^{-1}(x)$, so $y = \frac{1}{2}f^{-1}(x)$. Thus, the inverse of $f(2x)$ is $\boxed{\frac{1}{2}f^{-1}(x)}$, not $f^{-1}(2x)$. We also could have found the inverse using the fact that the inverse of $(f(g(x))$ is $g^{-1}(f^{-1}(x))$, which we proved in the text. Here, we let $g(x) = 2x$, so $g^{-1}(x) = \frac{1}{2}x$ and the inverse of $f(2x)$ is $g^{-1}(f^{-1}(x)) = \frac{1}{2}f^{-1}(x)$, as before.

Note that substituting $f(2x)$ into the function $\frac{1}{2}f^{-1}(x)$ gives x: we have $\frac{1}{2}f^{-1}(f(2x)) = \frac{1}{2}(2x) = x$, as expected. Also, we have $f\left(2 \cdot \frac{1}{2}f^{-1}(x)\right) = f(f^{-1}(x)) = x$.

1.37   For the functions f and g to be inverses, we also require that $g(f(x)) = x$ for all x in the domain of f. But $g(f(x)) = \sqrt{x^2} = |x|$, which is not equal to x when x is negative, so f and g are not inverse functions.

Since the function $f(x) = x^2$ fails the horizontal line test (for example, $f(-1) = f(1) = 1$), no function g can be an inverse of f.

Challenge Problems

1.38   If $f(x) = f^{-1}(x)$, then $f(f(x)) = f(f^{-1}(x)) = x$ for all x in the domain of f. We find an expression for $f(f(x))$ as follows:

$$f(f(x)) = f\left(\frac{2x + a}{bx - 2}\right) = \frac{2 \cdot (2x + a)/(bx - 2) + a}{b \cdot (2x + a)/(bx - 2) - 2} = \frac{2(2x + a) + a(bx - 2)}{b(2x + a) - 2(bx - 2)}$$

$$= \frac{4x + 2a + abx - 2a}{2bx + ab - 2bx + 4} = \frac{(ab + 4)x}{ab + 4}.$$

This simplifies to x as long as $ab + 4 \neq 0$. If $ab + 4 = 0$, then $a = -4/b$, and

$$f(x) = \frac{2x + a}{bx - 2} = \frac{2x - 4/b}{bx - 2} = \frac{2bx - 4}{b(bx - 2)} = \frac{2(bx - 2)}{b(bx - 2)} = \frac{2}{b},$$

which is a constant, so f does not have an inverse if $ab = -4$.

Hence, $f^{-1}(x)$ exists and $f^{-1}(x) = f(x)$ $\boxed{\text{for all } a \text{ and } b \text{ such that } ab \neq -4}$.

1.39

(a) The function $f(x + 1)$ is defined only when $-1 < x + 1 < 1$, or $-2 < x < 0$. Therefore, the domain of $f(x + 1)$ is $\boxed{(-2, 0)}$.

(b) The function $f(1/x)$ is defined only when $-1 < 1/x < 1$. If x is positive, then multiplying all three parts of the inequality chain by x gives $-x < 1 < x$. Since x is positive, $-x$ is negative, so the inequality $-x < 1$ is always satisfied. Hence, the solution in this case is $x > 1$.

   If x is negative, then multiplying all three parts of the inequality chain by x gives $-x > 1 > x$. Since x is negative, the inequality $1 > x$ is always satisfied. Multiplying the inequality $-x > 1$ by -1, we get $x < -1$. Hence, the solution in this case is $x < -1$.

   Therefore, the domain of $f(1/x)$ is $\boxed{(-\infty, -1) \cup (1, +\infty)}$.

(c) The function $f(\sqrt{x})$ is defined only when both $x \geq 0$ and $-1 < \sqrt{x} < 1$. Since $\sqrt{x}$ is always nonnegative, the inequality $-1 < \sqrt{x}$ is always satisfied. For the same reason, we can square both sides of the inequality $\sqrt{x} < 1$ to get $x < 1$. Therefore, the domain of the function $f(\sqrt{x})$ is $\boxed{[0, 1)}$.

(d) The function $f(\frac{x+1}{x-1})$ is defined only when $-1 < \frac{x+1}{x-1} < 1$. We take the cases where $x - 1 > 0$ and $x - 1 < 0$ separately.

   If $x - 1 > 0$, then multiplying all parts of the inequality $-1 < \frac{x+1}{x-1} < 1$ by $x - 1$, we get

   $$-x + 1 < x + 1 < x - 1.$$

   Subtracting $x + 1$ from all three parts gives $-2x < 0 < -2$. Therefore, there are no solutions in this case.

   If $x - 1 < 0$, then multiplying all parts of the inequality $-1 < \frac{x+1}{x-1} < 1$ by $x - 1$ (and changing the directions of the inequalities because $x - 1$ is negative), we get

   $$-x + 1 > x + 1 > x - 1.$$

   Subtracting $x + 1$ from all three parts gives $-2x > 0 > -2$, so $x < 0$. Therefore, the solution is $x < 0$ in this case.

   Hence, the domain of $f(\frac{x+1}{x-1})$ is $\boxed{(-\infty, 0)}$.

1.40 The function $f(x) = \sqrt{2 - x - x^2}$ is defined only when $2 - x - x^2 \geq 0$. Factoring gives $(2 + x)(1 - x) \geq 0$. Either both factors must be nonnegative, or both factors must be nonpositive.

If both factors are nonnegative, then $2 + x \geq 0$, so $x \geq -2$, and $1 - x \geq 0$, so $x \leq 1$. Hence, the solution in this case is $-2 \leq x \leq 1$. If both factors are nonpositive, then $2 + x \leq 0$, so $x \leq -2$, and $1 - x \leq 0$, so $x \geq 1$. There is no value of x such that $x \leq -2$ and $x \geq 1$ simultaneously, so there is no solution in this case. Therefore, the domain of $f(x)$ is $\boxed{[-2, 1]}$.

To find the range of $f(x)$, we complete the square inside the radical:

$$f(x) = \sqrt{2 - x - x^2} = \sqrt{\frac{9}{4} - \left(x + \frac{1}{2}\right)^2}.$$

As x varies from -2 to $-1/2$, the expression $9/4 - (x + 1/2)^2$ varies from 0 to 9/4, so $f(x)$ varies from 0 to 3/2. Then as x varies from $-1/2$ to 1, the expression $9/4 - (x + 1/2)^2$ varies from 9/4 to 0, so $f(x)$ varies from 3/2 to 0. Therefore, the range of $f(x)$ is $\boxed{[0, 3/2]}$.

1.41 Let $t = x^2 + 1$, so $x^2 = t - 1$ and $x^4 = (t - 1)^2$. Hence,

$$f(t) = f(x^2 + 1) = x^4 + 5x^2 + 3 = (t - 1)^2 + 5(t - 1) + 3 = t^2 - 2t + 1 + 5t - 5 + 3 = t^2 + 3t - 1.$$

Substituting $t = x^2 - 1$, we get $f(x^2 - 1) = (x^2 - 1)^2 + 3(x^2 - 1) - 1 = x^4 - 2x^2 + 1 + 3x^2 - 3 - 1 = \boxed{x^4 + x^2 - 3}$.

1.42 Consider a point $(x, f(|x|))$ on the graph of $y = f(|x|)$. If $x \geq 0$, then $|x| = x$, so $(x, f(|x|))$ is the same as $(x, f(x))$. On the other hand, if $x < 0$, then $|x| = -x$, and the point $(x, f(|x|)) = (x, f(-x))$ is the reflection of the point $(-x, f(-x))$ over the y-axis.

Thus, the graph of $y = f(|x|)$ consists of the portion of the graph of $y = f(x)$ that is on or to the right of the y-axis, together with the curve formed when reflecting this portion of the graph of $y = f(x)$ over the y-axis. An example is shown below.

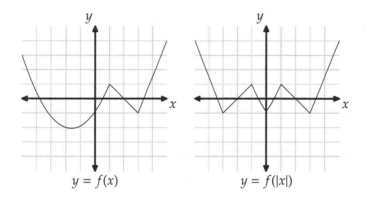

$$y = f(x) \qquad\qquad y = f(|x|)$$

1.43 Let x be a root of $f(x) = 0$. Applying f to both sides, we get $f(f(x)) = f(0) = 2003$. But $f(f(x)) = x$ for all values of x, so $x = 2003$. Therefore, the only root of the equation $f(x) = 0$ is $x = \boxed{2003}$.

1.44 To find the inverse of $f(x)$, set $x = f(y) = (ay + b)/(cy + d)$. Then

$$x(cy + d) = ay + b \quad \Rightarrow \quad cxy + dx = ay + b \quad \Rightarrow \quad cxy - ay = b - dx \quad \Rightarrow \quad y = \frac{b - dx}{cx - a}.$$

Thus, the inverse of $f(x)$, if it exists, must be $f^{-1}(x) = \dfrac{b - dx}{cx - a}$.

To check if this works, we substitute:

$$f(f^{-1}(x)) = f\left(\frac{b - dx}{cx - a}\right) = \frac{a \cdot (b - dx)/(cx - a) + b}{c \cdot (b - dx)/(cx - a) + d} = \frac{a(b - dx) + b(cx - a)}{c(b - dx) + d(cx - a)} = \frac{(bc - ad)x}{bc - ad}.$$

This simplifies to x as long as $ad - bc \neq 0$. Hence, the function f has an inverse if and only if $\boxed{ad - bc \neq 0}$.

Note that the function $f(x) = \frac{ax+b}{cx+d}$ is well-defined as long as c and d are not simultaneously 0, and the function

$$f^{-1}(x) = \frac{b - dx}{cx - a}.$$

is well-defined as long as a and c are not simultaneously 0. However, if $c = d = 0$, then $ad - bc = 0$, and if $a = c = 0$, then $ad - bc = 0$. Hence, the condition $ad - bc \neq 0$ also ensures that both of these functions are well-defined.

1.45 Let $(x, 0)$ be an x-intercept of the graph of f, so $f(x) = 0$. Then $g(x) = f(x)f(|x|) = 0$, so $(x, 0)$ is also an x-intercept of the graph of g. Hence, all x-intercepts of the graph of f must also be x-intercepts of the graph of g.

On the other hand, an x-intercept of the graph of g is not necessarily an x-intercept of the graph of f. For example, take $f(x) = x - 1$, so $g(x) = f(x)f(|x|) = (x - 1)(|x| - 1)$. Note that $g(-1) = 0$, but $f(-1) = -2 \neq 0$. Thus, the point $(-1, 0)$ is an x-intercept of the graph of g, but not an x-intercept of the graph of f.

1.46 To construct the graph of $y = \frac{3}{2}f(2x - 2) - 1$ from the graph of $y = f(x)$, we describe the construction one step at a time, working from the inside of the function out.

First, we look at the expression $f(2x - 2) = f(2(x - 1))$. As described in the solution to Problem 1.2.4, the graph of $y = f(2(x - 1))$ is the result of scaling the graph of $y = f(x)$ horizontally towards the y-axis by a factor of $1/2$ and shifting the ensuing graph to the right 1 unit. This produces the middle graph below.

Then, the graph of $y = \frac{3}{2}f(2x - 2)$ is the result of scaling the graph of $f(2x - 2)$ vertically away from the x-axis by a factor of $3/2$, and the graph of $y = \frac{3}{2}f(2x - 2) - 1$ results from shifting the graph of $y = \frac{3}{2}f(2x - 2)$ downward 1 unit. This produces the final graph at right below.

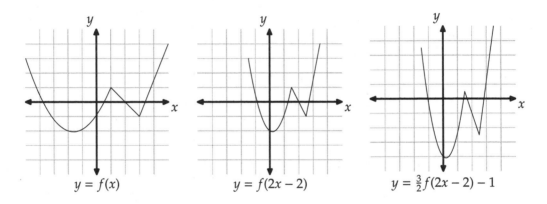

$$y = f(x) \qquad\qquad y = f(2x - 2) \qquad\qquad y = \tfrac{3}{2}f(2x - 2) - 1$$

1.47 To find the unique number that is not in the range of f, let $y = f(x) = (ax + b)/(cx + d)$. Then

$$y(cx + d) = ax + b \quad \Rightarrow \quad cxy + dy = ax + b \quad \Rightarrow \quad cxy - ax = b - dy \quad \Rightarrow \quad x = \frac{b - dy}{cy - a}.$$

Hence, for every value of y, there is a corresponding value of x such that $f(x) = y$, except when the denominator $cy - a$ is equal to 0. Therefore, the unique number that is not in the range of f is a/c.

Now we compute $f(f(x))$:

$$f(f(x)) = f\left(\frac{ax + b}{cx + d}\right) = \frac{a \cdot (ax + b)/(cx + d) + b}{c \cdot (ax + b)/(cx + d) + d} = \frac{a(ax + b) + b(cx + d)}{c(ax + b) + d(cx + d)}$$
$$= \frac{(a^2 + bc)x + (ab + bd)}{(ac + cd)x + (bc + d^2)}.$$

If $f(f(x)) = x$ for all x, then $\dfrac{(a^2 + bc)x + (ab + bd)}{(ac + cd)x + (bc + d^2)} = x$, so $(a^2 + bc)x + (ab + bd) = (ac + cd)x^2 + (bc + d^2)x$ for all x.

Then the coefficient of x^2 must be 0, so $ac + cd = 0$, which means $c(a + d) = 0$. Since c is nonzero, we have $a + d = 0$, so $d = -a$. Furthermore, if $d = -a$, then

$$f(f(x)) = \frac{(a^2 + bc)x + (ab + bd)}{(ac + cd)x + (bc + d^2)} = \frac{(a^2 + bc)x + (ab - ab)}{(ac - ac)x + (a^2 + bc)} = \frac{(a^2 + bc)x}{a^2 + bc} = x,$$

as desired. Therefore, $f(x) = \dfrac{ax+b}{cx+d} = \dfrac{ax+b}{cx-a}$.

Since $f(19) = 19$, we have

$$\frac{19a+b}{19c-a} = 19 \quad \Rightarrow \quad 19^2 c - 19a = 19a + b \quad \Rightarrow \quad 19^2 c = 2 \cdot 19a + b.$$

Also, $f(97) = 97$, so we have

$$\frac{97a+b}{97c-a} = 97 \quad \Rightarrow \quad 97^2 c - 97a = 97a + b \quad \Rightarrow \quad 97^2 c = 2 \cdot 97a + b.$$

Subtracting the first equation from the second, we get

$$(97^2 - 19^2)c = 2(97 - 19)a \quad \Rightarrow \quad (97+19)(97-19)c = 2(97-19)a \quad \Rightarrow \quad \frac{a}{c} = \frac{97+19}{2} = \boxed{58}.$$

Here is a faster way to derive the same equations. Since $f(f(x)) = x$ for all x, we have $f(f(0)) = 0$. Since $f(0) = b/d$, we have

$$f(f(0)) = f(b/d) = \frac{a \cdot b/d + b}{c \cdot b/d + d} = \frac{ab + bd}{bc + d^2},$$

so $f(f(0)) = 0$ gives $(ab + bd)/(bc + d^2) = 0$. Therefore, we have $ab + bd = 0$, so $b(a + d) = 0$. Since b is nonzero, we have $a + d = 0$, so $d = -a$.

Now, we know that $x = 19$ and $x = 97$ satisfy the equation $f(x) = x$. Writing this equation out, we get

$$\frac{ax+b}{cx+d} = x \quad \Rightarrow \quad x(cx+d) = ax + b \quad \Rightarrow \quad cx^2 + (d-a)x - b = 0.$$

The sum of the roots of this quadratic is $-(d-a)/c = 2a/c$. We know that the roots of this quadratic are 19 and 97, since these are the solutions to $f(x) = x$. Therefore, we have $2a/c = 19 + 97 = 116$, which gives us $a/c = 116/2 = 58$.

CHAPTER 2

Introduction to Trigonometric Functions

Exercises for Section 2.1

2.1.1

(a) Let P be the terminal point of $120°$ and let S be the foot of the perpendicular from P to the x-axis. From 30-60-90 triangle POS, we have $PS = \sqrt{3}/2$. Since $120°$ lies in the second quadrant, we have $\sin 120° = \boxed{\sqrt{3}/2}$.

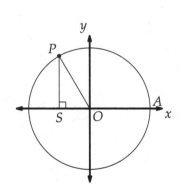

Figure 2.1: Diagram for Part (a)

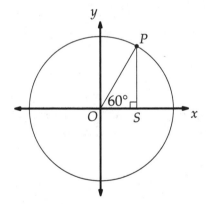

Figure 2.2: Diagram for Part (b)

(b) Let P be the terminal point of $60°$, and let S be the foot of the perpendicular from P to the x-axis. From 30-60-90 triangle POS, we have $PS = \sqrt{3}/2$ and $OS = 1/2$, so

$$\cot 60° = \frac{\cos 60°}{\sin 60°} = \frac{OS}{PS} = \frac{1/2}{\sqrt{3}/2} = \boxed{\frac{\sqrt{3}}{3}}.$$

(c) Since $\cos 45° = \sqrt{2}/2 = 1/\sqrt{2}$, we have $\sec 45° = \frac{1}{\cos 45°} = \frac{1}{1/\sqrt{2}} = \boxed{\sqrt{2}}$.

(d) Let P be the terminal point of $210°$ and let S be the foot of the altitude from P to the x-axis. From 30-60-90 triangle POS, we have $OS = \sqrt{3}/2$ and $PS = 1/2$. Since $210°$ lies in the third quadrant, we have $\cos 210° = -\sqrt{3}/2$ and $\sin 210° = -1/2$, so

$$\tan 210° = \frac{\sin 210°}{\cos 210°} = \frac{-1/2}{-\sqrt{3}/2} = \frac{1}{\sqrt{3}} = \boxed{\frac{\sqrt{3}}{3}}.$$

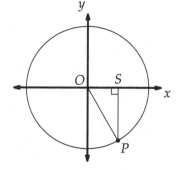

(e) The terminal point of $180°$ is $(-1, 0)$, so $\cos 180° = \boxed{-1}$.

(f) Let P be the terminal point of $300° = 360° - 60°$, and let S be the foot of the altitude from P to the x-axis. Then $\angle POS = 60°$, so $OS = 1/2$ and $PS = \sqrt{3}/2$. Since $300°$ lies in the fourth quadrant, we have $\sin 300° = -\sqrt{3}/2$, so

$$\csc 300° = \frac{1}{\sin 300°} = \frac{1}{-\sqrt{3}/2} = -\frac{2}{\sqrt{3}} = \boxed{-\frac{2\sqrt{3}}{3}}.$$

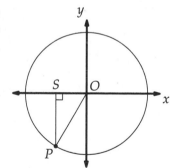

2.1.2

(a) The terminal point of $-90°$ is $(0, -1)$, so $\cos(-90°) = \boxed{0}$.

(b) Since $36000° = 100 \cdot 360°$, the terminal point of $36000°$ is the same as the terminal point of $0°$. Therefore, $\tan 36000° = \tan 0° = \boxed{0}$.

(c) Since $750° = 2 \cdot 360° + 30°$, the terminal point of $750°$ is the same as the terminal point of $30°$. Therefore, $\sin 750° = \sin 30° = \boxed{1/2}$.

(d) Since $-1200° = -4 \cdot 360° + 240°$, the terminal point of $-1200°$ is the same as the terminal point of $240° = 180° + 60°$. We let P be this terminal point, and let S be the foot of the altitude from P to the x-axis. Then $\angle POS = 60°$, so $OS = 1/2$ and $PS = \sqrt{3}/2$. Since $240°$ lies in the third quadrant, we have $\cos 240° = -1/2$, so

$$\sec(-1200°) = \sec 240° = \frac{1}{\cos 240°} = \frac{1}{-1/2} = \boxed{-2}.$$

2.1.3 The angle $293°$ lies in the fourth quadrant, so $\sin 293°$ is negative. The angle $-68°$ lies in the fourth quadrant, so $\cos(-68°)$ is positive. The angle $206°$ lies in the third quadrant, so $\cot 206° = \frac{\cos 206°}{\sin 206°}$ is positive. The angle $90.5°$ lies in the second quadrant, so $\csc 90.5° = \frac{1}{\sin 90.5°}$ is positive.

Thus, $\boxed{\cos(-68°), \cot 206°, \text{ and } \csc 90.5°}$ are positive.

2.1.4 If $\cos\theta = \sqrt{2}/2$, then $\cos^2\theta = 2/4 = 1/2$. We know that $\cos^2\theta + \sin^2\theta = 1$, so $\sin^2\theta = 1 - 1/2 = 1/2$, which means $\sin\theta = \pm\sqrt{1/2} = \pm\sqrt{2}/2$. Hence, $(\cos\theta, \sin\theta) = (\sqrt{2}/2, \sqrt{2}/2)$ or $(\sqrt{2}/2, -\sqrt{2}/2)$.

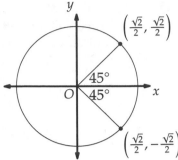

We know that $(\sqrt{2}/2, \sqrt{2}/2)$ is the terminal point of $45°$. However, the point $(\sqrt{2}/2, \sqrt{2}/2)$ is also the terminal point of $45° + 360° = 405°$, $405° + 360° = 765°$, and $765° + 360° = 1125°$. The angles in this list that are in the range $0° \le \theta \le 900°$ are $45°$, $405°$, and $765°$.

We know that $(\sqrt{2}/2, -\sqrt{2}/2)$ is the terminal point of $-45°$. However, the point $(\sqrt{2}/2, -\sqrt{2}/2)$ is also the terminal point of $-45° + 360° = 315°$, $315° + 360° = 675°$, and $675° + 360° = 1035°$. The angles in the range $0° \le \theta \le 900°$ are $315°$ and $675°$.

Therefore, the solutions such that $0° \le \theta \le 900°$ are $\theta = 45°, 315°, 405°, 675°$, and $765°$. There are $\boxed{5}$ such solutions.

2.1.5 The range of the function $\sin 3x$ is $[-1, 1]$. Then the range of the function $2\sin 3x$ is $[-2, 2]$, and so the range of the function $f(x) = 7 + 2\sin 3x$ is $\boxed{[5, 9]}$.

Exercises for Section 2.2

2.2.1

(a) $330° = 330° \cdot \frac{2\pi}{360°} = \boxed{\frac{11\pi}{6}}$.

(b) $270° = 270° \cdot \frac{2\pi}{360°} = \boxed{\frac{3\pi}{2}}$.

(c) $-315° = -315° \cdot \frac{2\pi}{360°} = \boxed{-\frac{7\pi}{4}}$.

(d) $2000° = 2000° \cdot \frac{2\pi}{360°} = \boxed{\frac{100\pi}{9}}$.

2.2.2

(a) $3\pi = 3\pi \cdot \frac{360°}{2\pi} = \boxed{540°}$.

(b) $\frac{2\pi}{3} = \frac{2\pi}{3} \cdot \frac{360°}{2\pi} = \boxed{120°}$.

(c) $-\frac{7\pi}{5} = -\frac{7\pi}{5} \cdot \frac{360°}{2\pi} = \boxed{-252°}$.

(d) $-3 = -3 \cdot \frac{360°}{2\pi} = \boxed{-\frac{540°}{\pi}}$.

2.2.3 Converting the angle measures to degrees, we find that:

$$30° = 30° \cdot \frac{\pi}{180°} = \frac{\pi}{6}, \qquad 45° = 45° \cdot \frac{\pi}{180°} = \frac{\pi}{4}, \qquad 60° = 60° \cdot \frac{\pi}{180°} = \frac{\pi}{3}, \qquad 90° = 90° \cdot \frac{\pi}{180°} = \frac{\pi}{2}.$$

Each leg of the 45-45-90 triangle has length $\frac{\sqrt{2}}{2}$, and the legs of the 30-60-90 triangle have length $\frac{1}{2}$ and $\frac{\sqrt{3}}{2}$, with the shorter leg opposite the smaller acute angle. The triangles are drawn below.

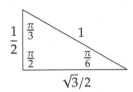

2.2.4

(a) Converting $\frac{\pi}{4}$ to degrees, we find $\frac{\pi}{4} = \frac{\pi}{4} \cdot \frac{360°}{2\pi} = 45°$. Then $\sin\frac{\pi}{4} = \sin 45° = \boxed{\sqrt{2}/2}$.

(b) The terminal point of 7π is the same as the terminal point of $7\pi - 3(2\pi) = \pi$, which is $(-1, 0)$, so $\sin\pi = 0$. Therefore, $\csc 7\pi = \csc\pi$, which is $\boxed{\text{undefined}}$.

(c) We let P be the terminal point of $5\pi/6$ and S the foot of the altitude from P to the x-axis. Then $\angle POS = \pi/6$, so $OS = \sqrt{3}/2$ and $PS = 1/2$. Since $5\pi/6$ lies in the second quadrant, we have $\cos(5\pi/6) = -\sqrt{3}/2$ and $\sin(5\pi/6) = 1/2$, so

$$\tan\frac{5\pi}{6} = \frac{\sin\frac{5\pi}{6}}{\cos\frac{5\pi}{6}} = \frac{1/2}{-\sqrt{3}/2} = -\frac{1}{\sqrt{3}} = \boxed{-\frac{\sqrt{3}}{3}}.$$

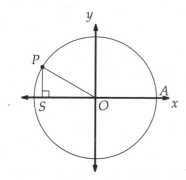

Figure 2.3: Diagram for Part (c)   Figure 2.4: Diagram for Part (d)

(d) We let P be the terminal point of $-\pi/3$ and S be the foot of the altitude from P to the x-axis. Then $\angle POS = \pi/3$, so $OS = 1/2$ and $PS = \sqrt{3}/2$. Since $-\pi/3$ lies in the fourth quadrant, we have $\cos\left(-\frac{\pi}{3}\right) = \boxed{1/2}$.

2.2.5 Since $\tan t = \frac{\sin t}{\cos t}$, we see that $\tan t$ is defined as long as $\cos t \neq 0$. But $\cos t = 0$ if and only if t is an odd multiple of $\frac{\pi}{2}$, that is, if $t = \frac{(2k+1)\pi}{2}$ for some integer k. So, $\tan t$ is defined if and only if t is not of the form $\frac{(2k+1)\pi}{2}$.

Hence, $f(x) = 3\tan(2x - \frac{\pi}{3})$ is defined if and only if $2x - \frac{\pi}{3}$ is not of the form $\frac{(2k+1)\pi}{2}$. Therefore, the values of x that we must omit from the domain are the values of x such that $2x - \frac{\pi}{3} = \frac{(2k+1)\pi}{2}$ where k is an integer. Solving this equation for x gives $x = \frac{(6k+5)\pi}{12}$. Therefore, the domain of $f(x) = 3\tan(2x - \frac{\pi}{3})$ is all real numbers except those numbers of the form $\frac{(6k+5)\pi}{12}$ where k is an integer.

2.2.6

(a) Since $40°$ lies in the first quadrant, $\tan 40°$ is positive.

(b) Jake probably computed $\tan 40$, where the angle 40 is in radians. Indeed, $\tan 40 = -1.11721$ to five decimal places. Jake should have either set his calculator to degrees mode, or computed $\tan(40° \cdot \frac{2\pi}{360°}) = \tan\frac{2\pi}{9}$ instead.

Exercises for Section 2.3

2.3.1

(a) The graph is shown at right.

(b) Since $\sec x = \frac{1}{\cos x}$, and $\cos x$ has period 2π, $\sec x$ also has period $\boxed{2\pi}$.

(c) Since $\sec x = \frac{1}{\cos x}$ and $\csc x = \frac{1}{\sin x}$, and the graph of $y = \cos x$ is the graph of $y = \sin x$ shifted $\frac{\pi}{2}$ to the left, the graph of $y = \sec x$ is the graph of $y = \csc x$ shifted $\frac{\pi}{2}$ to the left.

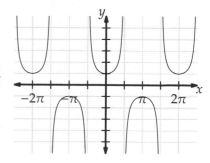

2.3.2

(a) The graph is shown at right.

(b) Since $\cot x = \frac{1}{\tan x}$, and $\tan x$ has period π, $\cot x$ also has period $\boxed{\pi}$.

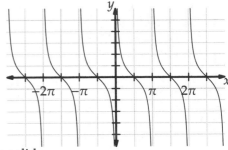

2.3.3 Below, the graph of $y = \cos x$ is dashed and the graph of $y = \sin x$ is solid.

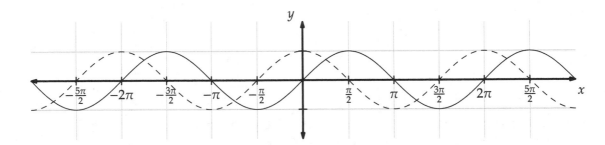

It appears that the two graphs intersect twice in each period. To see why this is the case, consider the unit circle. If $\cos\theta = \sin\theta$, then the coordinates of the terminal point of the angle θ are the same. All the points whose coordinates are the same lie on the graph of $x = y$. The intersection of this line with the unit circle gives us the two terminal points whose coordinates are the same, $\left(\frac{\sqrt{2}}{2}, \frac{\sqrt{2}}{2}\right)$ and $\left(-\frac{\sqrt{2}}{2}, -\frac{\sqrt{2}}{2}\right)$.

We know that $\left(\frac{\sqrt{2}}{2}, \frac{\sqrt{2}}{2}\right)$ is the terminal point of $\frac{\pi}{4}$. However, $\left(\frac{\sqrt{2}}{2}, \frac{\sqrt{2}}{2}\right)$ is also the terminal point of $\frac{\pi}{4} + 2\pi k$ for any integer k. The smallest number of this form in the interval $[0, 100]$ is $\frac{\pi}{4}$, and the largest number of this form is $\frac{\pi}{4} + 2\pi \cdot 15 \approx 95.03$. Hence, we can take $k = 0, 1, \ldots, 15$, for a total of 16 different values of k.

We know that $\left(-\frac{\sqrt{2}}{2}, -\frac{\sqrt{2}}{2}\right)$ is the terminal point of $\frac{5\pi}{4}$. However, $\left(-\frac{\sqrt{2}}{2}, -\frac{\sqrt{2}}{2}\right)$ is also the terminal point of $\frac{5\pi}{4} + 2\pi k$ for any integer k. The smallest number of this form in the interval $[0, 100]$ is $\frac{5\pi}{4}$, and the largest number of this form is $\frac{5\pi}{4} + 2\pi \cdot 15 \approx 98.17$. Hence, we can take $k = 0, 1, \ldots, 15$, for a total of 16 different values of k.

Therefore, the graphs of $y = \sin x$ and $y = \cos x$ intersect $16 + 16 = \boxed{32}$ times in the interval $[0, 100]$.

Exercises for Section 2.4

2.4.1 The range of $\sin x$ is $[-1, 1]$. If a is nonnegative, then the range of $f(x) = a\sin x$ is $[-a, a]$, so the amplitude of $f(x)$ is $[a - (-a)]/2 = a = |a|$. If a is negative, then the range of $f(x) = a\sin x$ is $[a, -a]$, so the amplitude of $f(x)$ is $(-a - a)/2 = -a = |a|$. In either case, the amplitude of $f(x)$ is $|a|$.

2.4.2 The graph of $y = 3+\sin x$ is obtained by shifting the graph of $y = \sin x$ up by 3 units. This graph is shown at right.

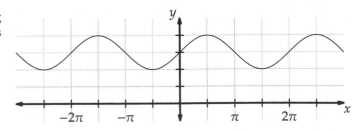

2.4.3

(a) The graph of $y = \cos 2x$ is obtained by compressing the graph of $y = \cos x$ horizontally towards the x-axis by a factor of 2. This graph is shown at right.

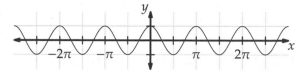

(b) The graph of $y = -3\cos 2x$ is obtained by stretching the graph of $y = \cos 2x$ vertically by a factor of 3, and then reflecting the result over the x-axis. This graph is shown at right.

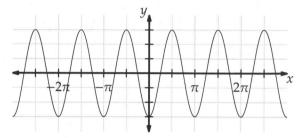

(c) We note that $-3\cos 2x$ is equal to -3 when $x = 0$. Correspondingly, the function $-3\cos(2x - \frac{\pi}{3})$ is equal to -3 when $2x - \frac{\pi}{3} = 0$, or $x = \frac{\pi}{6}$. Therefore, $\frac{\pi}{6}$ is the phase shift of the graph $y = -3\cos(2x - \frac{\pi}{3})$, and we find this graph by shifting the graph of $y = -3\cos 2x$ to the right by $\frac{\pi}{6}$. The resulting graph of $y = -3\cos(2x - \frac{\pi}{3})$ is shown at right.

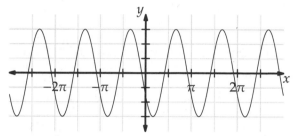

2.4.4 The amplitude of the function corresponding to the given graph is 3. The amplitude of the function $g(x) = t\sin ux$ is t when t is positive, so we have $t = 3$.

If $t = 3$, then $g(x) = 3\sin ux$, so the graph of $g(x)$ is obtained by stretching the graph of $y = \sin x$ vertically by a factor of 3 and compressing it horizontally by a factor of u. To find u, note that the graph completes 5 cycles in the interval $[0, 2\pi]$. Therefore, $u = 5$, and $g(x) = 3\sin 5x$, so $\boxed{(t, u) = (3, 5)}$.

2.4.5 We start with the function $\cos x$, which has period 2π and amplitude 1. The graph of $\cos 2x$ is obtained by compressing the graph of $\cos x$ horizontally by a factor of 2, so the function $\cos 2x$ has period $\frac{2\pi}{2} = \pi$ and amplitude 1.

The graph of $-3\cos 2x$ is obtained by stretching the graph of $\cos 2x$ vertically by a factor 3, and then reflecting it in the x-axis. The period is still π, but the amplitude is now 3.

The function $-3\cos 2x$ is equal to -3 when $x = 0$. Correspondingly, the function $-3\cos(2x + \pi)$ is equal to -3 when $2x + \pi = 0$, or $x = -\frac{\pi}{2}$. The function $-3\cos(2x + \pi)$ also equals -3 when $2x + \pi = 2\pi$, or $x = \frac{\pi}{2}$. So, we see that we can view the graph of $y = -3\cos(2x + \pi)$ as either a rightward or leftward shift of $y = -3\cos(2x)$ by $\frac{\pi}{2}$. Therefore, we can say that the phase shift of $-3\cos(2x + \pi)$ is either $-\frac{\pi}{2}$ or $\frac{\pi}{2}$. (Either is acceptable.) The period and amplitude are still π and 3, respectively.

Finally, adding 4 does not change the period, amplitude, or phase shift. Therefore, the function $f(x) = -3\cos(2x + \pi) + 4$ has $\boxed{\text{period } \pi, \text{ amplitude 3, and phase shift } \frac{\pi}{2} \text{ or } -\frac{\pi}{2}}$.

2.4.6 Because the distance between the block and the ceiling varies sinusoidally with time, the distance is $f(t) = a\sin(bt + c) + d$, where a, b, c, and d are constants. The middle point of the block's path is 1.5 meters from the ceiling, and it varies 0.3 meters above and below this point. So, the amplitude is 0.3, and $f(t)$ must equal 1.5 when $a\sin(bt + c)$ is 0 (when the block is at the midpoint of its path). Therefore, the function is $f(t) = 0.3\sin(bt + c) + 1.5$ for some values of b and c.

Suppose we let $t = 0$ at a point when the block is closest to the ceiling, so we must have $f(0) = 1.2$. Therefore, we must have $\sin(bt + c) = -1$ when $t = 0$, which means we can take $c = -\frac{\pi}{2}$, and we have

$$f(t) = 0.3\sin\left(bt - \frac{\pi}{2}\right) + 1.5.$$

Finally, we turn to the last piece of information we have: that the block goes from the high point to the low point in 1 second. This means that it goes from high to low and back to high again in 2 seconds, so its period is 2 seconds. Since the period of $f(t)$ is $\frac{2\pi}{b}$ when b is positive, we must have $b = \pi$, and we have

$$f(t) = 0.3\sin\left(\pi t - \frac{\pi}{2}\right) + 1.5.$$

So, 2.7 seconds after the block is at a high point, the distance between the block and the ceiling is

$$f(2.7) = 0.3\sin\left(2.7\pi - \frac{\pi}{2}\right) + 1.5 \approx \boxed{1.68 \text{ meters}}.$$

2.4.7

(a) The graph completes one cycle in the interval $[0, 3\pi]$, so the function has period 3π. But the function $\cos ax$ has period $\frac{2\pi}{|a|}$, so $\frac{2\pi}{|a|} = 3\pi$, which means $|a| = 2/3$. Since $a > 0$, we have $\boxed{a = 2/3}$.

(b) The function $\sin(bx + c)$ has the same period as the function $\sin bx$, which has period $\frac{2\pi}{|b|}$. Since the period must be 3π and $b > 0$, we have $\boxed{b = 2/3}$.

(c) If $f(x) = \sin(bx + c)$, then we have $f(0) = \sin c$. Since the graph passes through $(0, 1)$, we must have $\sin c = 1$. There are no other restrictions on c, so there are infinitely many possible values of c. Since the only restriction on c is that $\sin c = 1$, the permissible values of c are $\boxed{\text{the numbers of the form } \frac{\pi}{2} + 2k\pi, \text{ where } k \text{ is an integer}}$.

2.4.8 We claim that the function $f(x) = 2\sin x - \tan x$ has period 2π.

First, $\sin x$ has period 2π and $\tan x$ has period π, so $\sin(x + 2\pi) = \sin x$ and $\tan(x + 2\pi) = \tan(x + \pi) = \tan x$, which means

$$f(x + 2\pi) = 2\sin(x + 2\pi) - \tan(x + 2\pi) = 2\sin x - \tan x = f(x).$$

This tells us that f is periodic, and that the period of f is no greater than 2π.

Let d be the period of $f(x) = 2\sin x - \tan x$, so $f(x + d) = f(x)$. To show that the period of f is 2π, we must show that d cannot be less than 2π. Note that

$$f(x) = 2\sin x - \tan x = 2\sin x - \frac{\sin x}{\cos x} = \frac{2\cos x\sin x - \sin x}{\cos x},$$

so $f(x)$ is undefined whenever $\cos x = 0$. But if $f(x)$ is undefined for some value of x, then $f(x + d)$ must also be undefined. (Otherwise, the equality $f(x+d) = f(x)$ could not hold.) In other words, if $\cos x = 0$, then $\cos(x+d) = 0$. The values of x such that $\cos x = 0$ are the integer multiples of π. So, d must be an integer multiple of π. We have already seen that $f(x + 2\pi) = f(x)$ for all x, so the only other possibility for the period is π. To show the period is not π, we can note that $f\left(\frac{\pi}{4}\right) = 2\sin\frac{\pi}{4} - \tan\frac{\pi}{4} = \sqrt{2} - 1$ and $f\left(\frac{5\pi}{4}\right) = 2\sin\frac{5\pi}{4} - \tan\frac{5\pi}{4} = -\sqrt{2} - 1$. Therefore, we have $f\left(\frac{\pi}{4}\right) \neq f\left(\frac{\pi}{4} + \pi\right)$, which means the period is not π.

We conclude that the period of $f(x) = 2\sin x - \tan x$ is $\boxed{2\pi}$.

Exercises for Section 2.5

2.5.1

(a) Since $\sin\frac{\pi}{6} = 0.5$ and $-\frac{\pi}{2} \le \frac{\pi}{6} \le \frac{\pi}{2}$, we have $\arcsin 0.5 = \boxed{\frac{\pi}{6}}$.

(b) Since $\cos\frac{3\pi}{4} = -\frac{\sqrt{2}}{2}$ and $0 \le \frac{3\pi}{4} \le \pi$, we have $\arccos\left(-\frac{\sqrt{2}}{2}\right) = \boxed{\frac{3\pi}{4}}$.

(c) Since $\tan\left(-\frac{\pi}{4}\right) = -1$ and $-\frac{\pi}{2} < -\frac{\pi}{4} < \frac{\pi}{2}$, we have $\arctan(-1) = \boxed{-\frac{\pi}{4}}$.

(d) Since $\cos\frac{\pi}{2} = 0$ and $0 \le \frac{\pi}{2} \le \pi$, we have $\arccos 0 = \boxed{\frac{\pi}{2}}$.

(e) Since $\cot\frac{\pi}{3} = \frac{\sqrt{3}}{3}$ and $0 < \frac{\pi}{3} < \pi$, we have $\text{arccot}\,\frac{\sqrt{3}}{3} = \boxed{\frac{\pi}{3}}$.

(f) We seek an angle θ such that $\csc\theta = -2$. Since $\csc\theta = \frac{1}{\sin\theta}$, we have $\frac{1}{\sin\theta} = -2$, so $\sin\theta = -\frac{1}{2}$. Since $\sin(-\frac{\pi}{6}) = -\frac{1}{2}$, and $-\frac{\pi}{6} \in [-\frac{\pi}{2}, 0) \cup (0, \frac{\pi}{2}]$ (the range of $\csc^{-1} x$), we have $\csc^{-1}(-2) = \boxed{-\frac{\pi}{6}}$.

2.5.2 Let $\theta = \arcsin 0.3$, so $-\frac{\pi}{2} \le \theta \le \frac{\pi}{2}$ and $\sin\theta = 0.3 = 3/10$. Then

$$\cos^2\theta = 1 - \sin^2\theta = 1 - \left(\frac{3}{10}\right)^2 = 1 - \frac{9}{100} = \frac{91}{100}.$$

Since $-\frac{\pi}{2} \le \theta \le \frac{\pi}{2}$, we have $\cos\theta \ge 0$, so $\cos(\arcsin 0.3) = \cos\theta = \sqrt{\frac{91}{100}} = \boxed{\frac{\sqrt{91}}{10}}$.

2.5.3 Let $\theta = \arctan(\tan\frac{63\pi}{5})$, so $-\frac{\pi}{2} < \theta < \frac{\pi}{2}$ and $\tan\theta = \tan\frac{63\pi}{5}$. Since $\tan\theta$ has period π, and

$$\frac{63\pi}{5} = 13\pi - \frac{2\pi}{5},$$

we conclude that $\tan\frac{63\pi}{5} = \tan\left(-\frac{2\pi}{5}\right)$. Also, $-\frac{\pi}{2} < -\frac{2\pi}{5} < \frac{\pi}{2}$. Therefore, $\theta = \boxed{-\frac{2\pi}{5}}$.

2.5.4 Since $\tan\frac{\pi}{2}$ is undefined, the range of $\arctan$ cannot include $\frac{\pi}{2}$, so in particular, it cannot be the interval $[0, \pi)$. But even if we removed the value $\frac{\pi}{2}$ from this interval to make the range $[0, \frac{\pi}{2}) \cup (\frac{\pi}{2}, \pi)$, we would face another problem: this choice would introduce a discontinuity (that is, a gap) in the graph of $y = \arctan x$.

Note that $\tan 0 = \tan\pi = 0$. For small positive values of x, $\arctan x$ would be close to 0. However, for small negative values of x, $\arctan x$ would be close to π.

This kind of behavior is unnatural, and can easily be avoided by setting $(-\frac{\pi}{2}, \frac{\pi}{2})$ to be the range of $\arctan x$. The graph of $\arctan x$ is shown below.

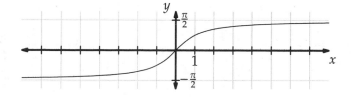

2.5.5 The range of the arccos function is $0 \le \theta \le \pi$. But $\sin\theta \ge 0$ for all $0 \le \theta \le \pi$, and $\sin\theta = 0$ only for $\theta = 0$ and $\theta = \pi$. Since $\cos 0 = 1$ and $\cos\pi = -1$, the only values x for which $\sin(\arccos x) \le 0$ are $\boxed{x = 1 \text{ and } -1}$.

The range of the arcsin function is $-\frac{\pi}{2} \le \theta \le \frac{\pi}{2}$. But $\cos\theta \ge 0$ for all $-\frac{\pi}{2} \le \theta \le \frac{\pi}{2}$, so in particular, there are $\boxed{\text{no values } x}$ for which $\cos(\arcsin x) < 0$.

2.5.6 If $y = \arctan x$, then $\tan y = x$ and $-\frac{\pi}{2} < y < \frac{\pi}{2}$. Then $x = \tan y = \frac{\sin y}{\cos y}$. Squaring, we get $x^2 = \frac{\sin^2 y}{\cos^2 y} = \frac{\sin^2 y}{1-\sin^2 y}$. Solving for $\sin^2 y$, we find $\sin^2 y = \frac{x^2}{x^2+1}$. Taking the square root of both sides, we get

$$\sin y = \pm \sqrt{\frac{x^2}{x^2 + 1}} = \pm \frac{x}{\sqrt{x^2 + 1}}.$$

The term $\sqrt{x^2 + 1}$ is always positive. If $0 \le y < \frac{\pi}{2}$, then $\sin y \ge 0$ and $x = \tan y \ge 0$, so in the equation above, the sign of $\sin y$ matches the sign of x. If $-\frac{\pi}{2} < y < 0$, then $\sin y < 0$ and $x = \tan y < 0$, so again, the sign of $\sin y$ matches the sign of x. Since $\sin y$ always matches the sign of x, we can always express $\sin y$ as the "+" result of the "$\pm$" in our expression above for $\sin y$. So, we have

$$\sin y = \boxed{\frac{x}{\sqrt{x^2 + 1}}}.$$

2.5.7 The solution in the text includes $t = \frac{25\pi}{36} + \frac{2k\pi}{3}$, while Kira's solution includes $t = \frac{\pi}{36} + \frac{2k\pi}{3}$. We will prove that these two expressions produce the same set of solutions as follows.

Suppose $k = a$ for some integer a in the text's solution, and $k = b$ for some integer in Kira's solution. We wish to show that for any integer a, there is a value of b such that these two solutions are the same. So, for a given a, we want to show that there is an integer b such that

$$\frac{25\pi}{36} + \frac{2a\pi}{3} = \frac{\pi}{36} + \frac{2b\pi}{3}.$$

Solving for b, we find $b = a + 1$. So, for every solution the form in the text produces, Kira's form also produces that solution. Conversely, we can see that for any integer b that we choose for Kira's form, setting $a = b - 1$ in the form in the text produces the same solution. Putting these observations together, we see that the two expressions produce the same set of solutions. So, Kira's answer is correct.

Review Problems

2.39

(a) Let P be the terminal point of $150° = 180° - 30°$ and let S be the foot of the perpendicular from P to the x-axis. Then $\angle POS = 30°$, so $OS = \sqrt{3}/2$ and $PS = 1/2$. Since $150°$ lies in the second quadrant, $\cos 150° = \boxed{-\sqrt{3}/2}$.

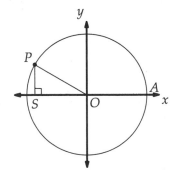

Figure 2.5: Diagram for Part (a)

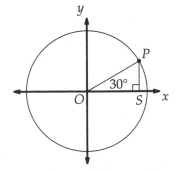

Figure 2.6: Diagram for Part (b)

(b) Let P be the terminal point of $30°$ and let S be the foot of the perpendicular from P to the x-axis. From 30-60-90 triangle POS, we have $OS = \sqrt{3}/2$ and $PS = 1/2$, so $\tan 30° = PS/OS = 1/\sqrt{3} = \boxed{\sqrt{3}/3}$.

(c) The terminal point of 90° is (0, 1), so sin 90° = 1, which means csc 90° = 1/ sin 90° = $\boxed{1}$.

(d) The terminal point of 270° is (0, −1), so cos 270° = 0 and sin 270° = −1. Therefore, tan 270° is $\boxed{\text{undefined}}$.

(e) Let P be the terminal point of 315° = 360° − 45° and let S be the foot of the perpendicular from P to the x-axis. Then $\angle POS = 45°$, so $OS = \sqrt{2}/2$ and $PS = \sqrt{2}/2$. Since 315° lies in the fourth quadrant, we have cos 315° = $\boxed{\sqrt{2}/2}$.

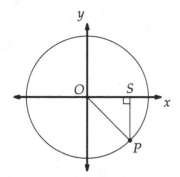

Figure 2.7: Diagram for Part (e)   Figure 2.8: Diagram for Part (f)

(f) Let P be the terminal point of 225° = 180° + 45° and let S be the foot of the perpendicular from P to the x-axis. Then $\angle POS = 45°$, so $OS = \sqrt{2}/2$ and $PS = \sqrt{2}/2$. Since 225° lies in the third quadrant, we have sin 225° = $\boxed{-\sqrt{2}/2}$.

2.40

(a) Let P be the terminal point of −120° = −180° + 60° and let S be the foot of the perpendicular from P to the x-axis. Then $\angle POS = 60°$, so $OS = 1/2$ and $PS = \sqrt{3}/2$. Since −120° lies in the third quadrant, we have sin(−120°) = $\boxed{-\sqrt{3}/2}$.

(b) Since −540° = −2 · 360° + 180°, the terminal point of −540° is the same as the terminal point of 180°, which is (−1, 0). Therefore, cos(−540°) = $\boxed{-1}$.

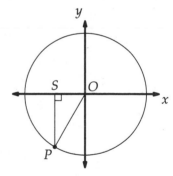

(c) Since 1320° = 3 · 360° + 240°, the terminal point of 1320° is the same as the terminal point of 240° = 180° + 60°. Let P be this terminal point and let S be the foot of the perpendicular from P to the x-axis. Then $\angle POS = 60°$, so $OS = 1/2$ and $PS = \sqrt{3}/2$. Since 240° lies in the third quadrant, we have cos 240° = −1/2 and sin 240° = $-\sqrt{3}/2$, so

$$\tan 1320° = \tan 240° = \frac{\sin 240°}{\cos 240°} = \frac{-\sqrt{3}/2}{-1/2} = \boxed{\sqrt{3}}.$$

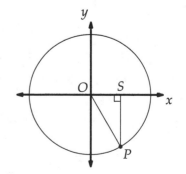

Figure 2.9: Diagram for Part (c)   Figure 2.10: Diagram for Part (d)

(d) Since $660° = 2 \cdot 360° - 60°$, the terminal point of $660°$ is the same as the terminal point of $-60°$. Let P be this terminal point and let S be the foot of the perpendicular from P to the x-axis. Then $\angle POS = 60°$, so $OS = 1/2$ and $PS = \sqrt{3}/2$. Since $-60°$ lies in the fourth quadrant, we have $\cos(-60°) = 1/2$, so

$$\sec 660° = \sec(-60°) = \frac{1}{\cos(-60°)} = \frac{1}{1/2} = \boxed{2}.$$

2.41 As θ increases from 0 to π, $\cos \theta$ varies from 1 to -1. Furthermore, it takes on each value in the interval $[-1, 1]$ exactly once, including -0.1. As θ increases from π to 2π, $\cos \theta$ varies from -1 to 1, in the same way. In general, for each integer n, as θ increases from $n\pi$ to $(n + 1)\pi$, $\cos \theta$ takes on each value in the interval $[-1, 1]$ exactly once.

Thus, we can divide the interval $[0, 7\pi]$ into the seven intervals $[0, \pi], [\pi, 2\pi], \dots, [6\pi, 7\pi]$. On each of these intervals, $\cos \theta$ takes on the value -0.1 exactly once. Therefore, there are $\boxed{7}$ values of θ in the interval $0 \le \theta \le 7\pi$ for which $\cos \theta = -0.1$.

We can also tackle this problem by thinking about the unit circle. The angles θ for which $\cos \theta = -0.1$ are those angles whose terminal points have x-coordinate equal to -0.1. The intersections of the graph of $x = -0.1$ with the unit circle give us the two such points. One is in the second quadrant and one is in the third quadrant. So, there are two values from 0 to 2π for which $\cos \theta = -0.1$, one between 0 and π and the other between π and 2π. Since the period of cosine is 2π, we have $\cos \theta = -0.1$ for exactly one value of θ between $k\pi$ and $(k + 1)\pi$ for every integer k. There are clearly 7 such intervals from 0 to 7π, so there are 7 values of θ in the interval $0 \le \theta \le 7\pi$ for which $\cos \theta = -0.1$.

2.42 The function $\tan x$ is defined for all x, except when x is an odd multiple of $\frac{\pi}{2}$; in other words, when $x = \frac{(2n+1)\pi}{2}$ for some integer n. Hence, in order for $f(t) = 2 \tan(3t - \pi)$ to be defined, we cannot have $3t - \pi = \frac{(2n+1)\pi}{2}$ for some integer n. Solving for t gives $t = \frac{(2n+3)\pi}{6}$. Therefore, $\tan(3t - \pi)$ is defined for all t except $t = \frac{(2n+3)\pi}{6}$, where n is an integer. So, the domain of $f(t)$ is all real numbers except numbers of the form $\frac{(2n+3)\pi}{6}$ where n is an integer.

The range of $f(t)$ is all real numbers, since tangent can output any real number.

2.43

(a) $144° = 144° \cdot \frac{2\pi}{360°} = \boxed{\frac{4\pi}{5}}$.   (b) $336° = 336° \cdot \frac{2\pi}{360°} = \boxed{\frac{28\pi}{15}}$.

2.44

(a) $\frac{5\pi}{2} = \frac{5\pi}{2} \cdot \frac{360°}{2\pi} = \boxed{450°}$.   (b) $\frac{11\pi}{3} = \frac{11\pi}{3} \cdot \frac{360°}{2\pi} = \boxed{660°}$.

2.45

(a) Let P be the terminal point of $\frac{3\pi}{4} = \pi - \frac{\pi}{4}$ and let S be the foot of the perpendicular from P to the x-axis. Then $\angle POS = \frac{\pi}{4}$, so $OS = \frac{\sqrt{2}}{2}$ and $PS = \frac{\sqrt{2}}{2}$. Since $\frac{3\pi}{4}$ lies in the second quadrant, we have $\cos \frac{3\pi}{4} = \boxed{-\frac{\sqrt{2}}{2}}$.

(b) The terminal point of $-\pi$ is $(-1, 0)$, so $\sin(-\pi) = \boxed{0}$.

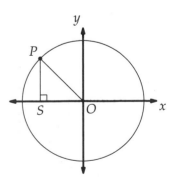

(c) Let P be the terminal point of $\frac{5\pi}{4} = \pi + \frac{\pi}{4}$ and let S be the foot of the perpendicular from P to the x-axis. Then $\angle POS = \frac{\pi}{4}$, so $OS = \frac{\sqrt{2}}{2}$ and $PS = \frac{\sqrt{2}}{2}$. Since $\frac{5\pi}{4}$ lies in the third quadrant, we have $\cos \frac{5\pi}{4} = -\frac{\sqrt{2}}{2}$ and $\sin \frac{5\pi}{4} = -\frac{\sqrt{2}}{2}$, so

$$\cot \frac{5\pi}{4} = \frac{\cos \frac{5\pi}{4}}{\sin \frac{5\pi}{4}} = \frac{-\sqrt{2}/2}{-\sqrt{2}/2} = \boxed{1}.$$

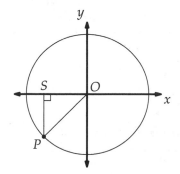

Figure 2.11: Diagram for Part (c)

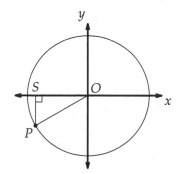

Figure 2.12: Diagram for Part (d)

(d) Let P be the terminal point of $\frac{7\pi}{6} = \pi + \frac{\pi}{6}$ and let S be the foot of the perpendicular from P to the x-axis. Then $\angle POS = \frac{\pi}{6}$, so $OS = \frac{\sqrt{3}}{2}$ and $PS = \frac{1}{2}$. Since $\frac{7\pi}{6}$ lies in the third quadrant, we have $\cos \frac{7\pi}{6} = -\frac{\sqrt{3}}{2}$, so

$$\sec \frac{7\pi}{6} = \frac{1}{-\sqrt{3}/2} = -\frac{2}{\sqrt{3}} = \boxed{-\frac{2\sqrt{3}}{3}}.$$

2.46 Since $\pi < 1.3\pi < \frac{3\pi}{2}$, the angle 1.3π lies in the third quadrant, so $\cos 1.3\pi$ is negative. Since $-2\pi < -6 < -\frac{3\pi}{2}$, the angle -6 radians lies in the first quadrant, so $\sin(-6)$ is positive. Since $\pi < \frac{7\pi}{5} < \frac{3\pi}{2}$, the angle $\frac{7\pi}{5}$ lies in the third quadrant, so $\tan \frac{7\pi}{5}$ is positive. Since $\frac{3\pi}{2} < 6 < 2\pi$, the angle 6 radians lies in the fourth quadrant, so $\sec 6 = 1/\cos 6$ is positive.

Therefore, $\boxed{\sin(-6), \tan \frac{7\pi}{5}, \text{ and } \sec 6}$ are positive.

2.47 Because 3 is between $\frac{\pi}{2}$ and π, the angle 3 radians is a second quadrant angle, which means that $\sin 3$ is positive. Since 4 is between π and $\frac{3\pi}{2}$, the angle 4 radians is a third quadrant angle, which means $\sin 4$ is negative. So, $\sin 3 + \sin 4$ is . . . Uh-oh, we need more information! Fortunately, we can get it. Since 3 is between $\frac{3\pi}{4}$ and π, we know that $0 < \sin 3 < \frac{\sqrt{2}}{2}$. Since 4 is between $\frac{5\pi}{4}$ and $\frac{3\pi}{2}$, we know that $-1 < \sin 4 < -\frac{\sqrt{2}}{2}$. Aha! Since $\sin 3 < \frac{\sqrt{2}}{2}$ and $\sin 4 < -\frac{\sqrt{2}}{2}$, we have $\sin 3 + \sin 4 < 0$. Therefore, $\sin 3 + \sin 4$ is $\boxed{\text{negative}}$. (Basically, all we're doing here is noticing that 3 radians is a second quadrant angle, 4 radians is a third quadrant angle, and 4 is farther from π than 3 is.)

2.48 In order to have $\sin(\sin x) = 1$, we must have $\sin x = \frac{\pi}{2} + 2k\pi$ for some integer k. The smallest positive value $\frac{\pi}{2} + 2k\pi$ can have is $\frac{\pi}{2}$. However, $\frac{\pi}{2}$ is greater than 1, so we cannot have $\sin x = \frac{\pi}{2}$. Similarly, the greatest negative value of $\frac{\pi}{2} + 2k\pi$ occurs when $k = -1$, which gives us $\sin x = -\frac{3\pi}{2}$. However, $\sin x$ cannot be less than -1, so this equation also has no solutions. Therefore, $\sin(\sin x)$ is never equal to 1.

2.49 The angle y in degrees is $y \cdot \frac{2\pi}{360} = \frac{y\pi}{180}$ in radians. Therefore, the line of code should be changed to $\boxed{\text{x = t * sin(y * pi/180)}}$.

2.50

(a) Since $\cot x = \frac{1}{\tan x} = \frac{\cos x}{\sin x}$, $\cot x$ is undefined whenever $\sin x = 0$. In turn, $\sin x = 0$ if and only if x is a multiple of π; in other words, $x = n\pi$ for some integer n. Therefore, the asymptotes of the graph of $y = \cot x$ are the lines $\boxed{x = n\pi, \text{ where } n \text{ is an integer}}$.

(b) By part (a), $\cot 3x$ is undefined whenever $\sin 3x = 0$. In turn, $\sin 3x = 0$ if and only if $3x$ is a multiple of π; in other words, $3x = n\pi$ for some integer n. Solving for x, we find $x = \frac{n\pi}{3}$. Therefore, the asymptotes of the graph of $y = \cot x$ are the lines $\boxed{x = \frac{n\pi}{3}, \text{ where } n \text{ is an integer}}$.

2.51   Note that $f(0) = a \sin 0 = 0$ and $g(x) = a \cos 0 = a$, so the graph of $f(x)$ passes through the origin and the graph of $g(x)$ does not pass through the origin. This gives us a quick way to identify which graph corresponds to which function.

2.52   We start with the function $\sin x$, which has period 2π and amplitude 1. The graph of $\sin \frac{x}{4}$ is obtained by stretching the graph of $\sin x$ horizontally away from the y-axis by a factor of 4, so the function $\sin \frac{x}{4}$ has period $4 \cdot 2\pi = 8\pi$ and amplitude 1.

The graph of $-2 \sin \frac{x}{4}$ is obtained by stretching the graph of $\sin \frac{x}{4}$ vertically away from the x-axis by a factor 2, and then reflecting it over the x-axis. The period is still 8π, but the amplitude is now 2.

Finally, the function $-2 \sin \frac{x}{4}$ is equal to 0 when $x = 0$. Correspondingly, the function $g(x) = -2 \sin \left(\frac{x}{4} - \frac{\pi}{3} \right)$ is equal to 0 when $\frac{x}{4} - \frac{\pi}{3} = 0$, or $x = \frac{4\pi}{3}$. Therefore, the phase shift of $g(x) = -2 \sin(\frac{x}{4} - \frac{\pi}{3})$ is $\boxed{\frac{4\pi}{3}}$. The period and amplitude are still $\boxed{8\pi \text{ and } 2}$, respectively.

2.53   We start with the function $\tan x$, which has period π. The graph of $\tan 3x$ is obtained by compressing the graph of $\tan x$ horizontally towards the y-axis by a factor of 3, so the function $\tan 3x$ has period $\frac{\pi}{3}$.

The graph of $-2 \tan 3x$ is obtained by stretching the graph of $\tan 3x$ vertically away from the x-axis by a factor of 2, and then reflecting it over the x-axis. The period is still $\frac{\pi}{3}$.

Finally, the graph of the function $g(x) = -2 \tan(3x - \pi)$ is obtained by translating the graph of $-2 \tan 3x$ by the phase shift, which does not affect the period. Therefore, the period of $g(x)$ is $\boxed{\frac{\pi}{3}}$.

2.54

(a)   The graph of $y = \sin(\frac{\pi}{2} + x)$ is a $\frac{\pi}{2}$ leftward shift of the graph of $y = \sin x$, the result of which is shown below.

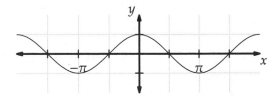

(b)   If $f(x) = \sin(\frac{\pi}{2} + x)$, then $f(-x) = \sin(\frac{\pi}{2} - x)$, so the graph of $y = \sin(\frac{\pi}{2} - x)$ is the graph of $y = \sin(\frac{\pi}{2} + x)$ reflected over the y-axis. Reflecting the graph from part (a) over the y-axis gives us

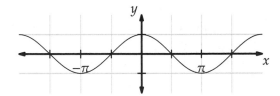

(c)   The graph of $y = \sin(\frac{\pi}{2} - x)$ appears to be exactly the same as the graph of $y = \cos x$, so it appears that $\sin(\frac{\pi}{2} - x) = \cos x$ for all x. We'll prove this relationship in the next chapter. See if you can do it on your own!

2.55   We produce the graph of $y = 2 \sin \frac{2x}{3}$ by scaling the graph of $y = \sin x$ vertically from the x-axis by a factor of 2, and horizontally away from the y-axis by a factor of 3/2, producing the graph below.

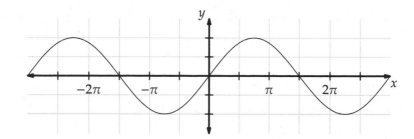

We note that if $f(x) = 2\sin\frac{2x}{3}$, then $f(x) = 0$ for $x = 0$. If $g(x) = 2\sin\left(\frac{2x}{3} - \pi\right)$, then we have $g(x) = 0$ for $x = \frac{3\pi}{2}$. Therefore, we can shift the graph of $y = 2\sin\frac{2x}{3}$ by $\frac{3\pi}{2}$ to the right to produce the graph of $y = 2\sin\left(\frac{2x}{3} - \pi\right)$. The resulting graph is shown below.

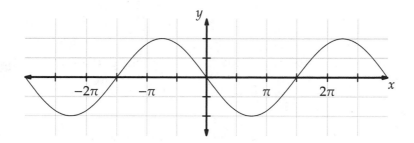

2.56 The given graph passes through the point $(0, -2)$, so $h(0) = -2$. But $h(x) = a\cos bx$, so $h(0) = a\cos 0 = a$. Hence, $\boxed{a = -2}$.

The graph completes 3 cycles between -2π and 2π, so the period is $[2\pi - (-2\pi)]/3 = \frac{4\pi}{3}$. But the period of $h(x) = a\cos bx$ is $\frac{2\pi}{|b|}$, so $\frac{2\pi}{|b|} = \frac{4\pi}{3}$. Solving for $|b|$, we find $|b| = 3/2$. Therefore, $\boxed{b = 3/2 \text{ or } -3/2}$.

2.57

(a) Since $\cos\frac{2\pi}{3} = -0.5$ and $0 \le \frac{2\pi}{3} \le \pi$, $\arccos(-0.5) = \boxed{\frac{2\pi}{3}}$.

(b) Since $\sin\left(-\frac{\pi}{2}\right) = -1$ and $-\frac{\pi}{2} \le -\frac{\pi}{2} \le \frac{\pi}{2}$, $\arcsin(-1) = \boxed{-\frac{\pi}{2}}$.

(c) Since $\tan 0 = 0$ and $-\frac{\pi}{2} < 0 < \frac{\pi}{2}$, $\tan^{-1} 0 = \boxed{0}$.

(d) Since $\cos\frac{\pi}{6} = \frac{\sqrt{3}}{2}$ and $0 \le \frac{\pi}{6} \le \pi$, $\arccos\left(\frac{\sqrt{3}}{2}\right) = \boxed{\frac{\pi}{6}}$.

2.58 Let $\theta = \arctan 2$, so $-\frac{\pi}{2} < \theta < \frac{\pi}{2}$ and $\tan\theta = 2$. Then we have $2 = \tan\theta = \frac{\sin\theta}{\cos\theta}$. Squaring both ends, we get $4 = \frac{\sin^2\theta}{\cos^2\theta} = \frac{\sin^2\theta}{1-\sin^2\theta}$. Solving for $\sin^2\theta$, we find $\sin^2\theta = \frac{4}{5}$. Taking the square root of both sides, we get $\sin\theta = \pm\frac{2}{\sqrt{5}} = \pm\frac{2\sqrt{5}}{5}$.

We know $\tan\theta = 2$. If $-\frac{\pi}{2} < \theta < \frac{\pi}{2}$ and $\tan\theta$ is positive, then $0 < \theta < \frac{\pi}{2}$, so $\sin\theta$ is also positive. Therefore, $\sin\theta = \boxed{\frac{2\sqrt{5}}{5}}$.

2.59 If $3\cot(4k - \pi) = \sqrt{3}$, then $\cot(4k - \pi) = \sqrt{3}/3$, so

$$\tan(4k - \pi) = \frac{1}{\cot(4k - \pi)} = \frac{3}{\sqrt{3}} = \sqrt{3}.$$

Since $\tan x$ has period π, this equation is equivalent to $\tan 4k = \sqrt{3}$.

We know that arctan $\sqrt{3} = \frac{\pi}{3}$. Since $\tan x$ has period π, we conclude that $\tan x = \sqrt{3}$ if and only if

$$x = \frac{\pi}{3} + n\pi = \frac{(3n+1)\pi}{3}$$

for some integer n. Therefore, if $\tan 4k = \sqrt{3}$, then $4k = \frac{(3n+1)\pi}{3}$ for some integer n. This means the solutions are of the form $k = \boxed{\frac{(3n+1)\pi}{12}}$, where n is an integer.

2.60   The graph of $y = \arctan x$ is shown below:

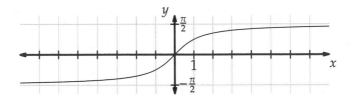

The graph of $y = \arctan x$ is the reflection over the line $y = x$ of the portion of the graph of $y = \tan x$ on the interval $-\frac{\pi}{2} < x < \frac{\pi}{2}$. Since the graph of $y = \tan x$ has the vertical asymptotes $x = -\frac{\pi}{2}$ and $x = \frac{\pi}{2}$, the graph of $y = \arctan x$ has the corresponding horizontal asymptotes $y = \frac{\pi}{2}$ and $y = -\frac{\pi}{2}$. And since $\arctan x$ is defined for all x, the graph of $y = \arctan x$ has no vertical asymptotes.

2.61   Since $\sec x = 1/\cos x$, and $\cos \frac{\pi}{2} = 0$, $\sec \frac{\pi}{2}$ is undefined. Hence, $\frac{\pi}{2}$ cannot be in the range of arcsec x.

2.62   Since $\tan \theta = \frac{\sin \theta}{\cos \theta}$, the equation $8\tan \theta = 3\cos \theta$ becomes $\frac{8\sin \theta}{\cos \theta} = 3\cos \theta$, so $8\sin \theta = 3\cos^2 \theta = 3(1 - \sin^2 \theta)$. Expanding and simplifying, we get $3\sin^2 \theta + 8\sin \theta - 3 = 0$, which factors as $(\sin \theta + 3)(3\sin \theta - 1) = 0$, so $\sin \theta = -3$ or $\sin \theta = \frac{1}{3}$. But $-1 \le \sin \theta \le 1$ for all θ, so $\sin \theta$ cannot be equal to -3. Therefore, $\sin \theta = \boxed{\frac{1}{3}}$.

Challenge Problems

2.63

(a)   First, we graph the functions $y = \sin(x - \frac{\pi}{2})$ (solid) and $y = \cos x$ (dashed).

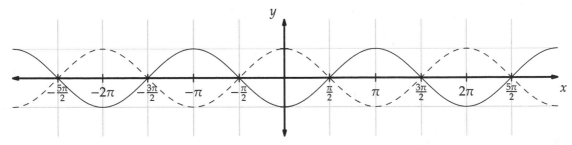

(b)   The two graphs appear to be reflections of each other over the x-axis, which suggests that we have the equation $\boxed{\sin\left(x - \frac{\pi}{2}\right) = -\cos x}$ for all x.

2.64   The period of the function $\sin(px + q)$ is the same as the period of the function $\sin px$, which is $\frac{2\pi}{|p|}$. But the period of the graphed function is $\frac{3\pi}{2}$, so $\frac{2\pi}{|p|} = \frac{3\pi}{2}$, which means $|p| = 4/3$, so $p = 4/3$ or $p = -4/3$.

First, we take the case $p = 4/3$. The graph of $y = \sin \frac{4x}{3}$ is shown below, which we must then translate to produce the graph given in the problem.

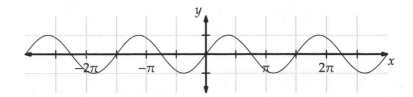

The graph of $y = \sin \frac{4x}{3}$ passes through the origin, and is increasing at the origin. The graph given in the problem passes through the point $(\frac{\pi}{4}, 0)$, and is increasing at this point. Hence, one possible equation for the given graph is

$$y = \sin\left[\frac{4}{3}\left(x - \frac{\pi}{4}\right)\right] = \sin\left(\frac{4}{3}x - \frac{\pi}{3}\right).$$

However, since the sine function has period 2π, the equation

$$y = \sin\left(\frac{4}{3}x - \frac{\pi}{3} + 2\pi n\right) = \sin\left(\frac{4}{3}x + \frac{(6n-1)\pi}{3}\right)$$

will also produce the same graph, for any integer n.

Next, we take the case $p = -4/3$. The graph of $y = \sin(-\frac{4x}{3})$ is shown below.

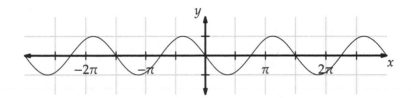

The graph of $y = \sin(-\frac{4x}{3})$ passes through the origin, and is decreasing at the origin. The graph given in the problem passes through the point $(-\frac{\pi}{2}, 0)$, and is decreasing at this point. Hence, one possible equation for the given graph is

$$y = \sin\left[-\frac{4}{3}\left(x + \frac{\pi}{2}\right)\right] = \sin\left(-\frac{4}{3}x - \frac{2\pi}{3}\right).$$

Just as in the previous case, the equation

$$y = \sin\left(-\frac{4}{3}x - \frac{2\pi}{3} + 2\pi n\right) = \sin\left(-\frac{4}{3}x + \frac{2(3n-1)\pi}{3}\right)$$

will also produce the same graph, for any integer n.

Therefore, the possible pairs (p, q) are

$$\boxed{\left(\frac{4}{3}, \frac{(6n-1)\pi}{3}\right) \quad \text{and} \quad \left(-\frac{4}{3}, \frac{2(3n-1)\pi}{3}\right),}$$

where n is any integer.

2.65 We claim that the period of $f(x) = \sin x \cos x$ is π.

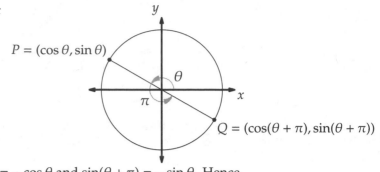

First, we prove that $\cos(\theta + \pi) = -\cos \theta$ and $\sin(\theta + \pi) = -\sin \theta$ for all θ. Let P and Q be the terminal points of θ and $\theta + \pi$, respectively, so $P = (\cos \theta, \sin \theta)$ and $Q = (\cos(\theta + \pi), \sin(\theta + \pi))$.

The point Q is obtained by rotating P around the origin counterclockwise by π, which means that P and Q are diametrically opposite points on the unit circle. Therefore, we have $\cos(\theta + \pi) = -\cos \theta$ and $\sin(\theta + \pi) = -\sin \theta$. Hence,

$$f(x + \pi) = \sin(x + \pi)\cos(x + \pi) = (-\sin x)(-\cos x) = \sin x \cos x = f(x).$$

To show that π is the period of $f(x)$, we must also show that there is no positive real number d smaller than π such that $f(x + d) = f(x)$ for all x.

If d is the period of f, then we have $f(x + d) = 0$ whenever $f(x) = 0$. So, finding all values of x such that $f(x) = 0$ will limit the possible values of d. We have $f(x) = 0$ if and only if $\cos x = 0$ or $\sin x = 0$, so we have $f(x) = 0$ for $x = \frac{n\pi}{2}$ for all integers n. Since the values of x for which $f(x) = 0$ are $\frac{\pi}{2}$ apart, the only positive real number d less than π that could possibly be the period of $f(x)$ is $\frac{\pi}{2}$. However,

$$f\left(\frac{\pi}{4}\right) = \sin\frac{\pi}{4}\cos\frac{\pi}{4} = \frac{1}{\sqrt{2}} \cdot \frac{1}{\sqrt{2}} = \frac{1}{2},$$

and

$$f\left(\frac{\pi}{4} + \frac{\pi}{2}\right) = f\left(\frac{3\pi}{4}\right) = \sin\frac{3\pi}{4}\cos\frac{3\pi}{4} = \frac{1}{\sqrt{2}} \cdot \left(-\frac{1}{\sqrt{2}}\right) = -\frac{1}{2},$$

so $f(\frac{\pi}{4}) \neq f(\frac{\pi}{4} + \frac{\pi}{2})$, which means that the period of $f(x)$ cannot be $\frac{\pi}{2}$. We conclude that the period of $f(x)$ is $\boxed{\pi}$.

2.66 Since $|\sin x| \leq 1$ for all x, if x satisfies the equation $\sin x = \frac{x^2}{625\pi^2}$, then

$$\left|\frac{x^2}{625\pi^2}\right| \leq 1 \quad \Rightarrow \quad |x^2| \leq 625\pi^2 \quad \Rightarrow \quad |x| \leq \sqrt{625\pi^2} = 25\pi.$$

Thus, all solutions x lie in the interval $-25\pi \leq x \leq 25\pi$.

We consider the graphs of $y = \sin x$ and $y = \frac{x^2}{625\pi^2}$ and count the number of times they intersect. The graph of $y = \frac{x^2}{625\pi^2}$ is an upward-opening parabola. When $|x| < 25\pi$, we have $\frac{x^2}{625\pi^2} < 1$, so for $|x| < 25\pi$, the parabola intersects the sine curve at two points in each period—once as $\sin x$ goes up from 0 to 1, and then again as $\sin x$ goes from 1 back down to 0. Therefore, in the 24 full periods from $x = -24\pi$ to $x = 24\pi$, there are $2 \cdot 24 = 48$ intersections. We have to be careful in the intervals $-25\pi \leq x < -24\pi$ and $24\pi < x \leq 25\pi$. The two intersections in each period are always in the left half of the period, since $\sin x$ is nonnegative in this half of the period. Therefore, there are two intersections for $24\pi < x \leq 25\pi$, but no intersections for $-25\pi \leq x < -24\pi$. This gives us a total of $48 + 2 = \boxed{50}$ intersections, and therefore 50 solutions to the given equation.

2.67 Let P be the terminal point of $\frac{6\pi}{7}$, and let S be the foot of the perpendicular from P to the x-axis. Then $\sin \frac{6\pi}{7} = PS$, so $\arccos(\sin \frac{6\pi}{7}) = \arccos PS$.

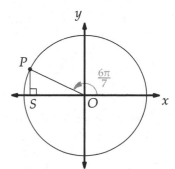

But $\angle POS = \pi - \frac{6\pi}{7} = \frac{\pi}{7}$, and $\angle OPS = \frac{\pi}{2} - \frac{\pi}{7} = \frac{5\pi}{14}$, so $\cos \frac{5\pi}{14} = \frac{PS}{OP} = PS$. Since $\frac{5\pi}{14}$ is in the range of the arccos function, we conclude that $\arccos PS = \boxed{\frac{5\pi}{14}}$.

2.68 The terminal point of $\frac{\pi}{6}$ is $\left(\cos \frac{\pi}{6}, \sin \frac{\pi}{6}\right) = \left(\frac{\sqrt{3}}{2}, \frac{1}{2}\right)$, and the terminal point of $\frac{5\pi}{6}$ is $\left(\cos \frac{5\pi}{6}, \sin \frac{5\pi}{6}\right) = \left(-\frac{\sqrt{3}}{2}, \frac{1}{2}\right)$. Hence, $\sin x \geq \frac{1}{2}$ for $\frac{\pi}{6} \leq x \leq \frac{5\pi}{6}$, and $\sin x < \frac{1}{2}$ for $\frac{5\pi}{6} < x < \frac{\pi}{6} + 2\pi = \frac{13\pi}{6}$.

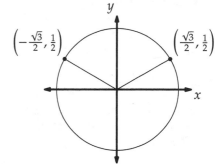

Since $\sin x$ has period 2π, we can also say that $\sin x < \frac{1}{2}$ for $\frac{5\pi}{6} + 2\pi k < x < \frac{13\pi}{6} + 2\pi k$ for any integer k, which simplifies as $\frac{12\pi k + 5\pi}{6} < x < \frac{12\pi k + 13\pi}{6}$. Then taking $x = 3\theta$, we find that $\sin x = \sin 3\theta$, and $\frac{12\pi k + 5\pi}{6} < 3\theta < \frac{12\pi k + 13\pi}{6}$. Therefore, $\sin 3\theta < \frac{1}{2}$ if and only if

$$\boxed{\frac{12\pi k + 5\pi}{18} < \theta < \frac{12\pi k + 13\pi}{18}}$$

for some integer k.

2.69

(a) We know that $\sin x$ and $\cos x$ have period 2π. Therefore,

$$g(x + 2\pi) = \sin[a(x + 2\pi)] + \cos[b(x + 2\pi)] = \sin(ax + 2\pi a) + \cos(bx + 2\pi b) = \sin ax + \cos bx = g(x),$$

so $g(x)$ is periodic. (However, the period of $g(x)$ is not necessarily 2π. For example, the function $\sin 2x$ has period π.)

(b) Let $a = a_1/a_2$ and $b = b_1/b_2$, where a_1, a_2, b_1, and b_2 are integers. Then $aa_2 = a_1$ and $bb_2 = b_1$, and

$$\begin{aligned}
g(x + 2\pi a_2 b_2) &= \sin[a(x + 2\pi a_2 b_2)] + \cos[b(x + 2\pi a_2 b_2)] \\
&= \sin(ax + 2\pi a a_2 b_2) + \cos(bx + 2\pi a_2 b b_2) \\
&= \sin(ax + 2\pi a_1 b_2) + \cos(bx + 2\pi a_2 b_1) \\
&= \sin ax + \cos bx \\
&= g(x),
\end{aligned}$$

so $g(x)$ is periodic.

2.70 We consider the sine function with respect to radians. Then the sine of t in radians is simply $\sin t$. However, to evaluate the sine of t in degrees, we must first convert t to radians, which becomes $\frac{\pi}{180}t$. Then the sine of t in degrees is $\sin \frac{\pi}{180}t$. Since the professor found the correct answer despite being in the wrong angular mode, we must have

$$\sin t = \sin \frac{\pi}{180}t.$$

In order for the sines of two angles to be equal, the terminal points of the two angles must have the same y-coordinate. So, we examine the terminal points of the angles t and $\frac{\pi}{180}t$.

For all positive t, the angle t is larger than the angle $\frac{\pi}{180}t$. Therefore, when $0 < t < \frac{\pi}{2}$, the terminal points of t and $\frac{\pi}{180}t$ cannot be the same. Moreover, as t goes from $\frac{\pi}{2}$ to π, the y-coordinate of its terminal point goes from 1 to 0. Meanwhile, the value of $\frac{\pi}{180}t$ goes from $\frac{\pi^2}{360}$ to $\frac{\pi^2}{180}$. The y-coordinate of the terminal point of $\frac{\pi}{180}t$ therefore is pretty close to 0, and increases slightly as t goes from $\frac{\pi}{2}$ to π. Putting these together, we see that for some value of t between $\frac{\pi}{2}$ and π, the two terminal points have the same y-coordinate.

Let these terminal points be P and Q, and let A and B be the feet of the altitudes from these points, as shown at right. (This figure is very much not to scale!) Since these terminal points have equal y-coordinates, we have $QB = PA$. Putting this together with $PO = QO$ gives us $\triangle POA \cong \triangle QOB$ by SA Congruence for right triangles, so we have $\angle QOB = \angle POA$. Since Q is the terminal point of $\angle QOA$, we have $\angle QOA = t$. Similarly, we have $\angle POA = \frac{\pi}{180}t$, so we have

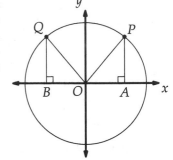

$$\pi = \angle QOB + \angle QOA = \angle POA + \angle QOA = \frac{\pi}{180}t + t.$$

Solving for t gives $t = \frac{180\pi}{180+\pi}$.

Fortunately, finding the next solution is easier. As t continues from the value we just found up to $t = \pi$, the y-coordinate of its terminal point decreases, while that of $\frac{\pi}{180}t$ increases, so there are no more positive solutions less than or equal to π. For $\pi < t < 2\pi$, the value of $\sin t$ is negative, but that of $\sin \frac{\pi}{180}t$ is positive, since $0 < t < \frac{\pi}{2}$ for values of t in this range. The next time the angles t and $\frac{\pi}{180}t$ will have equal sines is when their terminal points coincide. This occurs when the values of the angles themselves differ by 2π, so we have

$$t - 2\pi = \frac{\pi}{180}t.$$

Solving this equation gives $t = \frac{360\pi}{180-\pi}$.

Therefore, the two smallest positive values of x such that $\sin x = \sin \frac{\pi}{180}x$ are $\boxed{\frac{360\pi}{180-\pi}}$ and $\boxed{\frac{180\pi}{180+\pi}}$.

2.71 Since the price of the stock varies sinusoidally with time, we can express it as $p(t) = a\sin(bt + c) + d$ for some constants a, b, c and d, where t is in days. Since the price per share varies from \$40 to \$80, the amplitude is 20, and the price varies about an average price of \$60 per share. So, we can let $a = 20$ and $d = 60$, and we have $p(t) = 20\sin(bt + c) + 60$.

At the start of the year, the price is \$70, so we must have $p(0) = 70$, which means $20\sin(c) = 10$, so $\sin(c) = \frac{1}{2}$. Therefore, we can let $c = \frac{\pi}{6}$, and we have $p(t) = 20\sin\left(bt + \frac{\pi}{6}\right) + 60$. Finally, from the information that the price climbs to a peak of \$80 in 4 days, we have $p(4) = 80$, so $20\sin\left(4b + \frac{\pi}{6}\right) + 60 = 80$. From this, we have $\sin\left(4b + \frac{\pi}{6}\right) = 1$, so $4b + \frac{\pi}{6} = \frac{\pi}{2}$, which gives us $b = \frac{\pi}{12}$. Therefore, we have

$$p(t) = 20\sin\left(\frac{\pi}{12}t + \frac{\pi}{6}\right) + 60.$$

To maximize the amount of money we make, we buy as low as we can and sell as high as we can. So, we start by buying at \$70/share, since we know we'll be able to sell it at \$80/share. We can buy \$700/\$70 = 10 shares, and we'll sell them for \$80 per share, so we'll have \$800 after 4 days.

After that, we repeatedly buy at \$40 per share and sell at \$80 per share, doubling our money each time. (Note that our money is a multiple of \$40, so we'll always be buying a whole number of shares.) But how many times can we do so? The period of p is $\frac{2\pi}{\pi/12} = 24$. That is, the stock goes from \$80/share to \$40/share and back up to \$80/share in 24 days. So, every 24 days, we can double our money. We'll be able to do so $360/24 = 15$ times in the next 360 days, giving us a total of $\$800 \cdot 2^{15} = \$26{,}214{,}400$. At that point, we'll be 364 days into the year, and the price will go down on the last day, and we won't be able to make any more money. Since we started with \$700, our profit is $\$26{,}214{,}400 - \$700 = \boxed{\$26{,}213{,}700}$.

Trigonometric Identities

Exercises for Section 3.1

3.1.1 $\cot^2 \theta + 1 = \dfrac{\cos^2 \theta}{\sin^2 \theta} + 1 = \dfrac{\cos^2 \theta + \sin^2 \theta}{\sin^2 \theta} = \dfrac{1}{\sin^2 \theta} = \csc^2 \theta.$

3.1.2 The angle $360° - \theta$ has the same terminal point as the angle $-\theta$. Let P and Q be the terminal points of θ and $-\theta$, respectively. The actions of rotating $(1,0)$ an angle θ counterclockwise about the origin and rotating $(1,0)$ an angle θ *clockwise* about the origin are symmetric about the x-axis. That is, P and Q are reflections of each other over the x-axis, which means that their y-coordinates are opposites. Therefore, we have $\sin\theta = -\sin(-\theta) = -\sin(360° - \theta)$, which means that $\sin(360° - \theta) = -\sin\theta.$

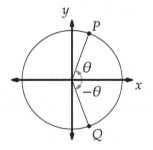

3.1.3 From the equation $\tan x + \sec x = \sqrt{3}$, we have $\sec x = \sqrt{3} - \tan x$. Substituting into the identity $\tan^2 x + 1 = \sec^2 x$, we get

$$\tan^2 x + 1 = (\sqrt{3} - \tan x)^2 \quad\Rightarrow\quad \tan^2 x + 1 = 3 - 2\sqrt{3}\tan x + \tan^2 x \quad\Rightarrow\quad 2\sqrt{3}\tan x = 2 \quad\Rightarrow\quad \tan x = \dfrac{2}{2\sqrt{3}} = \dfrac{\sqrt{3}}{3}.$$

The only values in the interval $0 \le x < 2\pi$ such that $\tan x = \frac{\sqrt{3}}{3}$ are $x = \frac{\pi}{6}$ and $x = \frac{7\pi}{6}$. Checking these values, we find that only $x = \boxed{\frac{\pi}{6}}$ satisfies the given equation.

3.1.4 We can combine the identities that we have derived so far. For example,

$$\cos\left(\frac{3\pi}{2} + \theta\right) = \cos\left(\frac{\pi}{2} + (\pi + \theta)\right).$$

In the text, we showed that $\cos\left(\frac{\pi}{2} + x\right) = -\sin x$ for all x, so

$$\cos\left(\frac{\pi}{2} + (\pi + \theta)\right) = -\sin(\pi + \theta),$$

which we can rewrite as $-\sin[\pi - (-\theta)]$.

We also showed that $\sin(\pi - x) = \sin x$ for all x, so $-\sin[\pi - (-\theta)] = -\sin(-\theta)$. Finally, we know that $\sin(-\theta) = -\sin\theta$, so we have $-\sin(-\theta) = \sin\theta$. Therefore, $\cos\left(\frac{3\pi}{2} + \theta\right) = \boxed{\sin\theta}$.

3.1.5 First, we note that $\tan\frac{3\pi}{12} = \tan\frac{\pi}{4} = 1$, and that we can pair off the other four angles conveniently: $\frac{\pi}{12} + \frac{5\pi}{12} = \frac{2\pi}{12} + \frac{4\pi}{12} = \frac{\pi}{2}$. In the text, we showed that $\tan\left(\frac{\pi}{2} - x\right) = \cot x$, so $\tan\left(\frac{\pi}{2} - x\right) = \frac{1}{\tan x}$, which means we have

$\tan\frac{\pi}{12}\tan\frac{5\pi}{12} = 1$ and $\tan\frac{2\pi}{12}\tan\frac{4\pi}{12} = 1$. Therefore, we have

$$\tan\frac{\pi}{12}\tan\frac{2\pi}{12}\tan\frac{3\pi}{12}\tan\frac{4\pi}{12}\tan\frac{5\pi}{12} = \boxed{1}.$$

3.1.6   The numerator is equal to

$$\sin A(3\cos^2 A + \cos^4 A + 3\sin^2 A + \sin^2 A\cos^2 A) = \sin A[(3\cos^2 A + 3\sin^2 A) + \cos^2 A\cos^2 A + \sin^2 A\cos^2 A]$$
$$= \sin A[3 + \cos^2 A(\cos^2 A + \sin^2 A)]$$
$$= \sin A(3 + \cos^2 A),$$

and, when A is not an integer multiple of $\frac{\pi}{2}$, the denominator is equal to

$$\tan A(\sec A - \sin A\tan A) = \frac{\sin A}{\cos A}\left(\frac{1}{\cos A} - \sin A\cdot\frac{\sin A}{\cos A}\right) = \frac{\sin A(1 - \sin^2 A)}{\cos^2 A} = \frac{\sin A\cos^2 A}{\cos^2 A} = \sin A.$$

Therefore,

$$f(A) = \frac{\sin A(3 + \cos^2 A)}{\sin A} = 3 + \cos^2 A.$$

The range of $\cos^2 A$ is the interval $[0, 1]$, so the range of $3 + \cos^2 A$ is the interval $[3, 4]$. However, A is not allowed to be any integer multiple of $\frac{\pi}{2}$, so $\cos A$ cannot be equal to 0, 1, or -1. Therefore, the range of $f(A)$ is the interval $\boxed{(3, 4)}$.

Exercises for Section 3.2

3.2.1

(a) $\sin(\pi + \theta) = \sin\pi\cos\theta + \sin\theta\cos\pi = \boxed{-\sin\theta}$.

(b) $\cos\left(\frac{\pi}{2} + \theta\right) = \cos\frac{\pi}{2}\cos\theta - \sin\frac{\pi}{2}\sin\theta = \boxed{-\sin\theta}$.

(c) $\cos(3\pi - \theta) = \cos 3\pi\cos\theta + \sin 3\pi\sin\theta = \boxed{-\cos\theta}$.

(d)   If we apply the angle sum identity for tangent, we obtain $\tan\left(\theta + \frac{3\pi}{2}\right) = \frac{\tan\theta + \tan\frac{3\pi}{2}}{1 - \tan\theta\tan\frac{3\pi}{2}}$. However, $\tan\frac{3\pi}{2}$ is not defined, so we try a different approach:

$$\tan\left(\theta + \frac{3\pi}{2}\right) = \frac{\sin\left(\theta + \frac{3\pi}{2}\right)}{\cos\left(\theta + \frac{3\pi}{2}\right)}.$$

By the angle sum identity for sine, we have

$$\sin\left(\theta + \frac{3\pi}{2}\right) = \sin\theta\cos\frac{3\pi}{2} + \sin\frac{3\pi}{2}\cos\theta = -\cos\theta,$$

and by the angle sum identity for cosine, we have

$$\cos\left(\theta + \frac{3\pi}{2}\right) = \cos\theta\cos\frac{3\pi}{2} - \sin\theta\sin\frac{3\pi}{2} = \sin\theta,$$

so $\dfrac{\sin\left(\theta + \frac{3\pi}{2}\right)}{\cos\left(\theta + \frac{3\pi}{2}\right)} = \dfrac{-\cos\theta}{\sin\theta} = -\cot\theta = \boxed{-\dfrac{1}{\tan\theta}}$.

3.2.2 By the angle sum identity for tangent, we have $\tan(\alpha + \gamma) = \dfrac{\tan \alpha + \tan \gamma}{1 - \tan \alpha \tan \gamma}$. Letting $\gamma = -\beta$, we have

$$\tan(\alpha - \beta) = \frac{\tan \alpha + \tan(-\beta)}{1 - \tan \alpha \tan(-\beta)} = \frac{\tan \alpha - \tan \beta}{1 + \tan \alpha \tan \beta}.$$

3.2.3 By the angle sum identity for tangent, we have

$$\tan 285° = \tan(240° + 45°) = \frac{\tan 240° + \tan 45°}{1 - \tan 240° \tan 45°} = \frac{\sqrt{3} + 1}{1 - \sqrt{3} \cdot 1} = \frac{\sqrt{3} + 1}{1 - \sqrt{3}}$$

$$= \frac{(\sqrt{3} + 1)(1 + \sqrt{3})}{(1 - \sqrt{3})(1 + \sqrt{3})} = \frac{4 + 2\sqrt{3}}{-2} = \boxed{-2 - \sqrt{3}}.$$

3.2.4 We have $\sin(\alpha - \beta) = \sin \alpha \cos \beta - \sin \beta \cos \alpha$ and $\sin(\alpha + \beta) = \sin \alpha \cos \beta + \sin \beta \cos \alpha$, so

$$\sin(\alpha - \beta) \sin(\alpha + \beta) = (\sin \alpha \cos \beta - \sin \beta \cos \alpha)(\sin \alpha \cos \beta + \sin \beta \cos \alpha)$$
$$= \sin^2 \alpha \cos^2 \beta + \sin \alpha \cos \alpha \sin \beta \cos \beta - \sin \alpha \cos \alpha \sin \beta \cos \beta - \sin^2 \beta \cos^2 \alpha$$
$$= \sin^2 \alpha \cos^2 \beta - \sin^2 \beta \cos^2 \alpha$$
$$= \sin^2 \alpha(1 - \sin^2 \beta) - \sin^2 \beta(1 - \sin^2 \alpha)$$
$$= \sin^2 \alpha - \sin^2 \alpha \sin^2 \beta - \sin^2 \beta + \sin^2 \alpha \sin^2 \beta$$
$$= \sin^2 \alpha - \sin^2 \beta.$$

3.2.5 If $\sin 2x \sin 3x = \cos 2x \cos 3x$, then

$$\cos 2x \cos 3x - \sin 2x \sin 3x = 0.$$

By the angle sum identity for cosine, we have $\cos 2x \cos 3x - \sin 2x \sin 3x = \cos(2x + 3x) = \cos 5x$, so $\cos 5x = 0$. Since x is an acute angle, we have $0 < x < \frac{\pi}{2}$, so $0 < 5x < \frac{5\pi}{2}$. The only angles between 0 and $\frac{5\pi}{2}$ whose cosine is 0 are $\frac{\pi}{2}$ and $\frac{3\pi}{2}$, so the solutions are $x = \boxed{\frac{\pi}{10}}$ and $\boxed{\frac{3\pi}{10}}$.

3.2.6 By the angle sum identity for tangent, we have $\tan(A + B) = \frac{\tan A + \tan B}{1 - \tan A \tan B}$. But we are given $\tan A + \tan B = 1 - \tan A \tan B$, so $\frac{\tan A + \tan B}{1 - \tan A \tan B} = 1$, which means $\tan(A + B) = 1$.

Since angles A and B are acute, we have $0° < A + B < 180°$. The only angle in this range whose tangent is 1 is $45°$, so $A + B = 45°$. Thus, we have the equations $A + B = 45°$ and $A - B = 41°$. Adding these two equations, we get $2A = 86°$, so $A = \boxed{43°}$.

3.2.7 Squaring the given equations, we get

$$\sin^2 a + 2 \sin a \sin b + \sin^2 b = \frac{5}{3},$$
$$\cos^2 a + 2 \cos a \cos b + \cos^2 b = 1.$$

Adding these equations, we get

$$\sin^2 a + 2 \sin a \sin b + \sin^2 b + \cos^2 a + 2 \cos a \cos b + \cos^2 b = \frac{8}{3}.$$

But $\sin^2 a + \cos^2 a = \sin^2 b + \cos^2 b = 1$, so $2 \sin a \sin b + 2 \cos a \cos b = \frac{8}{3} - 2 = \frac{2}{3}$, which means $\cos a \cos b + \sin a \sin b = \frac{1}{3}$. Then by the angle difference identity for cosine, we have $\cos(a - b) = \cos a \cos b + \sin a \sin b = \boxed{\frac{1}{3}}$.

3.2.8

(a) Since $0 < \alpha < \frac{\pi}{2}$ and $\gamma = \frac{\pi}{2} - \alpha$, we have $0 < \gamma < \frac{\pi}{2}$. Hence, γ is acute. Similarly, $0 < \beta < \frac{\pi}{2}$ and $\phi = \frac{\pi}{2} - \beta$, so $0 < \phi < \frac{\pi}{2}$. Hence, ϕ is acute.

(b) Substituting, we get $\sin(\alpha + \beta) = \sin\left[\left(\frac{\pi}{2} - \gamma\right) + \left(\frac{\pi}{2} - \phi\right)\right] = \sin[\pi - (\gamma + \phi)]$.

(c) Since $\sin(\pi - x) = \sin x$ for all x, we have $\sin(\alpha + \beta) = \sin[\pi - (\gamma + \phi)] = \sin(\gamma + \phi)$.

(d) We see that

$$\gamma + \phi = \left(\frac{\pi}{2} - \alpha\right) + \left(\frac{\pi}{2} - \beta\right) = \pi - (\alpha + \beta).$$

Since $\alpha + \beta$ is obtuse, $\gamma + \phi$ is acute. Furthermore, both γ and ϕ are acute, which is the case we have already proved (this is why it is important that γ and ϕ are acute). So, we may apply the angle sum identity to get

$$\sin(\gamma + \phi) = \sin \gamma \cos \phi + \sin \phi \cos \gamma.$$

But $\sin\left(\frac{\pi}{2} - x\right) = \cos x$ and $\cos\left(\frac{\pi}{2} - x\right) = \sin x$ for all x, so

$$\sin \gamma \cos \phi + \sin \phi \cos \gamma = \sin\left(\frac{\pi}{2} - \alpha\right)\cos\left(\frac{\pi}{2} - \beta\right) + \sin\left(\frac{\pi}{2} - \beta\right)\cos\left(\frac{\pi}{2} - \alpha\right)$$
$$= \cos \alpha \sin \beta + \cos \beta \sin \alpha.$$

By part (c), $\sin(\alpha + \beta) = \sin(\gamma + \phi)$. Therefore, the angle sum identity for sine holds when α and β are acute and $\alpha + \beta$ is obtuse.

(e) If α and β are obtuse, then $\pi - \alpha$ and $\pi - \beta$ are acute, so we can use the sine angle addition identity, which we have proved for acute angles, to find that

$$\sin\left((\pi - \alpha) + (\pi - \beta)\right) = \sin(\pi - \alpha)\cos(\pi - \beta) + \sin(\pi - \beta)\cos(\pi - \alpha).$$

Applying $\sin(\pi - \theta) = \sin \theta$ and $\cos(\pi - \theta) = -\cos \theta$, we have

$$\sin\left((\pi - \alpha) + (\pi - \beta)\right) = -\sin \alpha \cos \beta - \sin \beta \cos \alpha.$$

As Exercise 3.1.2, we showed that $\sin(2\pi - \theta) = -\sin \theta$, so

$$\sin\left((\pi - \alpha) + (\pi - \beta)\right) = \sin\left(2\pi - (\alpha + \beta)\right) = -\sin(\alpha + \beta),$$

which means we have $-\sin(\alpha+\beta) = -\sin \alpha \cos \beta - \sin \beta \cos \alpha$. Multiplying by -1 gives the desired $\sin(\alpha+\beta) = \sin \alpha \cos \beta + \sin \beta \cos \alpha$.

Exercises for Section 3.3

3.3.1 By the half-angle formula for sine, we have

$$\sin 22.5° = \sin\frac{45°}{2} = \pm\sqrt{\frac{1 - \cos 45°}{2}} = \pm\sqrt{\frac{1 - \sqrt{2}/2}{2}} = \pm\sqrt{\frac{2 - \sqrt{2}}{4}} = \pm\frac{\sqrt{2 - \sqrt{2}}}{2}.$$

Since the angle 22.5° lies in the first quadrant, its sine is positive, so $\sin 22.5° = \boxed{\frac{\sqrt{2-\sqrt{2}}}{2}}$.

3.3.2 By the double angle formula for cosine, we have $\cos 2\theta = 1 - 2\sin^2 \theta = 1 - 2\left(\frac{1}{4}\right)^2 = 1 - \frac{1}{8} = \boxed{\frac{7}{8}}$.

3.3.3 We have

$$\frac{2\tan x}{1 + \tan^2 x} = \frac{2 \cdot \frac{\sin x}{\cos x}}{1 + \frac{\sin^2 x}{\cos^2 x}} = \frac{2\sin x \cos x}{\cos^2 x + \sin^2 x} = 2\sin x \cos x,$$

and by the double angle formula for sine, we have $2\sin x \cos x = \sin 2x$. Therefore, we have $\sin 2x = \frac{2\tan x}{1+\tan^2 x}$, as desired.

3.3.4 We have

$$\tan x + \cot x = \frac{\sin x}{\cos x} + \frac{\cos x}{\sin x} = \frac{\sin^2 x + \cos^2 x}{\sin x \cos x} = \frac{1}{\sin x \cos x} = \frac{2}{2\sin x \cos x}.$$

By the double angle formula for sine, we have $2\sin x \cos x = \sin 2x$, so $\frac{2}{2\sin x \cos x} = \frac{2}{\sin 2x} = 2\csc 2x$.

3.3.5 By the double angle formula for cosine, we have $\cos 2A = 2\cos^2 A - 1 = 1 - 2\sin^2 A$, so

$$\frac{\sin 2A - \cos 2A + 1}{\sin 2A + \cos 2A + 1} = \frac{\sin 2A - (1 - 2\sin^2 A) + 1}{\sin 2A + (2\cos^2 A - 1) + 1} = \frac{\sin 2A + 2\sin^2 A}{\sin 2A + 2\cos^2 A}.$$

By the double angle formula for sine, we have $\sin 2A = 2\sin A \cos A$, so

$$\frac{\sin 2A + 2\sin^2 A}{\sin 2A + 2\cos^2 A} = \frac{2\sin A \cos A + 2\sin^2 A}{2\sin A \cos A + 2\cos^2 A} = \frac{2\sin A(\cos A + \sin A)}{2\cos A(\sin A + \cos A)} = \frac{\sin A}{\cos A} = \tan A.$$

3.3.6 Let $x = \cos\theta - \sin\theta$. Then

$$x^2 = \cos^2\theta - 2\cos\theta\sin\theta + \sin^2\theta = 1 - 2\cos\theta\sin\theta.$$

By the double angle formula for sine, we have $2\cos\theta\sin\theta = \sin 2\theta = \frac{21}{25}$, so $x^2 = 1 - \frac{21}{25} = \frac{4}{25}$, which means $x = \pm\sqrt{\frac{4}{25}} = \pm\frac{2}{5}$. We are given that $\cos\theta > \sin\theta$, so $x = \cos\theta - \sin\theta = \boxed{2/5}$.

3.3.7 Squaring the given equation, we get

$$\cos^2\theta + 2\cos\theta\sin\theta + \sin^2\theta = \frac{3}{2}.$$

But $\cos^2\theta + \sin^2\theta = 1$, so $2\cos\theta\sin\theta = \frac{1}{2}$. If $\cos\theta$ is negative, then $\sin\theta$ is also negative, because the product $\cos\theta\sin\theta$ is positive. This means $\cos\theta + \sin\theta$ is negative. But $\cos\theta + \sin\theta = \sqrt{3/2}$, so we cannot have $\cos\theta$ be negative. Therefore, $\cos\theta$ is positive, so $\sin\theta$ is also positive, which means θ lies in the first quadrant. In other words, $0 < \theta < \frac{\pi}{2}$.

Now, by the double angle formula for sine, we have $2\cos\theta\sin\theta = \sin 2\theta$, so $2\cos\theta\sin\theta = \frac{1}{2}$ gives us $\sin 2\theta = \frac{1}{2}$. We know that $0 < \theta \le \frac{\pi}{2}$, so $0 < 2\theta < \pi$. The only angles in this interval whose sine is $\frac{1}{2}$ are $\frac{\pi}{6}$ and $\frac{5\pi}{6}$. Hence, $2\theta = \frac{\pi}{6}$ or $2\theta = \frac{5\pi}{6}$, so the solutions are $\boxed{\theta = \frac{\pi}{12} \text{ and } \frac{5\pi}{12}}$.

We check these solutions: Note that $\frac{\pi}{12} = \frac{\pi}{3} - \frac{\pi}{4}$, so by the angle difference identity for cosine,

$$\cos\frac{\pi}{12} = \cos\left(\frac{\pi}{3} - \frac{\pi}{4}\right) = \cos\frac{\pi}{3}\cos\frac{\pi}{4} + \sin\frac{\pi}{3}\sin\frac{\pi}{4} = \frac{1}{2}\cdot\frac{\sqrt{2}}{2} + \frac{\sqrt{3}}{2}\cdot\frac{\sqrt{2}}{2} = \frac{\sqrt{2}+\sqrt{6}}{4},$$

and by the angle difference identity for sine,

$$\sin\frac{\pi}{12} = \sin\left(\frac{\pi}{3} - \frac{\pi}{4}\right) = \sin\frac{\pi}{3}\cos\frac{\pi}{4} - \sin\frac{\pi}{4}\cos\frac{\pi}{3} = \frac{\sqrt{3}}{2}\cdot\frac{\sqrt{2}}{2} - \frac{\sqrt{2}}{2}\cdot\frac{1}{2} = \frac{\sqrt{6}-\sqrt{2}}{4},$$

so $\cos\dfrac{\pi}{12} + \sin\dfrac{\pi}{12} = \dfrac{2\sqrt{6}}{4} = \dfrac{\sqrt{6}}{2} = \sqrt{\dfrac{6}{4}} = \sqrt{\dfrac{3}{2}}$. Also, $\dfrac{5\pi}{12} = \dfrac{\pi}{2} - \dfrac{\pi}{12}$, so

$$\cos\frac{5\pi}{12} + \sin\frac{5\pi}{12} = \cos\left(\frac{\pi}{2} - \frac{5\pi}{12}\right) + \sin\left(\frac{\pi}{2} - \frac{5\pi}{12}\right) = \sin\frac{\pi}{12} + \cos\frac{\pi}{12} = \sqrt{\frac{3}{2}}.$$

3.3.8

(a) By the angle sum identity for cosine, we have

$$\cos 3\theta = \cos(\theta + 2\theta) = \cos\theta\cos 2\theta - \sin\theta\sin 2\theta.$$

Then by the double angle formulas for cosine and sine, we have

$$\cos\theta\cos 2\theta - \sin\theta\sin 2\theta = \cos\theta(2\cos^2\theta - 1) - \sin\theta(2\sin\theta\cos\theta) = 2\cos^3\theta - \cos\theta - 2\sin^2\theta\cos\theta.$$

But $\sin^2\theta = 1 - \cos^2\theta$, so

$$\begin{aligned}
2\cos^3\theta - \cos\theta - 2\sin^2\theta\cos\theta &= 2\cos^3\theta - \cos\theta - 2(1 - \cos^2\theta)\cos\theta \\
&= 2\cos^3\theta - \cos\theta - 2\cos\theta + 2\cos^3\theta \\
&= \boxed{4\cos^3\theta - 3\cos\theta}.
\end{aligned}$$

(b) By the angle sum identity for sine, we have

$$\sin 3\theta = \sin(\theta + 2\theta) = \sin\theta\cos 2\theta + \sin 2\theta\cos\theta.$$

Then by the double angle formulas for cosine and sine, we have

$$\begin{aligned}
\sin\theta\cos 2\theta + \sin 2\theta\cos\theta &= \sin\theta(1 - 2\sin^2\theta) + (2\sin\theta\cos\theta)\cos\theta \\
&= \sin\theta - 2\sin^3\theta + 2\sin\theta\cos^2\theta.
\end{aligned}$$

But $\cos^2\theta = 1 - \sin^2\theta$, so

$$\begin{aligned}
\sin\theta - 2\sin^3\theta + 2\sin\theta\cos^2\theta &= \sin\theta - 2\sin^3\theta + 2\sin\theta(1 - \sin^2\theta) \\
&= \sin\theta - 2\sin^3\theta + 2\sin\theta - 2\sin^3\theta \\
&= \boxed{3\sin\theta - 4\sin^3\theta}.
\end{aligned}$$

(c) By part (a), $\cos 3x = 4\cos^3 x - 3\cos x$, so

$$\frac{\cos 3x}{\cos x} = \frac{4\cos^3 x - 3\cos x}{\cos x} = 4\cos^2 x - 3.$$

Hence, $4\cos^2 x - 3 = 1/3$, so $\cos^2 x = (3 + 1/3)/4 = 10/12 = 5/6$. Therefore, $\sin^2 x = 1 - \cos^2 x = 1 - 5/6 = 1/6$. By part (b), we have $\sin 3x = 3\sin x - 4\sin^3 x$, so

$$\frac{\sin 3x}{\sin x} = \frac{3\sin x - 4\sin^3 x}{\sin x} = 3 - 4\sin^2 x = 3 - 4\cdot\frac{1}{6} = 3 - \frac{2}{3} = \boxed{\frac{7}{3}}.$$

Exercises for Section 3.4

3.4.1 By the difference-to-product formula for cosine, we have $\dfrac{\cos 4x - \cos 2x}{2\sin 3x} = \dfrac{-2\sin 3x \sin x}{2\sin 3x} = -\sin x.$

3.4.2 By the sum-to-product formula for cosine, we have $\cos 27° + \cos 33° = 2\cos 30° \cos 3° = \sqrt{3}\cos 3°$, so $\theta = \boxed{3°}$.

3.4.3 Subtracting the second equation from the first equation and dividing by 2, we get

$$\sin\beta \cos\alpha = \frac{1}{2}[\sin(\alpha+\beta) - \sin(\alpha-\beta)].$$

Switching α and β and noting that $\sin(\beta - \alpha) = \sin(-(\alpha-\beta)) = -\sin(\alpha-\beta)$, we get

$$\sin\alpha \cos\beta = \frac{1}{2}[\sin(\alpha+\beta) - \sin(\beta-\alpha)] = \frac{1}{2}[\sin(\alpha+\beta) + \sin(\alpha-\beta)],$$

which agrees with our earlier formula.

3.4.4 By the sum-to-product formula for sine, we have

$$\sin 10° + \sin 80° = 2\sin 45° \cos 35° = \sqrt{2}\cos 35°,$$
$$\sin 20° + \sin 70° = 2\sin 45° \cos 25° = \sqrt{2}\cos 25°,$$
$$\sin 30° + \sin 60° = 2\sin 45° \cos 15° = \sqrt{2}\cos 15°,$$
$$\sin 40° + \sin 50° = 2\sin 45° \cos 5° = \sqrt{2}\cos 5°,$$

so

$$\sin 10° + \sin 20° + \sin 30° + \sin 40° + \sin 50° + \sin 60° + \sin 70° + \sin 80°$$
$$= \sqrt{2}(\cos 5° + \cos 15° + \cos 25° + \cos 35°).$$

By the sum-to-product formula for cosine, we have

$$\cos 5° + \cos 35° = 2\cos 20° \cos 15°,$$
$$\cos 15° + \cos 25° = 2\cos 20° \cos 5°,$$

so

$$\sqrt{2}(\cos 5° + \cos 15° + \cos 25° + \cos 35°) = \sqrt{2}(2\cos 20° \cos 15° + 2\cos 20° \cos 5°) = 2\sqrt{2}\cos 20°(\cos 5° + \cos 15°).$$

Finally, by the sum-to-product formula for cosine, we have $\cos 5° + \cos 15° = 2\cos 10° \cos 5°$, so

$$2\sqrt{2}\cos 20°(\cos 5° + \cos 15°) = 4\sqrt{2}\cos 5° \cos 10° \cos 20°.$$

Therefore,

$$\frac{\sin 10° + \sin 20° + \sin 30° + \sin 40° + \sin 50° + \sin 60° + \sin 70° + \sin 80°}{\cos 5° \cos 10° \cos 20°} = \frac{4\sqrt{2}\cos 5° \cos 10° \cos 20°}{\cos 5° \cos 10° \cos 20°} = \boxed{4\sqrt{2}}.$$

3.4.5 By the difference-to-product formula for sine,

$$\sin(B + C - A) - \sin(A + B + C) = 2\sin(-A)\cos(B + C) = -2\sin A \cos(B + C),$$

and by the sum-to-product formula for sine,

$$\sin(C + A - B) + \sin(A + B - C) = 2\sin A \cos(C - B).$$

Hence,

$$-\sin(A + B + C) + \sin(B + C - A) + \sin(C + A - B) + \sin(A + B - C) = -2\sin A \cos(B + C) + 2\sin A \cos(C - B)$$
$$= 2\sin A[\cos(C - B) - \cos(B + C)].$$

By the difference-to-product formula for cosine,

$$\cos(C - B) - \cos(B + C) = -2\sin C \sin(-B) = 2\sin B \sin C,$$

so $2\sin A[\cos(C - B) - \cos(B + C)] = 4\sin A \sin B \sin C.$

3.4.6 The sum in the numerator and difference in the denominator suggest we should try the sum-to-product and difference-to-product identities, particularly because the angles are the same, so the difference of the angles is $0°$, which we know how to handle. Unfortunately, all these identities involve the sum or difference of two of the same trig function; none of them has a sum or difference of one sine and one cosine. So, we try changing the problem into a form that involves the sum or difference of two sines or the sum or difference of two cosines.

There are many ways to turn a sine into a cosine. Clearly, $\sin^2 \theta + \cos^2 \theta = 1$ isn't what we're looking for, since this won't give us a sum of cosines. But $\sin \theta = \cos(90° - \theta)$ will. Applying this to both of the sines in the problem gives

$$\frac{\cos 96° + \sin 96°}{\cos 96° - \sin 96°} = \frac{\cos 96° + \cos(-6°)}{\cos 96° - \cos(-6°)}.$$

We then apply the sum-to-product and difference-to-product identities for cosine:

$$\frac{\cos 96° + \cos(-6°)}{\cos 96° - \cos(-6°)} = \frac{2\cos\left(\frac{96° + (-6°)}{2}\right)\cos\left(\frac{96° - (-6°)}{2}\right)}{-2\sin\left(\frac{96° + (-6°)}{2}\right)\sin\left(\frac{96° - (-6°)}{2}\right)} = -\frac{2\cos 45° \cos 51°}{2\sin 45° \sin 51°} = -\cot 51°,$$

so the equation we must solve now is $\tan 19x° = -\cot 51°$. We'd rather have tangent on both sides, so we convert cotangent to tangent. Since

$$\cot 51° = \frac{\cos 51°}{\sin 51°} = \frac{\sin(90° - 51°)}{\cos(90° - 51°)} = \tan(90° - 51°) = \tan 39°,$$

we can write $\tan 19x° = -\cot 51°$ as $\tan 19x = -\tan 39°$. We'd like to get rid of the negative sign so that we're equating two tangents. We have plenty of identities to do that. The two simplest are $\tan \theta = -\tan(-\theta)$ and $\tan \theta = -\tan(180° - \theta)$. The former gives $\tan 19x° = -\tan 39° = \tan(-39°)$ and the latter gives $\tan 19x° = -\tan 39° = \tan 141°$. Uh-oh! These look different! But they're not. Since the period of $\tan \theta$ is $180°$, we have $\tan(-39°) = \tan(-39° + 180°) = \tan 141°$. This observation also helps us see how to finish the problem.

We want $\tan 19x° = \tan 141°$, but we also need x to be a positive integer. If we let $19x = 141$, then x is not an integer. But we also know that $\tan 141° = \tan(141° + 180n°)$ for all integers n. So, we keep increasing $141°$ by multiples of $180°$ until we hit a multiple of $19°$, which will allow us to divide by $19°$ to get an integer answer for x.

Now we have a basic number theory problem. We want the smallest positive integer x such that there is some positive integer k with $19x = 141 + 180k$. Dividing by 19 gives

$$x = 7 + \frac{8}{19} + 9k + \frac{9k}{19} = 9k + 7 + \frac{9k + 8}{19}.$$

We could figure this out with trial and error, starting with $k = 1$ and increasing k by 1 until $9k + 8$ is divisible by 19. Instead, we'll use a little more clever algebra and number theory.

Since $9k + 7$ is an integer, we must have $(9k + 8)/19$ be an integer in order for x to be an integer. If $(9k + 8)/19$ is an integer, then there is a positive integer j such that $9k + 8 = 19j$. Solving for k gives

$$k = \frac{19j - 8}{9} = 2j + \frac{j - 8}{9}.$$

We want k to be an integer, so we let $j = 8$, since this is the smallest possible positive integer value of j that makes $(j - 8)/9$ an integer. When $j = 8$, we have $k = 16$, which gives

$$x = 9k + 7 + \frac{9k + 8}{19} = \boxed{159}.$$

Exercises for Section 3.5

3.5.1 From the identity $\cos\theta = \sin(90° - \theta)$, we have $\cos 41° = \sin 49°$, so $\cos 41° + \sin 41° = \sin 49° + \sin 41°$. Then by the sum-to-product formula for sine, we have

$$\sin 49° + \sin 41° = 2\sin 45° \cos 4° = 2 \cdot \frac{\sqrt{2}}{2} \cdot \cos 4° = \sqrt{2}\cos 4°.$$

Hence, we have $\cos 41° + \sin 41° = \sqrt{2}\cos 4°$. Comparing this to the given $\cos 41° + \sin 41° = \sqrt{2}\sin A$ gives $\sin A = \cos 4°$. Since $\cos 4° = \sin(90° - 4°) = \sin 86°$, we have $A = \boxed{86°}$.

3.5.2 We let $ad + bc = k$. We square this equation and the equation $ac - bd = \frac{1}{2}$, hoping to produce squares that we can combine with our other two equations. These squarings give us

$$a^2 d^2 + 2abcd + b^2 c^2 = k^2,$$
$$a^2 c^2 - 2abcd + b^2 d^2 = \frac{1}{4}.$$

Adding these two equations eliminates the $abcd$ terms and leaves $a^2 d^2 + b^2 c^2 + a^2 c^2 + b^2 d^2 = k^2 + \frac{1}{4}$. The left side of this equation is exactly what we get when we multiply the left sides of the other two equations. Since $(a^2 + b^2)(c^2 + d^2) = 1$ we have

$$1 = a^2 c^2 + a^2 d^2 + b^2 c^2 + b^2 d^2 = k^2 + \frac{1}{4},$$

which means $k^2 = \frac{3}{4}$. Since a, b, c, and d are positive, k must also be positive, and we have $k = \boxed{\frac{\sqrt{3}}{2}}$.

3.5.3 We can solve this problem with $\sin^2\theta + \cos^2\theta = 1$, but it's a little tricky—try it and see. Instead, we'll use the sum-to-product identities. Applying these identities gives

$$2\sin\left(\frac{A + B}{2}\right)\cos\left(\frac{A - B}{2}\right) = \sqrt{\frac{3}{2}},$$
$$2\cos\left(\frac{A + B}{2}\right)\cos\left(\frac{A - B}{2}\right) = \sqrt{\frac{1}{2}}.$$

We can eliminate the terms with $(A - B)/2$ by dividing the first equation by the second (which we can safely do since we know the left side of the second equation is not 0). This gives us

$$\frac{\sin\left(\frac{A+B}{2}\right)}{\cos\left(\frac{A+B}{2}\right)} = \sqrt{3}.$$

The angles are the same on the left side, so we can write this equation as $\tan\left(\frac{A+B}{2}\right) = \sqrt{3}$. Since we seek the value of $A + B$ such that $0° \le A, B \le 180°$, we know that $0 \le \frac{A+B}{2} \le 180°$, and the only angle in this range whose tangent equals $\sqrt{3}$ is $60°$. Therefore, we have $\frac{A+B}{2} = 60°$, so $A + B = \boxed{120°}$.

3.5.4 By the angle sum identity for tangent, we have $\tan(x + y) = \frac{\tan x + \tan y}{1 - \tan x \tan y}$. We are given that $\tan x + \tan y = 25$, so we only need to find $\tan x \tan y$.

We are also given that $\cot x + \cot y = 30$. But $\cot x + \cot y = \frac{1}{\tan x} + \frac{1}{\tan y} = \frac{\tan x + \tan y}{\tan x \tan y}$, so

$$\tan x \tan y = \frac{\tan x + \tan y}{\cot x + \cot y} = \frac{25}{30} = \frac{5}{6}.$$

Therefore, $\tan(x + y) = \frac{\tan x + \tan y}{1 - \tan x \tan y} = \frac{25}{1 - 5/6} = \boxed{150}$.

3.5.5 First, we determine $\cos B$ by noting that $\cos^2 B = 1 - \sin^2 B = 1 - \frac{1}{10} = \frac{9}{10}$, so $\cos B = \pm\sqrt{\frac{9}{10}} = \pm\frac{3}{\sqrt{10}}$. We are given that B is acute, so $\cos B = \frac{3}{\sqrt{10}}$, which means $\tan B = \frac{\sin B}{\cos B} = \frac{1/\sqrt{10}}{3/\sqrt{10}} = \frac{1}{3}$.

By the double angle formula for tangent, we have $\tan 2B = \frac{2\tan B}{1 - \tan^2 B} = \frac{2/3}{1 - (1/3)^2} = \frac{3}{4}$. Since B is acute, we have $0° < 2B < 180°$. Since $\tan 2B$ is positive, $2B$ is also acute.

Then by the angle sum identity for tangent, we have

$$\tan(A + 2B) = \frac{\tan A + \tan 2B}{1 - \tan A \tan 2B} = \frac{1/7 + 3/4}{1 - 1/7 \cdot 3/4} = 1.$$

Since both A and $2B$ are acute, we have $0° < A + 2B < 180°$. The only angle in this range whose tangent is 1 is $45°$, so $A + 2B = \boxed{45°}$.

3.5.6 Rearranging the terms, we get $\sin\theta - \cos\theta = \cos 2\theta - \sin 2\theta$. Squaring both sides, we get

$$\sin^2\theta - 2\sin\theta\cos\theta + \cos^2\theta = \cos^2 2\theta - 2\cos 2\theta \sin 2\theta + \sin^2 2\theta.$$

But $\sin^2\theta + \cos^2\theta = \cos^2 2\theta + \sin^2 2\theta = 1$, so the equation above becomes $1 - 2\sin\theta\cos\theta = 1 - 2\cos 2\theta \sin 2\theta$, which means $\sin\theta\cos\theta = \cos 2\theta \sin 2\theta$.

By the double angle formula for sine, we have $\sin 2\theta = 2\sin\theta\cos\theta$, so

$$\sin\theta\cos\theta = 2\cos 2\theta \sin\theta\cos\theta.$$

Moving all the terms to one side, we get

$$2\cos 2\theta \sin\theta\cos\theta - \sin\theta\cos\theta = 0,$$

which factors as

$$\sin\theta\cos\theta(2\cos 2\theta - 1) = 0.$$

Hence, $\sin\theta = 0$, $\cos\theta = 0$, or $\cos 2\theta = \frac{1}{2}$. We are given that θ is acute, so neither $\sin\theta$ nor $\cos\theta$ can be equal to 0. Therefore, $\cos 2\theta = \frac{1}{2}$. Since θ is acute, we have $0 < 2\theta < \pi$. The only angle in this range whose cosine is $\frac{1}{2}$ is $\frac{\pi}{3}$, so $2\theta = \frac{\pi}{3}$, and $\theta = \boxed{\frac{\pi}{6}}$. This solution works, because when $\theta = \frac{\pi}{6}$, we have $\sin\theta + \sin 2\theta = \cos\theta + \cos 2\theta = \frac{1}{2} + \frac{\sqrt{3}}{2}$.

3.5.7 By the angle sum identity for cosine and the angle difference identity for sine, we have

$$\cos(\alpha + \beta) = \cos\alpha\cos\beta - \sin\alpha\sin\beta,$$
$$\sin(\alpha - \beta) = \sin\alpha\cos\beta - \cos\alpha\sin\beta,$$

so the given equation becomes $\cos\alpha\cos\beta - \sin\alpha\sin\beta + \sin\alpha\cos\beta - \cos\alpha\sin\beta = 0$. Dividing by $\cos\alpha\cos\beta$, we get

$$1 - \frac{\sin\alpha\sin\beta}{\cos\alpha\cos\beta} + \frac{\sin\alpha}{\cos\alpha} - \frac{\sin\beta}{\cos\beta} = 0,$$

which becomes $1 - \tan\alpha\tan\beta + \tan\alpha - \tan\beta = 0$. This equation factors as $(1+\tan\alpha)(1-\tan\beta) = 0$, so $\tan\alpha = -1$ or $\tan\beta = 1$. We are given that $\tan\beta = 1/2000$, so $\tan\alpha = \boxed{-1}$.

3.5.8 Expanding the given equation, we get

$$1 + \sin t + \cos t + \sin t\cos t = \frac{5}{4}.$$

Let $x = (1-\sin t)(1-\cos t)$, so

$$1 - \sin t - \cos t + \sin t\cos t = x.$$

Adding these equations, we get $2 + 2\sin t\cos t = x + \frac{5}{4}$, so $\sin t\cos t = \frac{x+5/4-2}{2} = \frac{4x-3}{8}$. Taking the difference of the same two equations gives $2\sin t + 2\cos t = \frac{5}{4} - x$, so $\sin t + \cos t = \frac{5/4-x}{2} = \frac{5-4x}{8}$. Squaring this equation, we get

$$\sin^2 t + 2\sin t\cos t + \cos^2 t = \left(\frac{5-4x}{8}\right)^2.$$

However, $\sin^2 t + \cos^2 t = 1$ and $\sin t\cos t = (4x-3)/8$, so

$$1 + 2\cdot\frac{4x-3}{8} = \left(\frac{5-4x}{8}\right)^2.$$

Simplifying, we obtain the quadratic equation $16x^2 - 104x + 9 = 0$. By the quadratic formula, we have

$$x = \frac{104 \pm \sqrt{104^2 - 4\cdot16\cdot9}}{2\cdot16} = \frac{104 \pm 32\sqrt{10}}{32} = \frac{13}{4} \pm \sqrt{10}.$$

However, $-1 \le \sin t \le 1$ and $-1 \le \cos t \le 1$, so

$$x = (1-\sin t)(1-\cos t) \le 2\cdot2 = 4,$$

and $\frac{13}{4} + \sqrt{10} > \frac{13}{4} + 3 > 4$. Hence, we have $x = \boxed{\frac{13}{4} - \sqrt{10}}$.

3.5.9 Solving for y, z, and x in the given equations, respectively, we find

$$y = \frac{2x}{1-x^2}, \qquad z = \frac{2y}{1-y^2}, \qquad x = \frac{2z}{1-z^2}.$$

The right sides of these equations look like the double angle identity for tangent. Since x, y, and z are real numbers, we can let $\alpha = \arctan x$, $\beta = \arctan y$, and $\gamma = \arctan z$, so $x = \tan\alpha$, $y = \tan\beta$, and $z = \tan\gamma$. Then from $y = \frac{2x}{1-x^2}$, we have

$$\tan\beta = \frac{2\tan\alpha}{1-\tan^2\alpha}.$$

But from the double angle formula for tangent, we have

$$\frac{2\tan\alpha}{1-\tan^2\alpha} = \tan2\alpha,$$

so $\tan\beta = \tan2\alpha$. Similarly, from the other two initial equations, we have $\tan\gamma = \tan2\beta$ and $\tan\alpha = \tan2\gamma$.

We'd like to combine these equations in a way that allows us to produce an equation with only α. One tempting possibility is to combine $\tan\beta = \tan 2\alpha$ and $\tan\gamma = \tan 2\beta$ through the following argument: "Since $\tan\beta = \tan 2\alpha$, we have $\tan 2\beta = \tan 4\alpha$, and combining this with $\tan\gamma = \tan 2\beta$ gives $\tan\gamma = \tan 4\alpha$." But before we get too excited, we better check the claim that if the tangents of two angles are equal, then the tangents of double the angles are equal.

To do so, we let θ and ψ be two angles. Then

$$\tan 2\theta = \frac{2\tan\theta}{1-\tan^2\theta} \quad \text{and} \quad \tan 2\psi = \frac{2\tan\psi}{1-\tan^2\psi}.$$

Hence, if $\tan\theta = \tan\psi$, then $\tan 2\theta = \tan 2\psi$.

Returning to our problem, we know that $\tan\gamma = \tan 2\beta$, so $\tan 2\gamma = \tan 4\beta$. Also, $\tan\beta = \tan 2\alpha$, so $\tan 2\beta = \tan 4\alpha$, and $\tan 4\beta = \tan 8\alpha$. Therefore,

$$\tan\alpha = \tan 2\gamma = \tan 4\beta = \tan 8\alpha.$$

In particular, $\tan\alpha = \tan 8\alpha$. Two angles with equal tangents must differ by an integer multiple of π. To see why, suppose that $\tan\theta = m$. Then, by the definition of the tangent function, the terminal point of θ is one of the two points at which the line $y = mx$ intersects the unit circle. These two points are at the opposite ends of a diameter of the unit circle. That is, they are terminal points of angles that differ by an integer multiple of π.

Therefore, from $\tan\alpha = \tan 8\alpha$, we know that $8\alpha = \alpha + \pi k$, where k is an integer. Solving for α, we find $\alpha = \frac{\pi k}{7}$, so $x = \tan\alpha = \tan\frac{\pi k}{7}$.

We now check that this solution works. If $x = \tan\frac{\pi k}{7}$, then

$$y = \frac{2x}{1-x^2} = \frac{2\tan\frac{\pi k}{7}}{1-\tan^2\frac{\pi k}{7}} = \tan\frac{2\pi k}{7},$$

$$z = \frac{2y}{1-y^2} = \frac{2\tan\frac{2\pi k}{7}}{1-\tan^2\frac{2\pi k}{7}} = \tan\frac{4\pi k}{7},$$

and

$$\frac{2z}{1-z^2} = \frac{2\tan\frac{4\pi k}{7}}{1-\tan^2\frac{4\pi k}{7}} = \tan\frac{8\pi k}{7}.$$

But $\frac{8\pi k}{7} - \frac{\pi k}{7} = \frac{7\pi k}{7} = \pi k$, so $\tan\frac{8\pi k}{7} = \tan\frac{\pi k}{7} = x$. Hence, $x = \frac{2z}{1-z^2}$, and the system is satisfied.

Therefore, the solutions are of the form

$$\boxed{(x,y,z) = \left(\tan\frac{\pi k}{7}, \tan\frac{2\pi k}{7}, \tan\frac{4\pi k}{7}\right),}$$

where $k = 0, 1, \ldots, 6$. (We only take k up to 6 because the solutions for $k = 7$ and $k = 0$ are the same, as are the solutions for $k = 8$ and $k = 1$, and so on.)

Review Problems

3.33 $\tan(\pi - \theta) = \dfrac{\sin(\pi - \theta)}{\cos(\pi - \theta)} = \dfrac{\sin\theta}{-\cos\theta} = -\tan\theta.$

3.34 Since $\sin(-x) = -\sin x$ for all x, we have $\sin(\theta - \pi) = -\sin(\pi - \theta)$. We also have $\sin(\pi - x) = \sin x$ for all x, so $\sin(\theta - \pi) = -\sin(\pi - \theta) = \boxed{-\sin\theta}$.

3.35

(a) First, we have $\cos(x - y) = \sin[90° - (x - y)] = \sin(90° - x + y)$. Therefore, we have

$$\cos(x - y) = \sin(90° - x + y) = \sin[(90° - x) + y] = \sin(90° - x)\cos y + \sin y \cos(90° - x).$$

Letting $z = 90° - x$ in the identity $\sin(90° - z) = \cos z$ gives $\sin x = \cos(90° - x)$, so

$$\cos(x - y) = \sin(90° - x)\cos y + \sin y \cos(90° - x) = \cos x \cos y + \sin x \sin y.$$

(b) Taking $y = x$ in $\sin(x + y) = \sin x \cos y + \sin y \cos x$, we get $\sin 2x = 2 \sin x \cos x$.

(c) First, $\cos(x + y) = \sin[90° - (x + y)] = \sin(90° - x - y)$. Then

$$\sin(90° - x - y) = \sin[(90° - x) + (-y)] = \sin(90° - x)\cos(-y) + \sin(-y)\cos(90° - x) = \cos x \cos y - \sin x \sin y.$$

Hence, we have

$$\tan(x + y) = \frac{\sin(x + y)}{\cos(x + y)} = \frac{\sin x \cos y + \sin y \cos x}{\cos x \cos y - \sin x \sin y} = \frac{\frac{\sin x}{\cos x} + \frac{\sin y}{\cos y}}{1 - \frac{\sin x \sin y}{\cos x \cos y}} = \frac{\tan x + \tan y}{1 - \tan x \tan y}.$$

(d) Letting $y = -x$ in the identity we proved in part (a), we have

$$\cos 2x = \cos x \cos(-x) + \sin x \sin(-x) = \cos^2 x - \sin^2 x = \cos^2 x - (1 - \cos^2 x) = 2\cos^2 x - 1.$$

Letting $x = \frac{z}{2}$, we have $\cos z = 2\cos^2 \frac{z}{2} - 1$, so $\cos^2 \frac{z}{2} = \frac{1 + \cos z}{2}$. If $0 < z < \pi$, then $0 < \frac{z}{2} < \frac{\pi}{2}$, so $\cos \frac{z}{2} > 0$. Therefore, we take the positive square root of our equation above to find $\cos \frac{z}{2} = \sqrt{\frac{1 + \cos z}{2}}$.

(e) If part (a), we found

$$\cos(x - y) = \cos x \cos y + \sin x \sin y.$$

Replacing y with $-y$ in this equation, we have

$$\cos(x + y) = \cos x \cos(-y) + \sin x \sin(-y) = \cos x \cos y - \sin x \sin y.$$

Adding these two equations gives $\cos(x - y) + \cos(x + y) = 2 \cos x \cos y$, so we have the identity $\cos x \cos y = \frac{1}{2}(\cos(x + y) + \cos(x - y))$.

(f) Replacing y with $-y$ in $\sin(x + y) = \sin x \cos y + \sin y \cos x$ gives

$$\sin(x - y) = \sin x \cos(-y) + \sin(-y)\cos x = \sin x \cos y - \sin y \cos x.$$

Subtracting this from our equation for $\sin(x + y)$ gives

$$\sin(x + y) - \sin(x - y) = 2 \sin y \cos x.$$

Letting $\alpha = x + y$ and $\beta = x - y$, we then have $x = (\alpha + \beta)/2$ and $y = (\alpha - \beta)/2$, and we have $\sin \alpha - \sin \beta = 2 \sin\left(\frac{\alpha - \beta}{2}\right)\cos\left(\frac{\alpha + \beta}{2}\right)$.

3.36 By the angle sum formula for sine, we have $\sin(x - y)\cos y + \cos(x - y)\sin y = \sin[(x - y) + y] = \boxed{\sin x}$.

3.37 Applying the angle difference identity for cosine, we have

$$\sqrt{2}\cos\left(\frac{\pi}{4} - x\right) = \sqrt{2}\left(\cos\frac{\pi}{4}\cos x + \sin\frac{\pi}{4}\sin x\right) = \sqrt{2}\left(\frac{\sqrt{2}}{2}\cos x + \frac{\sqrt{2}}{2}\sin x\right) = \cos x + \sin x,$$

which is choice $\boxed{\text{(a)}}$.

3.38 Since $\cos^2\theta = 1 - \sin^2\theta$, we have

$$\frac{-3 + 4\cos^2\theta}{1 - 2\sin\theta} = \frac{-3 + 4(1 - \sin^2\theta)}{1 - 2\sin\theta} = \frac{1 - 4\sin^2\theta}{1 - 2\sin\theta} = \frac{(1 + 2\sin\theta)(1 - 2\sin\theta)}{1 - 2\sin\theta} = 1 + 2\sin\theta.$$

Thus, $\boxed{a = 1 \text{ and } b = 2}$.

3.39 We know how to find the range of a function of the form $f(t) = a\sin t$, so we try to write $g(t)$ in this form. The angle sum identity for sine has a single sine on one side and a sum of trig expressions on the other:

$$\sin(t + y) = \sin t \cos y + \sin y \cos t.$$

However, we have $g(t) = 2\sin t + 3\cos t$, and there is no angle y such that $\cos y = 2$ and $\sin y = 3$. But maybe we can find some constant c such that there is an angle y with $\cos y = \frac{2}{c}$ and $\sin y = \frac{3}{c}$. From $\cos^2 y + \sin^2 y = 1$, we have $\frac{4}{c^2} + \frac{9}{c^2} = 1$, from which we find that $c = \sqrt{13}$ fits the bill. That is, we have

$$g(t) = \sqrt{13}\left(\frac{2}{\sqrt{13}}\sin t + \frac{3}{\sqrt{13}}\cos t\right) = \sqrt{13}\sin(t + y),$$

where y is the angle such that $\cos y = \frac{2}{\sqrt{13}}$ and $\sin y = \frac{3}{\sqrt{13}}$. (We know that such an angle exists because $\left(\frac{2}{\sqrt{13}}\right)^2 + \left(\frac{3}{\sqrt{13}}\right)^2 = 1$, which means $\left(\frac{2}{\sqrt{13}}, \frac{3}{\sqrt{13}}\right)$ is on the unit circle.)

From $g(t) = \sqrt{13}\sin(t + y)$, where y is some constant angle, we see that the range of $g(t)$ is $\boxed{\left[-\sqrt{13}, \sqrt{13}\right]}$, because the range of sine is $[-1, 1]$.

3.40 We have

$$\frac{\tan^2 20° - \sin^2 20°}{\tan^2 20° \sin^2 20°} = \frac{\frac{\sin^2 20°}{\cos^2 20°} - \sin^2 20°}{\frac{\sin^2 20°}{\cos^2 20°} \cdot \sin^2 20°} = \frac{\sin^2 20° - \sin^2 20° \cos^2 20°}{\sin^4 20°} = \frac{\sin^2 20°(1 - \cos^2 20°)}{\sin^4 20°} = \frac{\sin^4 20°}{\sin^4 20°} = \boxed{1}.$$

3.41 By the angle sum identity for tangent, we have $\tan\alpha = \tan[(\alpha - \beta) + \beta] = \dfrac{\tan(\alpha - \beta) + \tan\beta}{1 - \tan(\alpha - \beta)\tan\beta}$.

3.42 By the angle sum formula for tangent, we have

$$\tan\left(\frac{3\pi}{2} + \theta\right) = \frac{\tan\frac{3\pi}{2} + \tan\theta}{1 - \tan\frac{3\pi}{2}\tan\theta}.$$

Uh-oh; $\tan\frac{3\pi}{2}$ is not defined! To get around this, we write tangent in terms of sine and cosine:

$$\tan\left(\frac{3\pi}{2} + \theta\right) = \frac{\sin\left(\frac{3\pi}{2} + \theta\right)}{\cos\left(\frac{3\pi}{2} + \theta\right)} = \frac{\sin\frac{3\pi}{2}\cos\theta + \cos\frac{3\pi}{2}\sin\theta}{\cos\frac{3\pi}{2}\cos\theta - \sin\frac{3\pi}{2}\sin\theta} = \frac{-\cos\theta}{\sin\theta} = -\cot\theta = \boxed{-\frac{1}{\tan\theta}}.$$

3.43 By the angle sum identity for cosine, we have $\cos(A + B) = \cos A\cos B - \sin A\sin B = 1$. Because A and B are angles of a triangle, we must have $0° < A + B < 180°$. However, there is no angle in this range whose cosine equals 1, so it is impossible for any $\triangle ABC$ to satisfy the equation $\cos A\cos B - \sin A\sin B = 1$.

3.44 By the angle sum formulas for sine and cosine, we have

$$\sin(x + 30°) = \sin x\cos 30° + \cos x\sin 30° = \frac{\sqrt{3}}{2}\sin x + \frac{1}{2}\cos x,$$

$$\cos(x + 30°) = \cos x\cos 30° - \sin x\sin 30° = \frac{\sqrt{3}}{2}\cos x - \frac{1}{2}\sin x.$$

Then

$$\frac{\sqrt{3}\sin(x+30°) - \cos(x+30°)}{4\cos x \sin(x+30°) - 4\sin x \cos(x+30°)} = \frac{\sqrt{3}\left(\frac{\sqrt{3}}{2}\sin x + \frac{1}{2}\cos x\right) - \left(\frac{\sqrt{3}}{2}\cos x - \frac{1}{2}\sin x\right)}{4\cos x \left(\frac{\sqrt{3}}{2}\sin x + \frac{1}{2}\cos x\right) - 4\sin x \left(\frac{\sqrt{3}}{2}\cos x - \frac{1}{2}\sin x\right)}$$

$$= \frac{2\sin x}{2\cos^2 x + 2\sin^2 x}$$

$$= \boxed{\sin x}.$$

3.45 We have $\arccos\left(-\frac{1}{2}\right) = 120°$ and $\arcsin\frac{\sqrt{2}}{2} = 45°$, so

$$\tan\left(\arccos\left(-\frac{1}{2}\right) + \arcsin\frac{\sqrt{2}}{2}\right) = \tan(120° + 45°).$$

Then by the angle sum identity for tangent, we have

$$\tan(120° + 45°) = \frac{\tan 120° + \tan 45°}{1 - \tan 120° \tan 45°} = \frac{-\sqrt{3} + 1}{1 - (-\sqrt{3})\cdot 1} = \frac{1 - \sqrt{3}}{1 + \sqrt{3}} = \frac{(1-\sqrt{3})(1-\sqrt{3})}{(1+\sqrt{3})(1-\sqrt{3})} = \boxed{\sqrt{3} - 2}.$$

3.46 We know that

$$\sin 15° = \sin(45° - 30°) = \sin 45° \cos 30° - \sin 30° \cos 45° = \frac{\sqrt{6} - \sqrt{2}}{4} \quad \text{and}$$

$$\sin 75° = \sin(45° + 30°) = \sin 45° \cos 30° + \sin 30° \cos 45° = \frac{\sqrt{6} + \sqrt{2}}{4},$$

so $(\sin 15° + \sin 75°)^6 = \left(\frac{\sqrt{6}}{2}\right)^6 = \frac{216}{64} = \boxed{\frac{27}{8}}.$

3.47 We have

$$\frac{\sin 2x}{\sin x} - \frac{\cos 2x}{\cos x} = \frac{\sin 2x \cos x - \sin x \cos 2x}{\sin x \cos x}.$$

By the angle difference identity for sine, we have

$$\sin 2x \cos x - \sin x \cos 2x = \sin(2x - x) = \sin x,$$

so $\dfrac{\sin 2x \cos x - \sin x \cos 2x}{\sin x \cos x} = \dfrac{\sin x}{\sin x \cos x} = \dfrac{1}{\cos x} = \boxed{\sec x}.$

3.48 Since $\cos^2\theta + \sin^2\theta = 1$, we have

$$\sin^2\theta = 1 - \cos^2\theta = 1 - (-0.6)^2 = 1 - 0.36 = 0.64,$$

so $\sin\theta = \pm\sqrt{0.64} = \pm 0.8$. Then by the double angle formula for sine, we have $\sin 2\theta = 2\sin\theta\cos\theta = \pm 0.96$.

Now we show that both 0.96 and -0.96 are achievable. There exists an angle θ in the second quadrant such that $\cos\theta = -0.6$ and $\sin\theta = 0.8$ (since $(-0.6)^2 + (0.8)^2 = 1$). For this angle, we have $\sin 2\theta = 2\sin\theta\cos\theta = -0.96$. There's an angle θ in the third quadrant for which $\cos\theta = -0.6$ and $\sin\theta = -0.8$. For this angle, we have $\sin 2\theta = 2\sin\theta\cos\theta = 0.96$. Thus, both $\boxed{0.96}$ and $\boxed{-0.96}$ are possible values of $\sin 2\theta$.

3.49 We have

$$\frac{\cot x - 1}{\cot x + 1} = \frac{\frac{\cos x}{\sin x} - 1}{\frac{\cos x}{\sin x} + 1} = \frac{\cos x - \sin x}{\cos x + \sin x}.$$

Looking at the other side of the identity we hope to prove, we have

$$\frac{\cos 2x}{1 + \sin 2x} = \frac{\cos^2 x - \sin^2 x}{1 + 2 \sin x \cos x} = \frac{(\cos x - \sin x)(\cos x + \sin x)}{1 + 2 \sin x \cos x}.$$

Aha! The factored numerator gives us a clue how to tie these together. We multiply the numerator and denominator of our fraction for $\frac{\cot x - 1}{\cot x + 1}$ by $\cos x + \sin x$, and we find

$$\frac{\cos x - \sin x}{\cos x + \sin x} = \frac{(\cos x - \sin x)(\cos x + \sin x)}{(\cos x + \sin x)(\cos x + \sin x)} = \frac{\cos^2 x - \sin^2 x}{\cos^2 x + 2 \sin x \cos x + \sin^2 x} = \frac{\cos 2x}{1 + 2 \sin x \cos x} = \frac{\cos 2x}{1 + \sin 2x},$$

where we have also used the identity $\cos^2 x + \sin^2 x = 1$ to simplify the denominator.

3.50 We start with the identity $\sin^2 x + \cos^2 x = 1$. Squaring both sides, we get

$$\sin^4 x + 2 \sin^2 x \cos^2 x + \cos^4 x = 1.$$

By the double angle formula for sine, we have $\sin 2x = 2 \sin x \cos x$. But we are given that $\sin 2x = \frac{24}{25}$, so $2 \sin x \cos x = \frac{24}{25}$. Then $\sin x \cos x = \frac{12}{25}$, so $2 \sin^2 x \cos^2 x = 2 \left(\frac{12}{25} \right)^2 = \frac{288}{625}$. Therefore, $\sin^4 x + \cos^4 x = 1 - \frac{288}{625} = \boxed{\frac{337}{625}}$.

3.51 Applying the half-angle formula for tangent, we have

$$\tan \frac{5\pi}{8} = \frac{\sin \frac{5\pi}{4}}{1 + \cos \frac{5\pi}{4}} = \frac{-\frac{\sqrt{2}}{2}}{1 - \frac{\sqrt{2}}{2}} = \frac{-\sqrt{2}}{2 - \sqrt{2}} = \frac{-\sqrt{2}(2 + \sqrt{2})}{(2 - \sqrt{2})(2 + \sqrt{2})} = \boxed{-1 - \sqrt{2}}.$$

3.52 By the double angle formula for sine, we have $\sin 2x = 2 \sin x \cos x$, so the inequality $\sin 2x \le \sin x$ becomes $\sin x - 2 \sin x \cos x \ge 0$, which means $\sin x(1 - 2 \cos x) \ge 0$. This inequality is satisfied if and only if $\sin x \ge 0$ and $\cos x \le \frac{1}{2}$, or $\sin x \le 0$ and $\cos x \ge \frac{1}{2}$.

For $0 \le x \le 2\pi$, $\sin x \ge 0$ if and only if $0 \le x \le \pi$ or $x = 2\pi$, and $\cos x \le \frac{1}{2}$ if and only if $\frac{\pi}{3} \le x \le \frac{5\pi}{3}$. Hence, $\sin x \ge 0$ and $\cos x \le \frac{1}{2}$ if and only if $\frac{\pi}{3} \le x \le \pi$.

Similarly, $\sin x \le 0$ if and only if $x = 0$ or $\pi \le x \le 2\pi$, and $\cos x \ge \frac{1}{2}$ if and only if $0 \le x \le \frac{\pi}{3}$ or $\frac{5\pi}{3} \le x \le 2\pi$. Hence, $\sin x \le 0$ and $\cos x \ge \frac{1}{2}$ if and only if $x = 0$ or $\frac{5\pi}{3} \le x \le 2\pi$.

Therefore, the solution is $x \in \boxed{\{0\} \cup [\frac{\pi}{3}, \pi] \cup [\frac{5\pi}{3}, 2\pi]}$.

3.53 Since $\cos^2 \theta = 1 - \sin^2 \theta$, the given equation becomes

$$2(\sin^2 \theta - 1 + \sin^2 \theta) = 8 \sin \theta - 5 \quad \Leftrightarrow \quad 4 \sin^2 \theta - 8 \sin \theta + 3 = 0 \quad \Leftrightarrow \quad (2 \sin \theta - 1)(2 \sin \theta - 3) = 0,$$

so $\sin \theta = \frac{1}{2}$ or $\sin \theta = \frac{3}{2}$.

Since $-1 \le \sin \theta \le 1$ for all θ, $\sin \theta$ can never be equal to $\frac{3}{2}$, so $\sin \theta = \frac{1}{2}$. The solutions are $\theta = \boxed{\frac{\pi}{6}}$ and $\boxed{\frac{5\pi}{6}}$.

3.54

(a) By the double angle formulas for sine and cosine, we have $\sin 2\theta = 2 \sin \theta \cos \theta$ and $\cos 2\theta = 2 \cos^2 \theta - 1$, so

$$\frac{\sin \theta + \sin 2\theta}{1 + \cos \theta + \cos 2\theta} = \frac{\sin \theta + 2 \sin \theta \cos \theta}{1 + \cos \theta + 2 \cos^2 \theta - 1} = \frac{\sin \theta + 2 \sin \theta \cos \theta}{\cos \theta + 2 \cos^2 \theta}$$
$$= \frac{\sin \theta(1 + 2 \cos \theta)}{\cos \theta(1 + 2 \cos \theta)} = \frac{\sin \theta}{\cos \theta} = \tan \theta.$$

(b) Our derivation in part (a) fails if $1 + 2\cos\theta = 0$, or $\cos\theta = -1/2$. The only value in the interval $\frac{\pi}{2} < \theta < \pi$ for which this occurs is $\theta = \boxed{\frac{2\pi}{3}}$.

3.55 By the product-to-difference identity, we have

$$2\sin 68° \sin 52° = \cos(68° - 52°) - \cos(68° + 52°) = \cos 16° - \cos 120° = \cos 16° + \frac{1}{2},$$

so we want to find the acute angle ϕ such that $\sin\phi = \cos 16° = \sin(90° - 16°) = \sin 74°$. Hence, $\phi = \boxed{74°}$.

3.56 To prove the identity, we express all the trigonometric functions in terms of $\sin x$ and $\cos x$. The left-hand side becomes

$$\frac{\tan x}{1 - \cot x} + \frac{\cot x}{1 - \tan x} = \frac{\sin x / \cos x}{1 - \cos x / \sin x} + \frac{\cos x / \sin x}{1 - \sin x / \cos x}$$

$$= \frac{\sin^2 x}{\cos x(\sin x - \cos x)} + \frac{\cos^2 x}{\sin x(\cos x - \sin x)}$$

$$= \frac{\sin^3 x}{\sin x \cos x(\sin x - \cos x)} - \frac{\cos^3 x}{\sin x \cos x(\sin x - \cos x)}$$

$$= \frac{\sin^3 x - \cos^3 x}{\sin x \cos x(\sin x - \cos x)}.$$

As a difference of cubes, we have $\sin^3 x - \cos^3 x = (\sin x - \cos x)(\sin^2 x + \sin x \cos x + \cos^2 x)$, so

$$\frac{\sin^3 x - \cos^3 x}{\sin x \cos x(\sin x - \cos x)} = \frac{(\sin x - \cos x)(\sin^2 x + \sin x \cos x + \cos^2 x)}{\sin x \cos x(\sin x - \cos x)}$$

$$= \frac{\sin^2 x + \sin x \cos x + \cos^2 x}{\sin x \cos x}$$

$$= \frac{1 + \sin x \cos x}{\sin x \cos x} = \frac{1}{\sin x \cos x} + 1 = \sec x \csc x + 1,$$

which is the right-hand side of the identity we wish to prove.

3.57 From the sum-to-product formulas, we have

$$\sin x + \sin 5x = 2\sin 3x \cos 2x,$$
$$\cos x + \cos 5x = 2\cos 3x \cos 2x,$$

so $\dfrac{\sin x + \sin 3x + \sin 5x}{\cos x + \cos 3x + \cos 5x} = \dfrac{2\sin 3x \cos 2x + \sin 3x}{2\cos 3x \cos 2x + \cos 3x} = \dfrac{\sin 3x(2\cos 2x + 1)}{\cos 3x(2\cos 2x + 1)} = \dfrac{\sin 3x}{\cos 3x} = \tan 3x.$

Challenge Problems

3.58 Squaring the given equation, we get $\sin^2 x + 2\sin x \cos x + \cos^2 x = \frac{1}{25}$. But $\sin^2 x + \cos^2 x = 1$, so we have $2\sin x \cos x = \frac{1}{25} - 1 = -\frac{24}{25}$, which means $\sin x \cos x = -\frac{12}{25}$. Substituting $\cos x = \frac{1}{5} - \sin x$ (this is a rearrangement of the given equation) into $\sin x \cos x = -\frac{12}{25}$ gives

$$\sin x\left(\frac{1}{5} - \sin x\right) = -\frac{12}{25}.$$

Multiplying both sides by 25, expanding the left side, and rearranging, gives $25 \sin^2 x - 5 \sin x - 12 = 0$. Factoring the left side as a quadratic in x gives $(5 \sin x - 4)(5 \sin x + 3) = 0$. Since we are given $\frac{\pi}{2} < x < \pi$, we have $\sin x > 0$, and the only positive solution to the quadratic for $\sin x$ is $\sin x = \frac{4}{5}$.

From $\cos^2 x = 1 - \sin^2 x = \frac{9}{25}$ and the fact that x is in the second quadrant, we have $\cos x = -\frac{3}{5}$, so

$$\tan x = \frac{\sin x}{\cos x} = \frac{4/5}{-3/5} = \boxed{-\frac{4}{3}}.$$

3.59 To find $f(\sec^2 \theta)$, we find the value of x such that $\frac{x}{x-1} = \sec^2 \theta$. Taking the reciprocal of both sides, we get $\frac{x-1}{x} = \frac{1}{\sec^2 \theta}$, which becomes $1 - \frac{1}{x} = \cos^2 \theta$. Hence, we have

$$\frac{1}{x} = 1 - \cos^2 \theta = \sin^2 \theta,$$

so $f(\sec^2 \theta) = f\left(\frac{x}{x-1}\right) = \frac{1}{x} = \boxed{\sin^2 \theta}$.

3.60 Since $\tan 4x = \frac{\sin 4x}{\cos 4x}$, the given equation becomes $\frac{\sin 4x}{\cos 4x} = \frac{\cos x - \sin x}{\cos x + \sin x}$. Cross-multiplying, we get the equation $\cos x \sin 4x + \sin x \sin 4x = \cos x \cos 4x - \sin x \cos 4x$, so

$$\cos x \sin 4x + \sin x \cos 4x = \cos x \cos 4x - \sin x \sin 4x.$$

By the angle sum formula for sine, we have $\cos x \sin 4x + \sin x \cos 4x = \sin(x + 4x) = \sin 5x$, and by the angle sum formula for cosine, we have $\cos x \cos 4x - \sin x \sin 4x = \cos(x + 4x) = \cos 5x$, so the equation above becomes $\sin 5x = \cos 5x$. Then $\tan 5x = \frac{\sin 5x}{\cos 5x} = 1$.

The smallest positive angle whose tangent is 1 is $45°$, so the smallest possible positive angle x is $45°/5 = \boxed{9°}$.

3.61 To find the intersections of the graphs, we must solve the equation

$$8 \cos x + 5 \tan x = 5 \sec x + 2 \sin 2x.$$

By the double angle formula for sine, we have $\sin 2x = 2 \sin x \cos x$, so the equation above becomes

$$8 \cos x + \frac{5 \sin x}{\cos x} = \frac{5}{\cos x} + 4 \sin x \cos x$$
$$\Rightarrow \quad 8 \cos^2 x + 5 \sin x = 5 + 4 \sin x \cos^2 x$$
$$\Leftrightarrow \quad 8(1 - \sin^2 x) + 5 \sin x = 5 + 4 \sin x (1 - \sin^2 x)$$
$$\Leftrightarrow \quad 8 - 8 \sin^2 x + 5 \sin x = 5 + 4 \sin x - 4 \sin^3 x$$
$$\Leftrightarrow \quad 4 \sin^3 x - 8 \sin^2 x + \sin x + 3 = 0.$$

We can factor the left side of the final equation as a cubic polynomial. To see why, let $t = \sin x$, so we have $4t^3 - 8t^2 + t + 3 = 0$. Clearly, $t = 1$ is one solution, and factoring out $t - 1$ gives us $(t - 1)(4t^2 - 4t - 3) = 0$, from which we have $(t - 1)(2t - 3)(2t + 1) = 0$. Substituting $t = \sin x$ back into this equation, we have $(\sin x - 1)(2 \sin x - 3)(2 \sin x + 1) = 0$.

Hence, $\sin x = 1$, $\sin x = \frac{3}{2}$, or $\sin x = -\frac{1}{2}$.

If $\sin x = 1$, then $\cos^2 x = 1 - \sin^2 x = 1 - 1 = 0$, so $\cos x = 0$. But if $\cos x = 0$, then $\tan x = \sin x / \cos x$ and $\sec x = 1 / \cos x$ are undefined, so there are no solutions in this case.

Since $-1 \leq \sin x \leq 1$ for all x, there are no solutions to $\sin x = \frac{3}{2}$.

If $\sin x = -\frac{1}{2}$, then $x = \frac{7\pi}{6}$ or $x = \frac{11\pi}{6}$. For $x = \frac{7\pi}{6}$, we have

$$8\cos x + 5\tan x = 8\cos\frac{7\pi}{6} + 5\tan\frac{7\pi}{6} = 8\cdot\left(-\frac{\sqrt{3}}{2}\right) + 5\cdot\frac{1}{\sqrt{3}} = -4\sqrt{3} + \frac{5\sqrt{3}}{3} = -\frac{7\sqrt{3}}{3}.$$

For $x = \frac{11\pi}{6}$, we have

$$8\cos x + 5\tan x = 8\cos\frac{11\pi}{6} + 5\tan\frac{11\pi}{6} = 8\cdot\frac{\sqrt{3}}{2} + 5\cdot\left(-\frac{1}{\sqrt{3}}\right) = 4\sqrt{3} - \frac{5\sqrt{3}}{3} = \frac{7\sqrt{3}}{3}.$$

Therefore, the points of intersection are $(x, y) = \boxed{\left(\frac{7\pi}{6}, -\frac{7\sqrt{3}}{3}\right)}$ and $\boxed{\left(\frac{11\pi}{6}, \frac{7\sqrt{3}}{3}\right)}$.

3.62 Since $\cos^2 x + \sin^2 x = 1$, we have $1 - \cos^2 x = \sin^2 x$, so

$$1 - \sin^4 x - \cos^2 x = \sin^2 x - \sin^4 x = \sin^2 x(1 - \sin^2 x) = \sin^2 x \cos^2 x.$$

By the double angle formula for sine, we have $\sin 2x = 2\sin x \cos x$, so $\sin^2 x \cos^2 x = \frac{\sin^2 2x}{4}$.

Hence, the given equation becomes $\frac{\sin^2 2x}{4} = \frac{1}{16}$, so $\sin^2 2x = \frac{1}{4}$, which means $\sin 2x = \pm\frac{1}{2}$.

We are given that $-\frac{\pi}{2} \le x \le \frac{\pi}{2}$, so $-\pi \le 2x \le \pi$. The only angles in this range whose sines are $\pm\frac{1}{2}$ are $-\frac{5\pi}{6}$, $-\frac{\pi}{6}$, $\frac{\pi}{6}$, and $\frac{5\pi}{6}$. Therefore, the solutions are $\boxed{x = -\frac{5\pi}{12}, -\frac{\pi}{12}, \frac{\pi}{12}, \text{ and } \frac{5\pi}{12}}$.

3.63 From the sum-to-product formula for sine and the difference-to-product formula for cosine, we have

$$\sin x + \sin(x+y) = 2\sin\frac{2x+y}{2}\cos\left(-\frac{y}{2}\right) = 2\sin\frac{2x+y}{2}\cos\frac{y}{2},$$

and

$$\cos x - \cos(x+y) = -2\sin\frac{2x+y}{2}\sin\left(-\frac{y}{2}\right) = 2\sin\frac{2x+y}{2}\sin\frac{y}{2},$$

so

$$f(x) = \frac{\sin x + \sin(x+y)}{\cos x - \cos(x+y)} = \frac{2\sin\dfrac{2x+y}{2}\cos\dfrac{y}{2}}{2\sin\dfrac{2x+y}{2}\sin\dfrac{y}{2}} = \cot\frac{y}{2}.$$

Thus, f is constant.

3.64 The given equation reminds us of the identity $1 + \tan^2 x = \sec^2 x$. Rearranging this gives $\sec^2 x - \tan^2 x = 1$, from which we have $(\sec x - \tan x)(\sec x + \tan x) = 1$. We are given $\sec x + \tan x = \frac{22}{7}$, so we have $\sec x - \tan x = \frac{7}{22}$. Adding this to the given equation gives $\sec x = \frac{533}{308}$, and subtracting it from the given equation yields $\tan x = \frac{435}{308}$. Therefore, we have $\cos x = \frac{308}{533}$ and $\sin x = \cos x \tan x = \frac{435}{533}$. This gives us $\csc x = \frac{533}{435}$ and $\cot x = \frac{1}{\tan x} = \frac{308}{435}$, so $\csc x + \cot x = \frac{841}{435} = \boxed{\frac{29}{15}}$.

Challenge: see if you can find a less computational solution starting from the fact that $\left(\frac{1+\sin x}{\cos x}\right)\left(\frac{1-\sin x}{\cos x}\right) = 1$.

3.65 The expression in x and y vaguely resembles some of our trig identities, particularly identities involving tangent. (We think of tangent instead of sine or cosine because x and y can be any real number.) So, we let

$x = \tan A$ and $y = \tan B$. Then

$$\begin{aligned}
\frac{(x+y)(1-xy)}{(1+x^2)(1+y^2)} &= \frac{(\tan A + \tan B)(1 - \tan A \tan B)}{(1 + \tan^2 A)(1 + \tan^2 B)} \\
&= \frac{\left(\frac{\sin A}{\cos A} + \frac{\sin B}{\cos B}\right)}{\sec^2 A \sec^2 B}(1 - \tan A \tan B) \\
&= \left(\frac{\sin A}{\cos A} + \frac{\sin B}{\cos B}\right)(\cos^2 A \cos^2 B)\left(1 - \frac{\sin A}{\cos A} \cdot \frac{\sin B}{\cos B}\right) \\
&= \left(\frac{\sin A}{\cos A} + \frac{\sin B}{\cos B}\right)\cos A \cos B \cdot \left(1 - \frac{\sin A}{\cos A} \cdot \frac{\sin B}{\cos B}\right)\cos A \cos B \\
&= (\sin A \cos B + \sin B \cos A)(\cos A \cos B - \sin A \sin B) \\
&= \sin(A + B)\cos(A + B) \\
&= \frac{1}{2}\sin[2(A+B)],
\end{aligned}$$

which is always between $-1/2$ and $1/2$.

3.66 Since $0 \le \theta \le \frac{\pi}{2}$, we have $0 \le \sin\theta \le 1$ and $0 \le \cos\theta \le 1$. Then $\sin^5\theta \le \sin^2\theta$ and $\cos^5\theta \le \cos^2\theta$, so

$$\sin^5\theta + \cos^5\theta \le \sin^2\theta + \cos^2\theta = 1.$$

But $\sin^5\theta + \cos^5\theta = 1$, so the inequalities above must be equalities. That is, $\sin^5\theta = \sin^2\theta$ and $\cos^5\theta = \cos^2\theta$. Moving all the terms to one side in the first equation, we get $\sin^5\theta - \sin^2\theta = 0$, which factors as $\sin^2\theta(\sin^3\theta - 1) = 0$. Therefore, $\sin\theta = 0$ or $\sin\theta = 1$, which means $\theta = \boxed{0}$ or $\boxed{\frac{\pi}{2}}$. Checking, we find that both of these solutions work. Rearranging and factoring $\cos^5\theta = \cos^2\theta$ leads to $\cos^2\theta(\cos^3\theta - 1) = 0$, which leads to the same values of θ.

3.67 We can rewrite the given equation as $\tan 7x + \cot 7x = \sin 6x + \cos 4x$. Let $t = \tan 7x$. Then $\cot 7x = \frac{1}{t}$, so $\tan 7x + \cot 7x = t + \frac{1}{t}$. If $t > 0$, then by the AM-GM inequality, we have

$$t + \frac{1}{t} \ge 2\sqrt{t \cdot \frac{1}{t}} = 2,$$

with equality if and only if $t = 1$. However, $\sin 6x \le 1$ and $\cos 4x \le 1$, so $\sin 6x + \cos 4x \le 2$. Therefore, we must have $\tan 7x = \sin 6x = \cos 4x = 1$.

If $t < 0$, then let $s = -t$, so $t + \frac{1}{t} = -\left(s + \frac{1}{s}\right)$. Again by the AM-GM inequality, $s + \frac{1}{s} \ge 2$, so $t + \frac{1}{t} \le -2$, with equality if and only if $t = -1$. However, $\sin 6x \ge -1$ and $\cos 4x \ge -1$, so $\sin 6x + \cos 4x \ge -2$. Therefore, we must have $\tan 7x = \sin 6x = \cos 4x = -1$.

Case 1: $\tan 7x = \sin 6x = \cos 4x = 1$.

Since $0 \le x \le 2\pi$, we have $0 \le 4x \le 8\pi$. The only angles in the interval $[0, 8\pi]$ whose cosines equal 1 are 0, 2π, 4π, 6π, and 8π, so x is equal to 0, $\frac{\pi}{2}$, π, $\frac{3\pi}{2}$, or 2π. None of these values satisfies $\tan 7x = 1$.

Case 2: $\tan 7x = \sin 6x = \cos 4x = -1$.

The only angles between 0 and 8π whose cosine is -1 are π, 3π, 5π, and 7π, so x is equal to $\frac{\pi}{4}$, $\frac{3\pi}{4}$, $\frac{5\pi}{4}$, or $\frac{7\pi}{4}$. The only angles that satisfy both $\tan 7x = -1$ and $\sin 6x = -1$ are $\frac{\pi}{4}$ and $\frac{5\pi}{4}$.

Therefore, the solutions are $x = \boxed{\frac{\pi}{4}}$ and $\boxed{\frac{5\pi}{4}}$.

3.68 First, we derive an angle sum identity for the cotangent function. From the angle sum identity for tangent,

$$\cot(x + y) = \frac{1}{\tan(x+y)} = \frac{1 - \tan x \tan y}{\tan x + \tan y}.$$

To obtain a formula in terms of $\cot x$ and $\cot y$, we divide both the numerator and denominator by $\tan x \tan y$:

$$\frac{1 - \tan x \tan y}{\tan x + \tan y} = \frac{\dfrac{1}{\tan x \tan y} - 1}{\dfrac{1}{\tan y} + \dfrac{1}{\tan x}} = \frac{\cot x \cot y - 1}{\cot x + \cot y}.$$

Hence, $\cot(x + y) = \dfrac{\cot x \cot y - 1}{\cot x + \cot y}$.

Let $\alpha = \operatorname{arccot} 3$ and $\beta = \operatorname{arccot} 7$. Then $\cot(\alpha + \beta) = \dfrac{\cot \alpha \cot \beta - 1}{\cot \alpha + \cot \beta} = \dfrac{3 \cdot 7 - 1}{3 + 7} = 2$.

Let $\gamma = \operatorname{arccot} 13$ and $\delta = \operatorname{arccot} 21$. Then $\cot(\gamma + \delta) = \dfrac{\cot \gamma \cot \delta - 1}{\cot \gamma + \cot \delta} = \dfrac{13 \cdot 21 - 1}{13 + 21} = 8$.

Then

$$10 \cot(\operatorname{arccot} 3 + \operatorname{arccot} 7 + \operatorname{arccot} 13 + \operatorname{arccot} 21) = 10 \cot\left((\alpha + \beta) + (\gamma + \delta)\right)$$
$$= 10 \cdot \frac{\cot(\alpha + \beta) \cot(\gamma + \delta) - 1}{\cot(\alpha + \beta) + \cot(\gamma + \delta)} = 10 \cdot \frac{2 \cdot 8 - 1}{2 + 8} = \boxed{15}.$$

3.69 Let $P = \cos \dfrac{\pi}{7} \cos \dfrac{2\pi}{7} \cos \dfrac{4\pi}{7}$. We note that in the sequence of angles $\dfrac{\pi}{7}$, $\dfrac{2\pi}{7}$, and $\dfrac{4\pi}{7}$, each angle is double the previous angle. We can take advantage of this observation by using the double angle formula for sine. Multiplying both sides by $\sin \dfrac{\pi}{7}$, we get

$$P \sin \frac{\pi}{7} = \sin \frac{\pi}{7} \cos \frac{\pi}{7} \cos \frac{2\pi}{7} \cos \frac{4\pi}{7}.$$

By double angle formula for sine, we have $2 \sin \dfrac{\pi}{7} \cos \dfrac{\pi}{7} = \sin \dfrac{2\pi}{7}$, so $P \sin \dfrac{\pi}{7} = \dfrac{1}{2} \sin \dfrac{2\pi}{7} \cos \dfrac{2\pi}{7} \cos \dfrac{4\pi}{7}$. Similarly, $2 \sin \dfrac{2\pi}{7} \cos \dfrac{2\pi}{7} = \sin \dfrac{4\pi}{7}$, so $P \sin \dfrac{\pi}{7} = \dfrac{1}{4} \sin \dfrac{4\pi}{7} \cos \dfrac{4\pi}{7}$. And again, $2 \sin \dfrac{4\pi}{7} \cos \dfrac{4\pi}{7} = \sin \dfrac{8\pi}{7}$, so $P \sin \dfrac{\pi}{7} = \dfrac{1}{8} \sin \dfrac{8\pi}{7}$. But $\dfrac{8\pi}{7} = \pi + \dfrac{\pi}{7}$, and $\sin(\pi + \theta) = -\sin \theta$ for all θ, so $\sin \dfrac{8\pi}{7} = -\sin \dfrac{\pi}{7}$. Therefore, $P = \boxed{-\dfrac{1}{8}}$.

3.70

(a) By the triple angle formulas for sine and cosine, we have $\sin 3x = 3 \sin x - 4 \sin^3 x$ and $\cos 3x = 4 \cos^3 x - 3 \cos x$, so

$$\tan 3x = \frac{\sin 3x}{\cos 3x} = \frac{3 \sin x - 4 \sin^3 x}{4 \cos^3 x - 3 \cos x} = \frac{\sin x(3 - 4 \sin^2 x)}{\cos x(4 \cos^2 x - 3)} = \tan x \cdot \frac{3 - 4 \sin^2 x}{4 \cos^2 x - 3}.$$

If we only had $\sin^2 x$ and $\cos^2 x$ terms in the numerator and denominator of the fraction, we could turn them into constants and $\tan^2 x$ terms by dividing the numerator and denominator by $\cos^2 x$. Therefore, we write 3 as $3(\sin^2 x + \cos^2 x)$:

$$\frac{3 - 4 \sin^2 x}{4 \cos^2 x - 3} = \frac{3(\sin^2 x + \cos^2 x) - 4 \sin^2 x}{4 \cos^2 x - 3(\sin^2 x + \cos^2 x)} = \frac{3 \cos^2 x - \sin^2 x}{\cos^2 x - 3 \sin^2 x}.$$

We can then divide the top and bottom by $\cos^2 x$, to get

$$\frac{3 \cos^2 x - \sin^2 x}{\cos^2 x - 3 \sin^2 x} = \frac{3 - \dfrac{\sin^2 x}{\cos^2 x}}{1 - \dfrac{3 \sin^2 x}{\cos^2 x}} = \frac{3 - \tan^2 x}{1 - 3 \tan^2 x}.$$

Therefore, $\tan 3x = \tan x \cdot \dfrac{3 - \tan^2 x}{1 - 3 \tan^2 x} = \boxed{\dfrac{3 \tan x - \tan^3 x}{1 - 3 \tan^2 x}}$.

(b) First, we determine $\sin\theta$. We have $\sin^2\theta = 1 - \cos^2\theta = 1 - \left(-\frac{8}{17}\right)^2 = 1 - \frac{64}{289} = \frac{225}{289}$, so $\sin\theta = \pm\sqrt{\frac{225}{289}} = \pm\frac{15}{17}$. Since θ lies in the second quadrant, we have $\sin\theta = \frac{15}{17}$.

By the half-angle formula for tangent, we have

$$\tan\frac{\theta}{2} = \frac{\sin\theta}{1+\cos\theta} = \frac{15/17}{1-8/17} = \frac{15}{9} = \frac{5}{3}.$$

Then by the triple angle formula for tangent, we have

$$\tan\frac{3\theta}{2} = \frac{3\tan(\theta/2) - \tan^3(\theta/2)}{1 - 3\tan^2(\theta/2)} = \frac{3\cdot(5/3) - (5/3)^3}{1 - 3(5/3)^2} = \boxed{-\frac{5}{99}}.$$

(c) The angles $40°$ and $80°$ are both $20°$ from $60°$, so we know that we can express $\tan 40°$ and $\tan 80°$ in terms of $\tan 20°$. So, we let $t = \tan 20°$. Then by the angle sum identity for tangent, we have

$$\tan 80° = \tan(60° + 20°) = \frac{\tan 60° + \tan 20°}{1 - \tan 60°\tan 20°} = \frac{\sqrt{3}+t}{1-t\sqrt{3}},$$

and by the angle difference identity for tangent, we have

$$\tan 40° = \tan(60° - 20°) = \frac{\tan 60° - \tan 20°}{1 + \tan 60°\tan 20°} = \frac{\sqrt{3}-t}{1+t\sqrt{3}},$$

so

$$\tan 20°\tan 40°\tan 80° = t\cdot\frac{\sqrt{3}+t}{1-t\sqrt{3}}\cdot\frac{\sqrt{3}-t}{1+t\sqrt{3}} = \frac{t(3-t^2)}{1-3t^2}.$$

This result resembles our earlier triple angle identity for tangent, and $60° = 3\cdot 20°$. So, we apply the triple angle identity and find:

$$\tan 60° = \tan(3\cdot 20°) = \frac{3\tan 20° - \tan^3 20°}{1 - 3\tan^2 20°} = \frac{3t - t^3}{1 - 3t^2} = \frac{t(3-t^2)}{1-3t^2}.$$

Hence, $\tan x = \tan 60°$, so $x = \boxed{60°}$.

3.71 Expressing all the trigonometric functions in terms of $\sin\theta$ and $\cos\theta$, we find

$$\sin\theta + \cos\theta + \tan\theta + \cot\theta + \sec\theta + \csc\theta = \sin\theta + \cos\theta + \frac{\sin\theta}{\cos\theta} + \frac{\cos\theta}{\sin\theta} + \frac{1}{\cos\theta} + \frac{1}{\sin\theta}$$
$$= \sin\theta + \cos\theta + \frac{\sin^2\theta + \cos^2\theta + \sin\theta + \cos\theta}{\cos\theta\sin\theta}$$
$$= \sin\theta + \cos\theta + \frac{1 + \sin\theta + \cos\theta}{\cos\theta\sin\theta}.$$

Let $s = \sin\theta + \cos\theta$ and $p = \sin\theta\cos\theta$. Then from the given equation, we have $s + \frac{1+s}{p} = 7$. Solving for s, we find $s = \frac{7p-1}{p+1}$. To solve for p and s, we must find another equation involving s and p.

Note that
$$s^2 = (\sin\theta + \cos\theta)^2 = \sin^2\theta + 2\sin\theta\cos\theta + \cos^2\theta = 1 + 2p.$$

Substituting $s = \frac{7p-1}{p+1}$ into this, we get $\left(\frac{7p-1}{p+1}\right)^2 = 1 + 2p$, which simplifies as $2p^3 - 44p^2 + 18p = 0$. This factors as $2p(p^2 - 22p + 9) = 0$. But $p \neq 0$ (because both $\sec\theta$ and $\csc\theta$ are defined), so $p^2 - 22p + 9 = 0$. By the quadratic formula, we have

$$p = \frac{22 \pm \sqrt{22^2 - 4\cdot 9}}{2} = \frac{22 \pm 8\sqrt{7}}{2} = 11 \pm 4\sqrt{7}.$$

Since $-1 \le \sin\theta \le 1$ and $-1 \le \cos\theta \le 1$, we have $p = \sin\theta\cos\theta \le 1$. But $11 + 4\sqrt{7} > 1$, so $p = 11 - 4\sqrt{7}$.

Finally, by the double angle formula for sine, we have $\sin 2\theta = 2\sin\theta\cos\theta = 2p = \boxed{22 - 8\sqrt{7}}$.

3.72 The right side of the recursion looks a lot like the angle sum identity for tangent. So, we let $a_n = \tan\theta_n$, and we note that

$$\tan(\theta_{n+2}) = a_{n+2} = \frac{a_n + a_{n+1}}{1 - a_n a_{n+1}} = \frac{\tan\theta_n + \tan\theta_{n+1}}{1 - \tan\theta_n \tan\theta_{n+1}} = \tan(\theta_n + \theta_{n+1}).$$

Therefore, if we let $\theta_{n+2} = \theta_{n+1} + \theta_n$, the recursion for a_n is satisfied when we let $a_n = \tan\theta_n$. Hence, we can find a_{2009} by finding θ_{2009}.

Since $a_1 = 1$ and $a_2 = \frac{1}{\sqrt{3}}$, we can let $\theta_1 = 45°$ and $\theta_2 = 30°$. We have $\theta_1 = 45° = 3 \cdot 15°$ and $\theta_2 = 30° = 2 \cdot 15°$, so each θ_n is an integer multiple of $15°$. Let $\theta_n = t_n \cdot 15°$, so $t_1 = 3$, $t_2 = 2$, and $t_{n+2} = t_{n+1} + t_n$ for all $n \ge 0$. Since $180° = 12 \cdot 15°$, and the tangent function has period $180°$, it suffices to compute t_{2009} modulo 12. (That is, we only have to find what the remainder is when we divide t_{2009} by 12.)

Computing the terms of the sequence $\{t_n\}$ modulo 12, we get

$$3, 2, 5, 7, 0, 7, 7, 2, 9, 11, 8, 7, 3, 10, 1, 11, 0, 11, 11, 10, 9, 7, 4, 11, 3, 2, 5, 7, 0, \ldots.$$

Once we see 3 immediately followed by 2 a second time, we know that the sequence is periodic, since each term is determined by the two terms before it. Hence, the sequence $\{t_n\}$ modulo 12 has period 24 (that is, it repeats every 24 terms) and we have $t_{2009} \equiv t_{17} \equiv 0 \pmod{12}$, so $a_{2009} = \tan(0 \cdot 15°) = \tan 0° = \boxed{0}$.

3.73 Let $45 \le k \le 133$. Then

$$\frac{\sin 1°}{\sin k° \sin(k+1)°} = \frac{\sin[(k+1)° - k°]}{\sin k° \sin(k+1)°}.$$

By the angle difference identity for sine, we have

$$\sin[(k+1)° - k°] = \sin(k+1)° \cos k° - \cos(k+1)° \sin k°,$$

so

$$
\begin{aligned}
\frac{\sin[(k+1)° - k°]}{\sin k° \sin(k+1)°} &= \frac{\sin(k+1)° \cos k° - \cos(k+1)° \sin k°}{\sin k° \sin(k+1)°} \\
&= \frac{\sin(k+1)° \cos k°}{\sin k° \sin(k+1)°} - \frac{\cos(k+1)° \sin k°}{\sin k° \sin(k+1)°} \\
&= \frac{\cos k°}{\sin k°} - \frac{\cos(k+1)°}{\sin(k+1)°} \\
&= \cot k° - \cot(k+1)°.
\end{aligned}
$$

Summing over $k = 45, 47, 49, \ldots, 133$, we get

$$\frac{\sin 1°}{\sin 45° \sin 46°} + \frac{\sin 1°}{\sin 47° \sin 48°} + \cdots + \frac{\sin 1°}{\sin 133° \sin 134°} = (\cot 45° - \cot 46°) + (\cot 47° - \cot 48°) + \cdots + (\cot 133° - \cot 134°).$$

But $\cot 46° = -\cot 134°$, $\cot 47° = -\cot 133°$, $\ldots$, $\cot 89° = -\cot 91°$, $\cot 90° = 0$, and $\cot 45° = 1$, so each cotangent on the right above cancels with another, except for $\cot 45°$. Therefore, we have

$$(\cot 45° - \cot 46°) + (\cot 47° - \cot 48°) + \cdots + (\cot 133° - \cot 134°) = 1,$$

which means

$$\frac{\sin 1°}{\sin 45° \sin 46°} + \frac{\sin 1°}{\sin 47° \sin 48°} + \cdots + \frac{\sin 1°}{\sin 133° \sin 134°} = 1.$$

Therefore,

$$\frac{1}{\sin 45° \sin 46°} + \frac{1}{\sin 47° \sin 48°} + \cdots + \frac{1}{\sin 133° \sin 134°} = \frac{1}{\sin 1°},$$

and the answer is $n = \boxed{1}$.

3.74 Note that $8\cos^3 4x \cos^3 x = (2\cos 4x \cos x)^3$. By the product-to-sum formula, $\cos 4x \cos x = (\cos 3x + \cos 5x)/2$, so $(2\cos 4x \cos x)^3 = (\cos 3x + \cos 5x)^3$. Hence, the given equation becomes

$$\cos^3 3x + \cos^3 5x = (\cos 3x + \cos 5x)^3.$$

Let $a = \cos 3x$ and $b = \cos 5x$. Then

$$a^3 + b^3 = (a+b)^3 \quad\Leftrightarrow\quad a^3 + b^3 = a^3 + 3a^2b + 3ab^2 + b^3 \quad\Leftrightarrow\quad 3a^2b + 3ab^2 = 0 \quad\Leftrightarrow\quad 3ab(a+b) = 0.$$

Hence, $\cos 3x = 0$, $\cos 5x = 0$, or $\cos 3x + \cos 5x = 0$. But $\cos 3x + \cos 5x = 2\cos x \cos 4x$, so $\cos 3x + \cos 5x = 0$ gives $\cos x = 0$ or $\cos 4x = 0$. So to summarize, x must satisfy one of the equations $\cos x = 0$, $\cos 3x = 0$, $\cos 4x = 0$, or $\cos 5x = 0$.

If $\cos x = 0$, then $x = 360°k + 90°$ or $x = 360°k + 270°$ for some integer k. There are no solutions such that $100° < x < 200°$. (Note that we can express "$x = 360°k + 90°$ or $x = 360°k + 270°$ for some integer k" as "$x = 180°k + 90°$ for some integer k.")

If $\cos 3x = 0$, then $3x = 360°k + 90°$ or $3x = 360°k + 270°$ for some integer k, so $x = 120°k + 30°$ or $x = 120°k + 90°$. The only solution such that $100° < x < 200°$ is $120° + 30° = 150°$.

If $\cos 4x = 0$, then $4x = 360°k + 90°$ or $4x = 360°k + 270°$ for some integer k, so $x = 90°k + \frac{45}{2}°$ or $x = 90°k + \frac{135}{2}°$. The only solutions such that $100° < x < 200°$ are $90° + \frac{45}{2}° = \frac{225}{2}°$ and $90° + \frac{135}{2}° = \frac{315}{2}°$.

If $\cos 5x = 0$, then $5x = 360°k + 90°$ or $5x = 360°k + 270°$ for some integer k, so $x = 72°k + 18°$ or $x = 72°k + 54°$. The only solutions such that $100° < x < 200°$ are $72° \cdot 2 + 18° = 162°$, $72° + 54° = 126°$, and $72° \cdot 2 + 54° = 198°$.

Therefore, the sum of the solutions such that $100° < x < 200°$ is

$$150° + \frac{225°}{2} + \frac{315°}{2} + 162° + 126° + 198° = \boxed{906°}.$$

3.75 Note that $\sin(90° - x) = \cos x$ and $\cos(90° - x) = \sin x$, so

$$\sin^2(90° - x) = \cos^2 x,$$
$$\cos^2(90° - x) = \sin^2 x,$$
$$\sin(90° - x)\cos(90° - x) = \cos x \sin x.$$

Hence, $\sin^2(90° - x)$, $\cos^2(90° - x)$, and $\sin(90° - x)\cos(90° - x)$ do not form the sides of a triangle if and only if $\sin^2 x$, $\cos^2 x$, and $\sin x \cos x$ do not form the sides of a triangle. Thus, the answer is the same if we restrict our attention to those values of x for which $0° < x \le 45°$.

For x such that $0° < x \le 45°$, we have $\cos x \ge \sin x > 0$, so $\cos^2 x \ge \sin x \cos x \ge \sin^2 x$. Hence, $\sin^2 x$, $\cos^2 x$, and $\sin x \cos x$ do not form the sides of a triangle if and only if $\sin x \cos x + \sin^2 x \le \cos^2 x$, which is equivalent to $\sin x \cos x \le \cos^2 x - \sin^2 x$.

By the double angle formulas for sine and cosine, we have $\sin 2x = 2\sin x \cos x$ and $\cos 2x = \cos^2 x - \sin^2 x$, so the inequality $\sin x \cos x \le \cos^2 x - \sin^2 x$ is equivalent to $\frac{\sin 2x}{2} \le \cos 2x$, which means $\frac{\sin 2x}{\cos 2x} \le 2$, or $\tan 2x \le 2$. Hence, $2x \le \arctan 2$ (since $\tan\theta$ is strictly increasing for $0 < \theta < 45°$), so

$$x \le \frac{\arctan 2}{2}.$$

Therefore, the probability that $\sin^2 x$, $\cos^2 x$, and $\sin x \cos x$ do not form the sides of a triangle is given by $\frac{\frac{\arctan 2}{2}}{45} = \boxed{\frac{\arctan 2}{90}}$.

3.76 The expressions in that terrifying equation we must prove strongly resemble the double angle formula for tangent. So, we let $x = \tan a$, $y = \tan b$, and $z = \tan c$. From the double angle formula for tangent, we have

$$\tan 2a = \frac{2 \tan a}{1 - \tan^2 a} = \frac{2x}{1 - x^2},$$
$$\tan 2b = \frac{2 \tan b}{1 - \tan^2 b} = \frac{2y}{1 - y^2},$$
$$\tan 2c = \frac{2 \tan c}{1 - \tan^2 c} = \frac{2z}{1 - z^2},$$

so we must show that

$$\tan 2a + \tan 2b + \tan 2c = (\tan 2a)(\tan 2b)(\tan 2c).$$

We must somehow use the given equation $x + y + z = xyz$, which we can write as $\tan a + \tan b + \tan c = (\tan a)(\tan b)(\tan c)$. We get expressions like $\tan a + \tan b$ and $(\tan a)(\tan b)$ from the identity for the tangent of a sum of two angles. So, we wonder if we get expressions like $\tan a + \tan b + \tan c$ and $(\tan a)(\tan b)(\tan c)$ in an identity for the tangent of a sum of three angles.

By applying the angle sum identity for tangent twice, we find

$$\tan(a + b + c) = \tan[(a + b) + c] = \frac{\tan(a + b) + \tan c}{1 - \tan(a + b) \tan c}$$
$$= \frac{\frac{\tan a + \tan b}{1 - \tan a \tan b} + \tan c}{1 - \frac{\tan a + \tan b}{1 - \tan a \tan b} \cdot \tan c} = \frac{\tan a + \tan b + \tan c - \tan a \tan b \tan c}{1 - (\tan a \tan b + \tan a \tan c + \tan b \tan c)}.$$

Aha! Since we are given $\tan a + \tan b + \tan c = (\tan a)(\tan b)(\tan c)$, we now know that $\tan(a + b + c) = 0$. Therefore, $a + b + c = \pi k$ for some integer k. Then $2a + 2b + 2c = 2\pi k$, so

$$\tan(2a + 2b + 2c) = 0.$$

But from our formula above for the tangent of the sum of three angles, we have

$$\tan(2a + 2b + 2c) = \frac{\tan 2a + \tan 2b + \tan 2c - \tan 2a \tan 2b \tan 2c}{1 - (\tan 2a \tan 2b + \tan 2a \tan 2c + \tan 2b \tan 2c)}.$$

Since this must equal 0, we must have

$$\tan 2a + \tan 2b + \tan 2c = \tan 2a \tan 2b \tan 2c,$$

so we have

$$\frac{2x}{1 - x^2} + \frac{2y}{1 - y^2} + \frac{2z}{1 - z^2} = \frac{2x}{1 - x^2} \cdot \frac{2y}{1 - y^2} \cdot \frac{2z}{1 - z^2}.$$

3.77 First, to deal with the 4^n term, we let $b_n = a_n/2^n$, so $a_n = 2^n b_n$. Substituting into the given formula, we get

$$2^{n+1} b_{n+1} = \frac{8}{5} \cdot 2^n b_n + \frac{6}{5} \sqrt{4^n - 4^n b_n^2} = \frac{8}{5} \cdot 2^n b_n + \frac{6}{5} \cdot 2^n \sqrt{1 - b_n^2}.$$

Dividing both sides by 2^{n+1}, we find

$$b_{n+1} = \frac{4}{5} b_n + \frac{3}{5} \sqrt{1 - b_n^2}.$$

CHAPTER 3. TRIGONOMETRIC IDENTITIES

The term $\sqrt{1 - b_n^2}$ strongly suggests a trigonometric substitution, as do the $\frac{4}{5}$ and $\frac{3}{5}$, since the sum of the squares of these is 1. Since each a_n is real, each b_n must be real. Therefore, we must have $1 - b_n^2 \geq 0$, so $-1 \leq b_n \leq 1$, which means we can let $b_n = \sin \alpha_n$ to take advantage of the identity $\sin^2 \alpha_n + \cos^2 \alpha_n = 1$. Letting $b_n = \sin \alpha_n$ therefore gives us

$$b_{n+1} = \frac{4}{5} \sin \alpha_n + \frac{3}{5} \sqrt{1 - \sin^2 \alpha_n} = \frac{4}{5} \sin \alpha_n + \frac{3}{5} \sqrt{\cos^2 \alpha_n} = \frac{4}{5} \sin \alpha_n + \frac{3}{5} |\cos \alpha_n|.$$

We can take further advantage of trigonometric identities by noting that because $\left(\frac{4}{5}\right)^2 + \left(\frac{3}{5}\right)^2 = 1$, there is an angle θ such that $\sin \theta = \frac{3}{5}$ and $\cos \theta = \frac{4}{5}$. Therefore, we can write the equation above as

$$\sin \alpha_{n+1} = \cos \theta \sin \alpha_n + \sin \theta |\cos \alpha_n|.$$

By the angle sum and angle difference identities for sine, we have

$$\sin(\alpha \pm \beta) = \sin \alpha \cos \beta \pm \sin \beta \cos \alpha,$$

so

$$b_{n+1} = \begin{cases} \sin(\theta + \alpha_n) & \text{if } \cos \alpha_n \geq 0, \\ \sin(\theta - \alpha_n) & \text{if } \cos \alpha_n < 0. \end{cases}$$

We have $b_0 = \sin 0°$, so we can take $\alpha_0 = 0°$. Then $\cos \alpha_0 = \cos 0° = 1 \geq 0$, so $b_1 = \sin \theta$. We can take $\alpha_1 = \theta$.

Then $\cos \alpha_1 = \cos \theta = \frac{4}{5} \geq 0$, so $b_2 = \sin 2\theta$. We can take $\alpha_2 = 2\theta$.

To find b_3, we must find the sign of $\cos 2\theta$. By the double angle formula for cosine, we have $\cos 2\theta = \cos^2 \theta - \sin^2 \theta = \frac{16}{25} - \frac{9}{25} = \frac{7}{25} > 0$, so $b_3 = \sin 3\theta$. We can take $\alpha_3 = 3\theta$.

To find b_4, we must find the sign of $\cos 3\theta$. By the triple angle formula for cosine,

$$\cos 3\theta = 4 \cos^3 \theta - 3 \cos \theta = 4 \cdot \left(\frac{4}{5}\right)^3 - 3 \cdot \frac{4}{5} = -\frac{44}{125} < 0,$$

so $b_4 = \sin(3\theta - \theta) = \sin 2\theta$.

But since b_{n+1} depends only on the value of b_n, and $b_4 = \sin 2\theta = b_2$, the sequence $\{b_n\}$ is periodic starting with the term b_4, with period 2. Therefore, by the double angle formula for sine, we have

$$b_{10} = b_2 = \sin 2\theta = 2 \sin \theta \cos \theta = 2 \cdot \frac{3}{5} \cdot \frac{4}{5} = \frac{24}{25},$$

so $a_{10} = 2^{10} b_{10} = \boxed{\frac{24576}{25}}$.

3.78 First, we derive an angle sum identity for arctangent. Let α and β be acute angles, and let $x = \tan \alpha$ and $y = \tan \beta$, so x and y are positive. Then by the angle sum identity for tangent, we have

$$\tan(\alpha + \beta) = \frac{\tan \alpha + \tan \beta}{1 - \tan \alpha \tan \beta} = \frac{x + y}{1 - xy}.$$

If $xy < 1$, then $(x + y)/(1 - xy)$ is positive. Furthermore, $0 < \alpha < \frac{\pi}{2}$ and $0 < \beta < \frac{\pi}{2}$, so $0 < \alpha + \beta < \pi$. Since $\tan(\alpha + \beta) = (x + y)/(1 - xy)$ is positive, $\alpha + \beta$ must be an acute angle. Therefore, we may take the arctan of both sides, to get

$$\alpha + \beta = \arctan \frac{x + y}{1 - xy}.$$

Since $\alpha = \arctan x$ and $\beta = \arctan y$, the equation becomes

$$\arctan x + \arctan y = \arctan \frac{x + y}{1 - xy}.$$

This formula holds for all positive real numbers x and y such that $xy < 1$.

Let $S_n = \displaystyle\sum_{k=1}^{n} \arctan \dfrac{1}{2k^2}$. We will prove that $S_n = \arctan \dfrac{n}{n+1}$ for all positive integers n by induction.

For $n = 1$, we have $S_1 = \arctan(1/2)$, so the result holds for the base case $n = 1$. Assume that the result holds for some positive integer $n = m$, so

$$S_m = \sum_{k=1}^{m} \arctan \frac{1}{2k^2} = \arctan \frac{m}{m+1}.$$

Then

$$S_{m+1} = \sum_{k=1}^{m+1} \arctan \frac{1}{2k^2} = \sum_{k=1}^{m} \arctan \frac{1}{2k^2} + \arctan \frac{1}{2(m+1)^2} = \arctan \frac{m}{m+1} + \arctan \frac{1}{2(m+1)^2}.$$

Both $m/(m+1)$ and $1/[2(m+1)^2]$ are positive, and

$$\frac{m}{m+1} \cdot \frac{1}{2(m+1)^2} = \frac{m}{2(m+1)^3} < 1,$$

since $m < 2(m+1)^3$. Therefore, by our angle sum identity for arctangent, we have

$$S_{m+1} = \arctan \frac{m}{m+1} + \arctan \frac{1}{2(m+1)^2} = \arctan \frac{\frac{m}{m+1} + \frac{1}{2(m+1)^2}}{1 - \frac{m}{m+1} \cdot \frac{1}{2(m+1)^2}} = \arctan \frac{(m+1)[2m(m+1)+1]}{2(m+1)^3 - m}$$

$$= \arctan \frac{(m+1)(2m^2 + 2m + 1)}{2(m^3 + 3m^2 + 3m + 1) - m} = \arctan \frac{(m+1)(2m^2 + 2m + 1)}{2m^3 + 6m^2 + 5m + 2} = \arctan \frac{(m+1)(2m^2 + 2m + 1)}{(m+2)(2m^2 + 2m + 1)}$$

$$= \arctan \frac{m+1}{m+2}.$$

Hence, the result holds for $n = m + 1$, and by induction, it holds for all positive integers n.

CHAPTER 4

Applications to Geometry

Exercises for Section 4.1

4.1.1 We have $\sin \angle PRQ = \frac{PQ}{PR}$, so $PR = \frac{PQ}{\sin \angle PRQ} = \frac{8}{\sin 42°} \approx \boxed{11.96}$, and $\tan \angle PRQ = \frac{PQ}{QR}$, so $QR = \frac{PQ}{\tan \angle PRQ} = \frac{8}{\tan 42°} \approx \boxed{8.88}$.

4.1.2 We have $\tan \angle XZY = \frac{XY}{YZ}$, so $\angle XZY = \arctan \frac{XY}{YZ} = \arctan 4 \approx \boxed{76°}$.

4.1.3 The line $2x + 3y = 7$ has x-intercept $(7/2, 0)$ and y-intercept $(0, 7/3)$. Therefore, if θ is the acute angle formed by the x-axis and the line $2x + 3y = 7$, then

$$\tan \theta = \frac{7/3}{7/2} = \frac{2}{3},$$

so $\theta = \arctan(2/3) \approx \boxed{34°}$.

We could also have tackled this problem by using the relationship between tangent and slope that we proved in the text. Letting ϕ be an angle between the line and the x-axis, we have $\tan \phi = -2/3$, from which we find $\phi = \arctan(-2/3) \approx -34°$, which tells us that the acute angle between the x-axis and the line is approximately $34°$.

4.1.4 We see that $BC = 12/4 = 3$, so the area of triangle ABC is $\frac{1}{2} AB \cdot BC \sin \angle ABC = \frac{1}{2} \cdot 12 \cdot 3 \cdot \frac{1}{2} = \boxed{9}$.

4.1.5 First, we calculate the horizontal distance between the two peaks (that is, the distance as viewed from above, without the vertical component). On the map, the two peaks are 1 inch apart, so in real life, the horizontal distance between them is

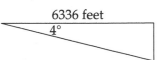

$$1 \text{ inch} \cdot \frac{1.2 \text{ miles}}{1 \text{ inch}} \cdot \frac{5280 \text{ feet}}{1 \text{ mile}} = 6336 \text{ feet.}$$

Then the vertical distance between the two peaks (in feet) is $6336 \cdot \tan 4° \approx 440$. Therefore, the second peak is $14000 - 440 \approx \boxed{13560 \text{ feet above sea level}}$.

4.1.6 Since $AO = 1$ and $\angle OCA$ is right, we immediately have $\boxed{CO = \cos \theta}$ and $\boxed{AC = \sin \theta}$. Since $AO = 1$ and $\angle OAD$ is right, we have $AD/AO = \tan \theta$, so $\boxed{AD = \tan \theta}$, and $AO/DO = \cos \theta$, so $\boxed{DO = \sec \theta}$.

Likewise, $\angle OAB$ is right. Also, $\angle AOB = \angle BOD - \angle AOD = 90° - \theta$, so $\angle ABO = 90° - \angle AOB = \theta$. Hence, $AO/AB = \tan \theta$ (and we still have $AO = 1$), so $\boxed{AB = \cot \theta}$, and $AO/BO = \sin \theta$, so $\boxed{BO = \csc \theta}$.

4.1.7 Let the side length of the cube be s. Since $\overline{AC}$ is the diagonal of a square with side length s, we have $AC = s\sqrt{2}$. Then by the Pythagorean Theorem on right triangle ACB, we have $AB = \sqrt{AC^2 + BC^2} = \sqrt{2s^2 + s^2} = \sqrt{3s^2} = s\sqrt{3}$.

Therefore,

$$\cos \angle ABC = \frac{BC}{AB} = \frac{s}{s\sqrt{3}} = \frac{1}{\sqrt{3}} = \boxed{\frac{\sqrt{3}}{3}}.$$

4.1.8 Let M be the midpoint of $\overline{AB}$, so $OM = AB/2 = 450$. Let $x = EM$ and $y = FM$, so $x + y = EM + FM = EF = 400$.

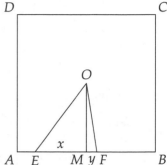

Let $\alpha = \angle EOM$ and $\beta = \angle FOM$. Then $\alpha + \beta = \angle EOM + \angle FOM = \angle EOF = 45°$, so $\tan(\alpha + \beta) = \tan 45° = 1$. But by the angle sum identity for tangent, we have $\tan(\alpha + \beta) = \frac{\tan\alpha + \tan\beta}{1 - \tan\alpha\tan\beta}$. Since $\angle EMO$ and $\angle FMO$ are right, we have $\tan\alpha = \frac{EM}{OM} = \frac{x}{450}$ and $\tan\beta = \frac{FM}{OM} = \frac{y}{450}$. Therefore, we have

$$\tan(\alpha + \beta) = \frac{\tan\alpha + \tan\beta}{1 - \tan\alpha\tan\beta} = \frac{\frac{x}{450} + \frac{y}{450}}{1 - \frac{x}{450}\cdot\frac{y}{450}} = \frac{450(x+y)}{202500 - xy} = \frac{180000}{202500 - xy},$$

where we used $x + y = 400$ in the final step. We also have $\tan(\alpha + \beta) = 1$, so $180000 = 202500 - xy$, which implies $xy = 22500$. Substituting $x = 400 - y$ from earlier into $xy = 22500$ gives $(400 - y)y = 22500$. Rearranging this equation gives $y^2 - 400y + 22500 = 0$. By the quadratic formula, we have

$$y = \frac{400 \pm \sqrt{400^2 - 4\cdot 22500}}{2} = \frac{400 \pm \sqrt{70000}}{2} = \frac{400 \pm 100\sqrt{7}}{2} = 200 \pm 50\sqrt{7}.$$

Since $AE < BF$, we have $x > y$, so $x = 200 + 50\sqrt{7}$ and $y = 200 - 50\sqrt{7}$. Therefore, we have

$$BF = BM - MF = 450 - (200 - 50\sqrt{7}) = \boxed{250 + 50\sqrt{7}}.$$

Exercises for Section 4.2

4.2.1 Let triangle ABC be obtuse, where $\angle C$ is obtuse. Let X be the foot of the altitude from B to $\overline{AC}$.

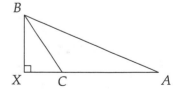

We see that $\angle BCX = 180° - \angle BCA = 180° - \angle C$, so $CX = BC\cos\angle BCX = a\cos(180° - \angle C) = -a\cos C$, and $BX = BC\sin\angle BCX = a\sin(180° - C) = a\sin C$. (Note that since $\angle C$ is obtuse, $\cos C$ is negative, so the expression $CX = -a\cos C$ makes sense.)

By the Pythagorean Theorem on right triangle BXA, we have $AB^2 = BX^2 + AX^2$. Since $AB = c$, $BX = a\sin C$, and $AX = AC + CX = b - a\cos C$, we have

$$c^2 = (a\sin C)^2 + (b - a\cos C)^2 = a^2\sin^2 C + b^2 - 2ab\cos C + a^2\cos^2 C$$
$$= (a^2\sin^2 C + a^2\cos^2 C) + b^2 - 2ab\cos C = a^2 + b^2 - 2ab\cos C.$$

Thus, the Law of Cosines holds for an obtuse triangle. The proof in the text suffices when $\angle B$ is obtuse or right. If $\angle A$ is obtuse or right, we can also use the proof in the text by simply swapping a and b, and swapping A and B, throughout the proof given in the text.

Next, let triangle ABC be right, where $\angle C$ is right. Then by the Pythagorean Theorem, $AB^2 = BC^2 + AC^2$, or $c^2 = a^2 + b^2$. But $\cos C = \cos 90° = 0$, so the equation $c^2 = a^2 + b^2 - 2ab\cos C$ holds. Thus, the Law of Cosines also holds for a right triangle.

4.2.2 Let A be the airport, W be the location of the westward-flying plane 90 minutes after takeoff, and E be the location of the eastward-flying plane 90 minutes after takeoff. Since the westward plane flies due west and the eastward plane flies $40°$ north of east, we have $\angle WAE = 90° + (90° - 40°) = 140°$. The westward plane flies at 200 miles per hour for 90 minutes, which is 1.5 hours, so $WA = 1.5(200) = 300$ miles. Similarly, we have $EA = 250 \cdot 1.5 = 375$ miles. Applying the Law of Cosines, we have

$$WE^2 = EA^2 + WA^2 - 2(WA)(EA)\cos \angle WAE \approx 403{,}000,$$

so $WE \approx \boxed{635 \text{ miles}}$.

4.2.3

(a) By the Pythagorean Theorem, $AB^2 = BC^2 + CA^2$ if and only if $\angle ACB$ is right.

(b) By the Law of Cosines, $\cos \angle ACB = \frac{AC^2 + BC^2 - AB^2}{2AC \cdot BC}$. Therefore, $AB^2 > BC^2 + CA^2$ if and only if $\cos \angle ACB < 0$. But $\angle ACB$ is an angle of a triangle, so $\cos \angle ACB < 0$ if and only if $\angle ACB$ is obtuse.

(c) By the Law of Cosines, $AB^2 < BC^2 + CA^2$ if and only if $\cos \angle ACB > 0$. But $\angle ACB$ is an angle of a triangle, so $\cos \angle ACB > 0$ if and only if $\angle ACB$ is acute.

4.2.4 Let the lengths of the sides be x, $2x$, and 6.

If the angle opposite the side of length x is $120°$, then by the Law of Cosines,

$$(2x)^2 + 6^2 - 2 \cdot 2x \cdot 6\cos 120° = x^2 \quad \Rightarrow \quad 4x^2 + 36 + 12x = x^2 \quad \Rightarrow \quad 3x^2 + 12x + 36 = 0.$$

This equation clearly has no positive solutions in x.

If the angle opposite the side of length $2x$ is $120°$, then

$$x^2 + 6^2 - 2 \cdot x \cdot 6\cos 120° = (2x)^2 \quad \Rightarrow \quad x^2 + 36 + 6x = 4x^2 \quad \Rightarrow \quad 3x^2 - 6x - 36 = 0 \quad \Rightarrow \quad x^2 - 2x - 12 = 0.$$

By the quadratic formula, $x = \frac{2 \pm \sqrt{2^2 + 4 \cdot 12}}{2} = \frac{2 \pm \sqrt{52}}{2} = \frac{2 \pm 2\sqrt{13}}{2} = 1 \pm \sqrt{13}$. Since x must be positive, the only solution in this case is $x = 1 + \sqrt{13}$.

If the angle opposite the side of length 6 is $120°$, then

$$x^2 + (2x)^2 - 2 \cdot x \cdot 2x\cos 120° = 6^2 \quad \Rightarrow \quad x^2 + 4x^2 + 2x^2 = 36 \quad \Rightarrow \quad 7x^2 = 36 \quad \Rightarrow \quad x^2 = \frac{36}{7},$$

so $x = 6/\sqrt{7} = (6\sqrt{7})/7$.

Therefore, the possible side lengths of the triangle are $\boxed{1 + \sqrt{13}, 2 + 2\sqrt{13}, \text{ and } 6}$ and $\boxed{6\sqrt{7}/7, 12\sqrt{7}/7, \text{ and } 6}$.

4.2.5 By the Law of Cosines, $\cos \angle XYZ = \frac{XY^2 + YZ^2 - XZ^2}{2XY \cdot YZ} = \frac{3^2 + 5^2 - 7^2}{2 \cdot 3 \cdot 5} = \frac{-15}{30} = -\frac{1}{2}$, so $\angle XYZ = \boxed{120°}$.

4.2.6 Let $\theta = \angle ABC$. Then by the Law of Cosines on triangle ABC,

$$AC^2 = AB^2 + BC^2 - 2AB \cdot BC\cos \theta = 2^2 + 3^2 - 2 \cdot 2 \cdot 3\cos \theta = 13 - 12\cos \theta.$$

Since quadrilateral $ABCD$ is inscribed in a circle, we have $\angle CDA = 180° - \angle ABC = 180° - \theta$. Therefore, by the Law of Cosines on triangle ADC, we have

$$AC^2 = CD^2 + DA^2 - 2CD \cdot DA\cos(180° - \theta) = 4^2 + 6^2 - 2 \cdot 4 \cdot 6(-\cos \theta) = 52 + 48\cos \theta.$$

Hence, we have

$$13 - 12\cos\theta = 52 + 48\cos\theta \quad\Rightarrow\quad 60\cos\theta = -39 \quad\Rightarrow\quad \cos\theta = -\frac{39}{60} = -\frac{13}{20}.$$

Then $AC^2 = 13 - 12\cos\theta = 13 + 12\cdot\frac{13}{20} = \frac{416}{20} = \frac{104}{5}$, so $AC = \sqrt{\frac{104}{5}} = \sqrt{\frac{520}{25}} = \boxed{\frac{2\sqrt{130}}{5}}$.

4.2.7 Let $\odot P$ denote the circle with center P for any point P. Let $\odot B$, $\odot C$, and $\odot D$ be tangent to $\odot A$ at X, Y, and Z, respectively. Let r be the radius of $\odot E$.

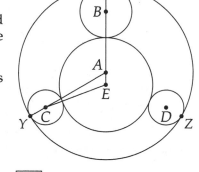

Since the radius of $\odot A$ is 10 and the radius of $\odot C$ is 2, we have $AC = 8$ and $CE = r + 2$. By symmetry, E lies on the diameter of $\odot A$ passing through X. Since the radius of $\odot B$ is 3, we have $AB = 7$ and $BE = r + 3$, so $AE = BE - AB = r - 4$.

Triangle XYZ is equilateral and A is its center, so $\angle XAY = 120°$, which means $\angle CAE = 60°$. Therefore, by the Law of Cosines on triangle CAE, we have

$$AC^2 + AE^2 - 2AC\cdot AE\cos 60° = CE^2$$
$$\Rightarrow\quad 8^2 + (r-4)^2 - 2\cdot 8\cdot(r-4)\cdot\frac{1}{2} = (r+2)^2$$
$$\Rightarrow\quad 64 + r^2 - 8r + 16 - 8r + 32 = r^2 + 4r + 4 \quad\Rightarrow\quad 20r = 108 \quad\Rightarrow\quad r = \frac{108}{20} = \boxed{\frac{27}{5}}.$$

4.2.8 Let $x = PC$. Since $\angle APB$, $\angle BPC$, and $\angle CPA$ are all equal and add up to $360°$, each angle is equal to $120°$. Then by the Law of Cosines on triangles PBC, PAB, and PAC, we have

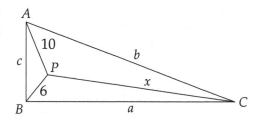

$$a^2 = x^2 + 6^2 - 2\cdot 6x\cos 120° = x^2 + 6x + 36,$$
$$c^2 = 6^2 + 10^2 - 2\cdot 6\cdot 10\cos 120° = 196,$$
$$b^2 = x^2 + 10^2 - 2\cdot 10x\cos 120° = x^2 + 10x + 100.$$

However, by the Pythagorean Theorem, we have $a^2 + c^2 = b^2$. Substituting the expressions above for a^2, b^2, and c^2 gives

$$x^2 + 6x + 36 + 196 = x^2 + 10x + 100 \quad\Rightarrow\quad 4x = 132 \quad\Rightarrow\quad x = \boxed{33}.$$

Exercises for Section 4.3

4.3.1 Let A, B, and W denote my house, my neighbor's house, and the well, respectively. Then from the given information, $AW = 150$ (in feet), $\angle BAW = 42°$, and $\angle AWB = 123°$. Then $\angle ABW = 180° - \angle BAW - \angle AWB = 180° - 42° - 123° = 15°$. Therefore, by the Law of Sines, we have

$$\frac{AB}{\sin\angle AWB} = \frac{AW}{\sin\angle ABW} \quad\Rightarrow\quad AB = \frac{AW\sin\angle AWB}{\sin\angle ABW} = \frac{150\sin 123°}{\sin 15°} \approx \boxed{486 \text{ feet}}.$$

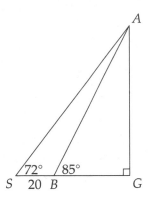

4.3.2 Let A and B denote the top and base of the tower, let S denote the tip of the shadow, and let G denote the point on the ground directly below A. Then from the given information, $BS = 20$, $\angle ABG = 85°$, and $\angle ASB = 72°$.

Then $\angle ABS = 180° - \angle ABG = 180° - 85° = 95°$, and $\angle BAS = 180° - \angle ASB - \angle ABS = 180° - 72° - 95° = 13°$, so by the Law of Sines on triangle ABS,

$$\frac{AB}{\sin \angle ASB} = \frac{BS}{\sin \angle BAS} \quad \Rightarrow \quad AB = \frac{BS \sin \angle ASB}{\sin \angle BAS} = \frac{20 \sin 72°}{\sin 13°} \approx 84.6.$$

Then from right triangle AGB, we have $AG = AB \sin \angle ABG = 84.6 \sin 85° \approx \boxed{84 \text{ feet}}$.

4.3.3 We claim that $AB > BC > CA$ if and only if $\angle C > \angle B > \angle A$. (In other words, the converse is also true.)

First, we cover the case of an acute or right triangle. By the Extended Law of Sines, we have $AB = 2R \sin C$, $AC = 2R \sin B$, and $BC = 2R \sin A$, so $AB > AC > BC$ if and only if $\sin C > \sin B > \sin A$. But the sine function is increasing on the interval $[0, \frac{\pi}{2}]$, so $\sin C > \sin B > \sin A$ if and only if $\angle C > \angle B > \angle A$.

Next, we cover the case of an obtuse triangle. Assume that $\angle C$ is obtuse, so $\angle C$ is the largest angle of the triangle. Furthermore, by the Law of Cosines,

$$AB^2 = AC^2 + BC^2 - 2AC \cdot BC \cos C.$$

Since $\angle C$ is obtuse, $\cos C$ is negative, so $-2AC \cdot BC \cos C$ is positive. Therefore, $AB^2 > AC^2$ and $AB^2 > BC^2$, which implies $AB > AC$ and $AB > BC$. Hence, AB is the largest side of the triangle.

Since $\angle C$ is obtuse, both $\angle A$ and $\angle B$ are acute. Hence, by the same argument as in the acute triangle case, $AC > BC$ if and only if $\angle B > \angle A$. Therefore, for an obtuse triangle, $AB > AC > BC$ if and only if $\angle C > \angle B > \angle A$.

4.3.4 From the given information, we have $\frac{\sin A + \sin B + \sin C}{3} = \frac{a+b+c}{12}$. By the Extended Law of Sines, we have $a = 2R \sin A$, $b = 2R \sin B$, and $c = 2R \sin C$, so

$$\frac{\sin A + \sin B + \sin C}{3} = \frac{a + b + c}{12} = \frac{2R \sin A + 2R \sin B + 2R \sin C}{12} = \frac{R}{6}(\sin A + \sin B + \sin C),$$

so $R = 2$.

Suppose that $a = 2$. Then $\sin A = \frac{a}{2R} = \frac{1}{2}$, so $\angle A$ is $\boxed{30° \text{ or } 150°}$.

4.3.5 Applying the Law of Sines gives $\frac{\sin \angle ACB}{AB} = \frac{\sin \angle CAB}{BC}$, so $\sin \angle ACB = \frac{AB}{BC} \sin \angle CAB = \frac{5}{7} \sin 30° = \frac{5}{14}$. We have $\arcsin \frac{5}{14} \approx 20.9°$, so the possible values of $\angle ACB$ are $20.9°$ or $180° - 20.9° = 159.1°$. However, since $\angle CAB = 30°$, we cannot possibly have $\angle ACB = 159.1°$, since the sum of these two angles is greater than $180°$, which we cannot have in $\triangle ABC$. So, there is only one possible value for $\angle ACB$. That is, if $AB = DE = 5$, $BC = EF = 7$, and $\angle CAB = \angle FDE = 30°$, then we *can* deduce that $\triangle ABC \cong \triangle DEF$. So, this wouldn't be an appropriate example to show why there isn't an SSA Congruence Theorem.

4.3.6 We cannot use the Law of Sines to prove AA Similarity because our very use of sine assumes that AA Similarity is true. For example, to show that every right triangle with a $23°$ angle has the same ratio

$$\frac{\text{leg opposite the } 23° \text{ angle}}{\text{hypotenuse}},$$

we use AA Similarity. In other words, we use AA Similarity to show that sine "works." So, we can't turn around and then use sine to show that AA Similarity works.

4.3.7 *Solution 1.* We have $\angle CAP = \angle CBP = 10°$ and $CA = CB$ (as radii of the same circle). The Law of Sines applied to $\triangle CAP$ and $\triangle CPB$ gives us

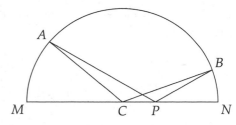

$$\frac{\sin \angle CPA}{AC} = \frac{\sin \angle CAP}{CP} \quad \text{and} \quad \frac{\sin \angle CPB}{CB} = \frac{\sin \angle CBP}{CP}.$$

Combining these with the equalities above, we have $\sin \angle CPA = \sin \angle CPB$. However, triangles ACP and BCP are NOT congruent, and $\angle CPA \neq \angle CPB$, so $\sin \angle CPA = \sin \angle CPB$ tells us that $\angle CPA + \angle CPB = 180°$.

Since $\angle ACM = 40°$, we have $\angle ACP = 180° - \angle ACM = 140°$. Then $\angle CPA = 180° - \angle CAP - \angle ACP = 180° - 10° - 140° = 30°$. Hence, $\angle CPB = 180° - \angle CPA = 180° - 30° = 150°$, so $\angle BCN = \angle BCP = 180° - \angle CBP - \angle CPB = 180° - 10° - 150° = 20°$. Therefore, we have $\widehat{BN} = \angle BCN = \boxed{20°}$.

Solution 2. As in Solution 1, we have $\angle CPA = 30°$. By the Law of Sines on triangle CPA,

$$\frac{AC}{\sin \angle CPA} = \frac{CP}{\sin \angle CAP} = \frac{CP}{\sin 10°}.$$

By the Law of Sines on triangle CPB,

$$\frac{BC}{\sin \angle CPB} = \frac{CP}{\sin \angle CBP} = \frac{CP}{\sin 10°}.$$

Since $AC = BC$, it follows that $\sin \angle CPB = \sin \angle CPA = \sin 30°$. Since $\angle CPB$ is obtuse, and $\sin 150° = \sin 30°$, we conclude that $\angle CPB = 150°$. Then again as in Solution 1, we have $\widehat{BN} = \angle BCN = \boxed{20°}$.

Exercises for Section 4.4

4.4.1 Let the incircle be tangent to $\overline{BC}$, $\overline{AC}$, and $\overline{AB}$ at D, E, and F, respectively, as shown at right. Since $\overline{AE}$ and $\overline{AF}$ are tangents to the same circle from the point, they are equal in length. Let $x = AE = AF$. Similarly, let $y = BD = BF$, and let $z = CD = CE$.

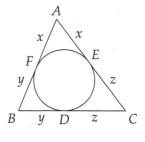

Hence,

$$x + y = AF + BF = AB = c,$$
$$x + z = AE + CE = AC = b,$$
$$y + z = BD + CD = BC = a.$$

Adding the first two equations, and subtracting the third equation, we get

$$(x + y) + (x + z) - (y + z) = b + c - a,$$

which simplifies as $2x = b + c - a$. Hence, $x = \frac{b+c-a}{2} = \frac{a+b+c-2a}{2} = \frac{a+b+c}{2} - a = s - a$.

Similarly, adding the first and third equations, and subtracting the second equation, we get

$$(x + y) + (y + z) - (x + z) = a + c - b,$$

which simplifies as $2y = a + c - b$. Hence, $y = \frac{a+c-b}{2} = \frac{a+b+c-2b}{2} = \frac{a+b+c}{2} - b = s - b$.

4.4.2 Let I be the incenter of triangle ABC. Then triangle IBC has height r and base $BC = a$, so

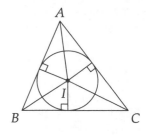

$$[IBC] = \frac{1}{2}ra.$$

Similarly, $[IAC] = \frac{1}{2}rb$ and $[IAB] = \frac{1}{2}rc$, so

$$[ABC] = [IBC] + [IAC] + [IAB] = \frac{1}{2}ra + \frac{1}{2}rb + \frac{1}{2}rc = r \cdot \frac{a+b+c}{2} = rs.$$

4.4.3 Let triangle ABC be obtuse, where $\angle C$ is obtuse. Let X be the foot of the altitude from B to $\overline{AC}$.

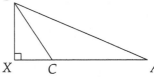

We see that $\angle BCX = 180° - \angle BCA = 180° - \angle C$, so $CX = BC \cos \angle BCX = a \cos(180° - \angle C) = -a \cos C$. Also, $AX = AB \cos \angle BAC = c \cos A$. Therefore, $b = AC = AX - CX = c \cos A + a \cos C$. This proof also addresses the case in which $\angle A$ is obtuse if swap the labels A and C, and swap a and c, throughout. The proof in the text suffices when $\angle B$ is obtuse. Thus, the formula holds for an obtuse triangle.

Next, let triangle ABC be right, where $\angle C$ is right. Then $b = AC = AB \cos \angle BAC = c \cos A$. But $\cos C = \cos 90° = 0$, so $b = c \cos A + a \cos C$. Thus, the formula also holds for a right triangle. As with the obtuse case, this proof addresses the case in which $\angle A$ is right, and the proof in the text covers the case in which $\angle B$ is right.

4.4.4 By the Law of Sines, we have $\sin A = \frac{a}{2R}$, $\sin B = \frac{b}{2R}$, and $\sin C = \frac{c}{2R}$, so

$$\sin A \sin B \sin C = \frac{a}{2R} \cdot \frac{b}{2R} \cdot \frac{c}{2R} = \frac{abc}{8R^3}.$$

Since $abc = 4R[ABC]$, we have $\sin A \sin B \sin C = \frac{abc}{8R^3} = \frac{4R[ABC]}{8R^3} = \frac{[ABC]}{2R^2}$.

4.4.5

(a) The desired formula resembles the formula we already found for $\cos \frac{A}{2}$, so we write $\sin \frac{A}{2}$ in terms of $\cos \frac{A}{2}$ and use the formula $\cos \frac{A}{2} = \sqrt{\frac{s(s-a)}{bc}}$ that we proved in the text:

$$\sin^2 \frac{A}{2} = 1 - \cos^2 \frac{A}{2} = 1 - \frac{s(s-a)}{bc} = \frac{bc - s(s-a)}{bc}.$$

We now must show that the numerator of that fraction equals $(s-b)(s-c)$. A little algebra does the trick:

$$bc - s(s-a) = bc - s^2 + as,$$
$$(s-b)(s-c) = s^2 - (b+c)s + bc = s^2 - (2s-a)s + bc = bc - s^2 + as.$$

So, we have $\sin \frac{A}{2} = \pm \sqrt{\frac{(s-b)(s-c)}{bc}}$. But $0° < A < 180°$, so $0° < \frac{A}{2} < 90°$, which means $\sin \frac{A}{2}$ is positive. Hence,

$$\sin \frac{A}{2} = \sqrt{\frac{(s-b)(s-c)}{bc}}.$$

(b) By the double angle formula for sine, we have $\sin A = 2 \sin \frac{A}{2} \cos \frac{A}{2}$, so

$$[ABC] = \frac{1}{2}bc \sin A = \frac{1}{2}bc \cdot 2 \sin \frac{A}{2} \cos \frac{A}{2} = bc \cdot \sqrt{\frac{(s-b)(s-c)}{bc}} \cdot \sqrt{\frac{s(s-a)}{bc}} = \sqrt{s(s-a)(s-b)(s-c)}.$$

(c) From the formula $[ABC] = \frac{abc}{4R}$, we have

$$R = \frac{abc}{4[ABC]} = \frac{abc}{4\sqrt{s(s-a)(s-b)(s-c)}} = \frac{abc}{\sqrt{16s(s-a)(s-b)(s-c)}} = \frac{abc}{\sqrt{2s(2s-2a)(2s-2b)(2s-2c)}}$$

$$= \frac{abc}{\sqrt{(a+b+c)(-a+b+c)(a-b+c)(a+b-c)}}.$$

4.4.6 Let $d = AM$. Then by Stewart's Theorem, we have

$$a\left(d^2 + \frac{a}{2} \cdot \frac{a}{2}\right) = \frac{a}{2} \cdot b^2 + \frac{a}{2} \cdot c^2 \quad \Rightarrow \quad d^2 + \frac{a^2}{4} = \frac{b^2 + c^2}{2} \quad \Rightarrow \quad d^2 = \frac{b^2 + c^2}{2} - \frac{a^2}{4}$$

$$\Rightarrow \quad d^2 = \frac{2b^2 + 2c^2 - a^2}{4} \quad \Rightarrow \quad d = \boxed{\frac{\sqrt{2b^2 + 2c^2 - a^2}}{2}}.$$

4.4.7 Let X, Y, and Z be the feet of the altitudes from A, B, and C, respectively.

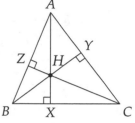

We can express AH in terms of other lengths and trig functions of angles with right triangle AYH, which gives us $\sin \angle AHY = \frac{AY}{AH}$, so $AH = \frac{AY}{\sin \angle AHY}$. Searching for other angles that equal $\angle AHY$, we note that

$$\angle AHY = 90^\circ - \angle YAH = 90^\circ - \angle CAX = 90^\circ - (90^\circ - C) = C,$$

so $AH = \frac{AY}{\sin C}$. Now we can use the Extended Law of Sines to introduce R. The Extended Law of Sines applied to $\triangle ABC$ gives $2R = \frac{c}{\sin C}$, so $AH = \frac{AY}{\sin C} = \frac{(2R)(AY)}{c}$. We wish to show that $AH = 2R\cos A$, so all we have left is to show that $\cos A = \frac{AY}{c}$. This follows immediately from $\triangle ABY$, in which we have $\cos A = \cos \angle YAB = \frac{AY}{AB} = \frac{AY}{c}$. Therefore, we have

$$AH = \frac{(2R)(AY)}{c} = 2R \cdot \frac{AY}{c} = 2R\cos A.$$

4.4.8 We have

$$\frac{\tan \frac{A-B}{2}}{\tan \frac{A+B}{2}} = \frac{\dfrac{\sin \frac{A-B}{2}}{\cos \frac{A-B}{2}}}{\dfrac{\sin \frac{A+B}{2}}{\cos \frac{A+B}{2}}} = \frac{\sin \frac{A-B}{2} \cos \frac{A+B}{2}}{\sin \frac{A+B}{2} \cos \frac{A-B}{2}}.$$

Applying product-to-sum identities gives

$$\sin \frac{A-B}{2} \cos \frac{A+B}{2} = \frac{1}{2}[\sin A + \sin(-B)] = \frac{1}{2}(\sin A - \sin B),$$

$$\sin \frac{A+B}{2} \cos \frac{A-B}{2} = \frac{1}{2}(\sin A + \sin B).$$

Therefore, $\dfrac{\tan \frac{A-B}{2}}{\tan \frac{A+B}{2}} = \dfrac{\sin \frac{A-B}{2} \cos \frac{A+B}{2}}{\sin \frac{A+B}{2} \cos \frac{A-B}{2}} = \dfrac{\sin A - \sin B}{\sin A + \sin B}.$

Finally, by the Extended Law of Sines, we have $\sin A = \frac{a}{2R}$ and $\sin B = \frac{b}{2R}$, so

$$\frac{\sin A - \sin B}{\sin A + \sin B} = \frac{\frac{a}{2R} - \frac{b}{2R}}{\frac{a}{2R} + \frac{b}{2R}} = \frac{a - b}{a + b}.$$

CHAPTER 4. APPLICATIONS TO GEOMETRY

Review Problems

4.26 Since $\angle U = 90°$, we have $\angle S + \angle T = 90°$. Combining this with the given $\angle S = 3\angle T$ produces $4\angle T = 90°$, so $\angle T = \frac{90°}{4} = 22.5°$. Since $\cos \angle T = \frac{TU}{ST}$, we have

$$\frac{ST}{TU} = \frac{1}{\cos 22.5°} \approx \boxed{1.08}.$$

4.27 The area of triangle PQR is given by

$$\frac{1}{2}PQ \cdot QR \sin \angle PQR = \frac{1}{2} \cdot 12 \cdot 4\sqrt{3} \sin \angle PQR = 24\sqrt{3} \sin \angle PQR.$$

We are given that the area of triangle PQR is $12\sqrt{6}$, so $\sin \angle PQR = \frac{12\sqrt{6}}{24\sqrt{3}} = \frac{\sqrt{2}}{2}$. Therefore, the possible values of $\angle PQR$ are $\boxed{45°}$ and $\boxed{135°}$.

4.28 Since $\angle JLK = 90°$, we have $\cos \angle JKL = \frac{KL}{JK} = \frac{7}{12}$, so $\angle JKL = \arccos \frac{7}{12} \approx \boxed{54°}$.

4.29

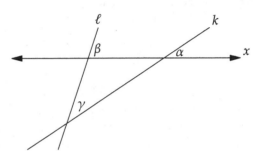

(a) Let α be the acute angle between k and the x-axis. The slope of line k is 2/3, so $\alpha = \arctan \frac{2}{3} \approx \boxed{34°}$.

(b) Let β be the acute angle between ℓ and the x-axis. The slope of line ℓ is 3, so $\beta = \arctan 3 \approx \boxed{72°}$.

(c) Let γ be the acute angle between k and ℓ, and consider the triangle formed by k, ℓ, and the x-axis. The angles of this triangle are α, $180° - \beta$, and γ, so $\alpha + (180° - \beta) + \gamma = 180°$. Therefore, $\gamma = \beta - \alpha \approx 72° - 34° \approx \boxed{38°}$.

(d) Let α be the angle between the line with slope m_1 and the x-axis such that $\tan \alpha = m_1$. Similarly, let β be the angle between the line with slope m_2 and the x-axis such that $\tan \beta = m_2$.

As in part (c), one of the angles between the two lines is $\beta - \alpha$. If we let $\gamma = \beta - \alpha$, then by the angle difference formula for tangent, we have

$$\tan \gamma = \tan(\beta - \alpha) = \frac{\tan \beta - \tan \alpha}{1 + \tan \beta \tan \alpha} = \boxed{\frac{m_2 - m_1}{1 + m_1 m_2}}.$$

We test the formula with lines k and ℓ. The slope of k is 2/3 and the slope of ℓ is 3, so taking $m_1 = 2/3$ and $m_2 = 3$, we get

$$\tan \gamma = \frac{3 - 2/3}{1 + 2/3 \cdot 3} = \frac{7/3}{3} = \frac{7}{9},$$

so $\gamma = \arctan(7/9) \approx 38°$, as before.

4.30 By symmetry, the overlap region is a rhombus. Let s be a side length of the rhombus. Then, we form a right triangle as shown. From this right triangle, we see that $\sin \alpha = \frac{1}{s}$, so $s = 1/\sin \alpha = \csc \alpha$. The area of the rhombus is simply the base times the height, which is $(1)(s) = \boxed{\csc \alpha}$.

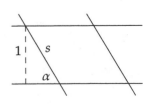

4.31 As the boat sails along the latitude 30° N, it follows the arc of a circle on the surface of Earth, so first we must find the length of that arc. Let P be denote the center of this circle, let A be the position of the boat, and let O denote the center of the Earth. We take a cross-section of the Earth that includes P, A, and O.

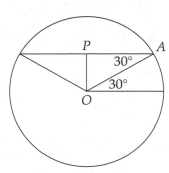

Since the boat is at 30° N latitude, we have $\angle OAP = 30°$. Since the radius of the Earth is 3950 (in miles), we have $PA = OA\cos 30° = 3950 \cdot \frac{\sqrt{3}}{2} \approx 3420$. The circumference of circle P is then $2\pi \cdot 3420 \approx 21500$.

As the boat sails from 19° W to 46° W, it sweeps out an arc of $46° - 19° = 27°$. Therefore, the boat sails a distance of $\frac{27°}{360°} \cdot 21500 \approx 1610$ miles. Finally, the boat sails at 9 miles per hour, so the travel time is $1610/9 = \boxed{179 \text{ hours}}$.

4.32 Let $x = CM$. Then $BM = CM\sin \angle MCB = 0.8x = \frac{4x}{5}$, and by the Pythagorean Theorem, we have

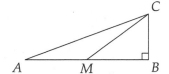

$$BC = \sqrt{CM^2 - BM^2} = \sqrt{x^2 - \frac{16x^2}{25}} = \sqrt{\frac{9x^2}{25}} = \frac{3x}{5}.$$

Since M is the midpoint of $\overline{AB}$, we have $AB = 2BM = \frac{8x}{5}$. Therefore, again by the Pythagorean Theorem,

$$AC = \sqrt{AB^2 + BC^2} = \sqrt{\left(\frac{8x}{5}\right)^2 + \left(\frac{3x}{5}\right)^2} = \sqrt{\frac{64x^2}{25} + \frac{9x^2}{25}} = \sqrt{\frac{73x^2}{25}} = \frac{x\sqrt{73}}{5}.$$

Hence, $\cos \angle ACB = \dfrac{BC}{AC} = \dfrac{\frac{3x}{5}}{\frac{x\sqrt{73}}{5}} = \dfrac{3}{\sqrt{73}} = \boxed{\dfrac{3\sqrt{73}}{73}}$.

4.33 By the Pythagorean Theorem, we have $AB = 5$ and $BD = 13$. We have $DE/DB = \sin \angle DBE$. Since $\angle DBE = 180° - \angle ABC - \angle ABD$, we have

$$\sin \angle DBE = \sin(180° - \angle ABC - \angle ABD) = \sin(\angle ABC + \angle ABD).$$

By the angle sum identity for sine, we have

$$\sin(\angle ABC + \angle ABD) = \sin \angle ABC \cos \angle ABD + \cos \angle ABC \sin \angle ABD$$

$$= \frac{AC}{AB} \cdot \frac{AB}{BD} + \frac{BC}{AB} \cdot \frac{AD}{BD} = \frac{3}{5} \cdot \frac{5}{13} + \frac{4}{5} \cdot \frac{12}{13} = \frac{15}{65} + \frac{48}{65} = \boxed{\frac{63}{65}}.$$

4.34

(a) We can build a right triangle with $\angle GIH$ as one of the acute angles by drawing altitude $\overline{HM}$ as shown. Because $GH = HI$, point M is the midpoint of $\overline{GI}$, so $GM = GI/2 = 3$. Then by the Pythagorean Theorem, we have $HM = \sqrt{GH^2 - GM^2} = \sqrt{72} = 6\sqrt{2}$, so $\tan \angle GIH = HM/IM = \boxed{2\sqrt{2}}$.

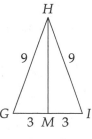

(b) We have

$$\tan \angle GHM = \frac{GM}{HM} = \frac{3}{6\sqrt{2}} = \frac{1}{2\sqrt{2}}.$$

Also, $\angle GHI = 2\angle GHM$, so by the double angle formula for tangent,

$$\tan \angle GHI = \tan(2\angle GHM) = \frac{2\tan \angle GHM}{1 - \tan^2 \angle GHM} = \frac{2 \cdot \frac{1}{2\sqrt{2}}}{1 - \left(\frac{1}{2\sqrt{2}}\right)^2} = \frac{\frac{1}{\sqrt{2}}}{1 - \frac{1}{8}} = \frac{\frac{1}{\sqrt{2}}}{\frac{7}{8}} = \frac{8}{7\sqrt{2}} = \frac{8\sqrt{2}}{14} = \boxed{\frac{4\sqrt{2}}{7}}.$$

(c)   By the Law of Cosines, we have $\cos \angle GHI = \dfrac{9^2 + 9^2 - 6^2}{2 \cdot 9 \cdot 9} = \dfrac{126}{162} = \boxed{\dfrac{7}{9}}$.

4.35 By the Law of Cosines, the distance between the two peaks is

$$\sqrt{4.6^2 + 5.8^2 - 2 \cdot 4.6 \cdot 5.8 \cos 37°} \approx \boxed{3.5 \text{ miles}}.$$

4.36 In the text, we used the Pythagorean Theorem to prove the Law of Cosines. In the "proof" offered in this problem, we then use the Law of Cosines to prove the Pythagorean Theorem. Because we used the Pythagorean Theorem to prove the Law of Cosines, we cannot use the Law of Cosines to prove the Pythagorean Theorem. (If we do, then we have essentially used the Law of Cosines to prove the Law of Cosines, which is clearly absurd!) We say this proof is guilty of "circular reasoning," and is therefore not a valid proof.

4.37 Label the vertices of the triangle A, B, and C, so that $AB = AC = 2$ and $BC = \sqrt{6} - \sqrt{2}$. Then by the Law of Cosines,

$$\cos A = \frac{2^2 + 2^2 - (\sqrt{6} - \sqrt{2})^2}{2 \cdot 2 \cdot 2} = \frac{4 + 4 - 6 + 4\sqrt{3} - 2}{8} = \frac{4\sqrt{3}}{8} = \frac{\sqrt{3}}{2}.$$

Hence, $A = 30°$. Since $AB = AC$, we have $\angle B = \angle C = (180° - 30°)/2 = 75°$. Therefore, the angles of the triangle are $\boxed{30°, 75°, \text{ and } 75°}$.

4.38 Let $x = AD$. By the Law of Cosines on triangle ABD, we have $\cos A = \frac{x^2 + 5^2 - 5^2}{2 \cdot x \cdot 5} = \frac{x}{10}$. (We can also see that $\cos A = \frac{x}{10}$ by noting that the altitude from B to $\overline{AD}$ bisects $\overline{AD}$, because $\triangle ABD$ is isosceles with $AB = BD$. So, we have $\cos A = \frac{AD/2}{AB} = \frac{x}{10}$.)

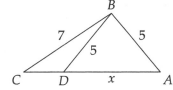

By the Law of Cosines on triangle ABC, we have $\cos A = \frac{5^2 + 9^2 - 7^2}{2 \cdot 5 \cdot 9} = \frac{19}{30}$, so $x/10 = 19/30$, which means $x = 19/3$. Then $CD = 9 - x = 8/3$, so $AD/DC = (19/3)/(8/3) = \boxed{19/8}$.

4.39 Let the side lengths of the triangle be $x - 1$, x, and $x + 1$. Since one angle is $60°$, the other two angles must add up to $120°$, so let them be $60° - \theta$ and $60° + \theta$, where θ is positive.

Since the angle with measure $60°$ is the "middle angle" (i.e., one other angle is larger and one is smaller), it must be opposite the side of with the "middle length," which is x. So by the Law of Cosines, we have

$$(x - 1)^2 + (x + 1)^2 - 2(x - 1)(x + 1) \cos 60° = x^2 \quad \Rightarrow \quad x^2 - 2x + 1 + x^2 + 2x + 1 - 2(x^2 - 1) \cdot \frac{1}{2} = x^2$$

$$\Rightarrow \quad 2x^2 + 2 - (x^2 - 1) = x^2 \quad \Rightarrow \quad 3 = 0,$$

which gives us a contradiction. That is, we have shown that if one angle of a triangle has measure $60°$ and the three side lengths of a triangle form an arithmetic sequence with common difference 1, then we have $3 = 0$. But it is impossible to have $3 = 0$, so we conclude that no such triangle exists.

4.40 Let $ABCD$ be a regular tetrahedron of side length 1, and let M be the midpoint of $\overline{AB}$, so $\overline{CM}$ and $\overline{DM}$ are altitudes of equilateral triangles ABC and ABD, respectively. Then the dihedral angle between faces ABC and ABD is equal to $\angle CMD$, since $\overline{CM}$ and $\overline{DB}$ are perpendicular to $\overline{AB}$ by symmetry.

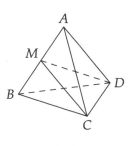

Since $AC = AD = 1$, we have $CM = DM = \sin 60° = \frac{\sqrt{3}}{2}$. Then by the Law of Cosines on triangle CMD,

$$\cos \angle CMD = \frac{\left(\frac{\sqrt{3}}{2}\right)^2 + \left(\frac{\sqrt{3}}{2}\right)^2 - 1^2}{2 \cdot \frac{\sqrt{3}}{2} \cdot \frac{\sqrt{3}}{2}} = \frac{\frac{3}{4} + \frac{3}{4} - 1}{\frac{3}{2}} = \frac{\frac{1}{2}}{\frac{3}{2}} = \frac{1}{3}.$$

Therefore, $\angle CMD = \arccos(\frac{1}{3}) \approx \boxed{70.5°}$.

4.41 By the Law of Sines, we have $\frac{PQ}{\sin R} = \frac{QR}{\sin P}$, so $\frac{4}{\sin R} = \frac{6}{\sin P} = \frac{6}{\sin 2R}$. By the double angle formula for sine, we have $\sin 2R = 2 \sin R \cos R$, so

$$\frac{4}{\sin R} = \frac{6}{2 \sin R \cos R} \quad \Rightarrow \quad \cos R = \frac{6}{2 \cdot 4} = \boxed{\frac{3}{4}}.$$

4.42 By the Law of Sines, we must have $\frac{AB}{\sin C} = \frac{BC}{\sin A}$, so $\frac{3}{\sin 57°} = \frac{8}{\sin 73°}$. However, $\frac{AB}{\sin C} = \frac{3}{\sin 57°} \approx 3.6$ and $\frac{BC}{\sin A} = \frac{8}{\sin 73°} \approx 8.4$, so no such triangle exists.

4.43 Let A, B, and C be the points where the plane starts, turns, and ends, respectively. Then $AB = 350$ (in miles), $AC = 800$, and $\angle B = 180° - 20° = 160°$. Then by the Law of Sines,

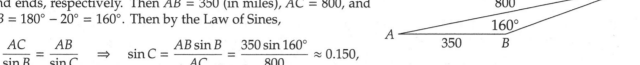

$$\frac{AC}{\sin B} = \frac{AB}{\sin C} \quad \Rightarrow \quad \sin C = \frac{AB \sin B}{AC} = \frac{350 \sin 160°}{800} \approx 0.150,$$

so $\angle C = \arcsin 0.150 \approx 8.6°$. Then $\angle A = 180° - \angle B - \angle C \approx 180° - 160° - 8.6° = 11.4°$.

Again by the Law of Sines, we have

$$\frac{BC}{\sin A} = \frac{AC}{\sin B} \quad \Rightarrow \quad BC = \frac{AC \sin A}{\sin B} = \frac{800 \sin 11.4°}{\sin 160°} \approx 462.$$

The speed of the plane then is 462 miles/2 hours = 231 miles per hour, so the total flight time is

$$\frac{350 \text{ miles} + 462 \text{ miles}}{231 \text{ miles per hour}} \approx \boxed{3.5 \text{ hours}}.$$

4.44 Label the vertices of the triangle A, B, and C, so that $AB = AC = 12$ and $BC = 6$. We want to find the circumradius of triangle ABC. From the Extended Law of Sines, the circumradius of triangle ABC is given by $R = \frac{AB}{2 \sin C}$. So, all we have to do is find $\sin C$. We can build a right triangle with $\angle C$ as one of the acute angles by drawing altitude $\overline{AD}$. The foot of this altitude is the midpoint of $\overline{BC}$ because $\triangle ABC$ is isosceles. The Pythagorean Theorem gives us $AD = \sqrt{AC^2 - CD^2} = 3\sqrt{15}$. Therefore, we have $\sin C = \frac{AD}{12} = \frac{\sqrt{15}}{4}$, so

$$R = \frac{AB}{2 \sin \angle C} = \frac{12}{\frac{\sqrt{15}}{2}} = \frac{24}{\sqrt{15}} = \frac{24\sqrt{15}}{15} = \boxed{\frac{8\sqrt{15}}{5}}.$$

4.45 *Solution 1.* By the Law of Cosines, the equation $AB \cos B = AC \cos C$ becomes

$$c \cdot \frac{a^2 + c^2 - b^2}{2ac} = b \cdot \frac{a^2 + b^2 - c^2}{2ab} \quad \Rightarrow \quad a^2 + c^2 - b^2 = a^2 + b^2 - c^2 \quad \Rightarrow \quad b^2 = c^2 \quad \Rightarrow \quad b = c.$$

Therefore, the triangle is isosceles with $AB = AC$, so the angles opposite these sides are equal: $\angle B = \angle C$.

Solution 2. The equation $AB \cos B = AC \cos C$ is equivalent to $\frac{\cos B}{\cos C} = \frac{AC}{AB}$. But by the Law of Sines, we have $\frac{AC}{AB} = \frac{\sin B}{\sin C}$, so $\frac{\cos B}{\cos C} = \frac{\sin B}{\sin C}$, which means $\sin B \cos C = \cos B \sin C$. Then by the angle difference formula for sine, we have

$$\sin(B - C) = \sin B \cos C - \cos B \sin C = 0,$$

so $B - C$ is an integer multiple of π. Since B and C are angles of a triangle, $B - C$ must be equal to 0, so $B = C$.

4.46 From the Law of Sines, we have $\frac{\sin \angle ABC}{AC} = \frac{\sin \angle BCA}{AB}$, so $\sin \angle ABC = \frac{AC}{AB} \sin \angle BCA = \frac{12}{6} \cdot \frac{1}{2} = 1$. Therefore, there is only one possible value of $\angle ABC$, which is $\angle ABC = 90°$. This means that we must have $\angle ABC = \angle DEF$, so $\triangle ABC \cong \triangle DEF$ by AAS Congruence.

4.47 If $\angle BAD = \angle CAD$, then both angles are equal to $A/2$. Then by the Law of Sines on triangle ABD, we have $\frac{AB}{BD} = \frac{\sin \angle ADB}{\sin(A/2)}$, and by the Law of Sines on triangle ACD, we have $\frac{AC}{CD} = \frac{\sin \angle ADC}{\sin(A/2)}$. But $\angle ADB + \angle ADC = 180°$, so $\sin \angle ADB = \sin \angle ADC$. Hence, we have $\frac{AB}{BD} = \frac{AC}{CD}$.

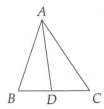

Challenge Problems

4.48 We have $BD = \tan \alpha$ and $CD = \tan \beta$, so $BC = BD + CD = \tan \alpha + \tan \beta$. Also, $\cos \alpha = AD/AB = 1/AB$ so $AB = 1/\cos \alpha$. Similarly, we have $AC = 1/\cos \beta$.

The area of triangle ABC is given by $\frac{1}{2}AD \cdot BC = \frac{1}{2}(\tan \alpha + \tan \beta)$. But the area of triangle ABC is also given by

$$\frac{1}{2}AB \cdot AC \sin \angle BAC = \frac{1}{2} \cdot \frac{1}{\cos \alpha} \cdot \frac{1}{\cos \beta} \sin(\alpha + \beta).$$

Therefore, $\frac{1}{2}(\tan \alpha + \tan \beta) = \frac{1}{2} \cdot \frac{1}{\cos \alpha} \cdot \frac{1}{\cos \beta} \sin(\alpha + \beta)$, which means

$$\sin(\alpha + \beta) = \cos \alpha \cos \beta (\tan \alpha + \tan \beta) = \cos \alpha \cos \beta \left(\frac{\sin \alpha}{\cos \alpha} + \frac{\sin \beta}{\cos \beta} \right) = \sin \alpha \cos \beta + \sin \beta \cos \alpha.$$

4.49 By the formula we found in the text for the cosine of half an angle of a triangle, we have

$$4 \cos \frac{A}{2} \cos \frac{B}{2} \cos \frac{C}{2} = 4 \sqrt{\frac{s(s-a)}{bc}} \cdot \sqrt{\frac{s(s-b)}{ac}} \cdot \sqrt{\frac{s(s-a)}{bc}} = 4 \sqrt{\frac{s^3(s-a)(s-b)(s-c)}{a^2 b^2 c^2}} = \frac{4\sqrt{s^2 \cdot s(s-a)(s-b)(s-c)}}{abc}.$$

The expression in the numerator looks like an expression in Heron's formula, which tells us that $[ABC] = \sqrt{s(s-a)(s-b)(s-c)}$. We also showed in the text that $abc = 4R[ABC]$. Making these substitutions above gives

$$4 \cos \frac{A}{2} \cos \frac{B}{2} \cos \frac{C}{2} = \frac{4\sqrt{s^2 \cdot s(s-a)(s-b)(s-c)}}{abc} = \frac{4s[ABC]}{4R[ABC]} = \frac{s}{R}.$$

4.50 Let Y and Z be the feet of the altitudes from B and C, as shown. From right triangle BHX, we have $\tan \angle HBX = \frac{HX}{BX}$, so $HX = BX \tan \angle HBX$. We now seek information about $\angle HBX$ and BX. From right triangle BYC, we have $\angle HBX = \angle YBC = 90° - \angle C$, so $\tan \angle HBX = \tan(90° - \angle C) = \cot C$. From right triangle AXB, we have $\cos B = \frac{BX}{AB}$, so $BX = AB \cos B$. Finally, we have

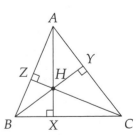

$$HX = BX \tan \angle HBX = BX \cot C = AB \cos B \cdot \frac{\cos C}{\sin C} = \frac{AB}{\sin C} \cos B \cos C = 2R \cos B \cos C,$$

where the Extended Law of Sines gives us the final step.

4.51 Let $x = AC$ and $\theta = \angle DCF$. Then from right triangle ACF, we have $\cos \theta = CF/AC = 1/x$.

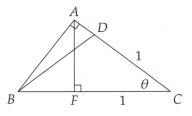

Since $BD = CD$, we have $\angle DBC = \angle DCB = \theta$. Then $\angle BDC = 180° - \angle DBC - \angle DCB = 180° - 2\theta$, so $\angle ADB = 180° - \angle BDC = 2\theta$. Then from right triangle BAD, we have $\cos 2\theta = AD/BD = AD = AC - CD = x - 1$.

By the double angle formula for cosine, we have $\cos 2\theta = 2\cos^2 \theta - 1$. Therefore,

$$x - 1 = \frac{2}{x^2} - 1 \quad \Rightarrow \quad x = \frac{2}{x^2} \quad \Rightarrow \quad x^3 = 2 \quad \Rightarrow \quad x = \boxed{\sqrt[3]{2}}.$$

4.52 Extend $\overline{AD}$ and $\overline{BC}$ to intersect at E. Since $\angle EAB = \angle EBA = 60°$, triangle ABE is equilateral. Let s be the side length of equilateral triangle ABE. Then $CE = BE - BC = s - 8$ and $DE = AE - AD = s - 10$.

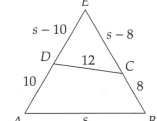

By the Law of Cosines on triangle CDE,

$$(s - 8)^2 + (s - 10)^2 - 2(s - 8)(s - 10)\cos 60° = 12^2$$
$$\Rightarrow \quad s^2 - 16s + 64 + s^2 - 20s + 100 - s^2 + 18s - 80 = 144$$
$$\Rightarrow \quad s^2 - 18s - 60 = 0.$$

By the quadratic formula,

$$s = \frac{18 \pm \sqrt{18^2 + 4 \cdot 60}}{2} = \frac{18 \pm \sqrt{564}}{2} = \frac{18 \pm 2\sqrt{141}}{2} = 9 \pm \sqrt{141}.$$

Since $9 - \sqrt{141} < 0$, we have $s = \boxed{9 + \sqrt{141}}$.

4.53 By the Pythagorean Theorem, the slant height of the cone is $\sqrt{600^2 + (200\sqrt{7})^2} = \sqrt{(200^2)(3^2 + (\sqrt{7})^2)} = 200\sqrt{9 + 7} = 800$. The circumference of the base is $2\pi \cdot 600 = 1200\pi$, so when the curved surface of the cone is unrolled, it becomes a sector with radius 800 and degree measure $\frac{1200\pi}{2\pi \cdot 800} \cdot 360° = 270°$.

Let O be the center of the sector (which corresponds to the vertex of the cone), and let A and B be the points where the ant starts and ends. Then the shortest path from A to B on the cone corresponds to a straight line in the sector. Since A and B are on opposite sides of the cone, we have $\angle AOB = 270°/2 = 135°$. Therefore, by the Law of Cosines,

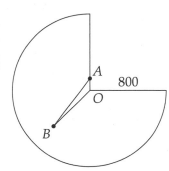

$$AB^2 = OA^2 + OB^2 - 2OA \cdot OB \cos 135° = 125^2 + (375\sqrt{2})^2 + 2 \cdot 125 \cdot 375\sqrt{2} \cdot \frac{\sqrt{2}}{2}$$
$$= 125^2(1^2 + (3\sqrt{2})^2 + 2 \cdot 1 \cdot 3) = 5^6(25) = 5^8,$$

so $AB = \sqrt{5^8} = \boxed{625}$.

4.54 By the angle difference formula for cosine, we have

$$\cos(A - B) = \cos A \cos B + \sin A \sin B,$$

so $\cos A \cos B = \cos(A - B) - \sin A \sin B$. Substituting into the given equation, we get

$$\cos(A - B) - \sin A \sin B + \sin A \sin B \sin C = 1,$$

so $\cos(A - B) = 1 + \sin A \sin B - \sin A \sin B \sin C = 1 + \sin A \sin B(1 - \sin C)$.

Since A, B, and C are the angles of a triangle, we have $0 < \sin A \le 1$, $0 < \sin B \le 1$, and $0 < \sin C \le 1$, so

$$1 + \sin A \sin B(1 - \sin C) \ge 1.$$

Furthermore, $\cos(A - B) \le 1$. Therefore, the equation above holds if and only if both sides are equal to 1, which means $\sin A \sin B(1 - \sin C) = 0$. Since $\sin A$ and $\sin B$ are positive, we have $\sin C = 1$, which means $C = \boxed{90°}$.

4.55 Since 2θ is an angle of a triangle, we have $0 < 2\theta < \pi$, so $0 < \theta < \frac{\pi}{2}$, which tells us that both $\sin \theta$ and $\cos \theta$ are positive. Let O be the center of square $ABCD$, and let M be the midpoint of $\overline{AB}$, so $AM = AB/2 = 1/2$.

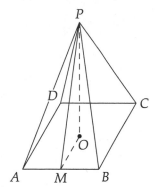

Then by symmetry, $\angle PMB$ and $\angle POM$ are right angles. Also, $\overline{PM}$ bisects $\angle APB$, so $\angle APM = \theta$. Then $AM = PM \tan \theta$, so

$$PM = \frac{AM}{\tan \theta} = \frac{1}{2 \tan \theta}.$$

Now, $OM = 1/2$, so by the Pythagorean Theorem on triangle POM, we have

$$OP^2 = PM^2 - OM^2 = \frac{1}{4 \tan^2 \theta} - \frac{1}{4} = \frac{\cos^2 \theta}{4 \sin^2 \theta} - \frac{1}{4} = \frac{\cos^2 \theta - \sin^2 \theta}{4 \sin^2 \theta}.$$

By the double angle formula for cosine, we have $\cos 2\theta = \cos^2 \theta - \sin^2 \theta$. Hence, we have $OP^2 = \frac{\cos^2 \theta - \sin^2 \theta}{4 \sin^2 \theta} = \frac{\cos 2\theta}{4 \sin^2 \theta}$, so $OP = \sqrt{\frac{\cos 2\theta}{4 \sin^2 \theta}} = \frac{\sqrt{\cos 2\theta}}{2 \sin \theta}$.

Finally, by the formula for the volume of a pyramid, the volume of pyramid $PABCD$ is

$$\frac{1}{3}[ABCD] \cdot OP = \frac{1}{3} \cdot \frac{\sqrt{\cos 2\theta}}{2 \sin \theta} = \boxed{\frac{\sqrt{\cos 2\theta}}{6 \sin \theta}}.$$

4.56 Let O be the center of the circle. The measure of $\angle BOC$ equals the measure of $\overarc{BC}$ in radians. Since the length of $\overarc{BC}$ is r, and the circumference of the circle is $2\pi r$, the arc is $\frac{1}{2\pi}$ of the circle, which means its measure in radians is simply 1. Therefore, we have $\angle BOC = 1$ in radians. (Indeed, we can see this from the very definition of a radian.) Then $\angle BAC = \overarc{BC}/2 = \angle BOC/2 = \frac{1}{2}$. Since $AB = AC$, we have $\angle ABC = \angle ACB = (\pi - \frac{1}{2})/2 = \frac{\pi}{2} - \frac{1}{4}$. Then by the Law of Sines,

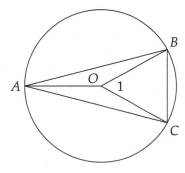

$$\frac{AB}{BC} = \frac{\sin \angle ACB}{\sin \angle BAC} = \frac{\sin(\frac{\pi}{2} - \frac{1}{4})}{\sin \frac{1}{2}} = \frac{\cos \frac{1}{4}}{\sin \frac{1}{2}}.$$

By the double angle formula for sine, we have $\sin \frac{1}{2} = 2 \sin \frac{1}{4} \cos \frac{1}{4}$, so

$$\frac{AB}{BC} = \frac{\cos \frac{1}{4}}{2 \sin \frac{1}{4} \cos \frac{1}{4}} = \frac{1}{2 \sin \frac{1}{4}} = \boxed{\frac{1}{2} \csc \frac{1}{4}}.$$

4.57 Let the sides of the triangle be $n - 1$, n, and $n + 1$. Let the smallest angle be θ, so the largest angle is 2θ. Therefore, the angle opposite $n - 1$ is θ, and the angle opposite $n + 1$ is 2θ, so by the Law of Sines,

$$\frac{n - 1}{\sin \theta} = \frac{n + 1}{\sin 2\theta} = \frac{n + 1}{2 \sin \theta \cos \theta}.$$

Solving for $\cos\theta$ gives $\cos\theta = \frac{n+1}{2(n-1)}$.

By the Law of Cosines, we have

$$\cos\theta = \frac{n^2 + (n+1)^2 - (n-1)^2}{2n(n+1)} = \frac{n^2 + n^2 + 2n + 1 - n^2 + 2n - 1}{2n(n+1)} = \frac{n^2 + 4n}{2n(n+1)} = \frac{n+4}{2(n+1)}.$$

Setting this equal to our earlier expression for $\cos\theta$ gives

$$\frac{n+1}{2(n-1)} = \frac{n+4}{2(n+1)} \quad\Rightarrow\quad (n+1)^2 = (n-1)(n+4) \quad\Rightarrow\quad n^2 + 2n + 1 = n^2 + 3n - 4 \quad\Rightarrow\quad n = 5,$$

so the cosine of the smallest angle is $\cos\theta = \frac{6}{2\cdot4} = \frac{6}{8} = \boxed{\frac{3}{4}}$.

4.58 Let $x = CD$. By the Angle Bisector Theorem, we have

$$\frac{CD}{AC} = \frac{BD}{AB} \quad\Rightarrow\quad BD = \frac{AB\cdot CD}{AC} = \frac{6x}{3} = 2x.$$

Let $d = AD$. By the Law of Cosines on triangle ACD, we have

$$3^2 + d^2 - 2\cdot3\cdot d\cos60° = x^2 \quad\Rightarrow\quad 9 + d^2 - 6d\cdot\frac{1}{2} = x^2 \quad\Rightarrow\quad d^2 - 3d + 9 = x^2,$$

and by the Law of Cosines on triangle ABD,

$$6^2 + d^2 - 2\cdot6\cdot d\cos60° = (2x)^2 \quad\Rightarrow\quad 36 + d^2 - 12d\cdot\frac{1}{2} = 4x^2 \quad\Rightarrow\quad d^2 - 6d + 36 = 4x^2.$$

Substituting the first equation into the second equation, we get

$$d^2 - 6d + 36 = 4(d^2 - 3d + 9) \quad\Rightarrow\quad d^2 - 6d + 36 = 4d^2 - 12d + 36 \quad\Rightarrow\quad 3d^2 - 6d = 0 \quad\Rightarrow\quad 3d(d-2) = 0.$$

Since d must be positive, we have $d = \boxed{2}$.

As an extra challenge, solve the problem without using the Law of Cosines or the Angle Bisector Theorem.

4.59 We wish to show that $AB > 2BC$. We can write this as $\frac{AB}{BC} > 2$. So, our goal is to find an expression for $\frac{AB}{BC}$ in terms of θ, and then show that this expression is greater than 2.

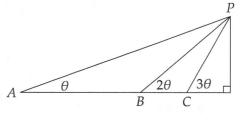

We see that $\angle PBA = 180° - \angle PBC = 180° - 2\theta$, so $\angle APB = 180° - \angle PAB - \angle PBA = 180° - \theta - (180° - 2\theta) = \theta$. Similarly, $\angle PCB = 180° - 3\theta$, so $\angle BPC = 180° - \angle PBC - \angle PCB = 180° - 2\theta - (180° - 3\theta) = \theta$. Since $\angle APB = \angle PAB = \theta$, we have $AB = PB$.

By the Law of Sines on triangle PBC, we have

$$\frac{PB}{BC} = \frac{\sin\angle PCB}{\sin\angle BPC} = \frac{\sin(180° - 3\theta)}{\sin\theta} = \frac{\sin3\theta}{\sin\theta},$$

so $\dfrac{AB}{BC} = \dfrac{PB}{BC} = \dfrac{\sin3\theta}{\sin\theta}$.

By the triple angle formula for sine, we have $\sin3\theta = 3\sin\theta - 4\sin^3\theta$, so

$$\frac{AB}{BC} = \frac{3\sin\theta - 4\sin^3\theta}{\sin\theta} = 3 - 4\sin^2\theta.$$

Since 3θ is an angle of a right triangle (other than the right angle), we have $3\theta < 90°$, so $\theta < 30°$. And since the sine function is increasing and positive on the interval $[0, \frac{\pi}{6}]$, we have $0 < \sin\theta < \sin 30° = \frac{1}{2}$. Therefore,

$$\frac{AB}{BC} = 3 - 4\sin^2\theta > 3 - 4 \cdot \left(\frac{1}{2}\right)^2 = 2,$$

so $AB > 2BC$.

4.60 Let $m = BD$ and $n = CD$, as shown in the diagram at right (the diagram is not to scale). Then $m + n = a$, and by the Angle Bisector Theorem, we have $m/c = n/b$, so $bm = cn$. Substituting $n = a - m$, we get $bm = c(a - m)$. Solving for m, we find

$$m = \frac{ac}{b + c},$$

so

$$n = \frac{bm}{c} = \frac{ab}{b + c}.$$

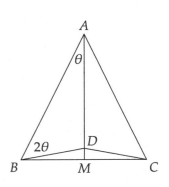

Let $t = AD$. By Stewart's Theorem, we have $a(t^2 + mn) = mb^2 + nc^2$. Substituting for m and n, the left side becomes

$$a(t^2 + mn) = a\left(t^2 + \frac{a^2bc}{(b + c)^2}\right),$$

and the right side becomes

$$mb^2 + nc^2 = \frac{ab^2c + abc^2}{b + c} = \frac{abc(b + c)}{b + c} = abc,$$

so $a\left(t^2 + \frac{a^2bc}{(b+c)^2}\right) = abc$, which means $t^2 + \frac{a^2bc}{(b+c)^2} = bc$. Isolating t^2, we find

$$t^2 = bc - \frac{a^2bc}{(b + c)^2} = \frac{bc(b + c)^2 - a^2bc}{(b + c)^2} = \frac{bc((b + c)^2 - a^2)}{(b + c)^2},$$

so $t = \boxed{\dfrac{\sqrt{bc((b + c)^2 - a^2)}}{b + c}}$.

4.61 *Solution 1.* Let $\theta = \angle BAM$, so $\angle BDM = 3\theta$. From right triangle BAM, we have $\tan\theta = \frac{BM}{AM} = \frac{BM}{11}$, and from right triangle BDM, we have $\tan 3\theta = \frac{BM}{DM} = \frac{BM}{AM - AD} = BM$. Therefore, we have $\frac{\tan 3\theta}{\tan\theta} = \frac{BM}{BM/11} = 11$. We also have

$$\tan 3\theta = \tan(2\theta + \theta) = \frac{\tan 2\theta + \tan\theta}{1 - \tan 2\theta \tan\theta} = \frac{\frac{2\tan\theta}{1 - \tan^2\theta} + \tan\theta}{1 - \frac{2\tan^2\theta}{1 - \tan^2\theta}}.$$

Multiplying the numerator and denominator of this last expression by $1 - \tan^2\theta$ gives

$$\tan 3\theta = \frac{2\tan\theta + \tan\theta(1 - \tan^2\theta)}{1 - 3\tan^2\theta} = \frac{3\tan\theta - \tan^3\theta}{1 - 3\tan^2\theta}.$$

Therefore, our earlier equation $\frac{\tan 3\theta}{\tan\theta} = 11$ gives us

$$11 = \frac{\tan 3\theta}{\tan\theta} = \frac{3 - \tan^2\theta}{1 - 3\tan^2\theta}.$$

Solving for $\tan\theta$, and noting that $\tan\theta > 0$ since θ is acute, we have $\tan\theta = \frac{1}{2}$. This gives us $BM = \frac{11}{2}$, and the Pythagorean Theorem applied to $\triangle AMB$ gives $AB = \frac{11\sqrt{5}}{2}$, so the perimeter of triangle ABC is $AB + AC + BC = 2AB + 2BM = \boxed{11 + 11\sqrt{5}}$.

Solution 2. Let $\theta = \angle BAM$. Then $\angle BAC = 2\angle BAM = 2\theta$, $\angle BDC = 3\angle BAC = 6\theta$, and $\angle BDM = \angle BDC/2 = 3\theta$. Then $\angle BDA = 180° - \angle BDM = 180° - 3\theta$, so $\angle ABD = 180° - \angle BAD - \angle BDA = 180° - \theta - (180° - 3\theta) = 2\theta$.

Let $x = BD$. By the Law of Sines on triangle ABD,

$$\frac{BD}{\sin \angle BAD} = \frac{AD}{\sin \angle ABD} \quad \Rightarrow \quad \frac{x}{\sin \theta} = \frac{10}{\sin 2\theta} = \frac{10}{2 \sin \theta \cos \theta} = \frac{5}{\cos \theta \sin \theta},$$

so $\cos \theta = \frac{5}{x}$.

Then from right triangle ABM, we have $AM = AB \cos \theta$, so $AB = \frac{AM}{\cos \theta} = \frac{11}{5/x} = \frac{11x}{5}$.

By the Pythagorean Theorem on right triangle ABM, we have

$$BM^2 = AB^2 - AM^2 = \left(\frac{11x}{5}\right)^2 - 11^2 = \frac{121(x^2 - 25)}{25}.$$

By the Pythagorean Theorem on right triangle DBM, we have $BM^2 = BD^2 - DM^2 = x^2 - 1$. Hence,

$$\frac{121(x^2 - 25)}{25} = x^2 - 1 \quad \Rightarrow \quad 121(x^2 - 25) = 25(x^2 - 1) \quad \Rightarrow \quad x = \sqrt{\frac{125}{4}} = \frac{5\sqrt{5}}{2}.$$

Then $AB = \frac{11x}{5} = \frac{11\sqrt{5}}{2}$, and $BM = \sqrt{x^2 - 1} = \sqrt{\frac{125}{4} - 1} = \sqrt{\frac{121}{4}} = \frac{11}{2}$, so the perimeter of triangle ABC is $AB + AC + BC = 2AB + 2BM = \boxed{11 + 11\sqrt{5}}$.

4.62 *Solution 1: Geometry.* We must create a diagram with an $18°$ angle, but where in geometry might we encounter such angles? We note that $18° = \frac{\pi}{10}$ radians. The factor of 5 in the denominator suggests investigating a regular pentagon, which we have drawn at right. We'd like to form a right triangle that we can use to find $\sin 18°$. We can build such a right triangle by drawing altitude $\overline{DG}$ from D to $\overline{AB}$, which bisects $\angle EDC$. We have $\angle EDG = 54°$, since each interior angle of a regular pentagon has measure $108°$. Then, drawing $\overline{AD}$ forms isosceles triangle AED in which we have $\angle ADE = \angle EAD = (180° - \angle E)/2 = 36°$, so $\angle ADG = 18°$. We now have a right triangle, $\triangle ADG$, with one angle equal to $18°$. So, if we let $AD = 1$, then we seek the length of AG, since $\sin 18° = \frac{AG}{AD}$.

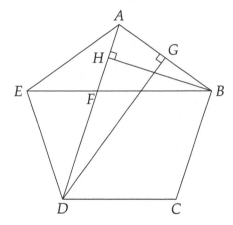

We let $AG = x$. We build another right triangle with one angle equal to $18°$ by drawing altitude $\overline{BH}$ from B to $\overline{AD}$. We have $\triangle BHA \sim \triangle DGA$ by AA, since these two right triangles share an acute angle. We have $AB = 2AG = 2x$. We can introduce yet another $18°$ angle in a right triangle by drawing $\overline{BE}$. Letting F be the intersection of $\overline{BE}$ and $\overline{AD}$, we see that $BFDC$ is a parallelogram. Furthermore, since $BC = CD$, this parallelogram is a rhombus, and we have $BF = FD = CD = AB = 2x$. Since $BF = AB$, altitude $\overline{AH}$ of isosceles triangle ABF bisects $\overline{AF}$, so $AH = \frac{AF}{2} = \frac{AD - FD}{2} = \frac{1 - 2x}{2} = \frac{1}{2} - x$.

Returning to similar triangles ABH and ADG, we have $\frac{AB}{AD} = \frac{AH}{AG}$, so $\frac{2x}{1} = \frac{\frac{1}{2} - x}{x}$. Cross-multiplying gives $2x^2 = \frac{1}{2} - x$. Multiplying by 2 and rearranging gives $4x^2 + 2x - 1 = 0$. The quadratic formula then gives $x = \frac{-2 \pm \sqrt{4 + 16}}{2(4)} = \frac{-1 \pm \sqrt{5}}{4}$. Since x must be positive, we have $\sin 18° = x = \boxed{\frac{-1 + \sqrt{5}}{4}}$.

Solution 2: Algebra Let $x = \cos 36°$ and $y = \cos 72° = \sin 18°$. Then by the double angle formula for cosine, we have $\cos 72° = 2 \cos^2 36° - 1$, so

$$y = 2x^2 - 1.$$

Again by the double angle formula for cosine, we have $\cos 36° = 1 - 2 \sin^2 18° = 1 - 2 \cos^2 72°$, so

$$x = 1 - 2y^2.$$

Adding these equations, we get $x + y = 2(x^2 - y^2) = 2(x + y)(x - y)$. Both x and y are positive, so $x + y$ is positive. Hence, we may divide both sides of $x + y = 2(x + y)(x - y)$ by $x + y$ to get $2(x - y) = 1$. Solving for x, we find $x = y + \frac{1}{2}$. Substituting into the equation $x = 1 - 2y^2$, we get $y + \frac{1}{2} = 1 - 2y^2$, which simplifies as $4y^2 + 2y - 1 = 0$. By the quadratic formula, we have $y = \frac{-1 \pm \sqrt{5}}{4}$. Since $\frac{-1 - \sqrt{5}}{4} < 0$, we conclude $\sin 18° = \boxed{\frac{-1+\sqrt{5}}{4}}$.

4.63 First, to evaluate the given expression, we express all the trigonometric functions in terms of sine and cosine:

$$\frac{\cot \gamma}{\cot \alpha + \cot \beta} = \frac{\dfrac{\cos \gamma}{\sin \gamma}}{\dfrac{\cos \alpha}{\sin \alpha} + \dfrac{\cos \beta}{\sin \beta}} = \frac{\sin \alpha \sin \beta \cos \gamma}{\sin \gamma(\sin \alpha \cos \beta + \cos \alpha \sin \beta)}.$$

By angle sum formula for sine, we have $\sin \alpha \cos \beta + \cos \alpha \sin \beta = \sin(\alpha + \beta)$ Furthermore, $\alpha + \beta = \pi - \gamma$, so $\sin(\alpha + \beta) = \sin(\pi - \gamma) = \sin \gamma$. Therefore,

$$\frac{\sin \alpha \sin \beta \cos \gamma}{\sin \gamma(\sin \alpha \cos \beta + \cos \alpha \sin \beta)} = \frac{\sin \alpha \sin \beta \cos \gamma}{\sin^2 \gamma}.$$

By the Law of Sines, we have $\frac{a}{\sin \alpha} = \frac{b}{\sin \beta} = \frac{c}{\sin \gamma}$, so $\frac{\sin \alpha}{\sin \gamma} = \frac{a}{c}$ and $\frac{\sin \beta}{\sin \gamma} = \frac{b}{c}$. Hence,

$$\frac{\sin \alpha \sin \beta \cos \gamma}{\sin^2 \gamma} = \frac{\sin \alpha}{\sin \gamma} \cdot \frac{\sin \beta}{\sin \gamma} \cdot \cos \gamma = \frac{ab}{c^2} \cos \gamma.$$

Finally, by the Law of Cosines, $\frac{ab}{c^2} \cos \gamma = \frac{ab}{c^2} \cdot \frac{a^2+b^2-c^2}{2ab} = \frac{a^2+b^2-c^2}{2c^2} = \frac{1989c^2-c^2}{2c^2} = \boxed{994}$.

4.64 We have $\tan \frac{A}{2} = \frac{r}{s-a}$, $\tan \frac{B}{2} = \frac{r}{s-b}$, and $\tan \frac{C}{2} = \frac{r}{s-c}$, so

$$\tan \frac{A}{2}\tan \frac{B}{2} + \tan \frac{A}{2}\tan \frac{C}{2} + \tan \frac{B}{2}\tan \frac{C}{2} = \frac{r}{s-a}\cdot\frac{r}{s-b} + \frac{r}{s-a}\cdot\frac{r}{s-c} + \frac{r}{s-b}\cdot\frac{r}{s-c} = \frac{r^2(s-c) + r^2(s-b) + r^2(s-a)}{(s-a)(s-b)(s-c)}$$

$$= \frac{r^2[3s - (a+b+c)]}{(s-a)(s-b)(s-c)} = \frac{r^2(3s - 2s)}{(s-a)(s-b)(s-c)} = \frac{r^2 s}{(s-a)(s-b)(s-c)}$$

$$= \frac{r^2 s^2}{s(s-a)(s-b)(s-c)} = \frac{[ABC]^2}{[ABC]^2} = 1.$$

4.65 We start with a cross-section of the Earth that includes Vancouver and the Earth's axis. In the diagram, Vancouver is point V and O is the center of the Earth. The Sun's rays come in from point S, putting half the Earth in night and half the Earth in daytime. $\overleftrightarrow{DO}$ is the dividing line between night and day, so it is perpendicular to the Sun's rays. $\overleftrightarrow{CO}$ is the axis of the Earth, point E is on the equator of the Earth, and $\overline{VY}$ is Vancouver's line of latitude (which means that V and Y are the same distance from the equator, so $\overline{VY}$ is perpendicular to the Earth's axis).

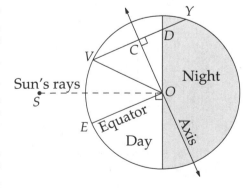

The length of daytime in Vancouver is maximized when we maximize the portion of $\overline{VY}$ that is in the "Day" section of the diagram. The axis of the Earth always makes a roughly 66.5° angle with the plane in which it orbits the Sun (we're simplifying here—the Earth "wobbles" a bit about this axis). However, as the Earth revolves about the sun, the angle between the Sun's rays and the upper half of the Earth's axis varies.

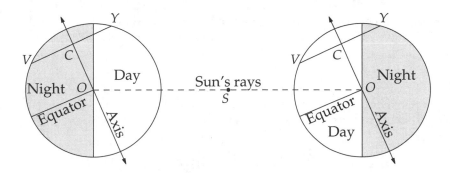

To get some intuition for why, consider the diagram above. The Sun is at the center of the diagram, and two positions of the Earth are shown. On the left is the Earth when Vancouver is in winter, and the days are short. On the right is Vancouver in summertime, when the days are long. The Earth's axis always points in the same direction (roughly at the "North Star") as it rotates around the Sun, so the axes in the two Earth positions shown are parallel.

The smallest that $\angle COS$ can be is when the upper half of the Earth's axis points towards the Sun, and the largest it can be is when this upper half of the axis points away from the Sun. Since the axis makes an angle of $66.5°$ with the plane in which the Earth orbits, this means that $\angle COS$ ranges from $66.5°$ to $180° - 66.5° = 113.5°$. As our diagram suggests, the smaller $\angle COS$ is, the more of Vancouver's latitude is included in daytime. Therefore, summer solstice, the longest day of the year in Vancouver, occurs when $\angle COS = 66.5°$.

Now we're ready to do some geometry. When the Earth rotates about its axis, Vancouver will move along its line of latitude one full turn about the axis. We wish to know for what portion of this trip Vancouver will be in sunlight. To measure this, we consider a second cross-section of the Earth, this time along Vancouver's latitude. This cross-section is at right. Points C and D are the same as in our initial diagram: C is where the axis of the Earth intersects the plane of Vancouver's latitude, and D is where the line that divides night and day in our first diagram intersects the line along Vancouver's latitude in that diagram. (Picture looking at the Earth from directly above the North Pole—C is on the axis and therefore at the center of your view, and D is the midpoint of the segment that divides night and day.) Points A and B are the boundary points between night and day.

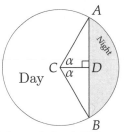

To determine how long Vancouver is in sunlight, we must determine what portion of the circumference of this cross-section is in sunlight. To do this, we must find $\angle ACB$. We start by drawing altitude $\overline{CD}$ to $\overline{AB}$, both because this forms right triangles and because we can learn more about $\overline{CD}$ from our first cross-section. Right triangles ACD and BCD are congruent, so our problem now is to determine $\angle ACD$, which we let be α. We can find α by first finding two side lengths of $\triangle ACD$, and then using trigonometry.

To find these side lengths, we return to our initial cross-section. $\overline{CD}$ is the same in this cross-section as in the latitude cross-section. $\overline{CA}$ in the latitude cross-section connects a point on Vancouver's latitude to the nearest point on the Earth's axis. Therefore, it has the same length as $\overline{CV}$ in the cross-section at right. Now, our problem is to determine CD and CV. Our diagram has many right triangles, and we know a lot of angles, so we reach for our trigonometric tools.

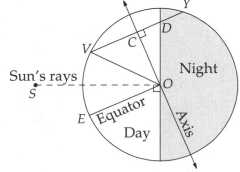

Because $\angle COS = 66.5°$, we have $\angle COD = 90° - 66.5° = 23.5°$. Since the latitude of Vancouver is $49°$ N, we have $\angle VOE = 49°$, which means that $\angle COV = 90° - \angle VOE = 41°$. Letting R be the radius of the Earth, we have $VO = R$. We now focus on the right triangles that contain our target lengths, CD and CV. Right triangle COV gives us $\sin \angle COV = \frac{CV}{VO}$, so $CV = VO \sin \angle COV = R \sin 41°$. In order to use $\triangle COD$ to build an expression for CD, we need an expression for CO or OD. Fortunately, we can use $\triangle COV$ to tackle CO. We have $\cos \angle COV = \frac{CO}{VO}$, so $CO = VO \cos \angle COV = R \cos 41°$. Finally, in $\triangle COD$, we have $\tan \angle COD = \frac{CD}{CO}$, so $CD = CO \tan \angle COD = R \cos 41° \tan 23.5°$.

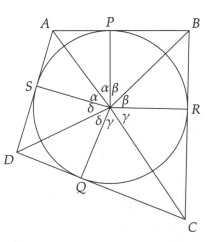

We're now ready to return to the latitude cross-section. Since CA in this cross-section is the same as CV in the axis cross-section, we have $CA = R \sin 41°$. We also found that $CD = R \cos 41° \tan 23.5°$, so we can write an expression for α:

$$\cos \alpha = \frac{CD}{CA} = \frac{R \cos 41° \tan 23.5°}{R \sin 41°} \approx 0.500.$$

(Yes, now you can see why we chose Vancouver for this example!) Since α is acute, $\cos \alpha \approx 0.5$ tells us that $\alpha \approx 60°$. Therefore, $\angle ACB \approx 120°$, which means that $\frac{1}{3}$ of the circumference of the circle is in the "night" section. This leaves the other $\frac{2}{3}$ of the circle in daytime, and gives Vancouver $\boxed{\frac{2}{3}(24) = 16}$ hours of sunlight on the summer solstice.

4.66 By the Pythagorean Theorem, we have $AB = 25$. Since M is the midpoint of hypotenuse $\overline{AB}$, M is the circumcenter of triangle ACB, so $CM = \frac{AB}{2} = \frac{25}{2}$. Since $\triangle ADB$ is isosceles with $AD = DB$ and point M is the midpoint of $\overline{AB}$, we have $\angle BMD = 90°$. Applying the Pythagorean Theorem to $\triangle DMB$ gives

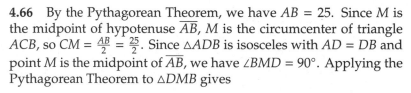

$$DM = \sqrt{BD^2 - BM^2} = \sqrt{15^2 - \left(\frac{25}{2}\right)^2} = \sqrt{\frac{275}{4}} = \frac{5\sqrt{11}}{2}.$$

Let $\theta = \angle CMD$, so $\angle CMB = \angle CMD + \angle DMB = \theta + 90°$. Then by the Law of Cosines on triangle BMC,

$$\cos(\theta + 90°) = \frac{BM^2 + CM^2 - BC^2}{2 BM \cdot CM} = \frac{\frac{625}{4} + \frac{625}{4} - 576}{2 \cdot \frac{25}{2} \cdot \frac{25}{2}} = \frac{-\frac{527}{2}}{\frac{625}{2}} = -\frac{527}{625}.$$

Since $\cos(\theta + 90°) = -\sin \theta$, we find $\sin \theta = \frac{527}{625}$. Therefore, the area of triangle CMD is

$$\frac{1}{2} CM \cdot DM \sin \theta = \frac{1}{2} \cdot \frac{25}{2} \cdot \frac{5\sqrt{11}}{2} \cdot \frac{527}{625} = \boxed{\frac{527\sqrt{11}}{40}}.$$

4.67 Let the circle be tangent to $\overline{BC}$ and $\overline{DA}$ at R and S, respectively, and let O and r be the center and radius of the circle, respectively. Let $\alpha = \angle POA$, $\beta = \angle POB$, $\gamma = \angle QOC$, and $\delta = \angle QOD$. Because $OP = OS = r$, $\angle APO = \angle ASO = 90°$, and right triangles APO and ASO share hypotenuse $\overline{AO}$, we have $\triangle APO \cong \triangle ASO$ by HL Congruence for right triangles. Therefore, we have $\angle SOA = \angle POA = \alpha$. Similarly, $\angle ROB = \angle POB = \beta$, $\angle ROC = \angle QOC = \gamma$, and $\angle SOD = \angle QOD = \delta$. Summing the angles around O, we get $2\alpha + 2\beta + 2\gamma + 2\delta = 2\pi$, so $\alpha + \beta + \gamma + \delta = \pi$.

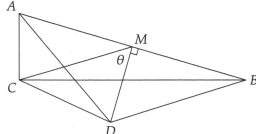

From right triangles POA and POB, we have $\tan \alpha = \frac{19}{r}$ and $\tan \beta = \frac{26}{r}$. Then by the angle sum identity for tangent, we have

$$\tan(\alpha + \beta) = \frac{\tan \alpha + \tan \beta}{1 - \tan \alpha \tan \beta} = \frac{19/r + 26/r}{1 - 19/r \cdot 26/r} = \frac{45r}{r^2 - 494}.$$

Similarly, from right triangles QOC and QOD, we have $\tan\gamma = \frac{37}{r}$ and $\tan\beta = \frac{23}{r}$, so

$$\tan(\gamma + \delta) = \frac{\tan\gamma + \tan\delta}{1 - \tan\gamma\tan\delta} = \frac{37/r + 23/r}{1 - 37/r \cdot 23/r} = \frac{60r}{r^2 - 851}.$$

Since $(\alpha + \beta) + (\gamma + \delta) = \pi$, we have $\tan(\alpha + \beta) + \tan(\gamma + \delta) = 0$. Therefore,

$$\frac{45r}{r^2 - 494} + \frac{60r}{r^2 - 851} = 0 \quad \Rightarrow \quad \frac{3}{r^2 - 494} + \frac{4}{r^2 - 851} = 0 \quad \Rightarrow \quad 3(r^2 - 851) + 4(r^2 - 494) = 0 \quad \Rightarrow \quad 7r^2 = 4529,$$

so $r^2 = \boxed{647}$.

4.68 By the double angle formula for cosine, we have

$$\cos^2 B = \frac{\cos 2B + 1}{2} \quad \text{and} \quad \cos^2 C = \frac{\cos 2C + 1}{2},$$

so

$$\cos^2 B + \cos^2 C = \frac{\cos 2B + 1}{2} + \frac{\cos 2C + 1}{2} = \frac{\cos 2B + \cos 2C}{2} + 1.$$

Then by the product-to-sum formula for cosine,

$$\frac{\cos 2B + \cos 2C}{2} + 1 = \cos(B + C)\cos(B - C) + 1.$$

But $B + C = 180° - A$, which means $\cos(B + C) = \cos(180° - A) = -\cos A$. Therefore,

$$\cos(B + C)\cos(B - C) + 1 = -\cos A \cos(B - C) + 1.$$

Hence,

$$\begin{aligned}
\cos^2 A + \cos^2 B + \cos^2 C &= \cos^2 A - \cos A \cos(B - C) + 1 \\
&= 1 - \cos A[\cos(B - C) - \cos A] \\
&= 1 - \cos A[\cos(B - C) - \cos(180° - B - C)] \\
&= 1 - \cos A[\cos(B - C) + \cos(B + C)].
\end{aligned}$$

Finally, by the sum-to-product formula for cosine, $\cos(B - C) + \cos(B + C) = 2\cos B \cos C$, so

$$1 - \cos A[\cos(B - C) + \cos(B + C)] = 1 - 2\cos A \cos B \cos C.$$

Hence, $\cos^2 A + \cos^2 B + \cos^2 C = 1 - 2\cos A \cos B \cos C$, or $\cos^2 A + \cos^2 B + \cos^2 C + 2\cos A \cos B \cos C = 1$.

4.69

(a) Let $a = AB$, $b = BC$, $c = CD$, and $d = DA$. By the Law of Cosines on triangles BAD and BCD, we have $BD^2 = a^2 + d^2 - 2ad\cos A$ and $BD^2 = b^2 + c^2 - 2bc\cos C$, respectively. Since quadrilateral $ABCD$ is inscribed in a circle, we have $C = 180° - A$, so $\cos C = \cos(180° - A) = -\cos A$, which means $BD^2 = b^2 + c^2 + 2bc\cos A$. Setting our two expressions for BD^2 equal gives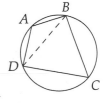

$$a^2 + d^2 - 2ad\cos A = b^2 + c^2 + 2bc\cos A \quad \Rightarrow \quad a^2 - b^2 - c^2 + d^2 = 2(ad + bc)\cos A$$

$$\Rightarrow \quad \cos A = \frac{a^2 - b^2 - c^2 + d^2}{2(ad + bc)}.$$

Then

$$\begin{aligned}
\sin^2 A = 1 - \cos^2 A &= 1 - \left[\frac{a^2 - b^2 - c^2 + d^2}{2(ad + bc)}\right]^2 = \frac{[2(ad + bc)]^2 - (a^2 - b^2 - c^2 + d^2)^2}{[2(ad + bc)]^2} \\
&= \frac{[2(ad + bc) + (a^2 - b^2 - c^2 + d^2)][2(ad + bc) - (a^2 - b^2 - c^2 + d^2)]}{[2(ad + bc)]^2}.
\end{aligned}$$

(♠)

The first factor in the numerator is equal to

$$2(ad + bc) + (a^2 - b^2 - c^2 + d^2) = (a^2 + 2ad + d^2) - (b^2 - 2bc + c^2) = (a + d)^2 - (b - c)^2 = (a + d + b - c)(a + d - b + c).$$

Since $2s = a + b + c + d$, we have $a + d + b - c = (a + b + c + d) - 2c = 2s - 2c = 2(s - c)$, and $a + d - b + c = (a + b + c + d) - 2b = 2s - 2b = 2(s - b)$, so

$$(a + d + b - c)(a + d - b + c) = 4(s - b)(s - c).$$

Similarly, the second factor in the numerator of ($\spadesuit$) is equal to

$$2(ad + bc) - (a^2 - b^2 - c^2 + d^2) = (b^2 + 2bc + c^2) - (a^2 - 2ad + d^2) = (b + c)^2 - (a - d)^2$$
$$= (b + c + a - d)(b + c - a + d)$$
$$= [(a + b + c + d) - 2d][(a + b + c + d) - 2a] = (2s - 2d)(2s - 2a) = 4(s - d)(s - a),$$

so

$$\sin^2 A = \frac{16(s - a)(s - b)(s - c)(s - d)}{[2(ad + bc)]^2} = \frac{4(s - a)(s - b)(s - c)(s - d)}{(ad + bc)^2}.$$

Since $0° < A < 180°$, $\sin A$ is positive, so

$$\sin A = \frac{2\sqrt{(s - a)(s - b)(s - c)(s - d)}}{ad + bc}.$$

Quadrilateral $ABCD$ is composed of the two triangles BAD and BCD. The area of triangle BAD is

$$[BAD] = \frac{1}{2} AB \cdot AD \sin \angle BAD = \frac{1}{2} ad \sin A.$$

The area of triangle BCD is

$$[BCD] = \frac{1}{2} BC \cdot DC \sin \angle BCD = \frac{1}{2} bc \sin C.$$

But $\sin C = \sin(180° - A) = \sin A$, so $[BCD] = \frac{1}{2} bc \sin A$. Therefore,

$$[ABCD] = [BAD] + [BCD] = \frac{1}{2} ad \sin A + \frac{1}{2} bc \sin A = \frac{1}{2}(ad + bc) \sin A = \sqrt{(s - a)(s - b)(s - c)(s - d)}.$$

(b) By the Law of Cosines on triangles BAD and BCD, we have $BD^2 = a^2 + d^2 - 2ad \cos A$ and $BD^2 = b^2 + c^2 - 2bc \cos C$, respectively, so

$$a^2 + d^2 - 2ad \cos A = b^2 + c^2 - 2bc \cos C.$$

Then $a^2 - b^2 - c^2 + d^2 = 2ad \cos A - 2bc \cos C$. Squaring both sides, we get

$$(a^2 - b^2 - c^2 + d^2)^2 = 4a^2d^2 \cos^2 A + 4b^2c^2 \cos^2 C - 8abcd \cos A \cos C. \tag{$*$}$$

Quadrilateral $ABCD$ is composed of the two triangles BAD and BCD, so

$$[ABCD] = [BAD] + [BCD] = \frac{1}{2} ad \sin A + \frac{1}{2} bc \sin C,$$

or $4[ABCD] = 2ad \sin A + 2bc \sin C$. Squaring both sides, we get

$$16[ABCD]^2 = 4a^2d^2 \sin^2 A + 4b^2c^2 \sin^2 C + 8abcd \sin A \sin C. \tag{$**$}$$

Adding equations ($*$) and ($**$), we get

$$16[ABCD]^2 + (a^2 - b^2 - c^2 + d^2)^2 = 4a^2d^2(\cos^2 A + \sin^2 A) + 4b^2c^2(\cos^2 C + \sin^2 C) - 8abcd(\cos A \cos C - \sin A \sin C)$$
$$= 4a^2d^2 + 4b^2c^2 - 8abcd(\cos A \cos C - \sin A \sin C).$$

By the angle sum formula for cosine, we have $\cos A \cos C - \sin A \sin C = \cos(A + C)$, and by the double angle formula for cosine, we have $\cos(A + C) = 2\cos^2\left(\frac{A+C}{2}\right) - 1$, so

$$4a^2d^2 + 4b^2c^2 - 8abcd(\cos A \cos C - \sin A \sin C) = 4a^2d^2 + 4b^2c^2 - 8abcd\left[2\cos^2\left(\frac{A+C}{2}\right) - 1\right]$$

$$= 4a^2d^2 + 8abcd + 4b^2c^2 - 16abcd\cos^2\left(\frac{A+C}{2}\right)$$

$$= [2(ad + bc)]^2 - 16abcd\cos^2\left(\frac{A+C}{2}\right).$$

Hence,

$$16[ABCD]^2 + (a^2 - b^2 - c^2 + d^2)^2 = [2(ad + bc)]^2 - 16abcd\cos^2\left(\frac{A+C}{2}\right),$$

which implies that

$$16[ABCD]^2 = [2(ad + bc)]^2 - (a^2 - b^2 - c^2 + d^2)^2 - 16abcd\cos^2\left(\frac{A+C}{2}\right).$$

We have already calculated in part (a) that

$$[2(ad + bc)]^2 - (a^2 - b^2 - c^2 + d^2)^2 = 16(s - a)(s - b)(s - c)(s - d),$$

so

$$16[ABCD]^2 = 16(s - a)(s - b)(s - c)(s - d) - 16abcd\cos^2\left(\frac{A+C}{2}\right).$$

Then $[ABCD]^2 = (s - a)(s - b)(s - c)(s - d) - abcd\cos^2\left(\frac{A+C}{2}\right)$, so

$$[ABCD] = \sqrt{(s - a)(s - b)(s - c)(s - d) - abcd\cos^2\left(\frac{A+C}{2}\right)}.$$

Note: This is called **Bretschneider's formula**, after someone who must have really liked algebra.

4.70 Let $\theta = \angle DBA$. Then $\angle CAB = \angle DBC = 2\theta$. Hence, $\angle AOB = 180° - \angle OAB - \angle OBA = 180° - 2\theta - \theta = 180° - 3\theta$, so $\angle BOC = 180° - \angle AOB = 3\theta$.

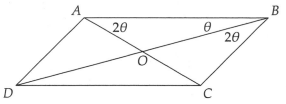

Then by the Law of Sines on $\triangle BOC$, we have $\frac{OC}{BC} = \frac{\sin 2\theta}{\sin 3\theta}$, and by the Law of Sines on $\triangle ABC$, we have $\frac{AC}{BC} = \frac{\sin 3\theta}{\sin 2\theta}$.

But quadrilateral $ABCD$ is a parallelogram, so O is the midpoint of $\overline{AC}$, which means $AC = 2OC$. Therefore,

$$\frac{\sin 3\theta}{\sin 2\theta} = \frac{AC}{BC} = \frac{2OC}{BC} = \frac{2\sin 2\theta}{\sin 3\theta},$$

so $2\sin^2 2\theta = \sin^2 3\theta$.

By the double angle formula for sine, we have $\sin 2\theta = 2\sin\theta\cos\theta$, and by the triple angle formula for sine, we have $\sin 3\theta = 3\sin\theta - 4\sin^3\theta = \sin\theta(3 - 4\sin^2\theta)$. Therefore,

$$8\sin^2\theta\cos^2\theta = \sin^2\theta(3 - 4\sin^2\theta)^2.$$

Since $\sin\theta$ is not equal to 0, we may divide both sides by $\sin^2\theta$, to get

$$8\cos^2\theta = (3 - 4\sin^2\theta)^2.$$

By the double angle formula for cosine, we have $\cos^2\theta = (1 + \cos 2\theta)/2$ and $\sin^2\theta = (1 - \cos 2\theta)/2$, so

$$4 + 4\cos 2\theta = (1 + 2\cos 2\theta)^2.$$

Expanding the right-hand side and simplifying, the equation above becomes $4\cos^2 2\theta = 3$, so $\cos^2 2\theta = \frac{3}{4}$.

Since 2θ and 3θ are angles of triangle BOC, we have $5\theta < 180°$, so $\theta < 36°$. Then $2\theta < 72°$, so 2θ is acute, which means $\cos 2\theta$ is positive, so $\cos 2\theta = \frac{\sqrt{3}}{2}$. Hence, $2\theta = 30°$, so $\theta = 15°$.

Finally,

$$r = \frac{\angle ACB}{\angle AOB} = \frac{180° - 5\theta}{180° - 3\theta} = \frac{105°}{135°} = \boxed{\frac{7}{9}}$$

4.71 Without loss of generality, assume that $\angle B \geq \angle C$. (The case $\angle B < \angle C$ is tackled similarly.) The differences of angles in the formula make us look for an angle with measure $B - C$. We construct one as follows: Let D be the foot of the altitude from A to side $\overline{BC}$, and let E be the reflection of B through D. This creates another angle with measure B, since $AE = AB$ implies that $\angle AEB = \angle B$. Now, since $\angle AEB$ is an exterior angle of $\triangle AEC$, we have $\angle AEB = \angle C + \angle EAC$. Therefore, $\angle EAC = \angle AEB - \angle C = \angle B - \angle C$.

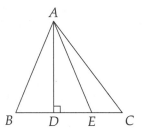

Turning to side lengths, from right triangle ABD, we have $BD = AB\cos\angle ABD = c\cos B$, and from right triangle ACD, we have $CD = AC\cos\angle ACD = b\cos C$. Then $DE = BD = c\cos B$, so $CE = CD - DE = b\cos C - c\cos B$. Then by the Law of Cosines,

$$CE = b\cos C - c\cos B = b\cdot\frac{a^2 + b^2 - c^2}{2ab} - c\cdot\frac{a^2 + c^2 - b^2}{2ac} = \frac{a^2 + b^2 - c^2}{2a} - \frac{a^2 + c^2 - b^2}{2a} = \frac{2b^2 - 2c^2}{2a} = \frac{b^2 - c^2}{a}.$$

Since $AE = AB = c$ and $AC = b$, applying the Law of Cosines to triangle ACE gives

$$\cos(B - C) = \cos\angle EAC = \frac{AC^2 + AE^2 - CE^2}{2AC\cdot AE} = \frac{b^2 + c^2 - (\frac{b^2 - c^2}{a})^2}{2bc} = \frac{a^2 b^2 + a^2 c^2 - (b^2 - c^2)^2}{2a^2 bc}$$
$$= \frac{a^2 b^2 + a^2 c^2 - b^4 + 2b^2 c^2 - c^4}{2a^2 bc}.$$

Hence,

$$a^3\cos(B-C) = \frac{a^3(a^2 b^2 + a^2 c^2 + 2b^2 c^2 - b^4 - c^4)}{2a^2 bc} = \frac{a^2(a^2 b^2 + a^2 c^2 + 2b^2 c^2 - b^4 - c^4)}{2abc} = \frac{a^4 b^2 + a^4 c^2 + 2a^2 b^2 c^2 - a^2 b^4 - a^2 c^4}{2abc}.$$

Similarly,

$$b^3\cos(C - A) = \frac{a^2 b^4 + b^4 c^2 + 2a^2 b^2 c^2 - a^4 b^2 - b^2 c^4}{2abc},$$
$$c^3\cos(A - B) = \frac{a^2 c^4 + b^2 c^4 + 2a^2 b^2 c^2 - a^4 c^2 - b^4 c^2}{2abc}.$$

Adding all three equations, we get

$$a^3\cos(B - C) + b^3\cos(C - A) + c^3\cos(A - B) = \frac{6a^2 b^2 c^2}{2abc} = 3abc.$$

5

CHAPTER

Parameterization and Trigonometric Coordinate Systems

Exercises for Section 5.1

5.1.1 We eliminate t from the two equations. Multiplying the equation for x by 8 and the equation for y by 3 gives $8x = 24t + 56$, $3y = 24t - 15$. Subtracting the second equation from the first gives $8x - 3y = 71$. Solving for y gives $y = \frac{8}{3}x - \frac{71}{3}$. The slope of the graph of this equation is $\boxed{\frac{8}{3}}$.

5.1.2 We have $x^2 + y^2 = 9\cos^2 2t + 9\sin^2 2t = 9(\cos^2 2t + \sin^2 2t) = 9$. Therefore, all points on the graph are on the circle centered at the origin with radius 3. As t varies from 0 to π, every point on this circle is produced, so the graph of the parametric equations is $\boxed{\text{a circle centered at the origin with radius 3}}$.

5.1.3 From the given equations, we have $x + 3 = 5\cos\theta$ and $y - 2 = 5\sin\theta$. Squaring both and adding them gives

$$(x + 3)^2 + (y - 2)^2 = (5\cos\theta)^2 + (5\sin\theta)^2 = 25\cos^2\theta + 25\sin^2\theta = 25.$$

Thus, the graph is $\boxed{\text{a circle with center } (-3, 2) \text{ and radius 5}}$. We can see that the graph of the parametric equations is the entire circle as follows. First, we note that the graph of $x = 5\cos\theta$, $y = 5\sin\theta$ is a dilation of the unit circle about the origin, so the graph of $x = 5\cos\theta$, $y = 5\sin\theta$ is a circle centered at the origin with radius 5. Each point on the graph of $x = -3 + 5\cos\theta$, $y = 2 + 5\sin\theta$ is 3 to the left and 2 above the corresponding point on the graph $x = 5\cos\theta$, $y = 5\sin\theta$. So, the graph of $x = -3 + 5\cos\theta$, $y = 2 + 5\sin\theta$ is the result of translating the graph of $x = 5\cos\theta$, $y = 5\sin\theta$ by 3 to the left and 2 up. Therefore, the graph is a full circle.

5.1.4 Solving $x = 4\cos 2t$ for $\cos 2t$ gives $\cos 2t = \frac{x}{4}$. Substituting this into $y = 1 + \cos^2 2t$ gives $y = 1 + \frac{x^2}{16}$. So, all points on the graph of the parametric equations are on the graph of $y = 1 + \frac{x^2}{16}$. However, the parametric equations do not produce the entire parabola described by $y = 1 + \frac{x^2}{16}$. Since $\cos 2t$ varies from -1 to 1, the value of $x = 4\cos 2t$ varies from -4 to 4. Therefore, graphing the parametric equations only produces the portion of the graph of $y = 1 + \frac{x^2}{16}$ with $-4 \le x \le 4$, as shown at right.

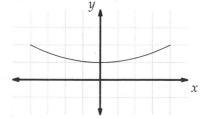

5.1.5 Dividing both sides by 144 gives $\frac{x^2}{36} + \frac{y^2}{16} = 1$. The graph of this equation is an ellipse. To parameterize it, we notice that the two terms on the left are perfect squares whose sum is 1. Therefore, we can take $\frac{x^2}{36} = \cos^2\theta$ and $\frac{y^2}{16} = \sin^2\theta$. One solution for x and y is $\boxed{x = 6\cos\theta, y = 4\sin\theta}$. Graphing these parametric equations as θ ranges from 0 to 2π traces out the full ellipse.

These are not the only parametric equations whose graph is this ellipse. We could also have taken $x = -6\cos\theta$, $y = 4\sin\theta$, or $x = -6\cos 2\theta$, $y = -4\sin 2\theta$, etc. We can find infinitely many other possibilities. We can let

x be 6 or -6 times either sine or cosine of some angle, and let y be 4 or -4 times the cosine or sine (we choose the trig function we didn't use for x) of the same angle we used to express x.

5.1.6   We can parameterize the unit circle with $x = \cos t$, $y = \sin t$, so we can parameterize a circle with radius 3 centered at the origin with $x = 3\cos t$, $y = 3\sin t$. The circle with radius 3 centered at $(4, 5)$ is a translation right 4 units and up 5 units of the circle with radius 3 centered at the origin. Therefore, if (x, y) is a point on the circle centered at $(4, 5)$, then there is some point (x', y') on the circle centered at the origin such that $x = 4 + x'$, $y = 5 + y'$. We can parameterize the circle centered at the origin with $x' = 3\cos t$, $y' = 3\sin t$, so the circle centered at $(4, 5)$ with radius 3 can be parameterized with $x = 4 + 3\cos t$, $y = 5 + 3\sin t$.

While the graph of the parameterization $x = 4 + 3\cos t$, $y = 5 + 3\sin t$ does produce the same circle that the particle traces, it does not accurately describe the location of the particle. At $t = 0$, the equations $x = 4 + 3\cos t$, $y = 5 + 3\sin t$ give $(x, y) = (7, 5)$, but the particle is at $(1, 5)$, which is diametrically opposite $(7, 5)$. We can account for this by introducing a phase shift of π, using the parameterization $x = 4 + 3\cos(t + \pi)$, $y = 5 + 3\sin(t + \pi)$. Applying the trig identities $\cos(t + \pi) = -\cos t$ and $\sin(t + \pi) = -\sin t$ makes this parameterization $x = 4 - 3\cos t$, $y = 5 - 3\sin t$.

We're not quite finished yet. We check if the the particle is moving in the right direction by noting that if $t = \frac{\pi}{2}$, then $x = 4 - 3\cos t$, $y = 5 - 3\sin t$ gives $(x, y) = (4, 2)$, which is indeed counterclockwise from $(1, 5)$. However, we have to take into account the speed of the particle. Since it travels around the whole circle in 8π seconds, we need the period of each of the trig functions to be 8π rather than 2π. We accomplish this by replacing t with $t/4$, and we have the parameterization $\boxed{x = 4 - 3\cos \frac{t}{4}, y = 5 - 3\sin \frac{t}{4}}$.

5.1.7   We can't easily eliminate the parameter, but we can at least eliminate the denominators of the fractions. Since x clearly cannot be 0, we can divide our equation for y by our equation for x, which gives $\frac{y}{x} = t$. Aha! Now, we can substitute for the parameter! Substituting into the equation for x gives

$$x = \frac{4}{1 + t^2} = \frac{4}{1 + \frac{y^2}{x^2}} = \frac{4x^2}{x^2 + y^2}.$$

Since x is nonzero, we can divide both sides by x, and multiply both sides by $x^2 + y^2$, to get $x^2 + y^2 = 4x$. The graph of this equation is a circle. We rearrange the equation and complete the square to find that $(x - 2)^2 + y^2 = 4$. The graph of this equation is a circle with center $(2, 0)$ and radius 2.

We can't yet conclude that the graph of the parametric equations *is* this circle. We have only shown that every point on the graph of the parametric equations is on this circle. Looking at the equation $x = \frac{4}{1+t^2}$, we see that the largest x can be is 4, which occurs when $t = 0$. For any other t, we have $1 + t^2 > 1$, so $x < 4$. We also see that x must be positive, and can take on any value in the interval $(0, 4]$. Specifically, we see that we must omit the origin from the graph, so the graph of the parametric equations is not the entire circle centered at $(2, 0)$ with radius 2.

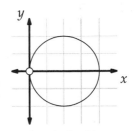

We're still not quite finished—for each value of x such that $0 < x < 4$, there are *two* points on the circle. Do our parametric equations produce both? Yes! The one above the x-axis is produced when $t > 0$ and the one below the x-axis is produced when $t < 0$. Therefore, the graph of the parametric equations is the circle centered at $(2, 0)$ with radius 2, except the origin, as shown at right above.

Exercises for Section 5.2

5.2.1

(a)   We have $x = -5$ and $y = 0$, so $r = \sqrt{x^2 + y^2} = \sqrt{(-5)^2 + 0^2} = \sqrt{25} = 5$. We seek an angle θ such that

$5\cos\theta = -5$ and $5\sin\theta = 0$, which means $\cos\theta = -1$ and $\sin\theta = 0$. We can take $\theta = 180°$. Hence, $\boxed{(5,180°)}$ are polar coordinates of the point.

(b) We have $x = 6\sqrt{2}$ and $y = 6$, so $r = \sqrt{x^2 + y^2} = \sqrt{(6\sqrt{2})^2 + 6^2} = \sqrt{108} = 6\sqrt{3}$. We seek an angle θ such that $6\sqrt{3}\cos\theta = 6\sqrt{2}$ and $6\sqrt{3}\sin\theta = 6$, which means $\cos\theta = \sqrt{\frac{2}{3}}$ and $\sin\theta = \frac{\sqrt{3}}{3}$. Since the point $(6\sqrt{2}, 6)$ lies in the first quadrant, we can take $\theta = \arccos\sqrt{\frac{2}{3}} \approx 35.3°$. Hence, $\boxed{(6\sqrt{3}, 35.3°)}$ are polar coordinates of the point.

(c) We have $x = -4$ and $y = -4\sqrt{3}$, so $r = \sqrt{x^2 + y^2} = \sqrt{(-4)^2 + (-4\sqrt{3})^2} = \sqrt{64} = 8$. We seek an angle θ such that $8\cos\theta = -4$ and $8\sin\theta = -4\sqrt{3}$, which means $\cos\theta = -\frac{1}{2}$ and $\sin\theta = -\frac{\sqrt{3}}{2}$. We can take $\theta = 240°$. Hence, $\boxed{(8, 240°)}$ are polar coordinates of the point.

5.2.2

(a) We have $r = 7$ and $\theta = 210°$, so $x = 7\cos 210° = 7\cdot\left(-\frac{\sqrt{3}}{2}\right) = -\frac{7\sqrt{3}}{2}$, and $y = 7\sin 210° = 7\cdot\left(-\frac{1}{2}\right) = -\frac{7}{2}$. Hence, the rectangular coordinates are $\boxed{\left(-\frac{7\sqrt{3}}{2}, -\frac{7}{2}\right)}$.

(b) We have $r = 9$ and $\theta = -180°$, so $x = 9\cos(-180°) = 9\cdot(-1) = -9$, and $y = 9\sin(-180°) = 9\cdot 0 = 0$. Hence, the rectangular coordinates are $\boxed{(-9, 0)}$.

(c) We have $r = -8$ and $\theta = \frac{9\pi}{4}$, so $x = -8\cos\frac{9\pi}{4} = -8\cdot\frac{\sqrt{2}}{2} = -4\sqrt{2}$, and $y = -8\sin\frac{9\pi}{4} = -8\cdot\frac{\sqrt{2}}{2} = -4\sqrt{2}$. Hence, the rectangular coordinates are $\boxed{(-4\sqrt{2}, -4\sqrt{2})}$.

5.2.3 Multiplying both sides by $2\cos\theta - 5\sin\theta$, we get $2r\cos\theta - 5r\sin\theta = 4$. We don't have worry about introducing extraneous solutions here, since clearly this resulting equation has no solutions for which we have $2r\cos\theta - 5r\sin\theta = 0$. Since $x = r\cos\theta$ and $y = r\sin\theta$, this equation becomes $2x - 5y = 4$. Thus, the graph is a $\boxed{\text{line}}$. (This line has x-intercept $(2, 0)$ and y-intercept $(0, -\frac{4}{5})$.)

5.2.4 We can express $r = 6\sin\theta$ in the form given for a circle in the text,

$$r^2 - 2rr_0\cos(\theta - \alpha) + r_0^2 = t^2,$$

with a little clever manipulation. We multiply both sides of $r = 6\sin\theta$ by r and move everything to the left side to get $r^2 - 6r\sin\theta = 0$. Then, we note that $\sin\theta = \cos(90° - \theta) = \cos(-(90° - \theta)) = \cos(\theta - 90°)$, so we can write the equation $r^2 - 6r\sin\theta = 0$ as $r^2 - 6r\cos(\theta - 90°) = 0$. Almost there! Comparing this to our form given for a circle, we see that we can let $r_0 = 3$ and $\alpha = 90°$. Finally, we must also have $t = 3$ to make the constants cancel in the form $r^2 - 2rr_0\cos(\theta - \alpha) + r_0^2 = t^2$. So, we write our equation as:

$$r^2 - 2(3)r\cos(\theta - 90°) + 3^2 = 3^2.$$

We thereby see that the graph of $r = 6\sin\theta$ is a circle with radius 3 and center $(3, 90°)$ (which is indeed $(0, 3)$ in rectangular coordinates).

5.2.5

(a) For segments $\overline{AO}$ and $\overline{BO}$ to be perpendicular, the angles θ and $137°$ must differ by an odd multiple of $90°$. Therefore, the possible values of θ are $137° + 90° = \boxed{227°}$ and $137° - 90° = \boxed{47°}$.

(b) If $\overline{AB} \perp \overline{AO}$, then $\angle OAB = 90°$. Then $\cos\angle AOB = \frac{OA}{OB} = \frac{6}{12} = \frac{1}{2}$, so $\angle AOB = 60°$. Therefore, the possible values of θ are $137° - 60° = \boxed{77°}$ and $137° + 60° = \boxed{197°}$.

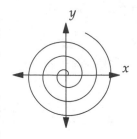

5.2.6 When we graph $r = 2 + 3\theta$, r increases as θ increases, so the graph gets farther and farther from the origin as θ increases. Because increasing θ when $r > 0$ means rotating around the origin, we obtain a $\boxed{\text{spiral}}$ as the graph. The graph of an equation of the form $r = a + b\theta$ is called an *Archimedean spiral*.

5.2.7 First, we determine the polar coordinates of the point $(3, 3)$. We have $x = 3$ and $y = 3$, so $r = \sqrt{x^2 + y^2} = \sqrt{3^2 + 3^2} = \sqrt{18} = 3\sqrt{2}$. We seek an angle θ such that $3\sqrt{2} \cos\theta = 3$ and $3\sqrt{2} \sin\theta = 3$, which means $\cos\theta = \frac{\sqrt{2}}{2}$ and $\sin\theta = \frac{\sqrt{2}}{2}$. We can take $\theta = 45°$. Hence, the polar coordinates are $(3\sqrt{2}, 45°)$.

To find the other vertices, we note that if we draw a circle centered at the origin such that the circle passes through the six vertices of the hexagon, then the vertices are equally spaced about the circle. So, in polar coordinates, the radius is the same for each, and we find the angle of successive vertex by adding $360°/6 = 60°$ to the angle of the previous vertex. Then the angles corresponding to the other five vertices of the hexagon are $45° + 60° = 105°$, $105° + 60° = 165°$, $165° + 60° = 225°$, $225° + 60° = 285°$, and $285° + 60° = 345°$. The two second quadrant angles are $105°$ and $165°$.

Now we must compute the coordinates. We have:

$$\cos 15° = \cos(45° - 30°) = \cos 45° \cos 30° + \sin 45° \sin 30° = \frac{\sqrt{6} + \sqrt{2}}{4}, \text{ and}$$

$$\sin 15° = \sin(45° - 30°) = \sin 45° \cos 30° - \sin 30° \cos 45° = \frac{\sqrt{6} - \sqrt{2}}{4}.$$

Then

$$\cos 105° = \cos(15° + 90°) = -\sin 15° = \frac{-\sqrt{6} + \sqrt{2}}{4}, \quad \sin 105° = \sin(15° + 90°) = \cos 15° = \frac{\sqrt{6} + \sqrt{2}}{4},$$

$$\cos 165° = \cos(180° - 15°) = -\cos 15° = \frac{-\sqrt{6} - \sqrt{2}}{4}, \quad \sin 165° = \sin(180° - 15°) = \sin 15° = \frac{\sqrt{6} - \sqrt{2}}{4}.$$

Therefore, the vertices of the hexagon that are in the second quadrant are

$$(3\sqrt{2} \cos 105°, 3\sqrt{2} \sin 105°) = \left(3\sqrt{2} \cdot \frac{-\sqrt{6} + \sqrt{2}}{4}, 3\sqrt{2} \cdot \frac{\sqrt{6} + \sqrt{2}}{4}\right) = \boxed{\left(\frac{3 - 3\sqrt{3}}{2}, \frac{3 + 3\sqrt{3}}{2}\right)},$$

and

$$(3\sqrt{2} \cos 165°, 3\sqrt{2} \sin 165°) = \left(3\sqrt{2} \cdot \frac{-\sqrt{6} - \sqrt{2}}{4}, 3\sqrt{2} \cdot \frac{\sqrt{6} - \sqrt{2}}{4}\right) = \boxed{\left(\frac{-3 - 3\sqrt{3}}{2}, \frac{-3 + 3\sqrt{3}}{2}\right)}.$$

Exercises for Section 5.3

5.3.1

(a) In two dimensions, the point with rectangular coordinates $(-2\sqrt{2}, 2\sqrt{2})$ has polar coordinates $\left(4, \frac{3\pi}{4}\right)$. Therefore, the cylindrical coordinates of the point with rectangular coordinates $(-2\sqrt{2}, 2\sqrt{2}, 4)$ are $\boxed{\left(4, \frac{3\pi}{4}, 4\right)}$.

Turning to spherical coordinates, we have $\rho = \sqrt{(-2\sqrt{2})^2 + (2\sqrt{2})^2 + 4^2} = 4\sqrt{2}$. The θ-coordinate is the same in spherical coordinates as in cylindrical coordinates. To find the ϕ-coordinate, we note that

$z = \rho \cos \phi$, so $4 = 4\sqrt{2} \cos \phi$. Therefore, we have $\cos \phi = \frac{\sqrt{2}}{2}$, which means $\phi = \frac{\pi}{4}$. The desired spherical coordinates then are $\boxed{\left(4\sqrt{2}, \frac{3\pi}{4}, \frac{\pi}{4}\right)}$.

(b) In two dimensions, the point with rectangular coordinates $(3, 0)$ has polar coordinates $(3, 0)$. Therefore, the point in space with rectangular coordinates $(3, 0, -3\sqrt{3})$ has cylindrical coordinates $\boxed{(3, 0, -3\sqrt{3})}$.

Turning to spherical coordinates, we have $\rho = \sqrt{(3)^2 + (0)^2 + (-3\sqrt{3})^2} = 6$. The θ-coordinate is the same in spherical and cylindrical coordinates, so $\theta = 0$. Finally, from $z = \rho \cos \phi$, we have $-3\sqrt{3} = 6 \cos \phi$, from which we have $\cos \phi = -\frac{\sqrt{3}}{2}$. This gives us $\phi = \frac{5\pi}{6}$, and the desired spherical coordinates are $\boxed{\left(6, 0, \frac{5\pi}{6}\right)}$.

5.3.2 Since $x^2 + y^2 = r^2$, we have $\rho = \sqrt{x^2 + y^2 + z^2} = \boxed{\sqrt{r^2 + z^2}}$.

5.3.3 We first tackle the case in which $\frac{\pi}{2} < \phi < \pi$. Let P be the point with spherical coordinates (ρ, θ, ϕ) and rectangular coordinates (x, y, z). Let O be the origin and let B be the foot of the altitude from P to the xy-plane. We have $\angle POB = \phi - \frac{\pi}{2}$, so

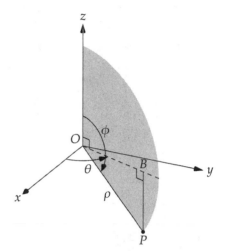

$$PB = OP \sin \angle POB = \rho \sin \left(\phi - \frac{\pi}{2}\right) = -\rho \cos \phi.$$

Since P is below the xy-plane, the z-coordinate of P is $-PB = \rho \cos \phi$. We also have

$$OB = \rho \cos \angle POB = \rho \cos \left(\phi - \frac{\pi}{2}\right) = \rho \sin \phi.$$

Therefore, the x- and y-coordinates of B are $x = \rho \sin \phi \cos \theta$ and $y = \rho \sin \phi \sin \theta$. The x- and y-coordinates of P are the same as those of P, so the rectangular coordinates of P are $(\rho \sin \phi \cos \theta, \rho \sin \phi \sin \theta, \rho \cos \phi)$.

If $\phi = \frac{\pi}{2}$, then $(\rho \sin \phi \cos \theta, \rho \sin \phi \sin \theta, \rho \cos \phi)$ becomes $(\rho \cos \theta, \rho \sin \theta, 0)$, which are indeed the correct coordinates. (To see why, recall that if (r, θ) are the polar coordinates of a point in the Cartesian plane, then the rectangular coordinates of the point are $(r \cos \theta, r \sin \theta)$.)

If $\phi = \pi$, then P is on the z-axis, and the point is ρ below the origin. When $\phi = \pi$, the rectangular coordinates $(\rho \sin \phi \cos \theta, \rho \sin \phi \sin \theta, \rho \cos \phi)$ are $(0, 0, -\rho)$, so our expressions relating rectangular to spherical coordinates work when $\phi = \pi$.

Finally, if $\phi = 0$, then P is on the z-axis, and the point is ρ above the origin. When $\phi = 0$, the rectangular coordinates $(\rho \sin \phi \cos \theta, \rho \sin \phi \sin \theta, \rho \cos \phi)$ are $(0, 0, \rho)$, so our expressions relating rectangular to spherical coordinates work when $\phi = 0$.

5.3.4 We have

$$
\begin{aligned}
\sqrt{(x-0)^2 + (y-0)^2 + (z-0)^2} &= \sqrt{(\rho \cos \theta \sin \phi)^2 + (\rho \sin \theta \sin \phi)^2 + (\rho \cos \phi)^2} \\
&= \sqrt{\rho^2 \cos^2 \theta \sin^2 \phi + \rho^2 \sin^2 \theta \sin^2 \phi + \rho^2 \cos^2 \phi} \\
&= \sqrt{\rho^2 (\cos^2 \theta + \sin^2 \theta) \sin^2 \phi + \rho^2 \cos^2 \phi} \\
&= \sqrt{\rho^2 \sin^2 \phi + \rho^2 \cos^2 \phi} = \sqrt{\rho^2 (\sin^2 \phi + \cos^2 \phi)} = \sqrt{\rho^2} = \rho.
\end{aligned}
$$

5.3.5 For each part, let O be the origin and P' be the image of the described transformation.

(a) If P is on the xy-plane, then P' is the same as P. Otherwise, the xy-plane is between P and P'. Since P' is the reflection of P through the xy-plane, the midpoint of $\overline{PP'}$ is on the xy-plane, and $\overline{PP'}$ is perpendicular to the xy-plane. Let A be intersection of $\overline{PP'}$ and the xy-plane, so A is the foot of the perpendiculars from P and from P' to the xy-plane. The polar coordinates of A in the xy-plane give us the r and θ coordinates of both P and P' in cylindrical coordinates. Therefore, the r- and θ-coordinates of P' are the same as those of P. However, because P and P' are on opposite sides of xy-plane, and are equidistant from the xy-plane, their z-coordinates are opposites. Therefore, the cylindrical coordinates of P' are $\boxed{(r, \theta, -z)}$. Note that these also hold when P is on the xy-plane, since then we have $z = -z = 0$.

(b) Rotation about the z-axis doesn't change the distance from the point to the xy-plane, nor does it change the distance between the origin and the foot of the altitude from the point to the xy-plane. Therefore, the r- and z-coordinates of P' are the same as those of P. The only change is in the θ-coordinate. The foot of the altitude from P' to the xy-plane is a rotation of π around the origin of the foot of the altitude from P to the xy-plane. Therefore, the cylindrical coordinates of P' are $\boxed{(r, \theta + \pi, z)}$.

(c) The reflection of P through the origin is the same as rotating P by π about the z-axis, and then reflecting the result through the xy-plane. Therefore, P' is the result of performing both of the transformations described in parts (a) and (b), which means its cylindrical coordinates are $\boxed{(r, \theta + \pi, -z)}$.

We also could have reasoned as follows. Let A and A' be the feet of the altitudes from P and P', respectively, to the xy-plane. Then, triangles PAO and $P'A'O$ are congruent. Since $PA = PA'$, but P and P' are on opposite sides of the xy-plane, the z-coordinates of P and P' are opposites. Because $AO = A'O$, the r-coordinates of P and P' are the same. Because the rays from O to A and A' are in opposite directions, the θ-coordinates of P and P' differ by π.

5.3.6 $\boxed{\text{Yes.}}$ The graph of $\phi = 0$ and the graph of $\phi = \pi$ are both the z-axis, and the graph of $\phi = \frac{\pi}{2}$ is the xy-plane. This is why lines and planes are sometimes considered "degenerate" cones.

5.3.7 We tackle parts (a) and (b) together. The r-coordinate is the distance from the point to the Earth's axis, and the z-coordinate is the distance from the point to the plane containing the equator. So, traveling with constant latitude holds both the r-coordinate and the z-coordinate constant. Since the θ-coordinate is the same in cylindrical coordinates as in spherical coordinates, traveling with constant longitude holds the θ-coordinate constant (while varying r and z). So, matters are a little better in cylindrical coordinates than in rectangular coordinates—we can describe east-west travel as varying just one coordinate. But they aren't as convenient as with with spherical coordinates, since north-south travel still requires varying two coordinates. (In summary, the answer to every question in this problem is "yes.")

5.3.8

(a) The graph of $\rho = 4$ is a sphere centered at the origin with radius 4. The graph of $\theta = \frac{2\pi}{5}$ in spherical coordinates is the same as the graph of $\theta = \frac{2\pi}{5}$ in cylindrical coordinates. In the text, we explained why the graph of $\theta = c$ in cylindrical coordinates, where c is a constant, is a plane through the origin. Therefore, the points such that $\rho = 4$ and $\theta = \frac{2\pi}{5}$ are the intersection of a sphere centered at the origin and a plane that passes through the origin. This intersection is a $\boxed{\text{circle}}$ with radius 4 centered at the origin such that the circle is in the plane described by $\theta = \frac{2\pi}{5}$.

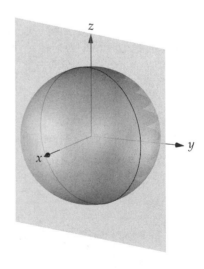

Another way to see this is to note that the desired graph consists of all points of the form $\left(4, \frac{2\pi}{5}, \phi\right)$. When $\phi = 0$, we have the point on the z-axis, 4 units above the origin. As we increase ϕ from 0 to $\frac{\pi}{2}$, we trace out a quarter circle centered at the origin from $(0,0,4)$ (in rectangular coordinates) to a point in the xy-plane that is 4 units from the origin. Continuing to increase ϕ from $\frac{\pi}{2}$ to π, we trace out another quarter circle from the point in the xy-plane to $(0,0,-4)$. As we increase ϕ from π to 2π, we trace out the remainder of the circle.

(b) When $\theta = \frac{\pi}{2}$, we have $x = \rho \cos \theta \sin \phi = 0$, $y = \rho \sin \theta \sin \phi = \rho \sin \phi$, and $z = \rho \cos \phi$. Therefore, the entire graph is in the yz-plane. Letting $\rho = 6 \sin \phi$ in our equations for y and z gives $y = 6 \sin^2 \phi$ and $z = 6 \cos \phi \sin \phi$. The double angle identities for sine and cosine give us $\sin 2\phi = 2 \cos \phi \sin \phi$ and $\cos 2\phi = 1 - 2 \sin^2 \phi$, so we can write y and z as $y = 3 - 3 \cos 2\phi$, $z = 3 \sin 2\phi$. We can now eliminate ϕ with $\cos^2 2\phi + \sin^2 2\phi = 1$. We have

$$\cos^2 2\phi + \sin^2 2\phi = \left(\frac{3-y}{3} \right)^2 + \left(\frac{z}{3} \right)^2,$$

so $\cos^2 2\phi + \sin^2 2\phi = 1$ gives us $(3 - y)^2 + z^2 = 9$. The graph of this equation is a circle in the yz-plane with center $(0, 3, 0)$ and radius 3.

Another way to approach this part is to note that when $\theta = \frac{\pi}{2}$, we have $x = 0$, $y = \rho \sin \phi$, and $z = \rho \cos \phi$, so $\rho^2 = y^2 + z^2$. Multiplying $\rho = 6 \sin \phi$ by ρ gives $\rho^2 = 6\rho \sin \phi$, so $y^2 + z^2 = 6y$. Completing the square leads to $(y - 3)^2 + z^2 = 9$. The graph of this equation is a circle in the yz-plane with center $(0, 3, 0)$ and radius 3.

5.3.9

(a) We can use the conversion from spherical to rectangular coordinates to generate our parameterization. The graph of the equation is a sphere with radius 2 and center $(-3, -5, 1)$. A sphere with radius 2 centered at the origin has parameterization $x = \rho \cos \theta \sin \phi = 2 \cos \theta \sin \phi$, $y = \rho \sin \theta \sin \phi = 2 \sin \theta \sin \phi$, $z = \rho \cos \phi = 2 \cos \phi$. If we add -3 to the x-coordinate, -5 to the y-coordinate, and 1 to the z-coordinate of every point on this sphere, we get the sphere with radius 2 and center $(-3, -5, 1)$. Therefore, the desired parameterization is

$$\boxed{x = -3 + 2 \cos \theta \sin \phi, y = -5 + 2 \sin \theta \sin \phi, z = 1 + 2 \cos \phi}.$$

(b) We must restrict our choices of θ and ϕ such that $z \geq 0$. Since z doesn't depend on θ, we don't restrict θ. However, $z \geq 0$ gives us $1 + 2 \cos \phi \geq 0$, from which we have $\cos \phi \geq -\frac{1}{2}$. Since $\cos \frac{2\pi}{3} = -\frac{1}{2}$, the inequality $\cos \phi \geq -\frac{1}{2}$ tells us that we must restrict ϕ to $0 \leq \phi \leq \frac{2\pi}{3}$. Therefore, the desired parameterization is

$$\boxed{x = -3 + 2 \cos \theta \sin \phi, y = -5 + 2 \sin \theta \sin \phi, z = 1 + 2 \cos \phi, 0 \leq \phi \leq \frac{2\pi}{3}}.$$

Review Problems

5.25 Solving $x = 3a - 1$ for a gives $a = (x+1)/3$. Substituting into $y = 2a^2 - a$ gives

$$y = 2 \left(\frac{x+1}{3} \right)^2 - \frac{x+1}{3} = \frac{2x^2 + 4x + 2}{9} - \frac{x+1}{3} = \frac{2x^2 + x - 1}{9}.$$

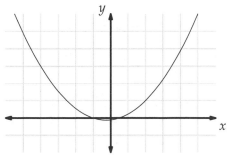

The graph of this equation is a parabola, as shown at right. Since x can take on any value, the full parabola is produced by graphing the parametric equations.

5.26 Completing the square in x and y gives $(x + 2)^2 + (y - 1)^2 = 16$. The graph of this equation is a circle centered at $(-2, 1)$ with radius 4. The unit circle can be parameterized by $x = \cos \theta$, $y = \sin \theta$, so the circle with radius 4 centered at the origin can be parameterized by $x = 4 \cos \theta$, $y = 4 \sin \theta$. Translating this circle 2 units left and 1 unit up produces the circle that is the graph of $(x + 2)^2 + (y - 1)^2 = 16$, so a parameterization of this circle is $\boxed{x = -2 + 4 \cos \theta, y = 1 + 4 \sin \theta}$.

Another way we could have found this parameterization is by noting that $\cos^2\theta + \sin^2\theta = 1$, so we have $(4\cos\theta)^2 + (4\sin\theta)^2 = 16$. Therefore, letting $x + 2 = 4\cos\theta$ and $y - 1 = 4\sin\theta$ gives $(x+2)^2 + (y-1)^2 = 16$. Solving these equations for x and y gives the parameterization $x = -2 + 4\cos\theta$, $y = 1 + 4\sin\theta$.

5.27 Applying the sine double angle identity, we have $x = \frac{1}{2}\sin 2t$. Therefore, we have

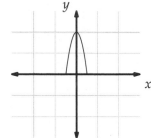

$$x^2 = \frac{1}{4}\sin^2 2t = \frac{1}{4}(1 - \cos^2 2t) = \frac{1}{4}\left(1 - \frac{y}{2}\right).$$

Solving for y gives $y = 2 - 8x^2$. The graph of this equation is a parabola. However, from $x = \frac{1}{2}\sin 2t$, we see that x is restricted to the interval $\left[-\frac{1}{2}, \frac{1}{2}\right]$, and that all values of x in this interval are attained. Therefore, the graph of these parametric equations is the portion of the parabola within this interval, as shown at right.

5.28 The right side of $x = \frac{1}{1-t^2}$ cannot equal zero, so x is nonzero and we can divide $y = \frac{t}{1-t^2}$ by $x = \frac{1}{1-t^2}$ to eliminate the denominators. This gives us $\frac{y}{x} = t$. Substituting this into the given equation for x gives $x = \frac{1}{1-(y/x)^2} = \frac{x^2}{x^2-y^2}$. Since we know that x is nonzero, we can divide both ends by x, and then multiply by $x^2 - y^2$ to give $x^2 - y^2 = x$. Rearranging and completing the square gives $\left(x - \frac{1}{2}\right)^2 - y^2 = \frac{1}{4}$. The graph of this equation is a hyperbola.

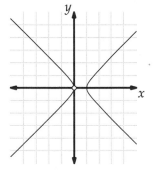

We know that all points in the graph of the parametric equations are on this hyperbola, but we still must check if every point on the hyperbola is in the graph of the parametric equations. First, we note that we must omit the origin from the graph of the parametric equations, since x cannot be 0 in $x = \frac{1}{1-t^2}$. Are there any other points we must omit?

We cannot have $0 < x < 1$, since $1 - t^2$ cannot be greater than 1. However, there are no points on the graph of $\left(x - \frac{1}{2}\right)^2 - y^2 = \frac{1}{4}$ with $0 < x < 1$, since $y^2 = \left(x - \frac{1}{2}\right)^2 - \frac{1}{4} < 0$ for $0 < x < 1$. Solving $x = \frac{1}{1-t^2}$ for t gives $t = \pm\sqrt{1 - \frac{1}{x}}$. For $x < 0$ or $x \geq 1$, we have $1 - \frac{1}{x} \geq 0$, so for all values of x except $0 \leq x < 1$, we can find a value of t such that $x = \frac{1}{1-t^2}$. Except when $t = 0$ (so $x = 1$), there are two such values of t, one negative and one positive. Since $y = \frac{t}{1-t^2}$, the positive value of t produces the corresponding point on the hyperbola above the x-axis, and the negative value of t produces the point below the x-axis. The graph is shown at right above; it consists of the entire hyperbola except the origin.

5.29

(a) We have $x = -7$ and $y = -7$, so $r = \sqrt{x^2 + y^2} = \sqrt{(-7)^2 + (-7)^2} = \sqrt{98} = 7\sqrt{2}$. We seek an angle θ such that $7\sqrt{2}\cos\theta = -7$ and $7\sqrt{2}\sin\theta = -7$, which means $\cos\theta = -\frac{\sqrt{2}}{2}$ and $\sin\theta = -\frac{\sqrt{2}}{2}$. We can take $\theta = \frac{5\pi}{4} \approx 3.93$. Hence, the polar coordinates are $\boxed{(7\sqrt{2}, 3.93)}$, where the angle is in radians.

(b) We have $x = 6\sqrt{3}$ and $y = -6\sqrt{2}$, so $r = \sqrt{x^2 + y^2} = \sqrt{(6\sqrt{3})^2 + (-6\sqrt{2})^2} = \sqrt{180} = 6\sqrt{5}$. We seek an angle θ such that $6\sqrt{5}\cos\theta = 6\sqrt{3}$ and $6\sqrt{5}\sin\theta = -6\sqrt{2}$, which means $\cos\theta = \sqrt{\frac{3}{5}}$ and $\sin\theta = -\sqrt{\frac{2}{5}}$. Since the point $(6\sqrt{3}, -6\sqrt{2})$ lies in the fourth quadrant, we can take $\theta = \arcsin\left(-\sqrt{\frac{2}{5}}\right) \approx -0.68$, which is equivalent to $-0.68 + 2\pi = 5.6$ radians. Hence, the polar coordinates are $\boxed{(6\sqrt{5}, 5.60)}$, where the angle is in radians. (The answer $(6\sqrt{5}, -0.68)$ is also acceptable.)

(c) We have $x = -5$ and $y = 8$, so $r = \sqrt{x^2 + y^2} = \sqrt{(-5)^2 + 8^2} = \sqrt{89}$. We seek an angle θ such that $\sqrt{89}\cos\theta = -5$ and $\sqrt{89}\sin\theta = 8$, which means $\cos\theta = -\frac{5\sqrt{89}}{89}$ and $\sin\theta = \frac{8\sqrt{89}}{89}$. Since the point $(-5, 8)$ lies in the second quadrant, we can take $\theta = \arccos\left(-\frac{5\sqrt{89}}{89}\right) \approx 2.13$ radians. Hence, the polar coordinates are $\boxed{(\sqrt{89}, 2.13)}$, where the angle is in radians.

5.30

(a) We have $r = 5$ and $\theta = 0°$, so $x = 5\cos 0° = 5 \cdot 1 = 5$, and $y = 5\sin 0° = 5 \cdot 0 = 0$. Hence, the rectangular coordinates are $\boxed{(5, 0)}$.

(b) We have $r = -4$ and $\theta = 210°$, so $x = -4\cos 210° = -4 \cdot \left(-\frac{\sqrt{3}}{2}\right) = 2\sqrt{3}$, and $y = -4\sin 210° = -4 \cdot \left(-\frac{1}{2}\right) = 2$. Hence, the rectangular coordinates are $\boxed{(2\sqrt{3}, 2)}$.

(c) We have $r = 8$ and $\theta = -\frac{\pi}{4}$, so $x = 8\cos\left(-\frac{\pi}{4}\right) = 8 \cdot \frac{\sqrt{2}}{2} = 4\sqrt{2}$, and $y = 8\sin\left(-\frac{\pi}{4}\right) = 8 \cdot \left(-\frac{\sqrt{2}}{2}\right) = -4\sqrt{2}$. Hence, the rectangular coordinates are $\boxed{(4\sqrt{2}, -4\sqrt{2})}$.

5.31 From the relations $r^2 = x^2 + y^2$, $x = r\cos\theta$, and $y = r\sin\theta$, the given equation becomes $x^2 + y^2 = 3x - 5y - 2$. Moving all the variable terms to one side, we get $x^2 - 3x + y^2 + 5y = -2$. Completing the square in x and y, we get $\left(x - \frac{3}{2}\right)^2 + \left(y + \frac{5}{2}\right)^2 = \frac{13}{2}$. Hence, the graph is a $\boxed{\text{circle with center } \left(\frac{3}{2}, -\frac{5}{2}\right) \text{ and radius } \sqrt{\frac{13}{2}}}$.

5.32

(a) In the text, we showed that the point that results from rotating (x, y) counterclockwise by the angle θ about the origin is the point $(x\cos\theta - y\sin\theta, y\cos\theta + x\sin\theta)$. A $60°$ clockwise rotation is a $300°$ counterclockwise rotation, so the point that results when the $(4, 5)$ is rotated $60°$ clockwise about the origin is

$$(4\cos 300° - 5\sin 300°, 5\cos 300° + 4\sin 300°) = \boxed{\left(2 + \frac{5\sqrt{3}}{2}, \frac{5}{2} - 2\sqrt{3}\right)}.$$

(b) Let P be the point that results when $(4, 5)$ is rotated $60°$ clockwise about $(2, 5)$. We know how to handle rotations about the origin. Since point $(4, 5)$ is 2 units to the right of $(2, 5)$, we consider the $60°$ clockwise rotation of point $(2, 0)$ about the origin (because $(2, 0)$ is 2 units to the right of the origin). Let the result of this rotation be P'. Since $(2, 0)$ has the same relationship to the origin that $(4, 5)$ has to $(2, 5)$, point P' will have the same relationship to the origin that P has to $(2, 5)$.

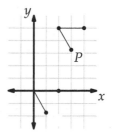

Applying the same formula we used in the previous part, we find that point P' is

$$(2\cos 300° - 0\sin 300°, 0\cos 300° + 2\sin 300°) = (1, -\sqrt{3}).$$

Since P' is 1 to the right of and $\sqrt{3}$ below the origin, point P is 1 to the right of and $\sqrt{3}$ below $(2, 5)$, which means its coordinates are $\boxed{(3, 5 - \sqrt{3})}$.

(c) We apply the same approach as in the previous parts. Let P be the desired point. Since $(4, 5)$ is 6 units to the right of and 2 units below $(-2, 7)$, we let point P' be the result of the $60°$ clockwise rotation of $(6, -2)$ about the origin. Therefore, point P' has coordinates

$$(6\cos 300° - (-2)\sin 300°, (-2)\cos 300° + 6\sin 300°) = (3 - \sqrt{3}, -1 - 3\sqrt{3}).$$

So, point P is $(-2 + (3 - \sqrt{3}), 7 + (-1 - 3\sqrt{3})) = \boxed{(1 - \sqrt{3}, 6 - 3\sqrt{3})}$.

5.33

(a) It is not always possible. For example, take $r_1 = r_2 = 1$, $\theta_1 = 0$, and $\theta_2 = 180°$. Then $(r_1, \theta_1) = (1, 0)$ and $(r_2, \theta_2) = (1, 180°)$ are the polar coordinates of the points with rectangular coordinates $(1, 0)$ and $(-1, 0)$, respectively. Their midpoint is the origin, which has magnitude 0. But any point of the form $\left(\frac{r_1 + r_2}{2}, \theta\right) = (1, \theta)$ always has magnitude 1.

(b) It is not always possible. For example, take $r_1 = 2$, $r_2 = 1$, $\theta_1 = 0$, and $\theta_2 = 180°$. Then $(r_1, \theta_1) = (2, 0)$ and $(r_2, \theta_2) = (1, 180°)$ are the polar coordinates of the points with rectangular coordinates $(2, 0)$ and $(-1, 0)$, respectively. Their midpoint is the point $\left(\frac{1}{2}, 0\right)$, which lies on the x-axis but not the y-axis. But any point of the form $\left(r, \frac{\theta_1+\theta_2}{2}\right) = (r, 90°)$ lies on the y-axis.

5.34

(a) Let $(r\cos\theta, r\sin\theta)$ be a point in the graph of $f(r, \theta) = 0$. Since $f(-r, \theta) = f(r, \theta)$, we have $f(-r, \theta) = 0$, so the point $(-r\cos\theta, -r\sin\theta)$ also lies on the graph. The point $(-r\cos\theta, -r\sin\theta)$ is a $180°$ rotation of $(r\cos\theta, r\sin\theta)$ about the origin is also in the graph. Therefore, the graph has $180°$ rotational symmetry about the origin. (In other words, for any point P in the graph, the $180°$ rotation of P about the origin is also in the graph.)

(b) Let $(r\cos\theta, r\sin\theta)$ be a point on the graph of $f(r, \theta) = 0$. Since $f(r, \theta) = f(r, -\theta)$, we have $f(r, -\theta) = 0$, so the point $(r\cos(-\theta), r\sin(-\theta)) = (r\cos\theta, -r\sin\theta)$ also lies on the graph. The point $(r\cos\theta, -r\sin\theta)$ is the reflection of the point $(r\cos\theta, r\sin\theta)$ over the x-axis. Hence, the graph must be $\boxed{\text{symmetric about the } x\text{-axis}}$.

5.35

(a) We have $r = \sqrt{x^2 + y^2} = 2\sqrt{2}$. From $x = r\cos\theta$ and $y = r\sin\theta$, we have $\cos\theta = \frac{\sqrt{2}}{2}$ and $\sin\theta = -\frac{\sqrt{2}}{2}$, so $\theta = \frac{7\pi}{4}$. Therefore, cylindrical coordinates of the point are $\boxed{\left(2\sqrt{2}, \frac{7\pi}{4}, 2\sqrt{2}\right)}$.

We have $\rho = \sqrt{x^2 + y^2 + z^2} = 4$, and the θ-coordinate in spherical coordinates is the same as the θ-coordinate in cylindrical coordinates. Finally, from $z = \rho\cos\phi$, we have $\cos\phi = \frac{z}{\rho} = \frac{\sqrt{2}}{2}$, so $\phi = \frac{\pi}{4}$. Therefore, spherical coordinates of the point are $\boxed{\left(4, \frac{7\pi}{4}, \frac{\pi}{4}\right)}$.

(b) We have $r = \sqrt{x^2 + y^2} = 2\sqrt{3}$. From $x = r\cos\theta$ and $y = r\sin\theta$, we have $\cos\theta = \frac{1}{2}$ and $\sin\theta = \frac{\sqrt{3}}{2}$, so $\theta = \frac{\pi}{3}$. Therefore, cylindrical coordinates of the point are $\boxed{\left(2\sqrt{3}, \frac{\pi}{3}, -2\sqrt{3}\right)}$.

We have $\rho = \sqrt{x^2 + y^2 + z^2} = 2\sqrt{6}$, and the θ-coordinate in spherical coordinates is the same as the θ-coordinate in cylindrical coordinates. Finally, from $z = \rho\cos\phi$, we have $\cos\phi = \frac{z}{\rho} = -\frac{\sqrt{2}}{2}$, so $\phi = \frac{3\pi}{4}$. Therefore, spherical coordinates of the point are $\boxed{\left(2\sqrt{6}, \frac{\pi}{3}, \frac{3\pi}{4}\right)}$.

5.36

(a) We have $x = r\cos\theta = 6\cos\frac{2\pi}{3} = 6\left(-\frac{1}{2}\right) = -3$ and $y = r\sin\theta = 6\sin\frac{2\pi}{3} = 6\left(\frac{\sqrt{3}}{2}\right) = 3\sqrt{3}$, so the rectangular coordinates are $\boxed{(-3, 3\sqrt{3}, -1)}$.

(b) We have
$$x = \rho\cos\theta\sin\phi = 12\cos\frac{\pi}{4}\sin\frac{3\pi}{4} = 12\left(\frac{\sqrt{2}}{2}\right)\left(\frac{\sqrt{2}}{2}\right) = 6,$$
$$y = \rho\sin\theta\sin\phi = 12\sin\frac{\pi}{4}\sin\frac{3\pi}{4} = 12\left(\frac{\sqrt{2}}{2}\right)\left(\frac{\sqrt{2}}{2}\right) = 6,$$
$$z = \rho\cos\phi = 12\cos\frac{3\pi}{4} = 12\left(-\frac{\sqrt{2}}{2}\right) = -6\sqrt{2},$$

so the rectangular coordinates are $\boxed{(6, 6, -6\sqrt{2})}$.

5.37 The portion of the graph of $y^2 + z^2 = 9$ that lies in the plane $x = c$ is a circle in the plane with radius 3 centered at the $(c, 0, 0)$. This is true for any value of c. So, every cross-section perpendicular to the x-axis is a circle with radius 3 and with center on the x-axis. Therefore, the graph is a

$$\boxed{\text{cylinder with radius 3 and the } x\text{-axis as its axis}}.$$

5.38 We have $r = 6 - z$. So, for any particular value of z, the value of r is fixed, and θ ranges from 0 to 2π. Therefore, each cross-section of the graph perpendicular to the z-axis is a circle centered at the point on the cross-section with $x = y = 0$, except when $z = 6$, for which the cross-section is the point $(0, 0, 6)$. The radius of the cross-section is directly proportional to the distance between the cross-section and the plane $z = 6$, so the graph is a pair of cones as shown at right. The cones have vertex $(0, 0, 6)$, and the z-axis is the axis of the cones.

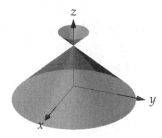

5.39

(a) The value of r in cylindrical coordinates is the distance from the point to the z-axis (if $r \geq 0$). Therefore, any point on the graph of $r = 5$ is 5 units from the z-axis, so the graph is a

$$\boxed{\text{cylinder with radius 5 and with the } z\text{-axis as its axis}}.$$

(b) The equation $\rho = 3$ describes all points that are 3 units from the origin. Therefore, the graph is a $\boxed{\text{sphere centered at the origin with radius 3}}$.

(c) In polar coordinates, the graph of $\theta = -\frac{\pi}{3}$ is a line through the origin that makes an angle of $\frac{\pi}{3}$ with the x-axis such that the line passes through the second and fourth quadrants. Since we can use any value for the z-coordinate in the graph of $\theta = -\frac{\pi}{3}$ in three dimensions, the graph of $\theta = -\frac{\pi}{3}$ consists of the aforementioned line in the plane $z = 0$, as well as every point "above" or "below" (i.e., with any z-coordinate) the line in space. In other words, the graph is a $\boxed{\text{plane}}$ that contains the z-axis and is perpendicular to the xy-plane.

(d) Let O be the origin and P be a point that has $\phi = \frac{3\pi}{4}$ in spherical coordinates. The ϕ-coordinate gives us the angle that $\overline{OP}$ makes with the positive portion of the z-axis. Therefore, any point on the ray $\overrightarrow{OP}$ has $\phi = \frac{3\pi}{4}$ when expressed in spherical coordinates. Moreover, we can rotate any point on this ray about the z-axis by any angle to produce another point with the same ϕ-coordinate. In other words, we can form a portion of the graph by rotating the whole ray around the z-axis, thereby sweeping out a cone. Since we can allow ρ to be negative, the graph consists of $\boxed{\text{two cones}}$ with the z-axis as the axis and the origin as the vertex. One cone is below the xy-plane (for $\rho > 0$) and the other is above the xy-plane (for $\rho < 0$).

5.40 Since $r^2 = x^2 + y^2$, we can write the equation as $x^2 + y^2 + z^2 = 2z$. Rearranging and completing the square in z gives $x^2 + y^2 + (z - 1)^2 = 1$. Therefore, the graph consists of all points that are 1 unit from $(0, 0, 1)$, which means the graph is the $\boxed{\text{sphere with center } (0, 0, 1) \text{ and radius 1}}$.

5.41 First, we note that the origin is in the graph, since $\phi = \frac{\pi}{2}$ gives $\rho = 0$. Multiplying both sides of $\rho = 6 \cos \phi$ by ρ gives $\rho^2 = 6\rho \cos \phi$, so we have $x^2 + y^2 + z^2 = 6z$. Rearranging and completing the square gives $x^2 + y^2 + (z - 3)^2 = 9$, so the graph is a sphere with radius 3 centered at $(0, 0, 3)$.

5.42 We present two solutions.

Solution 1: Reason geometrically. Let P be a point on the graph, A be the foot of the altitude from P to the z-axis, and O be the origin. Suppose we have $0 < \phi < \frac{\pi}{2}$. Then, we have $\sin \phi = \frac{PA}{PO} = \frac{PA}{\rho}$, so $PA = \rho \sin \phi$. The equation $\rho = \frac{3}{\sin \phi}$ gives $\rho \sin \phi = 3$, so $PA = 3$. This means that P is 3 units from the z-axis. Conversely, any point with $0 < \phi < \frac{\pi}{2}$ that is 3 units from the z-axis satisfies $\rho \sin \phi = 3$, so is on the graph.

If $\phi = \frac{\pi}{2}$, then $\sin \phi = 1$, and the graph is the circle in the xy-plane with radius 3 centered at the origin.

If $\frac{\pi}{2} < \phi < \pi$, then $\angle POA = \pi - \phi$. However, we still have $\sin \phi = \sin(\pi - \phi) = \sin \angle POA = \frac{PA}{PO} = \frac{PA}{\rho}$ as before. So, all the points that are 3 units from the z-axis and that have $\frac{\pi}{2} < \phi < \pi$ are also the graph.

Since $\sin \phi = 0$ for $\phi = 0$ and for $\phi = \pi$, no points with ϕ-coordinate 0 or π are on the graph. Combining all these observations, we see that the graph consists of all points that are 3 units from the z-axis. Therefore, the graph is a cylinder with radius 3 and the z-axis as its axis.

Solution 2: Algebra. That the graph is a cylinder suggests a connection to cylindrical coordinates. From $\rho = \frac{3}{\sin \phi}$, we have $\rho \sin \phi = 3$. Therefore, we have $x = \rho \cos \theta \sin \phi = 3 \cos \theta$ and $y = \rho \sin \theta \sin \phi = 3 \sin \theta$. Aha! Considering the equations $x = 3 \cos \theta$, $y = 3 \sin \theta$ in cylindrical coordinates gives $r = \sqrt{x^2 + y^2} = 3$. We have no restriction on z, so the graph is the same as the graph of $r = 3$ in cylindrical coordinates, and again we see that the graph is a cylinder with radius 3 and the z-axis as its axis.

5.43

(a) As indicated in the diagram, adding π to the θ-coordinate rotates the point by π about the z-axis. The θ-coordinate does not affect the distance from the point to the origin (ρ), or the distance from the point to the z-axis (r in cylindrical coordinates), or the distance from the point to the xy-plane (z in rectangular coordinates). Therefore, the transformation is a rotation of the point by π about the z-axis.

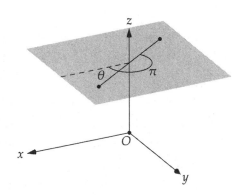

(b) Since $\sin(\pi - \phi) = \sin \phi$, the x- and y-coordinates of the point are unaffected. Since $\cos(\pi - \phi) = -\cos \phi$, the z-coordinate of the point with spherical coordinates (ρ, θ, ϕ) is the opposite of the z-coordinate of the point with spherical coordinates $(\rho, \theta, \pi - \phi)$. Therefore, the transformation is a reflection through the xy-plane.

Challenge Problems

5.44

(a) The graph of $r = \sin 2\theta$ is shown at right. To see how this is the shape formed, consider first the values $0 \le \theta \le \frac{\pi}{2}$. As θ increases from 0 to $\frac{\pi}{2}$, the value of r goes from 0 to 1 and then back to 0, and we form the petal in the first quadrant. Then, as θ goes from $\frac{\pi}{2}$ to π, the value of r goes from 0 to -1 and back to 0. This forms the petal in the *fourth* quadrant, not the second, because r is negative. As θ goes from π to $\frac{3\pi}{2}$, r goes from 0 to 1 and back to 0, forming the third quadrant petal. As θ goes from $\frac{3\pi}{2}$ to 2π, r again goes from 0 to -1 and back to 0, forming the *second* quadrant petal, because r is negative.

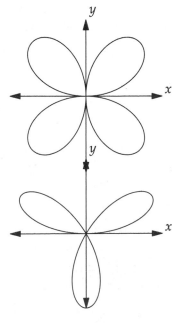

(b) The graph of $r = \sin 3\theta$ is shown at right. Here, the first quadrant petal corresponds to $0 \le \theta \le \frac{\pi}{3}$. The downward pointing petal corresponds to $\frac{\pi}{3} \le \theta \le \frac{2\pi}{3}$, since when θ is in this range, the value of 3θ is between π and 2π, which means $r = \sin 3\theta$ is negative. The second quadrant petal is formed for $\frac{2\pi}{3} \le \theta \le \pi$.

When $\pi \le \theta \le \frac{4\pi}{3}$, the value of 3θ is between 3π and 4π, which means r is negative. Therefore, the points produced by these values of θ are in the *first* quadrant, not the third. Moreover, each of the points has already been produced when $0 \le \theta \le \frac{\pi}{3}$. For example, when $\theta = \frac{7\pi}{6}$, we have $\sin 3\theta = -1$, and the point with polar coordinates $\left(-1, \frac{7\pi}{6}\right)$ is the same as the point with polar coordinates $\left(1, \frac{\pi}{6}\right)$, which is produced when $\theta = \frac{\pi}{6}$.

Similarly, when $\frac{4\pi}{3} \le \theta \le \frac{5\pi}{3}$, we reproduce the downward pointing petal, and when $\frac{5\pi}{3} \le \theta \le 2\pi$, we reproduce the second quadrant petal.

(c) The graph of $r = \sin 4\theta$ is shown below.

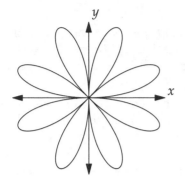

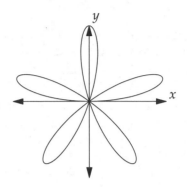

Figure 5.1: Diagram for Part (c)   Figure 5.2: Diagram for Part (d)

(d) The graph of $r = \sin 5\theta$ is shown at right above.

(e) The graph of an equation of the form $r = \sin k\theta$ is sometimes called a "rose." We claim that if k is even, then we obtain a graph with $2k$ "petals," and if k is odd, then the graph has k petals. To see why, we start by looking at the graph of $y = \sin 2x$ for $x \in [0, 2\pi]$, which is shown below.

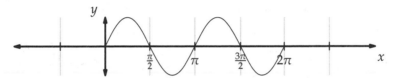

The crest (a portion of the graph above the x-axis) on the interval $[0, \frac{\pi}{2}]$ corresponds to the petal in the first quadrant in the graph of $r = \sin 2\theta$. The trough (a portion of the graph below the x-axis) on the interval $[\frac{\pi}{2}, \pi]$ corresponds to the petal in the fourth quadrant. (Even though an angle between $\frac{\pi}{2}$ and π usually indicates a point in the second quadrant, the petal lies in the fourth quadrant because $r = \sin 2\theta$ is negative on this interval.) More generally, each crest and trough in the graph of $y = \sin 2x$ (on the interval $[0, 2\pi]$) corresponds to a different petal. There are four crests and troughs in the graph of $y = \sin 2x$ on the interval $[0, 2\pi]$, hence there are four petals in the graph of $r = \sin 2\theta$.

In general, for even k, the high points of the crests occur at $x = \frac{\pi}{2k}, \frac{5\pi}{2k}, \ldots, \frac{(4k-3)\pi}{2k}$, and the low points of the troughs occur at $x = \frac{3\pi}{2k}, \frac{7\pi}{2k}, \ldots, \frac{(4k-1)\pi}{2k}$, which lead to the points $(1, \frac{\pi}{2k}), (1, \frac{5\pi}{2k}), \ldots, (1, \frac{(4k-3)\pi}{2k}), (-1, \frac{3\pi}{2k})$, $(-1, \frac{7\pi}{2k}), \ldots, (-1, \frac{(4k-1)\pi}{2k})$ in polar coordinates. These $2k$ points form the "tips" of the petals. To show that we obtain $2k$ petals, we must show that each petal tip is different.

Two petal tips can coincide if and only if they are of the form $(1, \phi)$ and $(-1, \phi + \pi)$. So, we must show that the r-coordinates are not opposites when the θ-coordinates differ by π. If $(1, \phi)$ is on the graph, then $1 = \sin k\phi$. When $\theta = \phi + \pi$, we have

$$\sin\big(k(\phi + \pi)\big) = \sin(k\phi + k\pi).$$

But k is even, which makes $k\pi$ an even multiple of π. Hence, $\sin(k\phi + k\pi) = \sin k\phi = 1$. This tells us that the point $(1, \phi + \pi)$ is also on the graph. So, for each petal in the graph of $r = \sin k\theta$, there is a petal on the opposite side of the origin. In particular, each petal tip is different, so the graph has $2k$ petals.

Now we look at the graph of $y = \sin 3x$.

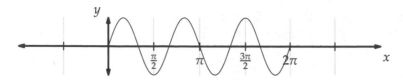

This graph exhibits 3 crests and 3 troughs, so we may expect 6 petals in the graph of $r = \sin 3\theta$. However, the crest at $x = \frac{\pi}{6}$ coincides with the trough at $x = \frac{7\pi}{6}$ in polar coordinates, because $\frac{\pi}{6} + \pi = \frac{7\pi}{6}$, $\sin(3 \cdot \frac{\pi}{6}) = \sin \frac{\pi}{2} = 1$, and $\sin(3 \cdot \frac{7\pi}{6}) = \sin \frac{7\pi}{2} = -1$. Hence, the 3 crests and 3 trough correspond to 3 petals in the graph of $r = \sin 3\theta$.

For general odd k, we expect that each petal corresponding to a crest overlaps with another petal corresponding to a trough. To show that this is the case, let $(1, \phi)$ be a point on the graph, so that $1 = \sin k\phi$. When $\theta = \phi + \pi$, we have $\sin[k(\phi + \pi)] = \sin(k\phi + k\pi)$. Since k is odd, $k\pi$ is an odd multiple of π, so $\sin(k\theta + k\pi) = -\sin k\theta = -1$. But the point $(-1, \phi + \pi)$ is the same point as $(1, \phi)$. Hence, each petal corresponding to a crest overlaps with another petal corresponding to a trough, so the graph of $r = \sin k\theta$ has k petals.

5.45

(a) The graph of the parametric equations $x = \cos \theta$, $y = \sin \theta$ is the unit circle centered at the origin. Dilating this circle by a factor of r, the graph of the parametric equations $x = r \cos \theta$, $y = r \sin \theta$ is the circle with radius r centered at the origin. Translating this circle horizontally by h (adding h to the x-coordinate) and vertically by y (adding k to the y-coordinate) gives us the circle with radius r and center (h, k). Therefore, parametric equations for this circle are $x = h + r \cos \theta$, $y = k + r \sin \theta$.

(b) The equations we use to convert spherical coordinates to rectangular coordinates give us parametric equations for a sphere centered at the origin with radius ρ. These are $x = \rho \cos \theta \sin \phi$, $y = \rho \sin \theta \sin \phi$, $z = \rho \cos \phi$. Adding a to the x-coordinate, b to the y-coordinate, and c to the z-coordinate of each point on this sphere is equivalent to translating the sphere centered at the origin to the desired sphere. Therefore, parametric equations for the sphere are $x = a + \rho \cos \theta \sin \phi$, $y = b + \rho \sin \theta \sin \phi$, $z = c + \rho \cos \phi$, where θ and ϕ are the parameters and ρ is a constant.

(c) Since x can be anything, we can let $x = t$ be one of the parametric equations. For any t, the point (t, y, z) must be on the circle centered at $(t, 0, 0)$ with radius r such that the circle is in the plane $x = t$. Therefore, we must have $(y - 0)^2 + (z - 0)^2 = r^2$, so $y^2 + z^2 = r^2$. This means we can let $y = r \cos \theta$ and $z = r \sin \theta$. Parametric equations for the cylinder then are $x = t$, $y = r \cos \theta$, $z = r \sin \theta$, where t and θ are the parameters.

5.46 Completing the square in x and y gives $(x - 2)^2 + (y - 1)^2 = 4$. The graph of this equation is a circle with center $(2, 1)$ and radius 2. We can parameterize a circle with radius 2 as $x = 2 \cos t$, $y = 2 \sin t$, so we can parameterize the circle with radius 2 and center $(2, 1)$ as $x = 2 + 2 \cos t$, $y = 1 + 2 \sin t$. However, graphing these equations produces the whole circle; we only want the portion above the x-axis. We must only include those values of t for which $y > 0$. From $1 + 2 \sin t > 0$, we have $\sin t > -\frac{1}{2}$. Since $\sin t \le -\frac{1}{2}$ for $t \in \left[\frac{7\pi}{6}, \frac{11\pi}{6}\right]$, we can exclude the portion of the circle below the x-axis and include the rest of the circle by limiting t to those values in the intervals $\left[0, \frac{7\pi}{6}\right)$ and $\left(\frac{11\pi}{6}, 2\pi\right]$. We can write this as a single interval by noting that for $t \in \left(-\frac{\pi}{6}, 0\right]$, we produce the same values for x and y as we do for $t \in \left(\frac{11\pi}{6}, 2\pi\right]$. Therefore, we can parameterize the portion of the graph of $x^2 - 4x + y^2 - 2y + 1 = 0$ that is above the x-axis with $x = 2 + 2 \cos t$, $y = 1 + 2 \sin t$, $t \in \left(-\frac{\pi}{6}, \frac{7\pi}{6}\right)$.

5.47 Let $\theta' = 2 \arctan \frac{y}{x + \sqrt{x^2 + y^2}}$. Because the range of arctangent is $\left(-\frac{\pi}{2}, \frac{\pi}{2}\right)$, we know that $\frac{\theta'}{2}$ is the unique angle in the interval $\left(-\frac{\pi}{2}, \frac{\pi}{2}\right)$ that has tangent $\frac{y}{x + \sqrt{x^2 + y^2}}$. Therefore, θ' is the unique angle in the interval $(-\pi, \pi)$ such that $\tan \frac{\theta'}{2} = \frac{y}{x + \sqrt{x^2 + y^2}}$. Our goal is to show that this angle is indeed the θ such that (r, θ), with $r > 0$ and $-\pi < \theta < \pi$, are the polar coordinates of the point (x, y).

We know that $r = \sqrt{x^2 + y^2}$, $x = r \cos \theta$, and $y = r \sin \theta$, so

$$\frac{y}{x + \sqrt{x^2 + y^2}} = \frac{r \sin \theta}{r \cos \theta + r} = \frac{\sin \theta}{1 + \cos \theta}.$$

By the half-angle formula for tangent, we have $\frac{\sin \theta}{1 + \cos \theta} = \tan \frac{\theta}{2}$, so $\tan \frac{\theta}{2} = \frac{y}{x + \sqrt{x^2 + y^2}}$. We showed above that only one angle θ' in the interval $(-\pi, \pi)$ satisfies $\tan \frac{\theta'}{2} = \frac{y}{x + \sqrt{x^2 + y^2}}$, so θ must be this angle. Therefore $\theta = \theta' = 2 \arctan \frac{y}{x + \sqrt{x^2 + y^2}}$.

5.48 The distance between $(1, 2, 3)$ and $(2 - t, 4 + t, 3 + 2t)$ is

$$\sqrt{(1 - (2 - t))^2 + (2 - (4 + t))^2 + (3 - (3 + 2t))^2} = \sqrt{(t - 1)^2 + (-2 - t)^2 + (2t)^2} = \sqrt{6t^2 + 2t + 5}.$$

Completing the square gives

$$\sqrt{6t^2 + 2t + 5} = \sqrt{6\left(t^2 + \frac{t}{3}\right) + 5} = \sqrt{6\left(t + \frac{1}{6}\right)^2 + 5 - 6\left(\frac{1}{6}\right)^2} = \sqrt{6\left(t + \frac{1}{6}\right)^2 + \frac{29}{6}}.$$

Since $6\left(t + \frac{1}{6}\right)^2$ is nonnegative for all real t, the minimum distance is $\boxed{\sqrt{\frac{29}{6}}}$, which occurs when $t = -\frac{1}{6}$.

5.49

(a) Let C be the center of the circle, and let θ be the angle that $\overline{OX}$ makes with the positive x-axis. Since $\overline{XT}$ is parallel to the x-axis, we have $\angle TXO = \theta$ as well. Therefore, we have $\frac{TX}{OT} = \cot \theta$. Since $OT = 1$, we have $TX = \cot \theta$, so the x-coordinate of A is $\cot \theta$.

The y-coordinate of A matches that of point P. We have $\angle POT = \frac{\pi}{2} - \theta$, and from isosceles triangle COP, we have $\angle CPO = \frac{\pi}{2} - \theta$, so $\angle OCP = \pi - \angle POC - \angle OPC = 2\theta$. We can find the y-coordinate of P by drawing an altitude from P to the y-axis. Or, we can consider the parameterization of P as it travels around the circle.

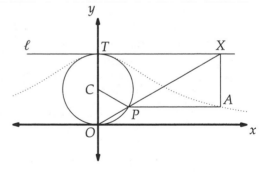

When $\theta = 0$, point P is at the origin, and when $\theta = \frac{\pi}{2}$, point P is at T. For any value of θ between 0 and $\frac{\pi}{2}$, we have $\angle OCP = 2\theta$. One parameterization of a circle with radius $\frac{1}{2}$ centered at $\frac{1}{2}$ is $\left(\frac{1}{2} \cos t, \frac{1}{2} + \frac{1}{2} \sin t\right)$. Here, we want the parameterization in terms of θ, the angle between $\overline{XO}$ and the x-axis. We have seen that $\angle OCP = 2\theta$, so we let 2θ be the angle in the parameterization. We also require P to be at $(0, 0)$ when $\theta = 0$, so our parameterization is $\left(\frac{1}{2} \cos\left(2\theta - \frac{\pi}{2}\right), \frac{1}{2} + \frac{1}{2} \sin\left(2\theta - \frac{\pi}{2}\right)\right)$. Applying the identity $\sin\left(x - \frac{\pi}{2}\right) = -\cos x$, the y-coordinate of P then is $\frac{1}{2} - \frac{1}{2} \cos 2\theta$. The y-coordinate of A matches the y-coordinate of P, so parametric equations for A are $\boxed{x = \cot \theta, y = \frac{1}{2} - \frac{1}{2} \cos 2\theta}$. As we'll see in the next part, we can also write this as $x = \cot \theta, y = \sin^2 \theta$.

(b) We have one trig identity that links $\cot \theta$ to sine or cosine, namely, $\cot^2 \theta + 1 = \csc^2 \theta$. We can introduce a $\sin^2 \theta$ to our parameterization $x = \cot \theta, y = \frac{1}{2} - \frac{1}{2} \cos 2\theta$ by using the double angle identity for cosine. We have $y = \frac{1}{2} - \frac{1}{2} \cos 2\theta = \frac{1}{2} - \frac{1}{2}(1 - 2 \sin^2 \theta) = \sin^2 \theta$. Substituting $x = \cot \theta$ and $y = \sin^2 \theta$ into $\cot^2 \theta + 1 = \csc^2 \theta$ gives $x^2 + 1 = \frac{1}{y}$, so the witch curve is the graph of $\boxed{y = \frac{1}{x^2 + 1}}$.

5.50 Parameterizing the red surface prior to the drilling is easy; we just borrow from spherical coordinates to produce the parameterization $x = 2\sin\phi\cos\theta$, $y = 2\sin\phi\sin\theta$, $z = 2\cos\phi$. The hole complicates matters. If the hole were along the z-axis, as shown at right, we could simply modify the parameterization we found for the whole sphere. To see why, let P be a point on the "top" of the drilled sphere, let A be the foot of the altitude from P to the z-axis, and let O be the origin. We have $AP = 1$ because the radius of the hole is 1, and right triangle OAP gives us $OA = \sqrt{3}$. Drilling the hole along the z-axis therefore has the effect of removing all points with z-coordinate above $\sqrt{3}$ or less than $-\sqrt{3}$. We then must have $-\sqrt{3} \le 2\cos\phi \le \sqrt{3}$, so $-\frac{\sqrt{3}}{2} \le \cos\phi \le \frac{\sqrt{3}}{2}$, which means $\frac{\pi}{6} \le \phi \le \frac{5\pi}{6}$. This gives us the parameterization $x = 2\sin\phi\cos\theta$, $y = 2\sin\phi\sin\theta$, $z = 2\cos\phi$, where $\frac{\pi}{6} \le \phi \le \frac{5\pi}{6}$.

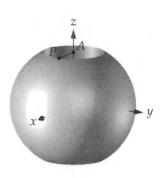

Turning to the surface formed when we drill along the x-axis, we see that we can't simply limit ϕ or θ alone in our initial parameterization for an un-drilled sphere. However, we can reassign the parameterizations that we used for x, y, and z when the hole was along the z-axis to produce a parameterization for the surface formed when the hole is along the x-axis. We simply swap the parameterizations for x and z, and we have

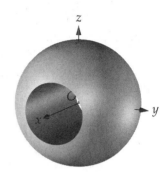

$$\boxed{x = 2\cos\phi,\, y = 2\sin\phi\sin\theta,\, z = 2\sin\phi\cos\theta,\ \text{where}\ \frac{\pi}{6} \le \phi \le \frac{5\pi}{6}.}$$

5.51

(a) The red point travels in a circle around the center of $\mathcal{B}$, but the center of $\mathcal{B}$ moves! So, we first find parametric equations for the center of $\mathcal{B}$. Let A be the center of $\mathcal{A}$, point B be the center of $\mathcal{B}$, and R be the red point. Point B is always 2 units from A, so it traces out a circle of radius 2 as $\mathcal{B}$ rolls around $\mathcal{A}$. We know how to parameterize a circle. Since the circle has radius 2, we might think the parameterization is $(2\cos t, 2\sin t)$. However, point B starts at $(0, 2)$, so we can be more precise by using the parameterization $\left(2\cos\left(t + \frac{\pi}{2}\right), 2\sin\left(t + \frac{\pi}{2}\right)\right)$. Applying trig identities, we can write this parameterization as $(-2\sin t, 2\cos t)$.

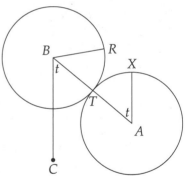

Next, we must figure out where R is relative to B. Let X be the point on $\mathcal{A}$ at which $\mathcal{A}$ and $\mathcal{B}$ are originally tangent. We draw vertical line $\overline{BC}$ as shown through B, and we have $\angle CBA = \angle XAB = t$ because $\overline{BC} \parallel \overline{AX}$. Let T be the point of tangency after $\mathcal{B}$ has rotated about $\mathcal{A}$ by angle t, as shown. Since $\mathcal{B}$ rolls along $\mathcal{A}$, the lengths of arcs $\overset{\frown}{RT}$ and $\overset{\frown}{TX}$ must be the same. The radii of the circles are the same, which means that $\angle RBT = \angle TAX = t$, as well. Therefore, we have $\angle CBR = 2t$. So, as point B rotates counterclockwise by an angle t about point A, point R rotates counterclockwise by $2t$ about point B. Since R is moving in a unit circle about the center of $\mathcal{B}$, and starts directly below the center of $\mathcal{B}$, we can parameterize its position *relative to B* as $\cos\left(2t - \frac{\pi}{2}\right)$ to the right of B and $\sin\left(2t - \frac{\pi}{2}\right)$ above B. The "$-\frac{\pi}{2}$" accounts for the fact that R begins directly below B. Combining this with the parameterization of B, the location of R is given by $\left(-2\sin t + \cos\left(2t - \frac{\pi}{2}\right), 2\cos t + \sin\left(2t - \frac{\pi}{2}\right)\right)$. Applying trig identities simplifies these coordinates to $\boxed{(-2\sin t + \sin 2t, 2\cos t - \cos 2t)}$.

(b) We saw in part (a) that point R is $-\cos 2t$ above point B. At $t = 0$, point R is -1 above point B (or, "1 below point B"). As $\mathcal{B}$ is rolled around $\mathcal{A}$, the value of t ranges from 0 to 2π, so the value of $2t$ ranges from 0 to 4π. We have $-\cos 2t = -1$ whenever $\cos 2t = 1$, which is whenever $2t$ is an integer multiple of 2π. Therefore, as t ranges from 0 to 4π, point R is directly below point B at $t = 0$, $t = 2\pi$, and $t = 4\pi$, so it makes $\boxed{2}$ full revolutions about the center of $\mathcal{B}$. We also could have seen this by using the fact noted above that as B moves an angle of t about A, point R moves $2t$ about B. So, as B goes around A once, point R goes around B twice.

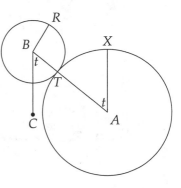

(c)   The situation is only slightly different than in the first two parts, so we give all the point labels the same meaning they had in part (a). We again find a parameterization for B first. Point B moves in a circle of radius $n + 1$ about point A and starts at $(0, n + 1)$, so, by an argument similar to that in part (a), we can parameterize point B with $(-(n + 1) \sin t, (n + 1) \cos t)$.

The main difference comes in our calculation of $\angle RBT$. Arcs $\overset{\frown}{RT}$ and $\overset{\frown}{XT}$ still have the same length, because $\mathcal{B}$ is being rolled along the circumference of $\mathcal{A}$. However, now the radii of the circles are different, so the angular measures of the arcs are different. The length of an arc equals the radius of the circle times the central angle that cuts off the arc. So, because arcs $\overset{\frown}{RT}$ and $\overset{\frown}{XT}$ have the same length, we have $(BT)(\angle RBT) = (AT)(\angle XAT)$. This gives us $\angle RBT = nt$, so $\angle RBC = (n + 1)t$. Now, following the argument of part (a), the parameterization of R is $\boxed{(-(n + 1) \sin t + \sin(n + 1)t, (n + 1) \cos t - \cos(n + 1)t)}$, and point R makes $\boxed{n + 1}$ revolutions about B as $\mathcal{B}$ makes one full revolution about $\mathcal{A}$.

6

CHAPTER

Basics of Complex Numbers

Exercises for Section 6.1

6.1.1 Isolating z^2 gives $z^2 = -3$, and "taking the square root of both sides" gives $z = \pm\sqrt{-3} = \pm\sqrt{3}\sqrt{-1} = \boxed{\pm i\sqrt{3}}$.

6.1.2

(a) $a - b = (3 + 4i) - (12 - 5i) = 3 - 12 + 4i + 5i = \boxed{-9 + 9i}$.

(b) We have $ab = (3 + 4i)(12 - 5i) = 3(12 - 5i) + 4i(12 - 5i) = 36 - 15i + 48i - 20i^2 = \boxed{56 + 33i}$.

(c) Factoring first makes this calculation easier:

$$a^2 + 3a + 2 = (a + 1)(a + 2) = (4 + 4i)(5 + 4i) = 20 + 16i + 20i - 16 = \boxed{4 + 36i}.$$

(d)

$$\frac{a}{\overline{b}} + \frac{\overline{a}}{b} = \frac{3 + 4i}{12 - 5i} + \frac{3 - 4i}{12 + 5i} = \frac{3 + 4i}{12 + 5i} \cdot \frac{12 - 5i}{12 - 5i} + \frac{3 - 4i}{12 - 5i} \cdot \frac{12 + 5i}{12 + 5i}$$

$$= \frac{(3 + 4i)(12 - 5i)}{(12 + 5i)(12 - 5i)} + \frac{(3 - 4i)(12 + 5i)}{(12 - 5i)(12 + 5i)} = \frac{56 + 33i}{169} + \frac{56 - 33i}{169} = \boxed{\frac{112}{169}}.$$

Is it a coincidence that the result is real?

6.1.3

(a) $\overline{2i} + \overline{7 - 2i} = -2i + (7 + 2i) = \boxed{7}$.

(b) $\dfrac{1}{\overline{2 + 3i} + \overline{1 + 2i}} = \dfrac{1}{(2 - 3i) + (1 - 2i)} = \dfrac{1}{3 - 5i} = \dfrac{1}{3 - 5i} \cdot \dfrac{3 + 5i}{3 + 5i} = \dfrac{3 + 5i}{(3 - 5i)(3 + 5i)} = \boxed{\dfrac{3}{34} + \dfrac{5}{34}i}$.

6.1.4

(a) Since $i^2 = -1$, we have $i^4 = (i^2)^2 = (-1)^2 = 1$ and $i^6 = (i^2)^3 = (-1)^3 = -1$. Therefore,

$$f(i) = \frac{i^6 + i^4}{i + 1} = \frac{-1 + 1}{i + 1} = \boxed{0}.$$

(b) Since $(-i)^2 = -1$, we have $(-i)^4 = [(-i)^2]^2 = (-1)^2 = 1$, and $(-i)^6 = [(-i)^2]^3 = (-1)^3 = -1$. Therefore,

$$f(-i) = \frac{(-i)^6 + (-i)^4}{-i + 1} = \frac{-1 + 1}{-i + 1} = \boxed{0}.$$

(c) First, we compute $(i - 1)^2 = i^2 - 2i + 1 = -1 - 2i + 1 = -2i$. Then $(i - 1)^4 = [(i - 1)^2]^2 = (-2i)^2 = 4i^2 = -4$, and $(i - 1)^6 = [(1 - i)^2]^3 = (-2i)^3 = -8i^3 = 8i$. Therefore,

$$f(i - 1) = \frac{(i - 1)^6 + (i - 1)^4}{(i - 1) + 1} = \frac{8i - 4}{i} = \frac{(8i - 4)(-i)}{i(-i)} = \boxed{8 + 4i}.$$

6.1.5 Note that $i^{-1} = 1/i = i/i^2 = i/(-1) = -i$. Therefore, $(i - i^{-1})^{-1} = (i + i)^{-1} = \dfrac{1}{2i} = \dfrac{i}{2i^2} = \boxed{-\dfrac{1}{2}i}$.

6.1.6

(a) Multiplying both sides of the equation by $z - 3$, we get $z + 3i = 2z - 6$. Subtracting z from both sides, and adding 6 to both sides, we find that $z = \boxed{6 + 3i}$.

(b) Multiplying both sides of the equation by z, we get $\frac{1+2i}{3} = (4 + 5i)z$. Then dividing both sides by $4 + 5i$, we get

$$z = \frac{1 + 2i}{3(4 + 5i)} = \frac{(1 + 2i)(4 - 5i)}{3(4 + 5i)(4 - 5i)} = \frac{14 + 3i}{3(4^2 + 5^2)} = \boxed{\frac{14}{123} + \frac{1}{41}i}.$$

(c) Multiplying both sides of the equation by $1 + i$, we get $3(1 + i)z + z = (10 - 4i)(1 + i)$, which simplifies to $(4 + 3i)z = 14 + 6i$. Dividing both sides by $4 + 3i$, we get

$$z = \frac{14 + 6i}{4 + 3i} = \frac{(14 + 6i)(4 - 3i)}{(4 + 3i)(4 - 3i)} = \frac{74 - 18i}{4^2 + 3^2} = \boxed{\frac{74}{25} - \frac{18}{25}i}.$$

6.1.7 Let $w = a + bi$, $z_1 = c_1 + d_1i$, and $z_2 = c_2 + d_2i$. Then, by the definition of complex number addition, we have $z_1 + z_2 = (c_1 + c_2) + (d_1 + d_2)i$. By the definition of complex number multiplication, we have

$$w(z_1 + z_2) = (a + bi)((c_1 + c_2) + (d_1 + d_2)i) = (a)(c_1 + c_2) - b(d_1 + d_2) + (b(c_1 + c_2) + a(d_1 + d_2))i.$$

We then apply the distributive property for real numbers to find

$$w(z_1 + z_2) = ac_1 + ac_2 - bd_1 - bd_2 + (bc_1 + bc_2 + ad_1 + ad_2)i.$$

Next, we compute $wz_1 + wz_2$. By the definition of complex number multiplication, we have

$$wz_1 + wz_2 = ((ac_1 - bd_1) + (ad_1 + bc_1)i) + ((ac_2 - bd_2) + (ad_2 + bc_2)i).$$

Then, applying the definition of complex number addition, followed by the associative property for real numbers, we have

$$wz_1 + wz_2 = ac_1 + ac_2 - bd_1 - bd_2 + (bc_1 + bc_2 + ad_1 + ad_2)i.$$

Therefore, we have $w(z_1 + z_2) = wz_1 + wz_2$, so complex numbers are distributive.

6.1.8 We have $i^{-n} = (i^{-1})^n = \left(\frac{1}{i}\right)^n$. Since $\frac{1}{i} = \frac{1}{i} \cdot \frac{i}{i} = \frac{i}{i^2} = -i$, we have $i^{-n} = (-i)^n$. Just as the powers of i repeat every four terms, the powers of $-i$ also repeat every four terms. So, we only have to check $i^n + (-i)^n$ for $n = 0, 1, 2$, and 3, to determine all possible values of $i^n + (-i)^n$. We have

$$i^0 + (-i)^0 = 2,$$
$$i + (-i)^1 = i - i = 0,$$
$$i^2 + (-i)^2 = -1 - 1 = -2,$$
$$i^3 + (-i)^3 = -i + i = 0.$$

Hence, the possible values of S are $\boxed{-2, 0, \text{ and } 2}$.

Exercises for Section 6.2

6.2.1 In the diagram at right, we see each of the points $4 + 7i$, $-6 - 2i$, and $(3 + i)(-2 + 5i) = -11 + 13i$ plotted in the complex plane, similar to the way the points $(4, 7)$, $(-6, -2)$, and $(-11, 13)$ would be plotted in the Cartesian coordinate plane.

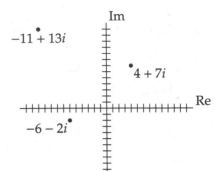

6.2.2

(a) The magnitude of $24 - 7i$ is $\sqrt{24^2 + (-7)^2} = \sqrt{625} = \boxed{25}$.

(b) The magnitude of $2 + 2\sqrt{3}i$ is $\sqrt{2^2 + (2\sqrt{3})^2} = \sqrt{16} = \boxed{4}$.

(c) The magnitude of $(1 + 2i)(2 + i) = 5i$ is $\sqrt{5^2} = \boxed{5}$.

6.2.3 As we see in the diagram at right, the complex numbers w, z, $\overline{w}$, and $\overline{z}$ form an isosceles trapezoid. The lengths of the bases are $|w - \overline{w}| = |10i| = 10$ and $|z - \overline{z}| = |4i| = 4$. The height is $12 - 3 = 9$, so the area of the trapezoid is $9 \cdot \frac{10+4}{2} = \boxed{63}$.

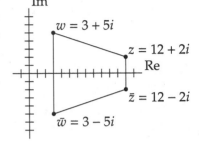

6.2.4 The distance between $4+7i$ and $-3-17i$ is the magnitude of their difference:

$$|4 + 7i - (-3 - 17i)| = |7 + 24i| = \sqrt{7^2 + 24^2} = \sqrt{625} = \boxed{25}.$$

6.2.5 Let $z_1 = a_1 + b_1 i$ and $z_2 = a_2 + b_2 i$. So, we have

$$\frac{z_1 + z_2}{2} = \frac{a_1 + b_1 i + a_2 + b_2 i}{2} = \frac{a_1 + a_2}{2} + \frac{b_1 + b_2}{2} \cdot i.$$

The midpoint of the segment connecting z_1 and z_2 on the complex plane corresponds to the midpoint of the segment connecting (a_1, b_1) and (a_2, b_2) on the Cartesian plane. This midpoint on the Cartesian plane is $\left(\frac{a_1+a_2}{2}, \frac{b_1+b_2}{2}\right)$, which does indeed correspond to the point $(z_1 + z_2)/2$ on the complex plane as shown by our calculation above.

6.2.6

(a) The magnitude of $\dfrac{1 + 2i}{2 + i} = \dfrac{(1 + 2i)(2 - i)}{(2 + i)(2 - i)} = \dfrac{4 + 3i}{5} = \dfrac{4}{5} + \dfrac{3}{5}i$ is $\sqrt{(4/5)^2 + (3/5)^2} = \sqrt{25/5^2} = \boxed{1}$.

(b) The magnitude of

$$\frac{6 + 11i}{11 + 6i} = \frac{(6 + 11i)(11 - 6i)}{(11 + 6i)(11 - 6i)} = \frac{132 + 85i}{157} = \frac{132}{157} + \frac{85}{157}i$$

is $\sqrt{(132/157)^2 + (85/157)^2} = \sqrt{24649/157^2} = \boxed{1}$.

(c) From parts (a) and (b), the magnitude of $(x + yi)/(y + xi)$ appears to be 1. We can prove this as follows. The magnitude of

$$\frac{x + yi}{y + xi} = \frac{(x + yi)(y - xi)}{(y + xi)(y - xi)} = \frac{2xy + (y^2 - x^2)i}{x^2 + y^2} = \frac{2xy}{x^2 + y^2} + \frac{y^2 - x^2}{x^2 + y^2}i$$

is

$$\sqrt{\left(\frac{2xy}{x^2 + y^2}\right)^2 + \left(\frac{y^2 - x^2}{x^2 + y^2}\right)^2} = \sqrt{\frac{4x^2y^2}{(x^2 + y^2)^2} + \frac{x^4 - 2x^2y^2 + y^4}{(x^2 + y^2)^2}}$$

$$= \sqrt{\frac{x^4 + 2x^2y^2 + y^4}{(x^2 + y^2)^2}} = \sqrt{\frac{(x^2 + y^2)^2}{(x^2 + y^2)^2}} = \boxed{1}.$$

6.2.7   Every square is also a parallelogram, and the diagonals of a parallelogram bisect each other. Plotting the points $1+2i$, $-2+i$, and $-1-2i$ in the complex plane, we see that the numbers $1+2i$ and $-1-2i$ must lie at opposite corners of the square. Hence, the midpoint of the diagonal between these points is $[(1+2i)+(-1-2i)]/2 = 0$.

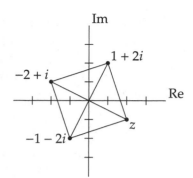

Therefore, the midpoint of z and $-2+i$ is also 0. In other words, $[z+(-2+i)]/2 = 0$, so $z = \boxed{2-i}$.

There are many other ways we might have approached this problem; we could have used slopes, distances, or even congruent triangles.

Exercises for Section 6.3

6.3.1   Instead of expanding the product, we can use the fact that the magnitude of the product is equal to the product of the magnitudes:

$$|(10+24i)(8-6i)| = |10+24i| \cdot |8-6i| = \sqrt{10^2+24^2} \cdot \sqrt{8^2+(-6)^2} = \sqrt{676} \cdot \sqrt{100} = 26 \cdot 10 = \boxed{260}.$$

6.3.2   Let $z = a+bi$, where a and b are real numbers. Then $|z| = |a+bi| = \sqrt{a^2+b^2}$, and $|\bar{z}| = |a-bi| = \sqrt{a^2+(-b)^2} = \sqrt{a^2+b^2}$. Hence, $|z| = |\bar{z}|$. We also could have noted that $|\bar{z}| = \sqrt{(\bar{z})(\bar{\bar{z}})} = \sqrt{(\bar{z})(z)} = |z|$.

6.3.3   Let $z = a+bi$ for some real numbers a and b. Then $\bar{z} = a-bi$, and the given equation becomes

$$3(a+bi)+4(a-bi) = 12-5i \quad \Rightarrow \quad 7a-bi = 12-5i.$$

Two complex numbers are equal if and only if their real parts are equal and their imaginary parts are equal, so we get $7a = 12$ and $b = 5$. Therefore, $z = a+bi = \boxed{\dfrac{12}{7}+5i}$.

We also could have solved for z without writing out the real and imaginary parts as follows. Taking the conjugate both sides of the given equation $3z+4\bar{z} = 12-5i$, we get $\overline{3z+4\bar{z}} = \overline{12-5i}$, so $3\bar{z}+4z = 12+5i$. This gives us the system of equations

$$3z+4\bar{z} = 12-5i,$$
$$4z+3\bar{z} = 12+5i.$$

Multiplying the first equation by 3 and the second equation by 4, we get

$$9z+12\bar{z} = 36-15i,$$
$$16z+12\bar{z} = 48+20i.$$

Subtracting the first equation from the second equation, we get $7z = 12+35i$, so $z = \frac{12}{7}+5i$.

6.3.4   Let $z = a+bi$, for some real numbers a and b.

(a)   If $z^2 = 2i$, then $2i = (a+bi)^2 = a^2-b^2+2abi$. Equating the real parts and equating the imaginary parts, we get the system of equations $a^2-b^2 = 0$ and $2ab = 2$. From the second equation, we have $b = 1/a$. Substituting this expression for b into the first equation, we get

$$a^2-\frac{1}{a^2} = 0 \quad \Rightarrow \quad a^4 = 1.$$

Since a is real, we have $a = \pm 1$. The corresponding values of $b = 1/a$ are ± 1. Therefore, the two square roots of $2i$ are $\boxed{1 + i \text{ and } -1 - i}$.

(b) If $z^2 = -5 + 12i$, then $-5 + 12i = (a + bi)^2 = a^2 - b^2 + 2abi$. Equating the real parts and equating the imaginary parts, we get the system of equations $a^2 - b^2 = -5$ and $2ab = 12$. From the second equation, we have $b = 6/a$. Substituting this expression for b into the first equation, we get

$$a^2 - \frac{36}{a^2} = -5 \quad \Rightarrow \quad a^4 + 5a^2 - 36 = 0 \quad \Rightarrow \quad (a^2 + 9)(a^2 - 4) = 0.$$

Since a is real, we have $a = \pm 2$. The corresponding values of b are $6/a = \pm 3$. Therefore, the two square roots of $-5 + 12i$ are $\boxed{2 + 3i \text{ and } -2 - 3i}$.

(c) If $z^2 = 24 - 10i$, then $24 - 10i = (a + bi)^2 = a^2 - b^2 + 2abi$. Equating the real parts and equating the imaginary parts, we get the system of equations $a^2 - b^2 = 24$ and $2ab = -10$. From the second equation, we have $b = -5/a$. Substituting this expression for b into the first equation, we get

$$a^2 - \frac{25}{a^2} = 24 \quad \Rightarrow \quad a^4 - 24a^2 - 25 = 0 \quad \Rightarrow \quad (a^2 - 25)(a^2 + 1) = 0.$$

Since a is real, we have $a = \pm 5$. The corresponding values of b are $b = -5/a = \mp 1$. (The $\mp$ symbol indicates that $b = -1$ goes with $a = 5$ and $b = 1$ goes with $a = -5$.) Therefore, the two square roots of $24 - 10i$ are $\boxed{5 - i \text{ and } -5 + i}$.

6.3.5

(a) The magnitude of

$$\frac{1}{z} = \frac{1}{3 + 4i} = \frac{3 - 4i}{(3 + 4i)(3 - 4i)} = \frac{3 - 4i}{25} = \frac{3}{25} - \frac{4}{25}i$$

is $|1/z| = \sqrt{(3/25)^2 + (4/25)^2} = \sqrt{25/25^2} = \boxed{1/5}$.

(b) The magnitude of z is $|z| = \sqrt{3^2 + 4^2} = \sqrt{25} = 5$, so $1/|z| = \boxed{1/5}$.

(c) First, we make the denominator of z/w real:

$$\frac{z}{w} = \frac{3 + 4i}{5 - 12i} = \frac{(3 + 4i)(5 + 12i)}{(5 - 12i)(5 + 12i)} = \frac{(3 + 4i)(5 + 12i)}{169}.$$

We leave the numerator as a product, because it makes the magnitude easier to compute:

$$\left|\frac{z}{w}\right| = \left|\frac{1}{169} \cdot (3 + 4i) \cdot (5 - 12i)\right| = \left|\frac{1}{169}\right| \cdot |3 + 4i| \cdot |5 - 12i|$$

$$= \frac{1}{169} \cdot \sqrt{3^2 + 4^2} \cdot \sqrt{5^2 + 12^2} = \frac{1}{169} \cdot \sqrt{25} \cdot \sqrt{169} = \frac{1}{169} \cdot 5 \cdot 13 = \boxed{\frac{5}{13}}.$$

(d) As computed in part (b), the magnitude of z is $|z| = 5$. The magnitude of w is $|w| = \sqrt{5^2 + (-12)^2} = \sqrt{169} = 13$. Therefore, $|z|/|w| = \boxed{5/13}$.

It appears that $|z/w| = |z|/|w|$. This holds for all complex numbers z and w, where $w \neq 0$. This is because

$$\left|\frac{z}{w}\right| \cdot |w| = \left|\frac{z}{w} \cdot w\right| = |z|.$$

Dividing both sides by $|w|$, we get $\left|\dfrac{z}{w}\right| = \dfrac{|z|}{|w|}$.

6.3.6 Let $z = a + bi$, where a and b are real numbers. Then

$$\frac{z}{\bar{z}} = \frac{a + bi}{a - bi} = \frac{(a + bi)(a + bi)}{(a - bi)(a + bi)} = \frac{(a^2 - b^2) + 2abi}{a^2 + b^2} = \frac{a^2 - b^2}{a^2 + b^2} + \frac{2ab}{a^2 + b^2}i.$$

(a) If $z/\bar{z}$ is real, then the imaginary part $2ab/(a^2 + b^2)$ must be 0. This is zero if and only if $a = 0$ or $b = 0$. Since $z = a + bi$ is real if and only if $b = 0$ and z is imaginary if and only if $a = 0$, we conclude that $z/\bar{z}$ is real if and only if $\boxed{z \text{ is real or imaginary (and not equal to 0)}}$.

(b) If $z/\bar{z}$ is imaginary, then the real part $(a^2 - b^2)/(a^2 + b^2)$ must be 0. This is zero if and only if $a^2 = b^2$, or $a = \pm b$. Therefore, $z/\bar{z}$ is imaginary if and only if z is of the form $a \pm ai = a(1 \pm i)$ for some nonzero real number a.

6.3.7 Let k be a real number such that $x = ki$ is an imaginary solution to the given equation. Substituting $x = ki$ into the left side gives

$$(ki)^4 - 3(ki)^3 + 5(ki)^2 - 27(ki) - 36 = k^4 + 3k^3 i - 5k^2 - 27ki - 36 = (k^4 - 5k^2 - 36) + (3k^3 - 27k)i.$$

If this expression equals 0, then the real part and the imaginary part equal 0, so we have the system of equations

$$k^4 - 5k^2 - 36 = 0, \qquad 3k^3 - 27k = 0.$$

The first equation factors as $(k^2 - 9)(k^2 + 4) = 0$, so $(k - 3)(k + 3)(k^2 + 4) = 0$. The second equation factors as $3k(k - 3)(k + 3) = 0$. The common roots are $k = 3$ and $k = -3$. Therefore, the imaginary solutions of the given equation are $x = \boxed{\pm 3i}$. (What are the other two roots?)

Exercises for Section 6.4

6.4.1 Note that all the coefficients are positive, so if there are any real roots, then they must be negative. Checking negative integers that divide 18, we find $g(-1) = 6$ and $g(-2) = -8$, so there's a root between -1 and -2. If a/b is a root of $g(x)$ (in lowest terms), then by the Rational Root Theorem, a must divide 18 and b must divide 2. Our only option between -1 and -2 is $-3/2$, so we try that. We find $g(-3/2) = 0$, so $2x + 3$ is a factor of $g(x)$ and $-3/2$ is a root of g. Dividing by $2x + 3$, we find $2x^3 + 5x^2 + 15x + 18 = (2x + 3)(x^2 + x + 6)$. Applying the quadratic formula to $x^2 + x + 6$ gives us the roots $x = (-1 \pm i\sqrt{23})/2$, so the roots of $g(x)$ are

$$x = \boxed{-\frac{3}{2}, \quad \frac{-1 + i\sqrt{23}}{2}, \quad \text{and} \quad \frac{-1 - i\sqrt{23}}{2}}.$$

6.4.2 First, we look for integer roots. If there are any integer roots, then they must be among the factors of 36. Checking these values, we find $f(3) = f(-4) = 0$, so $(t - 3)(t + 4)$ is a factor of $f(t)$. Dividing by $(t - 3)(t + 4)$, we find

$$2t^4 - 23t^2 + 27t - 36 = (t - 3)(t + 4)(2t^2 - 2t + 3).$$

Therefore, the roots are $t = \boxed{3, \quad -4, \quad \frac{1 + i\sqrt{5}}{2}, \quad \text{and} \quad \frac{1 - i\sqrt{5}}{2}}$.

6.4.3 The coefficients of g are all real, so nonreal solutions are present in conjugate pairs. Since $2 + i$ and $3 + i$ are roots of g, then so are $\boxed{2 - i}$ and $\boxed{3 - i}$. That gives a total of 4 roots, which is the maximum possible, so there are no other roots of g.

6.4.4 Since the nonreal roots of a polynomial with real coefficients come in conjugate pairs, the number of nonreal roots of the polynomial is even. Since the polynomial is of degree 4, the total number of roots is 4, so the number of real roots must also be even. We are given that the polynomial has at least one real root, so the polynomial must have at least two real roots.

6.4.5 Let the integer roots be r and s. We can then write the polynomial in the form

$$(x - r)(x - s)(x^2 + ax + b),$$

for some real numbers a and b. This expands as

$$x^4 + (a - r - s)x^3 + (b - ar - as + rs)x^2 + (ars - br - bs)x + brs.$$

We know that all the coefficients are integers. The coefficient of x^3 is $a - r - s$, and r and s are both integers, so a is an integer. Then the coefficient of x^2 is $b - ar - as + rs$, and ar, as, and rs are all integers, so b is also an integer.

Each of the roots listed is nonreal, so each must be a root of the quadratic $x^2 + ax + b$, if it is to be a root at all. Furthermore, the other root must be the conjugate. For example, if one root is $\frac{1+i\sqrt{11}}{2}$, then the other root must be $\frac{1-i\sqrt{11}}{2}$. The sum of these roots is 1, and their product is

$$\frac{1+i\sqrt{11}}{2} \cdot \frac{1-i\sqrt{11}}{2} = \frac{1+11}{4} = 3.$$

Hence, $\frac{1+i\sqrt{11}}{2}$ and $\frac{1-i\sqrt{11}}{2}$ are the roots of $x^2 - x + 3 = 0$. This shows $\frac{1+i\sqrt{11}}{2}$ can be a root of the original polynomial, so it must be the answer. We check the other choices to be sure:

Root	Quadratic
$\frac{1+i\sqrt{11}}{2}$	$x^2 - x + 3$
$\frac{1+i}{2}$	$x^2 - x + \frac{1}{2}$
$\frac{1}{2} + i$	$x^2 - x + \frac{5}{4}$
$1 + \frac{i}{2}$	$x^2 - 2x + \frac{5}{4}$
$\frac{1+i\sqrt{13}}{2}$	$x^2 - x + \frac{7}{2}$

The only quadratic of the form $x^2 + ax + b$ where a and b are integers is $x^2 - x + 3$, so the only possible root is $\boxed{\frac{1+i\sqrt{11}}{2}}$.

6.4.6 Note that all the coefficients are positive, so if there are any real roots, then they must be negative. First, we look for integer roots (which must be negative factors of 5), and find that there are none. Since the quartic is monic, all rational roots must be integers, so there are no rational roots either.

We know that the nonreal roots come in conjugate pairs. We also know that roots of quadratics with real coefficients come in conjugate pairs. So, we hope that we can factor the quartic as the product of quadratics. Maybe the quadratics will have integer coefficients that we can find. Let $x^4 + 5x^2 + 4x + 5 = (x^2 + ax + b)(x^2 + cx + d)$, where a, b, c, and d are integers. Expanding the right side, we get

$$x^4 + 5x^2 + 4x + 5 = x^4 + (a + c)x^3 + (b + d + ac)x^2 + (ad + bc)x + bd.$$

Equating the corresponding coefficients on both sides, we obtain the system of equations

$$a + c = 0,$$
$$b + d + ac = 5,$$
$$ad + bc = 4,$$
$$bd = 5.$$

Now, we have a number game much like factoring a quadratic. From the first equation, we have $c = -a$. The equation $bd = 5$ suggests that we try the values $b = 1$ and $d = 5$. Then from the second equation, we have $a = \pm 1$

and $c = \mp 1$. Only $a = 1, c = -1$ (together with $b = 1, d = 5$) satisfies the third equation, so we discard $a = -1, c = 1$. Hence, we have the factorization

$$x^4 + 5x^2 + 4x + 5 = (x^2 + x + 1)(x^2 - x + 5).$$

Therefore, the roots are $x = \boxed{\dfrac{-1 + i\sqrt{3}}{2}, \quad \dfrac{-1 - i\sqrt{3}}{2}, \quad \dfrac{1 + i\sqrt{19}}{2}, \quad \text{and} \quad \dfrac{1 - i\sqrt{19}}{2}}$.

Review Problems

6.29

(a) $2w - 3z = 2(2 + 3i) - 3(4 - 5i) = 4 + 6i - 12 + 15i = \boxed{-8 + 21i}$.

(b) $\dfrac{1}{w} = \dfrac{1}{2 + 3i} = \dfrac{(2 - 3i)}{(2 + 3i)(2 - 3i)} = \dfrac{2 - 3i}{2^2 + 3^2} = \dfrac{2 - 3i}{13} = \boxed{\dfrac{2}{13} - \dfrac{3}{13}i}$.

(c) We begin by computing the denominator: $\overline{w} + z = (2 - 3i) + (4 - 5i) = 6 - 8i$. Then

$$\frac{2}{\overline{w} + z} = \frac{2}{6 - 8i} = \frac{1}{3 - 4i} = \frac{3 + 4i}{(3 - 4i)(3 + 4i)} = \frac{3 + 4i}{3^2 + 4^2} = \boxed{\frac{3}{25} + \frac{4}{25}i}.$$

(d) First, we can factor the numerator and denominator:

$$\frac{w^3 + 2w^2z + wz^2}{w^2z + wz^2} = \frac{w(w^2 + 2wz + z^2)}{wz(w + z)} = \frac{w(w + z)^2}{wz(w + z)} = \frac{w + z}{z}.$$

Then

$$\frac{w + z}{z} = \frac{(2 + 3i) + (4 - 5i)}{4 - 5i} = \frac{6 - 2i}{4 - 5i} = \frac{(6 - 2i)(4 + 5i)}{(4 - 5i)(4 + 5i)} = \frac{34 + 22i}{41} = \boxed{\frac{34}{41} + \frac{22}{41}i}.$$

(e) $\dfrac{-y + xi}{x + yi} = \dfrac{(-y + xi)(x - yi)}{(x + yi)(x - yi)} = \dfrac{-xy + x^2i + y^2i + xy}{x^2 + y^2} = \dfrac{(x^2 + y^2)i}{x^2 + y^2} = \boxed{i}$. We could have also observed that $-y + xi = i(x + yi)$.

(f) Taking a cue from part (e), we can observe that $1 - i = -i(1 + i)$. Therefore,

$$\frac{(1 - i)^4}{(1 + i)^3} = \frac{(-i)^4(1 + i)^4}{(1 + i)^3} = \boxed{1 + i}.$$

6.30 Grouping the x^2 terms on one side and constants on the other gives $2x^2 = -12$. Dividing by 2 gives $x^2 = -6$. Taking the square root of both sides gives $x = \pm\sqrt{-6} = \pm\sqrt{6}\sqrt{-1} = \boxed{\pm i\sqrt{6}}$.

6.31 We evaluate the expression one step at a time, starting from the bottom. We have

$$1 - \frac{1}{1 + i} = \frac{1 + i - 1}{1 + i} = \frac{i}{1 + i},$$

but we don't bother rationalizing the denominator because we're going to take the reciprocal of this expression in the next step:

$$1 + \cfrac{1}{1 - \cfrac{1}{1 + i}} = 1 + \cfrac{1}{\cfrac{i}{1 + i}} = 1 + \frac{1 + i}{i} = \frac{i + 1 + i}{i} = \frac{1 + 2i}{i}.$$

Then

$$\cfrac{1}{1 + \cfrac{1}{1 - \cfrac{1}{1 + i}}} = \frac{1}{(1 + 2i)/i} = \frac{i}{1 + 2i} = \frac{i}{1 + 2i} \cdot \frac{1 - 2i}{1 - 2i} = \frac{i - 2i^2}{5} = \boxed{\frac{2}{5} + \frac{1}{5}i}.$$

6.32 Multiplying both sides by $1 + z$, we get $z = (-1 + i)(1 + z) = (-1 + i) + (-1 + i)z$. Then $(2 - i)z = -1 + i$, so

$$z = \frac{-1 + i}{2 - i} = \frac{(-1 + i)(2 + i)}{(2 - i)(2 + i)} = \frac{-3 + i}{5} = \boxed{-\frac{3}{5} + \frac{1}{5}i}.$$

6.33 Let $z = a + bi$, where a and b are real numbers. Then $\bar{z} = a - bi$, which has the same real part as z, but opposite imaginary part. Hence, $\bar{z}$ is the reflection of z over the real axis. Similarly, $-\bar{z} = -a + bi$ has the same imaginary part as z, but opposite real part. Hence, $-\bar{z}$ is the reflection of z over the imaginary axis. Finally, $-z = -a - bi$ has both opposite real part and opposite imaginary part as z, so $-z$ is the reflection of z through the origin. We can then label $-z$, $\bar{z}$, and $-\bar{z}$ as shown at the right.

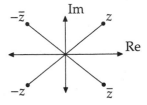

6.34 We present two solutions.

Solution 1: The faster way.

(a) We showed in the text that a complex number is real if and only if it equals its conjugate. We have

$$\overline{w\bar{z} + \bar{w}z} = \overline{w\bar{z}} + \overline{\bar{w}z} = \bar{w} \cdot \bar{\bar{z}} + \bar{\bar{w}} \cdot \bar{z} = \bar{w}z + w\bar{z} = w\bar{z} + \bar{w}z,$$

so $w\bar{z} + \bar{w}z$ is real.

(b) Similarly, we showed that a complex number is imaginary if and only if it equals the negative of its conjugate. We have

$$-(\overline{w\bar{z} - \bar{w}z}) = -\overline{w\bar{z}} + \overline{\bar{w}z} = -\bar{w} \cdot \bar{\bar{z}} + \bar{\bar{w}} \cdot \bar{z} = -\bar{w}z + w\bar{z} = w\bar{z} - \bar{w}z,$$

so $w\bar{z} - \bar{w}z$ is imaginary.

We could also have noted that the conjugate of $w\bar{z}$ is $\overline{w\bar{z}} = \bar{w} \cdot \bar{\bar{z}} = \bar{w}z$. Hence, $w\bar{z} + \bar{w}z$ is the sum of $w\bar{z}$ and its conjugate, which we know to be real. (In particular, it is twice the real part of $w\bar{z}$.)

Similarly, $w\bar{z} - \bar{w}z$ is the difference between $w\bar{z}$ and its conjugate, which we know to be imaginary. (In particular, it is $2i$ times the imaginary part of $w\bar{z}$.)

Solution 2: The slower way. Let $w = a + bi$ and $z = c + di$, where a, b, c, and d are all real.

(a) In terms of a, b, c, and d, we have

$$w\bar{z} + \bar{w}z = (a + bi)(c - di) + (a - bi)(c + di)$$
$$= ac + bd + (-ad + bc)i + ac + bd + (ad - bc)i = 2ac + 2bd,$$

which is real.

(b) In terms of a, b, c, and d, we have

$$w\bar{z} - \bar{w}z = (a + bi)(c - di) - (a - bi)(c + di)$$
$$= ac + bd + (-ad + bc)i - (ac + bd) - (ad - bc)i = (-2ad + 2bc)i,$$

which is imaginary.

6.35   Let $z = a + bi$ for some real numbers a and b. If $z^2 = 5 - 12i$, then $5 - 12i = (a + bi)^2 = a^2 - b^2 + 2abi$. Equating the real parts and equating the imaginary parts, we get the system of equations $a^2 - b^2 = 5$ and $2ab = -12$. From the second equation, we have $b = -6/a$. Substituting this expression for b into the first equation, we get

$$a^2 - \frac{36}{a^2} = 5 \quad \Rightarrow \quad a^4 - 5a^2 - 36 = 0 \quad \Rightarrow \quad (a^2 - 9)(a^2 + 4) = 0.$$

Since a is real, we have $a = \pm 3$. The corresponding values of $b = -6/a$ are ∓ 2. (The $\mp$ symbol indicates that $b = -2$ goes with $a = 3$ and $b = 2$ goes with $a = -3$.) Therefore, the square roots of $5 - 12i$ are $\boxed{3 - 2i \text{ and } -3 + 2i}$.

6.36   If $7 + i$ is 5 units from $10 + ci$, then from the distance formula, we have

$$\sqrt{(7 - 10)^2 + (1 - c)^2} = 5 \quad \Rightarrow \quad 9 + (c - 1)^2 = 25 \quad \Rightarrow \quad (c - 1)^2 = 16 \quad \Rightarrow \quad c - 1 = \pm 4,$$

so $c = \boxed{5 \text{ or } -3}$.

6.37   The value of $|z - (4 - 5i)|$ is the distance between z and $4 - 5i$ in the complex plane. The equation $|z - (4 - 5i)| = 2\sqrt{3}$ tells us that this distance is $2\sqrt{3}$. Therefore, the graph of $|z - (4 - 5i)| = 2\sqrt{3}$ is a circle centered at $4 - 5i$ with radius $2\sqrt{3}$, which has area $(2\sqrt{3})^2\pi = \boxed{12\pi}$.

6.38   Let $z = a + bi$ be the complex number corresponding to the point F, where a and b are real numbers. From the diagram, we can say two things about z. First, since F lies to the right of the imaginary axis and above the real axis, $a > 0$ and $b > 0$. Second, F lies outside the unit circle, so $|z| > 1$. We must deduce similar facts about $1/z$.

First we express $\frac{1}{z}$ as a complex number:

$$\frac{1}{z} = \frac{1}{a + bi} = \frac{a - bi}{(a + bi)(a - bi)} = \frac{a - bi}{a^2 + b^2} = \frac{a}{a^2 + b^2} - \frac{b}{a^2 + b^2}i.$$

Since a and b are positive, the real part of $1/z$ is positive, and the imaginary part is negative. Therefore, $1/z$ lies to the right of the imaginary axis, and below the real axis. This means $1/z$ must be A or C.

Next, we see that $\left|\dfrac{1}{z}\right| = \dfrac{1}{|z|} < 1$. Hence, $1/z$ lies inside the unit circle. Therefore, the reciprocal of F is $\boxed{C}$.

6.39   Let $z = a + bi$, where a and b are real numbers, so the equation becomes

$$|a + 1 + (b - 1)i| = |a - 2 + bi|$$
$$\Rightarrow \quad \sqrt{(a + 1)^2 + (b - 1)^2} = \sqrt{(a - 2)^2 + b^2}$$
$$\Rightarrow \quad (a + 1)^2 + (b - 1)^2 = (a - 2)^2 + b^2$$
$$\Rightarrow \quad a^2 + 2a + 1 + b^2 - 2b + 1 = a^2 - 4a + 4 + b^2$$
$$\Rightarrow \quad 6a = 2b + 2$$
$$\Rightarrow \quad b = 3a - 1.$$

Hence, all solutions are of the form $z = \boxed{a + (3a - 1)i}$, where a is any real number.

Note that we could also have solved this problem by noting that $|z + 1 - i|$ can be written as $|z - (-1 + i)|$, so this expression represents the distance from z to $-1 + i$ on the complex plane. The expression $|z - 2|$ equals the distance from z to 2, so the solutions to the equation $|z + 1 - i| = |z - 2|$ represent all z equidistant from $-1 + i$ and 2 on the complex plane. Therefore, z must be on the perpendicular bisector of the segment connecting $-1 + i$ and 2 on the complex plane. This segment has slope $-\frac{1}{3}$, and its midpoint is $\frac{1}{2} + \frac{i}{2}$. So, z is on the line through $\frac{1}{2} + \frac{i}{2}$ with slope 3. See if you can finish from here to get the same answer as above.

6.40   We are given that $(a + bi)(c + di) = ac - bd + (ad + bc)i$ is real, so $ad + bc = 0$. Then

$$(b + ai)(d + ci) = bd - ac + (bc + ad)i = bd - ac + (ad + bc)i = bd - ac,$$

so the product of $b + ai$ and $d + ci$ is also real.

6.41

(a) Since $(z - \bar{z})/(2i)$ equals $\operatorname{Im}(z)$ for all z, we divide the given equation by $2i$ to get $(z - \bar{z})/(2i) = -4$, which we can write as $\operatorname{Im}(z) = -4$. The graph is a horizontal line 4 units below the real axis, shown to the left below.

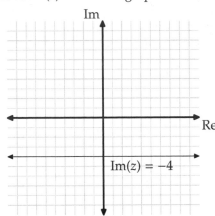

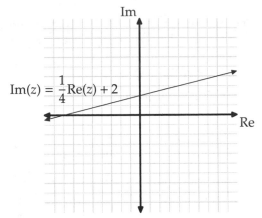

(b) Let $a = \operatorname{Re}(z)$ and $b = \operatorname{Im}(z)$. Then the given equation becomes

$$(4 - i)(a + bi) - (4 + i)(a - bi) = 16i$$
$$\Rightarrow \quad 4a + b - ai + 4bi - 4a - b - ai + 4bi = 16i$$
$$\Rightarrow \quad -2ai + 8bi = 16i$$
$$\Rightarrow \quad -a + 4b = 8$$
$$\Rightarrow \quad b = \frac{a}{4} + 2.$$

Thus, the graph is a line with slope $1/4$ that intersects the imaginary axis at $2i$, shown to the right above.

(c) The left side can be rewritten as $|7 + i - 2z| = |-(7 + i - 2z)| = |2z - 7 - i|$, so $|2z - 7 - i| = 4$. Dividing both sides by 2, we get

$$\left| z - \frac{7 + i}{2} \right| = 2.$$

This equation tells us that the distance from z to $(7 + i)/2$ on the complex plane is 2. Therefore, the graph is a circle centered at $\frac{7}{2} + \frac{1}{2}i$ with radius 2.

6.42 For a complex number z, $|z - 5 + 2i| = |z - (5 - 2i)|$ is the distance between z and $5 - 2i$. Hence, the graph of $|z - (5 - 2i)| \le 4$ is the set of all complex numbers z that are within 4 units of $5 - 2i$. This is the closed disk (which is both the circle and its interior) centered at $5 - 2i$ with radius 4, shown at right.

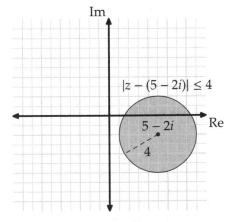

6.43 To deal with such large exponents, we compute the first few powers of $i + 1$ to see if we can find a pattern:

$$(i + 1)^1 = i + 1,$$
$$(i + 1)^2 = i^2 + 2i + 1 = 2i,$$
$$(i + 1)^3 = (i + 1)^2 \cdot (i + 1) = 2i(i + 1) = -2 + 2i,$$
$$(i + 1)^4 = (i + 1)^3 \cdot (i + 1) = (-2 + 2i)(i + 1) = -4.$$

We note that $(1 + i)^4$ is a real number, which makes it easy to work with. Similarly, we find that $(i - 1)^4 = -4$. Therefore,

$$(i + 1)^{3200} - (i - 1)^{3200} = [(i + 1)^4]^{800} - [(i - 1)^4]^{800} = (-4)^{800} - (-4)^{800} = \boxed{0}.$$

6.44 Let $z = a + bi$, for some real numbers a and b. We are given that $|z| = 1$, so $|z| = |a + bi| = \sqrt{a^2 + b^2} = 1$, or $a^2 + b^2 = 1$. Then

$$|z - 1|^2 + |z + 1|^2 = |(a - 1) + bi|^2 + |(a + 1) + bi|^2 = (a - 1)^2 + b^2 + (a + 1)^2 + b^2$$
$$= a^2 - 2a + 1 + b^2 + a^2 + 2a + 1 + b^2 = 2a^2 + 2b^2 + 2$$
$$= 2(a^2 + b^2) + 2 = 2 + 2 = 4.$$

Alternatively, we could have used the fact that $z\bar{z} = |z|^2$ for all complex numbers z:

$$|z - 1|^2 + |z + 1|^2 = (z - 1) \cdot \overline{(z - 1)} + (z + 1) \cdot \overline{(z + 1)} = (z - 1)(\bar{z} - 1) + (z + 1)(\bar{z} + 1)$$
$$= z\bar{z} - z - \bar{z} + 1 + z\bar{z} + z + \bar{z} + 1 = 2z\bar{z} + 2 = 2|z|^2 + 2 = 4.$$

Finally, can you find a geometric interpretation of the result?

6.45

(a) We have

$$|w - z|^2 + |w + z|^2 = (w - z)(\bar{w} - \bar{z}) + (w + z)(\bar{w} + \bar{z})$$
$$= w\bar{w} - w\bar{z} - z\bar{w} + z\bar{z} + w\bar{w} + w\bar{z} + z\bar{w} + z\bar{z}$$
$$= 2w\bar{w} + 2z\bar{z} = 2(|w|^2 + |z|^2).$$

(b) Consider the parallelogram in the complex plane with vertices at 0, w, z, and $w + z$. The distance from w to z is $|w - z|$, so the expression $|w - z|^2 + |w + z|^2$ is the sum of the squares of the diagonals. The sides of the parallelogram from the origin have lengths $|w|$ and $|z|$, and opposite sides of any parallelogram are equal, so $2(|w|^2 + |z|^2)$ is the sum of the squares of the sides of the parallelogram. Therefore, the equation $2(|w|^2 + |z|^2) = |w - z|^2 + |w + z|^2$ expresses the fact that the sum of the squares of the sides of a parallelogram equals the sum of the squares of the diagonals.

6.46 Since the coefficients of the polynomial are real, the complex conjugate of $-3 + 5i$, namely $-3 - 5i$, must also be a root. The sum of $-3 + 5i$ and $-3 - 5i$ is -6, and their product is $(-3 + 5i)(-3 - 5i) = 9 + 25 = 34$. Therefore, the quadratic whose roots are $-3 + 5i$ and $-3 - 5i$ is $x^2 + 6x + 34$. Since this quadratic must be a factor of the given quartic, we have

$$x^4 + 9x^3 + 48x^2 + 78x - 136 = (x^2 + 6x + 34)(ax^2 + bx + c)$$

for some constants a, b, and c. By considering the x^4 term in the expansion of the right side, we find $a = 1$. By considering the constant term in the expansion on the right side, we have $34c = -136$, from which we find $c = -4$. Finally, we consider the x^3 term on both sides. We have $6ax^3 + bx^3$ on the right, which must equal the $9x^3$ term on the left. We already have $a = 1$, so $b = 3$. We must check that the other coefficients match up. Checking reveals that we do indeed have

$$(x^2 + 6x + 34)(x^2 + 3x - 4) = x^4 + 9x^3 + 48x^2 + 78x - 136.$$

Therefore, the other two roots of the quartic are the roots of the quadratic $x^2 + 3x - 4 = (x - 1)(x + 4)$. So, the roots of the quartic other than $-3 + 5i$ are $\boxed{-3 - 5i, 1, \text{ and } -4}$.

6.47

(a) We are given that the polynomial has five distinct roots, but a polynomial of degree 6 has six roots (counting repeated roots as different roots), so one of the roots must be repeated.

(b) Since the coefficients of the polynomial are real, the nonreal roots must come in conjugate pairs. The nonreal roots $2 - i$ and $2 + i$ already form a conjugate pair, as do the nonreal roots $1 - i$ and $1 + i$. We also know that 1 is a root, so the sixth and final root must be real. Therefore, the double root is $\boxed{1}$.

6.48   We have $r_1 i = -b + ai$ and $r_2 i = b + ai$. Since the polynomial has real coefficients, the complex conjugate of $r_1 i = -b + ai$, namely $-b - ai$, must also be a root. Since a and b are nonzero, $-b - ai$ cannot be equal to $r_2 i = b + ai$. Similarly, the conjugate of $r_2 i = b + ai$, namely $b - ai$, must also be a root. We have shown that the polynomial must have at least four roots. Furthermore, the polynomial

$$(x + b - ai)(x + b + ai)(x - b - ai)(x - b + ai) = [(x + b)^2 + a^2][(x - b)^2 + a^2]$$

has real coefficients, and its roots are $-b + ai$, $-b - ai$, $b + ai$, and $b - ai$. Therefore, the minimum degree of the polynomial is $\boxed{4}$.

Challenge Problems

6.49   The powers of i repeat every four terms: $i^0 = 1$, $i^1 = i$, $i^2 = -1$, $i^3 = -i$, $i^4 = 1$, $i^5 = i$, $i^6 = -1$, $i^7 = -i$, $i^8 = 1$, and so on. Hence, the given sum becomes

$$i + 2i^2 + 3i^3 + 4i^4 + \cdots + 64i^{64}$$
$$= i - 2 - 3i + 4 + 5i - 6 - 7i + 8 + 9i - 10 - 11i + 12 + \cdots + 61i - 62 - 63i + 64$$
$$= (-2 + 4 - 6 + 8 - 10 + 12 + \cdots - 62 + 64) + (1 - 3 + 5 - 7 + 9 - 11 + \cdots + 61 - 63)i$$
$$= [(-2 + 4) + (-6 + 8) + (-10 + 12) + \cdots + (-62 + 64)] + [(1 - 3) + (5 - 7) + (9 - 11) + \cdots + (61 - 63)]i$$
$$= 2 \cdot 16 - 2 \cdot 16i = \boxed{32 - 32i}.$$

6.50

(a)   Taking the conjugate of the given equation, we get $\bar{b}z + \bar{a}\bar{z} = \bar{c}$. We can view the equations

$$az + b\bar{z} = c,$$
$$\bar{b}z + \bar{a}\bar{z} = \bar{c}$$

as a system of equations in z and $\bar{z}$, and solve for z accordingly. Multiplying the first equation by $\bar{a}$ and the second equation by b, we get

$$a\bar{a}z + \bar{a}b\bar{z} = \bar{a}c,$$
$$b\bar{b}z + \bar{a}b\bar{z} = b\bar{c}.$$

Subtracting the second equation from the first gives us

$$(a\bar{a} - b\bar{b})z = \bar{a}c - b\bar{c}.$$

We have $a\bar{a} - b\bar{b} = |a|^2 - |b|^2$, and we are given that $|a| \neq |b|$. Hence, we may divide both sides by $a\bar{a} - b\bar{b}$, to get

$$z = \frac{\bar{a}c - b\bar{c}}{a\bar{a} - b\bar{b}}.$$

We multiplied by $\bar{a}$ and by b, and one of these might be 0. So, we have to check our solution and make sure it isn't extraneous. If $z = \dfrac{\bar{a}c - b\bar{c}}{a\bar{a} - b\bar{b}}$, then $\bar{z} = \dfrac{a\bar{c} - \bar{b}c}{a\bar{a} - b\bar{b}}$, so

$$az + b\bar{z} = \frac{a\bar{a}c - ab\bar{c}}{a\bar{a} - b\bar{b}} + \frac{ab\bar{c} - b\bar{b}c}{a\bar{a} - b\bar{b}} = \frac{a\bar{a}c - b\bar{b}c}{a\bar{a} - b\bar{b}} = c.$$

(b) Let $z = x + yi$, where x and y are real numbers. Then the equation $az + b\overline{z} = c$ becomes

$$(-1 + 3i)(x + yi) + (1 + 3i)(x - yi) = i$$
$$\Leftrightarrow \quad -x - yi + 3xi - 3y + x - yi + 3xi + 3y = i$$
$$\Leftrightarrow \quad -2yi + 6xi = i.$$

Solving for y, we find $y = 3x - \frac{1}{2}$. Thus, the solutions are of the form $x + \left(3x - \frac{1}{2}\right)i$ for all real numbers x. This represents a line passing through the point corresponding to $-\frac{1}{2}i$ with slope 3.

6.51 Let $z = a + bi$, where a and b are real numbers. From the equation $|z - 1| = |z + 3|$, we obtain

$$|(a - 1) + bi| = |(a + 3) + bi|$$
$$\Rightarrow \quad \sqrt{(a - 1)^2 + b^2} = \sqrt{(a + 3)^2 + b^2}$$
$$\Rightarrow \quad (a - 1)^2 + b^2 = (a + 3)^2 + b^2$$
$$\Rightarrow \quad a^2 - 2a + 1 + b^2 = a^2 + 6a + 9 + b^2$$
$$\Rightarrow \quad 8a = -8$$
$$\Rightarrow \quad a = -1.$$

Then from the equations $|z - 1| = |z - i|$ and $a = -1$, we obtain

$$|-2 + bi| = |-1 + (b - 1)i|$$
$$\Rightarrow \quad \sqrt{(-2)^2 + b^2} = \sqrt{(-1)^2 + (b - 1)^2}$$
$$\Rightarrow \quad 4 + b^2 = 1 + (b - 1)^2$$
$$\Rightarrow \quad 4 + b^2 = 1 + b^2 - 2b + 1$$
$$\Rightarrow \quad 2b = -2$$
$$\Rightarrow \quad b = -1.$$

Therefore, $z = a + bi = \boxed{-1 - i}$.

We can interpret the problem geometrically as follows. The equations $|z - 1| = |z + 3| = |z - i|$ state that the complex number z is equidistant from the numbers 1, -3, and i. Hence, z is the circumcenter of the triangle formed by the complex numbers 1, -3, and i in the complex plane. From $|z - 1| = |z + 3|$, we find that z is on the perpendicular bisector of the segment connecting 1 and -3 on the complex plane. This perpendicular bisector is $\text{Re}(z) = -1$, so the real part of z is -1. From $|z - 1| = |z - i|$, we find that z is on the perpendicular bisector of the segment connecting 1 and i on the complex plane. This line is simply $\text{Re}(z) = \text{Im}(z)$. Combining this with $\text{Re}(z) = -1$, we have $z = -1 - i$, as before.

6.52 The first inequality is satisfied when z is *inside or on* the circle centered at $-4 + 4i$ with radius $4\sqrt{2}$, and the second inequality is satisfied when z is *outside or on* the circle centered at $4 + 4i$ with radius $4\sqrt{2}$. Hence, the region of interest is the set of points that lie inside or on the first circle but outside or on the second circle, as shown below.

Let A be the point $-4 + 4i$, the center of the first circle, and let B be the point $4 + 4i$, the center of the second circle. Then

$$AB = |(4 + 4i) - (-4 + 4i)| = |8| = 8.$$

Let M be the midpoint of $\overline{AB}$, so $AM = BM = AB/2 = 4$.

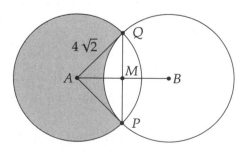

Let the two circles intersect at P and Q. Since $\overline{PQ}$ is a common chord of the two circles, $\overline{PQ}$ is perpendicular to $\overline{AB}$. Furthermore, M is also the midpoint of $\overline{PQ}$.

By the Pythagorean Theorem, $AM^2 + QM^2 = AQ^2$, so $16 + QM^2 = (4\sqrt{2})^2 = 32$. Therefore, $QM^2 = 16$, so $QM = 4$. Then $PQ = 2QM = 8$.

Now, to find the area of the shaded region, we can take the area of the first circle and subtract two times the area of the circular segment cut off by $\overline{PQ}$. To find the area of the circular segment cut off by $\overline{PQ}$, we can find the area of sector PAQ, and subtract the area of triangle PAQ.

Since $AM = QM$ and $\angle AMQ = 90°$, we have $\angle QAM = 45°$. Also, $AM = PM$ and $\angle AMP = 90°$, so $\angle PAM = 45°$. Therefore, $\angle PAQ = 45° + 45° = 90°$. Hence, the area of sector PAQ is one quarter the area of the circle, or $\frac{1}{4}\pi(4\sqrt{2})^2 = 8\pi$. The area of triangle PAQ is $\frac{1}{2}AM \cdot PQ = \frac{1}{2} \cdot 4 \cdot 8 = 16$. Hence, the area of the circular segment cut off by $\overline{PQ}$ is $8\pi - 16$.

Finally, the area of the shaded region is $\pi(4\sqrt{2})^2 - 2(8\pi - 16) = \boxed{16\pi + 32}$.

6.53 To deal with the appearances of i in the coefficients, we let $y = ix$. Then the given cubic becomes

$$iy^3 - 8y^2 - 22iy + 21 = i(ix)^3 - 8(ix)^2 - 22i(ix) + 21 = i(-ix^3) - 8(-x^2) - 22i(ix) + 21 = x^3 + 8x^2 + 22x + 21.$$

We look for roots of this new polynomial. First, we look for integer roots. If there are any integer roots, then they must be negative (because all the coefficients are positive) and among the factors of 21 (due to the Rational Root Theorem). Checking these values, we find that the cubic becomes 0 for $x = -3$. So by the Factor Theorem, $x + 3$ is a factor of the cubic. Dividing by $x + 3$, we find

$$x^3 + 8x^2 + 22x + 21 = (x + 3)(x^2 + 5x + 7).$$

The roots of the quadratic are $\frac{-5 \pm i\sqrt{3}}{2}$. Since $y = ix$, the solutions of the original cubic are

$$y = \boxed{-3i, \quad \frac{\sqrt{3} - 5i}{2}, \quad \text{and} \quad \frac{-\sqrt{3} - 5i}{2}}.$$

6.54 Let w, x, y, and z be the vertices of a parallelogram in the complex plane, where w and y are opposite vertices, so x and z are also opposite vertices. In every parallelogram, the diagonals bisect each other; in other words, the midpoints of each diagonal coincide. Therefore,

$$\frac{w + y}{2} = \frac{x + z}{2} \quad \Rightarrow \quad w + y = x + z.$$

Thus, the sum of w and y is equal to the sum of x and z.

Conversely, let the sum of two of w, x, y, and z be equal to the sum of the other two. Suppose $w + y = x + z$. Then $(w + y)/2 = (x + z)/2$. This means the midpoint of the line segment between w and y and the midpoint of the line segment between x and z coincide. In other words, in the quadrilateral formed by the complex numbers w, x, y, and z, the diagonals bisect each other. Hence, the quadrilateral formed by these four points is a parallelogram.

Make sure you see why we need to include both parts of this proof.

6.55 Let $z = \frac{4uv}{(u+v)^2}$. To show that z is real, it suffices to show that $z = \bar{z}$ (as described in the text), so we compute the conjugate of z. Since $|u| = 1$, we have $u\bar{u} = |u|^2 = 1$, so $\bar{u} = 1/u$. Similarly, $\bar{v} = 1/v$. Then

$$\bar{z} = \overline{\left(\frac{4uv}{(u+v)^2}\right)} = \frac{\overline{4uv}}{(u+v)^2} = \frac{4\bar{u}\cdot\bar{v}}{(\bar{u}+\bar{v})^2} = \frac{4\cdot\frac{1}{u}\cdot\frac{1}{v}}{\left(\frac{1}{u}+\frac{1}{v}\right)^2}.$$

Multiplying the numerator and denominator by u^2v^2, this becomes $\bar{z} = \frac{4uv}{(u+v)^2} = z$. Therefore, $z = 4uv/(u + v)^2$ is real.

Now, let $w = \dfrac{u + v}{u - v}$. To show that w is imaginary, it suffices to show that $w = -\bar{w}$. We have

$$\bar{w} = \overline{\left(\frac{u+v}{u-v}\right)} = \frac{\overline{u+v}}{\overline{u-v}} = \frac{\bar{u}+\bar{v}}{\bar{u}-\bar{v}} = \frac{\frac{1}{u}+\frac{1}{v}}{\frac{1}{u}-\frac{1}{v}} = \frac{\frac{1}{u}+\frac{1}{v}}{\frac{1}{u}-\frac{1}{v}} \cdot \frac{uv}{uv} = \frac{v+u}{v-u} = -w.$$

Therefore, $w = (u + v)/(u - v)$ is imaginary.

6.56 We have seen the expression $x^2 + y^2$ arise when computing the reciprocal of the complex number $z = x + yi$, where x and y are real numbers:
$$\frac{1}{z} = \frac{1}{x + yi} = \frac{x - yi}{(x + yi)(x - yi)} = \frac{x - yi}{x^2 + y^2}.$$

Thus, it may be possible to express the given equations in terms of the complex number $z = x + yi$. We are given that $x/(x^2 + y^2) = 33x - 56y$ and $56x + 33y = -y/(x^2 + y^2)$, so
$$\frac{1}{x + yi} = \frac{x}{x^2 + y^2} - \frac{y}{x^2 + y^2}i = (33x - 56y) + (56x + 33y)i.$$

But $(x + yi)(33 + 56i) = 33x + 56xi + 33yi - 56y = (33x - 56y) + (56x + 33y)i$, so using our expression for $\frac{1}{x+yi}$ from above, we have
$$\frac{1}{x + yi} = (33x - 56y) + (56x + 33y)i = (x + yi)(33 + 56i).$$

Hence,
$$(x + yi)^2 = \frac{1}{33 + 56i} = \frac{33 - 56i}{(33 + 56i)(33 - 56i)} = \frac{33 - 56i}{4225} = \frac{33 - 56i}{65^2}.$$

Therefore, $[65(x + yi)]^2 = (65x + 65yi)^2 = 33 - 56i$.

Let $a = 65x$ and $b = 65y$, so $(a + bi)^2 = 33 - 56i$. Expanding, we get $a^2 - b^2 + 2abi = 33 - 56i$. Equating real parts and equating imaginary parts, we obtain the system of equations
$$a^2 - b^2 = 33,$$
$$2ab = -56.$$

From the second equation, we have $b = -28/a$. Substituting into the first equation, we get $a^2 - \frac{784}{a^2} = 33$. Rearranging, we get $a^4 - 33a^2 - 784 = 0$, which factors as $(a^2 - 49)(a^2 + 16) = 0$, so $(a - 7)(a + 7)(a^2 + 16) = 0$. Since a is real, we have $a = 7$ or $a = -7$.

If $a = 7$, then $b = -28/7 = -4$, $x = a/65 = 7/65$, and $y = b/65 = -4/65$, so $|x| + |y| = 11/65$. If $a = -7$, then $b = -28/(-7) = 4$, $x = a/65 = -7/65$, and $y = b/65 = 4/65$, so $|x| + |y| = 11/65$. In either case, $|x| + |y| = \boxed{11/65}$.

6.57 We compute the first few terms of the sequence:
$$z_2 = z_1^2 + i = i,$$
$$z_3 = z_2^2 + i = -1 + i,$$
$$z_4 = z_3^2 + i = (-1 + i)^2 + i = 1 - 2i - 1 + i = -i,$$
$$z_5 = z_4^2 + i = (-i)^2 + i = -1 + i,$$
$$z_6 = z_5^2 + i = (-1 + i)^2 + i = 1 - 2i - 1 + i = -i,$$

and so on.

We observe that the sequence repeats every two terms beginning with z_3. We know that this pattern persists, because each term in the sequence depends only on the previous term. We conclude that $z_{111} = z_3 = -1 + i$. Therefore, the distance between z_{111} and the origin is $|-1 + i| = \sqrt{(-1)^2 + 1^2} = \boxed{\sqrt{2}}$.

6.58 We can produce $w_1^2 + w_2^2 + w_3^2$ by squaring the first given equation, which gives
$$w_1^2 + w_2^2 + w_3^2 + 2(w_1w_2 + w_2w_3 + w_1w_3) = 0.$$

Solving for $w_1^2 + w_2^2 + w_3^2$ gives

$$w_1^2 + w_2^2 + w_3^2 = -2(w_1w_2 + w_2w_3 + w_1w_3).$$

Now, we wish to show that $w_1w_2 + w_2w_3 + w_1w_3 = 0$. We let $t = w_1w_2 + w_2w_3 + w_1w_3$. We know we'll need to use the facts that $|w_1| = |w_2| = |w_3| = 1$, which we can write as $w_1\overline{w_1} = w_2\overline{w_2} = w_3\overline{w_3} = 1^2 = 1$, to get rid of the magnitudes. To make use of these equations, we consider $t\bar{t}$:

$$t\bar{t} = (w_1w_2 + w_2w_3 + w_1w_3)\overline{(w_1w_2 + w_2w_3 + w_1w_3)} = (w_1w_2 + w_2w_3 + w_1w_3)(\overline{w_1w_2} + \overline{w_2w_3} + \overline{w_1w_3})$$

$$= (w_1\overline{w_1})(w_2\overline{w_2}) + w_1(w_2\overline{w_2})\overline{w_3} + (w_1\overline{w_1})w_2\overline{w_3}$$
$$+ (w_2\overline{w_2})\overline{w_1}w_3 + (w_2\overline{w_2})(w_3\overline{w_3}) + \overline{w_1}w_2(w_3\overline{w_3})$$
$$+ (w_1\overline{w_1})\overline{w_2}w_3 + w_1\overline{w_2}(w_3\overline{w_3}) + (w_1\overline{w_1})(w_3\overline{w_3}).$$

We have $w_1\overline{w_1} = w_2\overline{w_2} = w_3\overline{w_3} = 1$, so

$$t\bar{t} = 3 + w_1\overline{w_3} + w_2\overline{w_3} + \overline{w_1}w_3 + \overline{w_1}w_2 + \overline{w_2}w_3 + w_1\overline{w_2}$$
$$= 3 + \overline{w_3}(w_1 + w_2) + \overline{w_1}(w_2 + w_3) + \overline{w_2}(w_1 + w_3).$$

Since $w_1 + w_2 + w_3 = 0$, we can write the three expressions in parentheses as $-w_3$, $-w_1$, and $-w_2$, respectively, and we have

$$t\bar{t} = 3 - \overline{w_3}w_3 - \overline{w_1}w_1 - \overline{w_2}w_2 = 3 - 1 - 1 - 1 = 0.$$

Therefore, we have $|t| = 0$, so $t = 0$, and we conclude that $w_1^2 + w_2^2 + w_3^2 = 0$.

6.59 From the equation $|a + bi| = 8$, we get $\sqrt{a^2 + b^2} = 8$, so $a^2 + b^2 = 64$.

We are also given that the image of each point in the complex plane is equidistant from that point and the origin. In other words, for all complex numbers z, we have $|f(z) - z| = |f(z)|$. Substituting $f(z) = (a + bi)z$, we get $|(a - 1 + bi)z| = |(a + bi)z|$, which we can rewrite as $|a - 1 + bi| \cdot |z| = |a + bi| \cdot |z|$.

Since this relation holds for all complex numbers z, the coefficients of $|z|$ must be equal, so

$$|a - 1 + bi| = |a + bi| \quad \Rightarrow \quad \sqrt{(a-1)^2 + b^2} = \sqrt{a^2 + b^2} \quad \Rightarrow \quad (a-1)^2 + b^2 = a^2 + b^2 \quad \Rightarrow \quad a = \frac{1}{2}.$$

Hence, $b^2 = 64 - a^2 = 64 - \left(\frac{1}{2}\right)^2 = \boxed{\frac{255}{4}}$.

Can you find a geometric approach to this problem?

6.60

(a) If $z \neq 0$, then $f(f(z)) = f\left(\dfrac{1}{z}\right) = \dfrac{1}{1/\bar{z}} = \dfrac{1}{1/z} = z$.

(b) The problem asks us to find what curve $f(z)$ traces, so we let $w = f(z)$ and try to find an equation for w that doesn't include z. Since $w = f(z) = 1/\bar{z}$, we have $\bar{z} = 1/w$. Taking the conjugate of both sides, we get $\bar{\bar{z}} = 1/\overline{w}$, so $z = 1/\overline{w}$. Substituting these expressions into the equation $(1 + 2i)z - (1 - 2i)\bar{z} = i$, we get

$$\frac{1 + 2i}{\overline{w}} - \frac{1 - 2i}{w} = i \quad \Rightarrow \quad (1 + 2i)w - (1 - 2i)\overline{w} = iw\overline{w}.$$

Let $w = a + bi$, where a and b are real numbers. Then the equation above becomes

$$(1 + 2i)(a + bi) - (1 - 2i)(a - bi) = i(a + bi)(a - bi)$$
$$\Rightarrow \quad (a - 2b) + (2a + b)i - (a - 2b) + (2a + b)i = (a^2 + b^2)i$$
$$\Rightarrow \quad (4a + 2b)i = (a^2 + b^2)i$$
$$\Rightarrow \quad 4a + 2b = a^2 + b^2$$
$$\Rightarrow \quad a^2 - 4a + b^2 - 2b = 0$$
$$\Rightarrow \quad a^2 - 4a + 4 + b^2 - 2b + 1 = 5$$
$$\Rightarrow \quad (a - 2)^2 + (b - 1)^2 = 5.$$

Hence, it appears that w traces a circle centered at $2 + i$ with radius $\sqrt{5}$. However, we must be careful to note that w cannot be 0, since there is no complex number z for which $0 = 1/\bar{z}$. So, w traces the entire circle centered at $2 + i$ with radius $\sqrt{5}$, except the origin.

6.61 We have seen the expression $a^2 + b^2$ arise when computing the reciprocal of the complex number $z = a + bi$, where a and b are real numbers:

$$\frac{1}{z} = \frac{1}{a + bi} = \frac{a - bi}{(a + bi)(a - bi)} = \frac{a - bi}{a^2 + b^2}.$$

Thus, it may be possible to express the given equations in terms of the complex number $z = a + bi$.

We see that the first equation contains the term a, which is the real part of $a + bi$. The second equation contains the term b, which is the imaginary part of $a + bi$. Furthermore, the first equation contains the term $\frac{a}{a^2+b^2}$, which is the real part of $1/z$, as seen above, and the second equation contains the term $-\frac{b}{a^2+b^2}$, which is the imaginary part of $1/z$.

Hence, by taking the expression in the first equation as the real part of a complex number, and the expression in the second equation as the imaginary part, we get

$$2 + 0i = \left(a + \frac{a + 8b}{a^2 + b^2}\right) + \left(b + \frac{8a - b}{a^2 + b^2}\right)i = a + bi + \frac{a - bi}{a^2 + b^2} + \frac{8(b + ai)}{a^2 + b^2} = z + \frac{1}{z} + \frac{8(b + ai)}{a^2 + b^2}, \qquad (\spadesuit)$$

where $z = a + bi$. This leaves us with the task of expressing $\dfrac{8(b + ai)}{a^2 + b^2}$ in terms of z.

Since $i(a - bi) = b + ai$ and $\dfrac{1}{z} = \dfrac{a - bi}{a^2 + b^2}$, we have $\dfrac{8(b + ai)}{a^2 + b^2} = \dfrac{8i(a - bi)}{a^2 + b^2} = \dfrac{8i}{z}$. Thus, we can write $(\spadesuit)$ as $z + \dfrac{1}{z} + \dfrac{8i}{z} = 2$.

Multiplying both sides by z, we get $z^2 + 1 + 8i = 2z$. Rearranging this equation gives $z^2 - 2z + 1 = -8i$, so $(z - 1)^2 = -8i$. Hence, we must find the numbers whose squares are $-8i$.

Let $w = c + di$, where c and d are real numbers and $w^2 = -8i$. Then, we have $w^2 = (c + di)^2$, so $c^2 - d^2 + 2cdi = -8i$. Equating the real parts and equating the imaginary parts, we get the system of equations $c^2 - d^2 = 0$ and $2cd = -8$. From the second equation, we have $d = -4/c$. Substituting this expression for d into the first equation, we get

$$c^2 = \frac{16}{c^2} \quad \Rightarrow \quad c^4 = 16 \quad \Rightarrow \quad c = \pm 2.$$

(We cannot have $c = 2i$ or $c = -2i$ because c is real.) The corresponding values of $d = -4/c$ are ∓ 2. (The $\mp$ symbol indicates that $d = -2$ goes with $c = 2$ and $d = 2$ goes with $c = -2$.) Therefore, the square roots of $-8i$ are $2 - 2i$ and $-2 + 2i$.

Hence, $z - 1 = 2 - 2i$ or $z - 1 = -2 + 2i$, so $z = 3 - 2i$ or $z = -1 + 2i$. The corresponding values of (a, b) are $(3, -2)$ and $(-1, 2)$. We check that both of these solutions work, so the solutions are $(a, b) = \boxed{(3, -2) \text{ and } (-1, 2)}$.

7

CHAPTER

Trigonometry and Complex Numbers

Exercises for Section 7.1

7.1.1

(a) The real part is 0 and the imaginary part is -32, so $r = \sqrt{0^2 + (-32)^2} = 32$. We seek an angle θ such that $32\cos\theta = 0$ and $32\sin\theta = -32$, which means $\cos\theta = 0$ and $\sin\theta = -1$. We can take $\theta = \frac{3\pi}{2}$, so $-32i = \boxed{32\left(\cos\frac{3\pi}{2} + i\sin\frac{3\pi}{2}\right)}$.

(b) The real part is 2 and the imaginary part is -2, so $r = \sqrt{2^2 + (-2)^2} = \sqrt{8} = 2\sqrt{2}$. We seek an angle θ such that $2\sqrt{2}\cos\theta = 2$ and $2\sqrt{2}\sin\theta = -2$, which means $\cos\theta = \frac{\sqrt{2}}{2}$ and $\sin\theta = -\frac{\sqrt{2}}{2}$. We can take $\theta = \frac{7\pi}{4}$, so $2 - 2i = \boxed{2\sqrt{2}\left(\cos\frac{7\pi}{4} + i\sin\frac{7\pi}{4}\right)}$.

(c) The real part is $2\sqrt{3}$ and the imaginary part is 6, so $r = \sqrt{(2\sqrt{3})^2 + 6^2} = \sqrt{48} = 4\sqrt{3}$. We seek an angle θ such that $4\sqrt{3}\cos\theta = 2\sqrt{3}$ and $4\sqrt{3}\sin\theta = 6$, which means $\cos\theta = \frac{1}{2}$ and $\sin\theta = \frac{\sqrt{3}}{2}$. We can take $\theta = \frac{\pi}{3}$, so $2\sqrt{3} + 6i = \boxed{4\sqrt{3}\left(\cos\frac{\pi}{3} + i\sin\frac{\pi}{3}\right)}$.

(d) The real part is $\frac{1}{4}$ and the imaginary part is $-\frac{\sqrt{3}}{4}$, so $r = \sqrt{(1/4)^2 + (-\sqrt{3}/4)^2} = \sqrt{1/4} = \frac{1}{2}$. We seek an angle θ such that $\frac{1}{2}\cos\theta = \frac{1}{4}$ and $\frac{1}{2}\sin\theta = -\frac{\sqrt{3}}{4}$, which means $\cos\theta = \frac{1}{2}$ and $\sin\theta = -\frac{\sqrt{3}}{2}$. We can take $\theta = \frac{5\pi}{3}$, so $\frac{1}{4} - \frac{\sqrt{3}}{4}i = \boxed{\frac{1}{2}\left(\cos\frac{5\pi}{3} + i\sin\frac{5\pi}{3}\right)}$.

7.1.2

(a) $6\operatorname{cis}150° = 6\cos150° + 6i\sin150° = 6\cdot\left(-\frac{\sqrt{3}}{2}\right) + 6i\cdot\frac{1}{2} = \boxed{-3\sqrt{3} + 3i}$.

(b) $8\operatorname{cis}3780° = 8\operatorname{cis}180° = 8\cos180° + 8i\sin180° = 8\cdot(-1) + 8i\cdot0 = \boxed{-8}$.

(c) $2\operatorname{cis}\frac{2\pi}{3} = 2\cos\frac{2\pi}{3} + 2i\sin\frac{2\pi}{3} = 2\cdot\left(-\frac{1}{2}\right) + 2i\cdot\frac{\sqrt{3}}{2} = \boxed{-1 + \sqrt{3}i}$.

(d) $9\operatorname{cis}\frac{3\pi}{4} = 9\cos\frac{3\pi}{4} + 9i\sin\frac{3\pi}{4} = 9\cdot\left(-\frac{\sqrt{2}}{2}\right) + 9i\cdot\frac{\sqrt{2}}{2} = \boxed{-\frac{9\sqrt{2}}{2} + \frac{9\sqrt{2}}{2}i}$.

7.1.3 We showed in the text that $\operatorname{cis}x\operatorname{cis}y = \operatorname{cis}(x + y)$ for any x and y, so $\operatorname{cis}\beta\operatorname{cis}(\alpha - \beta) = \operatorname{cis}[\beta + (\alpha - \beta)] = \operatorname{cis}\alpha$, which means $\frac{\operatorname{cis}\alpha}{\operatorname{cis}\beta} = \operatorname{cis}(\alpha - \beta)$. Hence, we have $\frac{w}{z} = \frac{r\operatorname{cis}\alpha}{s\operatorname{cis}\beta} = \frac{r}{s}\operatorname{cis}(\alpha - \beta)$.

We also could have tackled this problem as follows:

$$\frac{w}{z} = \frac{r(\cos \alpha + i \sin \alpha)}{s(\cos \beta + i \sin \beta)} = \frac{r(\cos \alpha + i \sin \alpha)}{s(\cos \beta + i \sin \beta)} \cdot \frac{\cos \beta - i \sin \beta}{\cos \beta - i \sin \beta}$$

$$= \frac{r}{s} \cdot \frac{(\cos \alpha \cos \beta + \sin \alpha \sin \beta) + (\sin \alpha \cos \beta - \cos \alpha \sin \beta)i}{\cos^2 \beta - i^2 \sin^2 \beta}$$

$$= \frac{r}{s} \cdot \frac{\cos(\alpha - \beta) + i \sin(\alpha - \beta)}{\cos^2 \beta + \sin^2 \beta} = \frac{r}{s} \operatorname{cis}(\alpha - \beta)$$

7.1.4

(a) For the first factor, we can visualize a right triangle with legs $6\sqrt{3}$ and 18. Since one leg is $\sqrt{3}$ times the other, this triangle is a 30-60-90 triangle. Its hypotenuse is twice the shorter side, so the hypotenuse is $12\sqrt{3}$. Therefore, the magnitude of $6\sqrt{3} + 18i$ is $12\sqrt{3}$, and we have $6\sqrt{3} + 18i = 12\sqrt{3}\left(\frac{1}{2} + \frac{\sqrt{3}}{2}i\right) = 12\sqrt{3}\left(\cos\frac{\pi}{3} + i\sin\frac{\pi}{3}\right)$.

Similarly, a right triangle with legs $2\sqrt{3}$ and 6 is a 30-60-90 triangle with hypotenuse $4\sqrt{3}$, and we have $-2\sqrt{3} + 6i = 4\sqrt{3}\left(-\frac{1}{2} + \frac{\sqrt{3}}{2}i\right) = 4\sqrt{3}\left(\cos\frac{2\pi}{3} + i\sin\frac{2\pi}{3}\right)$.

So, we have

$$(6\sqrt{3} + 18i)(-2\sqrt{3} + 6i) = (12\sqrt{3})\left(\cos\frac{\pi}{3} + i\sin\frac{\pi}{3}\right)(4\sqrt{3})\left(\cos\frac{2\pi}{3} + i\sin\frac{2\pi}{3}\right)$$

$$= 48 \cdot 3 \left(\cos\frac{\pi}{3} + i\sin\frac{\pi}{3}\right)\left(\cos\frac{2\pi}{3} + i\sin\frac{2\pi}{3}\right)$$

$$= 144 \left(\cos\left(\frac{\pi}{3} + \frac{2\pi}{3}\right) + i\sin\left(\frac{\pi}{3} + \frac{2\pi}{3}\right)\right)$$

$$= \boxed{-144}.$$

All these computations can be done in your head—find the magnitudes using 30-60-90 triangles, and the arguments using your understanding of cosine and sine. Then, multiply the magnitudes and add the arguments.

(b) As in part (a), we use our understanding of special right triangles to find the magnitudes, and our understanding of cosine and sine to find the arguments. Considering 45-45-90 right triangles, we see that the magnitudes of the three factors are $4\sqrt{2}$, $3\sqrt{2}$, and $10\sqrt{2}$, respectively. Each has an argument of $\frac{k\pi}{4}$ for some odd k. Since the argument of $4 + 4i$ must be in the first quadrant, it is $\frac{\pi}{4}$. The argument of $-3 - 3i$ is in the third quadrant, so it is $\frac{5\pi}{4}$. The argument of $-10 + 10i$ is in the second quadrant, so it is $\frac{3\pi}{4}$. Therefore, the magnitude of the product is $(4\sqrt{2})(3\sqrt{2})(10\sqrt{2}) = 240\sqrt{2}$, and the argument of the product is $\frac{\pi}{4} + \frac{5\pi}{4} + \frac{3\pi}{4} = \frac{9\pi}{4}$, and the product is

$$240\sqrt{2}\left(\cos\frac{9\pi}{4} + i\sin\frac{9\pi}{4}\right) = \boxed{240 + 240i}.$$

7.1.5 Since $\operatorname{cis}\theta = \cos\theta + i\sin\theta$ and $\operatorname{cis}(-\theta) = \cos(-\theta) + i\sin(-\theta) = \cos\theta - i\sin\theta$, we see that $\operatorname{cis}\theta = \operatorname{cis}(-\theta)$ if and only if $\sin\theta = -\sin\theta$. But $\sin\theta = -\sin\theta$ if and only if $\sin\theta = 0$. Finally, $\sin\theta = 0$ if and only if θ is a $\boxed{\text{multiple of } \pi}$.

See if you can also solve this problem geometrically. How are $\operatorname{cis}\theta$ and $\operatorname{cis}(-\theta)$ related to the point corresponding to 1 in the complex plane?

7.1.6   For $\theta = 30°$, we have $\cos 5\theta = \cos 150° = -\frac{\sqrt{3}}{2}$, and

$$16\cos^5\theta - 20\cos^3\theta + 5\cos\theta = 16\left(\frac{\sqrt{3}}{2}\right)^5 - 20\left(\frac{\sqrt{3}}{2}\right)^3 + 5\cdot\frac{\sqrt{3}}{2}$$

$$= 16\cdot\frac{3^2\sqrt{3}}{32} - 20\cdot\frac{3\sqrt{3}}{8} + \frac{5\sqrt{3}}{2} = \frac{9\sqrt{3}}{2} - \frac{15\sqrt{3}}{2} + \frac{5\sqrt{3}}{2} = -\frac{\sqrt{3}}{2}.$$

Thus, the identity holds for $\theta = 30°$.

7.1.7   By de Moivre's Theorem, $\cos 7\theta + i\sin 7\theta = (\cos\theta + i\sin\theta)^7$. By the Binomial Theorem,

$$(\cos\theta + i\sin\theta)^7 = \cos^7\theta + 7i\cos^6\theta\sin\theta - 21\cos^5\theta\sin^2\theta - 35i\cos^4\theta\sin^3\theta$$
$$+ 35\cos^3\theta\sin^4\theta + 21i\cos^2\theta\sin^5\theta - 7\cos\theta\sin^6\theta - i\sin^7\theta.$$

We know that this equals $\cos 7\theta + i\sin 7\theta$, so equating the real part of this to the real part of the expansion above, we find

$$\cos 7\theta = \cos^7\theta - 21\cos^5\theta\sin^2\theta + 35\cos^3\theta\sin^4\theta - 7\cos\theta\sin^6\theta.$$

We want to express this only in terms of $\cos\theta$, not $\sin\theta$. Substituting $\sin^2\theta = 1 - \cos^2\theta$, we get

$$\cos^7\theta - 21\cos^5\theta\sin^2\theta + 35\cos^3\theta\sin^4\theta - 7\cos\theta\sin^6\theta$$
$$= \cos^7\theta - 21\cos^5\theta(1 - \cos^2\theta) + 35\cos^3\theta(1 - \cos^2\theta)^2 - 7\cos\theta(1 - \cos^2\theta)^3$$
$$= \cos^7\theta - 21\cos^5\theta + 21\cos^7\theta + 35\cos^3\theta(1 - 2\cos^2\theta + \cos^4\theta)$$
$$- 7\cos\theta(1 - 3\cos^2\theta + 3\cos^4\theta - \cos^6\theta)$$
$$= \boxed{64\cos^7\theta - 112\cos^5\theta + 56\cos^3\theta - 7\cos\theta}.$$

7.1.8   Since $\cos(90° - \theta) = \sin\theta$ and $\sin(90° - \theta) = \cos\theta$, we have

$$\sin t + i\cos t = \cos(90° - t) + i\sin(90° - t) = \operatorname{cis}(90° - t),$$

so by de Moivre's Theorem,

$$(\sin t + i\cos t)^n = \operatorname{cis}^n(90° - t) = \operatorname{cis}[n(90° - t)] = \operatorname{cis}(90°n - nt).$$

Likewise, $\sin nt + i\cos nt = \operatorname{cis}(90° - nt)$. Thus, the given equation becomes $\operatorname{cis}(90°n - nt) = \operatorname{cis}(90° - nt)$.

This equation holds for all t if and only if the two arguments differ by a multiple of 360°. In other words, if and only if $90°n - 90° = 360°k$ for some integer k. Hence, $n - 1 = 4k$, so $n = 4k + 1$. The positive integers less than or equal to 1000 that are of this form are $1, 5, 9, \ldots, 997$, of which there are $\boxed{250}$.

Exercises for Section 7.2

7.2.1

(a)   The real part is 0 and the imaginary part is -32, so $r = \sqrt{0^2 + (-32)^2} = 32$. We seek an angle θ such that $32\cos\theta = 0$ and $32\sin\theta = -32$, which means $\cos\theta = 0$ and $\sin\theta = -1$. We can take $\theta = \frac{3\pi}{2}$, so $-32i = \boxed{32e^{3\pi i/2}}$.

(b) The real part is -8 and the imaginary part is $8\sqrt{3}$, so $r = \sqrt{(-8)^2 + (8\sqrt{3})^2} = \sqrt{256} = 16$. We seek an angle θ such that $16\cos\theta = -8$ and $16\sin\theta = 8\sqrt{3}$, which means $\cos\theta = -\frac{1}{2}$ and $\sin\theta = \frac{\sqrt{3}}{2}$. We can take $\theta = \frac{2\pi}{3}$, so $-8 + 8\sqrt{3}i = \boxed{16e^{2\pi i/3}}$.

(c) The real part is $-\frac{\sqrt{2}}{6}$ and the imaginary part is $\frac{\sqrt{2}}{6}$, so $r = \sqrt{\left(-\frac{\sqrt{2}}{6}\right)^2 + \left(\frac{\sqrt{2}}{6}\right)^2} = \sqrt{\frac{1}{9}} = \frac{1}{3}$. We seek an angle θ such that $\frac{1}{3}\cos\theta = -\frac{\sqrt{2}}{6}$ and $\frac{1}{3}\sin\theta = \frac{\sqrt{2}}{6}$, which means $\cos\theta = -\frac{\sqrt{2}}{2}$ and $\sin\theta = \frac{\sqrt{2}}{2}$. We can take $\theta = \frac{3\pi}{4}$, so $-\frac{\sqrt{2}}{6} + \frac{\sqrt{2}}{6}i = \boxed{\frac{1}{3}e^{3\pi i/4}}$.

(d) The real part is $2\sqrt{15}$ and the imaginary part is $2\sqrt{5}$, so $r = \sqrt{(2\sqrt{15})^2 + (2\sqrt{5})^2} = \sqrt{80} = 4\sqrt{5}$. We seek an angle θ such that $4\sqrt{5}\cos\theta = 2\sqrt{15}$ and $4\sqrt{5}\sin\theta = 2\sqrt{5}$, which means $\cos\theta = \frac{\sqrt{3}}{2}$ and $\sin\theta = \frac{1}{2}$. We can take $\theta = \frac{\pi}{6}$, so $2\sqrt{15} + 2\sqrt{5}i = \boxed{4\sqrt{5}e^{\pi i/6}}$.

7.2.2

(a) $e^{4\pi i} = \cos 4\pi + i\sin 4\pi = \boxed{1}$.

(b) $e^{5\pi i/4} = \cos\frac{5\pi}{4} + i\sin\frac{5\pi}{4} = \boxed{-\frac{\sqrt{2}}{2} - \frac{\sqrt{2}}{2}i}$.

(c) $e^{-5\pi i/3} = e^{-5\pi i/3 + 2\pi i} = e^{\pi i/3} = \cos\frac{\pi}{3} + i\sin\frac{\pi}{3} = \boxed{\frac{1}{2} + \frac{\sqrt{3}}{2}i}$.

(d) $e^{17\pi i/3} = e^{17\pi i/3 - 4\pi i} = e^{5\pi i/3} = \cos\frac{5\pi}{3} + i\sin\frac{5\pi}{3} = \boxed{\frac{1}{2} - \frac{\sqrt{3}}{2}i}$.

7.2.3 We have $e^{i\theta} = \cos\theta + i\sin\theta$ and $e^{-i\theta} = \cos(-\theta) + i\sin(-\theta) = \cos\theta - i\sin\theta$. Subtracting the second equation from the first equation, we get $e^{i\theta} - e^{-i\theta} = 2i\sin\theta$, so $\sin\theta = \frac{e^{i\theta} - e^{-i\theta}}{2i}$.

7.2.4 We have

$$\frac{e^{ix}}{e^{iy}} = \frac{\cos x + i\sin x}{\cos y + i\sin y} = \frac{(\cos x + i\sin x)(\cos y - i\sin y)}{(\cos y + i\sin y)(\cos y - i\sin y)}$$
$$= \frac{\cos x\cos y + \sin x\sin y + (\sin x\cos y - \cos x\sin y)i}{\cos^2 y + \sin^2 y} = \cos x\cos y + \sin x\sin y + (\sin x\cos y - \cos x\sin y)i.$$

We also have $\frac{e^{ix}}{e^{iy}} = e^{i(x-y)} = \cos(x - y) + i\sin(x - y)$. Equating the real part of this to the real part of our expression above for $\frac{e^{ix}}{e^{iy}}$ gives $\cos(x - y) = \cos x\cos y + \sin x\sin y$, and equating the imaginary parts gives $\sin(x - y) = \sin x\cos y - \cos x\sin y$.

7.2.5

(a) By considering 45-45-90 right triangles and our knowledge of sine and cosine, we can see that $8 + 8i = 8\sqrt{2}e^{\pi i/4}$ and $-3 + 3i = 3\sqrt{2}e^{3\pi i/4}$, so their quotient is $\frac{8}{3}e^{(\pi i/4)-(3\pi i/4)} = \frac{8}{3}e^{-\pi i/2} = \boxed{-\frac{8}{3}i}$.

(b) A right triangle with legs $4\sqrt{3}$ and 12 has one leg that is $\sqrt{3}$ times the other, so it is a 30-60-90 triangle with hypotenuse twice the smaller leg, which means the hypotenuse is $2 \cdot 4\sqrt{3} = 8\sqrt{3}$. Therefore, $-12 - 4i\sqrt{3}$ has magnitude $8\sqrt{3}$. Its argument is the third quadrant angle whose sine equals $-\frac{1}{2}$, so its argument is $\frac{7\pi}{6}$. Similarly, $3 + 3i\sqrt{3}$ has magnitude 6 and argument $\frac{\pi}{3}$. So, we have

$$\frac{-12 - 4i\sqrt{3}}{3 + 3i\sqrt{3}} = \frac{8\sqrt{3}e^{7\pi i/6}}{6e^{\pi i/3}} = \frac{4\sqrt{3}}{3}e^{(7\pi i/6)-(\pi i/3)} = \frac{4\sqrt{3}}{3}e^{5\pi i/6} = \frac{4\sqrt{3}}{3}\left(-\frac{\sqrt{3}}{2} + \frac{1}{2}i\right) = \boxed{-2 + \frac{2\sqrt{3}}{3}i}.$$

7.2.6

(a) First we express $1 - i$ in exponential form. The real part is 1 and the imaginary part is -1, so $r = \sqrt{1^2 + (-1)^2} = \sqrt{2}$. We seek an angle θ such that $\sqrt{2}\cos\theta = 1$ and $\sqrt{2}\sin\theta = -1$, which means $\cos\theta = \frac{\sqrt{2}}{2}$ and $\sin\theta = -\frac{\sqrt{2}}{2}$. We can take $\theta = \frac{7\pi}{4}$, so $1 - i = \sqrt{2}e^{7\pi i/4}$. Hence, $(1 - i)^{16} = (\sqrt{2}e^{7\pi i/4})^{16} = 2^8 e^{28\pi i} = \boxed{256}$.

(b) First we express $-2\sqrt{3} + 6i$ in exponential form. The real part is $-2\sqrt{3}$ and the imaginary part is 6, so $r = \sqrt{(-2\sqrt{3})^2 + 6^2} = \sqrt{48} = 4\sqrt{3}$. We seek an angle θ such that $4\sqrt{3}\cos\theta = -2\sqrt{3}$ and $4\sqrt{3}\sin\theta = 6$, which means $\cos\theta = -\frac{1}{2}$ and $\sin\theta = \frac{\sqrt{3}}{2}$. We can take $\theta = \frac{2\pi}{3}$, so $-2\sqrt{3} + 6i = 4\sqrt{3}e^{2\pi i/3}$. Hence,

$$(-2\sqrt{3} + 6i)^8 = (4\sqrt{3}e^{2\pi i/3})^8 = 4^8 \cdot 3^4 e^{16\pi i/3} = 4^8 \cdot 3^4 \left(-\frac{1}{2} - \frac{\sqrt{3}}{2}i\right) = \boxed{-2654208 - 2654208i\sqrt{3}}.$$

7.2.7 We express $-3 + \sqrt{3}i$ in exponential form. The real part is -3 and the imaginary part is $\sqrt{3}$, so $r = \sqrt{(-3)^2 + (\sqrt{3})^2} = \sqrt{12} = 2\sqrt{3}$. We seek an angle θ such that $2\sqrt{3}\cos\theta = -3$ and $2\sqrt{3}\sin\theta = \sqrt{3}$, which means $\cos\theta = -\frac{\sqrt{3}}{2}$ and $\sin\theta = \frac{1}{2}$. We can take $\theta = \frac{5\pi}{6}$, so $-3 + \sqrt{3}i = 2\sqrt{3}e^{5\pi i/6}$. Hence,

$$(-3 + \sqrt{3}i)^n = (2\sqrt{3})^n e^{5n\pi i/6} = (2\sqrt{3})^n \left(\cos\frac{5n\pi}{6} + i\sin\frac{5n\pi}{6}\right).$$

This is a real number if and only if $\sin\frac{5n\pi}{6} = 0$. But $\sin\frac{5n\pi}{6} = 0$ if and only if $\frac{5n\pi}{6}$ is a multiple of π, i.e. if $\frac{5n\pi}{6} = k\pi$ for some integer k, which means $k = 5n/6$. Since k is an integer, n must be divisible by 6.

Conversely, if n is a multiple of 6, say $n = 6m$, then

$$(-3 + \sqrt{3}i)^n = (2\sqrt{3}e^{5\pi i/6})^{6m} = (2\sqrt{3})^{6m}e^{5m\pi i} = (2\sqrt{3})^{6m}(-1)^{5m},$$

which is a real number. Therefore, $(-3 + \sqrt{3}i)^n$ is a real number if and only if $\boxed{n \text{ is a multiple of } 6}$.

7.2.8 Since dividing one exponential expression by another is easy, we try to show that

$$\frac{e^{i\alpha} + e^{i\beta}}{2e^{i(\alpha+\beta)/2}} = \cos\left(\frac{\alpha - \beta}{2}\right).$$

Multiplying both sides of this by $2e^{i(\alpha+\beta)/2}$ will give the desired equation. (This expression cannot be 0 because $e^{i\theta} \neq 0$ for all angles θ.)

We see that

$$\frac{e^{i\alpha} + e^{i\beta}}{2e^{i(\alpha+\beta)/2}} = \frac{1}{2}\left(\frac{e^{i\alpha}}{e^{i(\alpha+\beta)/2}} + \frac{e^{i\beta}}{e^{i(\alpha+\beta)/2}}\right) = \frac{e^{i(\alpha-\beta)/2} + e^{i(\beta-\alpha)/2}}{2} = \frac{e^{i(\alpha-\beta)/2} + e^{-i(\alpha-\beta)/2}}{2} = \cos\left(\frac{\alpha - \beta}{2}\right),$$

so $e^{i\alpha} + e^{i\beta} = 2\cos\left(\frac{\alpha - \beta}{2}\right)e^{i(\alpha+\beta)/2}$.

Exercises for Section 7.3

7.3.1 The solutions of $x^8 = 1$ are the eighth roots of unity, namely

$$\boxed{e^0,\ e^{2\pi i/8},\ e^{4\pi i/8},\ e^{6\pi i/8},\ e^{8\pi i/8},\ e^{10\pi i/8},\ e^{12\pi i/8},\ \text{and } e^{14\pi i/8}}.$$

The primitive eighth roots of unity are the eighth roots of unity of the form $e^{2k\pi i/8}$ for which $\gcd(k, 8) = 1$. Since $8 = 2^3$, we have $\gcd(k, 8) = 1$ if and only if k is odd, so the primitive eighth roots of unity are $\boxed{e^{2\pi i/8}, e^{6\pi i/8}, e^{10\pi i/8}, \text{ and } e^{14\pi i/8}}$.

7.3.2 Writing -64 in exponential form, we get $-64 = 64e^{\pi i}$. Therefore, the roots of the equation $x^6 = -64 = 2^6 e^{\pi i}$ are $2e^{\pi i/6}, 2e^{3\pi i/6}, 2e^{5\pi i/6}, 2e^{7\pi i/6}, 2e^{9\pi i/6}$, and $2e^{11\pi i/6}$.

The real part of a complex number is positive if and only if its argument is in the first or fourth quadrant (or if the argument is a multiple of 2π). The only roots of $x^6 = -64$ that have such an argument are $2e^{\pi i/6}$ and $2e^{11\pi i/6}$, and their product is $4e^{\pi i/6 + 11\pi i/6} = 4e^{2\pi i} = \boxed{4}$.

7.3.3 The sixth roots of unity are $e^0 = 1$, $e^{2\pi i/6} = e^{\pi i/3}$, $e^{4\pi i/6} = e^{2\pi i/3}$, $e^{6\pi i/6} = e^i = -1$, $e^{8\pi i/6} = e^{4\pi i/3}$, and $e^{10\pi i/6} = e^{5\pi i/3}$. We compute $\omega + \omega^2$ for each such root:

$$1 + 1^2 = 2,$$

$$e^{\pi i/3} + e^{2\pi i/3} = \cos\frac{\pi}{3} + i\sin\frac{\pi}{3} + \cos\frac{2\pi}{3} + i\sin\frac{2\pi}{3} = \frac{1}{2} + i\cdot\frac{\sqrt{3}}{2} - \frac{1}{2} + i\cdot\frac{\sqrt{3}}{2} = \sqrt{3}i,$$

$$e^{2\pi i/3} + e^{4\pi i/3} = \cos\frac{2\pi}{3} + i\sin\frac{2\pi}{3} + \cos\frac{4\pi}{3} + i\sin\frac{4\pi}{3} = -\frac{1}{2} + i\cdot\frac{\sqrt{3}}{2} - \frac{1}{2} - i\cdot\frac{\sqrt{3}}{2} = -1,$$

$$-1 + (-1)^2 = 0,$$

$$e^{4\pi i/3} + e^{8\pi i/3} = \cos\frac{4\pi}{3} + i\sin\frac{4\pi}{3} + \cos\frac{8\pi}{3} + i\sin\frac{8\pi}{3} = -\frac{1}{2} - i\cdot\frac{\sqrt{3}}{2} - \frac{1}{2} + i\cdot\frac{\sqrt{3}}{2} = -1,$$

$$e^{5\pi i/3} + e^{10\pi i/3} = \cos\frac{5\pi}{3} + i\sin\frac{5\pi}{3} + \cos\frac{10\pi}{3} + i\sin\frac{10\pi}{3} = \frac{1}{2} - i\cdot\frac{\sqrt{3}}{2} - \frac{1}{2} - i\cdot\frac{\sqrt{3}}{2} = -\sqrt{3}i.$$

Thus, the possible values of $\omega + \omega^2$ are $\boxed{2, 0, -1, \sqrt{3}i, \text{ and } -\sqrt{3}i}$.

7.3.4 We tackle both parts at once.

Solution 1: Consider each of the 12th roots of unity. The 12th roots of unity are the values of $e^{2\pi ik/12}$ for $k = 0, 1, 2, \ldots, 11$. As explained in the text, if $\gcd(k, 12) = 1$, then $e^{2\pi ik/12}$ is a primitive 12th root of unity. That covers $k = 1, 5, 7, 11$, so there are 4 primitive 12th roots of unity.

If $k = 2$, we have $e^{2\pi ik/12} = e^{2\pi i/6}$, which is a primitive 6th root of unity, since $(e^{2\pi i/6})^6 = 1$ but the argument of $(e^{2\pi i/6})^n$ is between 0 and 2π for $1 \le n \le 5$. The same holds for $k = 10$, which we can see by noting that $e^{2\pi i \cdot 10/12} = e^{2\pi i \cdot 12/12}/e^{2\pi i \cdot 2/12} = 1/e^{2\pi i/6}$. So, $(e^{2\pi i \cdot 10/12})^6 = 1$, and no smaller positive integer power of $e^{2\pi i \cdot 10/12}$ equals 1. So, there are 2 primitive 6th roots of unity among the 12th roots of unity.

Similarly, we find that when $k = 3$ or $k = 9$, then we have $(e^{2\pi ik/12})^4 = 1$, but no smaller positive power of $e^{2\pi ik/12}$ equals 1, so there are 2 primitive 4th roots of unity among the 12th roots of unity.

We also find that if $k = 4$ or $k = 8$, then $(e^{2\pi ik/12})^3 = 1$, but no smaller positive power of $e^{2\pi ik/12}$ equals 1, so there are 2 primitive 3rd roots of unity among the 12th roots of unity.

Finally, if $k = 6$, we have $(e^{2\pi ik/12})^2 = 1$ but $(e^{2\pi ik/12})^1 \ne 1$, and if $k = 0$, then $(e^{2\pi ik/12})^1 = 1$, so there is 1 primitive first root of unity and 1 primitive square root of unity among the 12th roots of unity.

Solution 2: Use a little number theory. Let ω be a 12th root of unity. Since $\omega^{12} = 1$, there is at least one positive integer n such that $\omega^n = 1$. Let k be the smallest such positive integer. Then $\omega^k = 1$ and ω is a primitive k^{th} root of unity.

Let q and r denote the quotient and remainder when 12 is divided by k, respectively, so $12 = kq + r$ and $0 \le r \le k - 1$. Then $\omega^{12} = \omega^{kq+r} = 1$. But $\omega^{kq+r} = (\omega^k)^q \cdot \omega^r = 1^q \cdot \omega^r = \omega^r$, so $\omega^r = 1$. But $r \le k - 1$, so r cannot be positive. (By definition, k is the smallest positive integer such that $\omega^k = 1$.) Therefore, $r = 0$, which means $12 = kq$.

Hence, k must be a divisor of 12.

Conversely, if ω is a primitive k^{th} root of unity, where k is a divisor of 12, then clearly ω is a 12th root of unity. Therefore, the possible values of k are $\boxed{1, 2, 3, 4, 6, \text{ and } 12}$.

The number of primitive k^{th} roots of unity is the number of positive integers less than or equal to k that have no positive factors in common with k besides 1. Thus, we only have to compute this for each quantity for each divisor of 12. Doing so, we find that among the 12th roots of unity, we have $\boxed{1}$ primitive first root of unity, $\boxed{1}$ primitive square root of unity, $\boxed{2}$ primitive cube roots of unity, $\boxed{2}$ primitive fourth roots of unity, $\boxed{2}$ primitive sixth roots of unity, and $\boxed{4}$ primitive 12th roots of unity. (Note that these account for all $1 + 1 + 2 + 2 + 2 + 4 = 12$ 12th roots of unity.)

7.3.5 Let ω be a primitive $(2m)^{\text{th}}$ root of unity, so $\omega^{2m} = 1$ and $\omega^k \neq 1$ for $k = 1, 2, \ldots, 2m-1$. Then $(\omega^2)^m = \omega^{2m} = 1$, so ω^2 is an mth root of unity. Furthermore, $(\omega^2)^k = \omega^{2k} \neq 1$ for $k = 1, 2, \ldots, m-1$, because of our observation above. Therefore, ω^2 is a primitive m^{th} root of unity.

7.3.6

(a) Let $\omega = \alpha\beta$. Then $\omega^{15} = (\alpha\beta)^{15} = \alpha^{15}\beta^{15} = (\alpha^3)^5(\beta^5)^3 = 1$, so ω is a 15th root of unity.

(b) Let $\omega = \alpha\beta$. Let k be an integer such that $\omega^k = 1$. Then $(\alpha\beta)^k = \alpha^k\beta^k = 1$. Raising both sides to the power of 3, we get $\alpha^{3k}\beta^{3k} = 1$. Since α is a cube root of unity, we have $\alpha^{3k} = 1$, so $\beta^{3k} = 1$. But β is a primitive fifth root of unity, so $3k$ must be divisible by 5. Since $\gcd(3,5) = 1$, we know that k must be divisible by 5.

Similarly, from the equation $\alpha^k\beta^k = 1$, we get $\alpha^{5k}\beta^{5k} = 1$. Since β is a fifth root of unity, it follows that $\alpha^{5k} = 1$. But α is a primitive cube root of unity, so $5k$ must be divisible by 3. Since $\gcd(3,5) = 1$, we know that k must be divisible by 3.

We have established that k must be divisible by both 3 and 5. Since $\gcd(3,5) = 1$, we know that k must be divisible by $3 \cdot 5 = 15$. Since this holds for any k such that $\omega^k = 1$, we conclude that k must be a multiple of 15. Since $\omega^{15} = 1$ and 15 is the smallest multiple of 15, we know that ω is a primitive 15th root of unity.

(We can show similarly that if α is a primitive n^{th} root of unity, and β is a primitive m^{th} root of unity, and $\gcd(m, n) = 1$, then $\alpha\beta$ is a primitive $(mn)^{\text{th}}$ root of unity.)

(c) We see that $e^{2\pi i/4} = e^{\pi i/2}$ is a primitive fourth root of unity, and that $e^{2\pi i/6} = e^{\pi i/3}$ is a primitive sixth root of unity. So, letting $\alpha = e^{\pi i/2}$ and $\beta = e^{\pi i/3}$, we have

$$\alpha\beta = e^{\pi i/2} \cdot e^{\pi i/3} = e^{\pi i/2 + \pi i/3} = e^{5\pi i/6}.$$

Since $(\alpha\beta)^{12} = (e^{5\pi i/6})^{12} = e^{10\pi i} = 1$, the product $\alpha\beta$ is not a primitive 24th root of unity.

7.3.7 We need a polynomial whose roots are the twelfth roots of unity. So, we start with the polynomial whose roots are all of the 12th roots of unity, namely $x^{12} - 1 = 0$. As a difference of squares, we have $x^{12} - 1 = (x^6 - 1)(x^6 + 1)$. We can factor $x^6 - 1$ as a difference of squares: $x^6 - 1 = (x^3 - 1)(x^3 + 1)$. As a difference of cubes, we have $x^3 - 1 = (x - 1)(x^2 + x + 1)$, and as a sum of cubes, we have $x^3 + 1 = (x + 1)(x^2 - x + 1)$.

We can also factor $x^6 + 1$ as a sum of cubes, $x^6 + 1 = (x^2 + 1)(x^4 - x^2 + 1)$. Thus, we have the factorization

$$x^{12} - 1 = (x - 1)(x^2 + x + 1)(x + 1)(x^2 - x + 1)(x^2 + 1)(x^4 - x^2 + 1).$$

The only quartic factor that could possibly fit the problem is therefore $\boxed{x^4 - x^2 + 1}$.

Proving that this quartic is irreducible is a good deal harder than finding it. Suppose that $x^4 - x^2 + 1$ can be factored over the rational numbers. First, we identify the roots of this polynomial. To do so, we must identify the roots of the other polynomials in the factorization above.

From our work above, we see that $x^4 - x^2 + 1$ is a factor of $x^6 + 1$. Since $x^{12} - 1 = (x^6 - 1)(x^6 + 1)$, the roots of $x^6 + 1 = 0$ are the 12th roots of unity that are not 6th roots of unity.

The 12th roots of unity are $e^0, e^{2\pi i/12}, e^{4\pi i/12}, \ldots, e^{22\pi i/12}$, and the 6th roots of unity are $e^0, e^{2\pi i/6}, e^{4\pi i/6}, \ldots, e^{10\pi i/6}$. Note that $e^0 = e^0$, $e^{4\pi i/12} = e^{2\pi i/6}$, $e^{8\pi i/12} = e^{4\pi i/6}$, and so on. Thus, the 12th roots of unity that are not 6th roots of unity are $e^{2\pi i/12}, e^{6\pi i/12}, e^{10\pi i/12}, e^{14\pi i/12}, e^{18\pi i/12}$, and $e^{22\pi i/12}$.

Furthermore, $x^6 + 1 = (x^2 + 1)(x^4 - x^2 + 1)$. The roots of $x^2 + 1 = 0$ are $i = e^{\pi i/2} = e^{6\pi i/12}$ and $-i = e^{3\pi i/2} = e^{18\pi i/12}$. Therefore, the roots of $x^4 - x^2 + 1 = 0$ are $e^{2\pi i/12}, e^{10\pi i/12}, e^{14\pi i/12}$, and $e^{22\pi i/12}$.

If $x^4 - x^2 + 1$ factors, then it either factors (at least) as the product of a linear polynomial and a cubic polynomial, or as the product of two quadratic polynomials. But each root is nonreal, so there cannot be any linear factors with rational coefficients. Therefore, it must factor as the product of two quadratic polynomials.

If a quadratic polynomial has real coefficients, and it has nonreal roots, then those roots must be conjugate pairs. Therefore, the roots of one of the quadratic factors must be $e^{2\pi i/12}$ and $e^{22\pi i/12}$. But

$$e^{2\pi i/12} = e^{\pi i/6} = \cos\frac{\pi}{6} + i\sin\frac{\pi}{6} = \frac{\sqrt{3}}{2} + \frac{1}{2}i,$$

$$e^{22\pi i/12} = e^{11\pi i/6} = \cos\frac{11\pi}{6} + i\sin\frac{11\pi}{6} = \frac{\sqrt{3}}{2} - \frac{1}{2}i,$$

which makes the quadratic

$$\left(x - \frac{\sqrt{3}}{2} - \frac{1}{2}i\right)\left(x - \frac{\sqrt{3}}{2} + \frac{1}{2}i\right) = \left(x - \frac{\sqrt{3}}{2}\right)^2 - \left(\frac{1}{2}i\right)^2 = x^2 - \sqrt{3}x + \frac{3}{4} + \frac{1}{4} = x^2 - \sqrt{3}x + 1.$$

Even if we multiplied this quadratic by a nonzero rational number, we couldn't make all the coefficients rational. Therefore, the polynomial $x^4 - x^2 + 1$ is irreducible over the rational numbers.

Exercises for Section 7.4

7.4.1

(a) Since $(z + 1)(z^2 - z + 1) = z^3 + 1$, the roots of $z^2 - z + 1 = 0$ are the roots of $z^3 = -1$ other than -1. Since $-1 = e^{\pi i}$ is one root of $z^3 = -1$, the other roots are $e^{\pi i + 2\pi i/3} = \boxed{e^{5\pi i/3}}$ and $e^{\pi i/3 + 2 \cdot 2\pi i/3} = e^{7\pi i/3} = \boxed{e^{\pi i/3}}$.

(b) Since $(z^5 - 1)(z^5 + 1) = z^{10} - 1$, the roots of $z^5 + 1$ are the roots of $z^{10} - 1$ that are not roots of $z^5 - 1$. The roots of $z^{10} - 1$ are $e^0, e^{2\pi i/10}, e^{4\pi i/10}, \ldots, e^{18\pi i/10}$. The roots of $z^5 - 1$ are $e^0, e^{2\pi i/5}, e^{4\pi i/5}, e^{6\pi i/5}$, and $e^{8\pi i/5}$.

Note that $e^0 = e^0$, $e^{2\pi i/5} = e^{4\pi i/10}$, $e^{4\pi i/5} = e^{8\pi i/10}$, $e^{6\pi i/5} = e^{12\pi i/10}$, and $e^{8\pi i/5} = e^{16\pi i/10}$. Therefore, the roots of $z^5 + 1$ are $e^{2\pi i/10} = \boxed{e^{\pi i/5}}$, $e^{6\pi i/10} = \boxed{e^{3\pi i/5}}$, $e^{10\pi i/10} = \boxed{e^{\pi i}}$, $e^{14\pi i/10} = \boxed{e^{7\pi i/5}}$, and $e^{18\pi i/10} = \boxed{e^{9\pi i/5}}$.

(c) The polynomial $z^5 + 1$ factors as $(z + 1)(z^4 - z^3 + z^2 - z + 1)$. The only root of $z^5 + 1$ that is real is -1, so the degree 4 polynomial whose roots are the nonreal roots of $z^5 + 1$ is $\boxed{z^4 - z^3 + z^2 - z + 1}$.

(d) First, the polynomial $x^7 + 1$ factors as $(x + 1)(x^6 - x^5 + x^4 - x^3 + x^2 - x + 1)$, so the roots of $x^6 - x^5 + x^4 - x^3 + x^2 - x + 1$ are the roots of $x^7 + 1$ other than -1.

From here, we can either find the roots of $x^7 + 1$ using the the same technique as in part (a), or we can be a little clever. We'll show the clever approach here. Since $(x^7 - 1)(x^7 + 1) = x^{14} - 1$, the roots of $x^7 + 1$ are the roots of $x^{14} - 1$ that are not also roots of $x^7 - 1$. The roots of $z^{14} - 1$ are $e^0, e^{2\pi i/14}, e^{4\pi i/14}, \ldots, e^{26\pi i/14}$. The roots of $z^7 - 1$ are $e^0, e^{2\pi i/7}, e^{4\pi i/7}, \ldots, e^{12\pi i/7}$. Note that $e^0 = e^0$, $e^{2\pi i/7} = e^{4\pi i/14}$, $e^{4\pi i/7} = e^{8\pi i/14}, \ldots, e^{12\pi i/7} = e^{24\pi i/14}$. Therefore, the roots of $z^7 + 1$ are $e^{2\pi i/14} = e^{\pi i/7}$, $e^{6\pi i/14} = e^{3\pi i/7}$, $e^{10\pi i/14} = e^{5\pi i/7}, \ldots, e^{26\pi i/14} = e^{13\pi i/7}$.

Finally, $-1 = e^{\pi i} = e^{7\pi i/7}$, so the roots of $x^6 - x^5 + x^4 - x^3 + x^2 - x + 1 = 0$ are

$$\boxed{e^{\pi i/7}, e^{3\pi i/7}, e^{5\pi i/7}, e^{9\pi i/7}, e^{11\pi i/7}, \text{ and } e^{13\pi i/7}.}$$

7.4.2 The roots of $x^{13} - 1 = 0$ are the 13th roots of unity, namely $1, \omega, \omega^2, \ldots, \omega^{12}$. Also, the polynomial $x^{13} - 1$ factors as $x^{13} - 1 = (x - 1)(x^{12} + x^{11} + \cdots + x + 1)$. Hence, the roots of $x^{12} + x^{11} + \cdots + x + 1 = 0$ are $\omega, \omega^2, \ldots, \omega^{12}$, which means

$$(x - \omega)(x - \omega^2) \cdots (x - \omega^{12}) = x^{12} + x^{11} + \cdots + x + 1.$$

Taking $x = 1$, we find $(1 - \omega)(1 - \omega^2) \cdots (1 - \omega^{12}) = \boxed{13}$.

7.4.3 Putting the expression over a common denominator, we get

$$z^{10} + \frac{1}{z^{10}} + z^{30} + \frac{1}{z^{30}} + z^{50} + \frac{1}{z^{50}} = \frac{z^{60} + z^{40} + z^{80} + z^{20} + z^{100} + 1}{z^{50}} = \frac{1 + z^{20} + z^{40} + z^{60} + z^{80} + z^{100}}{z^{50}}.$$

From the formula for a geometric series, the numerator is equal to

$$1 + z^{20} + z^{40} + z^{60} + z^{80} + z^{100} = \frac{z^{120} - 1}{z^{20} - 1}.$$

Since $z^7 = 1$, we have $z^{120} = z^{17 \cdot 7 + 1} = (z^7)^{17} \cdot z = z$, $z^{20} = z^{2 \cdot 7 + 6} = (z^7)^2 \cdot z^6 = z^6$, and $z^{50} = z^{7 \cdot 7 + 1} = (z^7)^7 \cdot z = z$, so

$$\frac{1 + z^{20} + z^{40} + z^{60} + z^{80} + z^{100}}{z^{50}} = \frac{z^{120} - 1}{z^{50}(z^{20} - 1)} = \frac{z - 1}{z(z^6 - 1)} = \frac{z - 1}{z^7 - z} = \frac{z - 1}{1 - z} = \frac{z - 1}{-(z - 1)} = \boxed{-1}.$$

We also could have made reductions in the powers of z before summing the geometric series. We have $z^{20} = (z^7)^2 z^6 = z^6$. Treating the other powers of z similarly gives

$$\frac{1 + z^{20} + z^{40} + z^{60} + z^{80} + z^{100}}{z^{50}} = \frac{1 + z^6 + z^5 + z^4 + z^3 + z^2}{z}.$$

Because z is a seventh root of unity besides 1, we have $z^6 + z^5 + z^4 + z^3 + z^2 + z + 1 = 0$, so $z^6 + z^5 + z^4 + z^3 + z^2 + 1 = -z$, which means the fraction above equals $-z/z$, which is -1.

7.4.4 Let $k = \frac{a}{b} = \frac{b}{c} = \frac{c}{a}$. Then $k^3 = \frac{a}{b} \cdot \frac{b}{c} \cdot \frac{c}{a} = 1$, so k is a cube root of unity, i.e. $k = 1$, $e^{2\pi i/3}$, or $e^{4\pi i/3}$.

We can write b and c in terms of a by noting that $c = ka$ and $b = kc$, so $b = k^2 a$. We then have $(a + b - c)/(a - b + c) = (a + k^2 a - ka)/(a - k^2 a + ka) = (1 + k^2 - k)/(1 - k^2 + k)$. This is easy to compute for one cube root of unity; for $k = 1$, this expression equals 1. We could pound it out for each of the other cube roots of unity, but instead we can make a clever simplification. If k is a cube root of unity besides 1, then we have $1 + k + k^2 = 0$, so $1 + k = -k^2$ and $1 + k^2 = -k$. Making these substitutions in the numerator and denominator of the fraction give

$$\frac{1 + k^2 - k}{1 - k^2 + k} = \frac{(1 + k^2) - k}{(1 + k) - k^2} = \frac{-k - k}{-k^2 - k^2} = \frac{1}{k}.$$

Since k is a cube root of unity, we have $k^3 = 1$, so $\frac{1}{k} = k^2$. Therefore, the other two possible values are the squares of the primitive cube roots of unity. But $(e^{2\pi i/3})^2 = e^{4\pi i/3}$ and $(e^{4\pi i/3})^2 = e^{8\pi i/3} = e^{2\pi i/3}$, which means that our other two possible values are the cube roots of unity.

Therefore, the possible values of $\frac{a+b-c}{a-b+c}$ are $\boxed{1, -\frac{1}{2} + \frac{\sqrt{3}}{2}i, \text{ and } -\frac{1}{2} - \frac{\sqrt{3}}{2}i}$.

We also could have simplified the solution even more with the observation that $a = kb$, $b = kc$, and $c = ka$, so

$$\frac{a + b - c}{a - b + c} = \frac{a + b - c}{kb - kc + ka} = \frac{a + b - c}{k(a + b - c)} = \frac{1}{k} = \frac{k^3}{k} = k^2.$$

7.4.5   We compare the polynomials $x^7 + x^4 + x^3 + x + 1$ and $x^4 + x^3 + x^2 + x + 1$, because both polynomials are the sum of five powers of x, and both have several terms in common. In fact, their difference is $(x^7 + x^4 + x^3 + x + 1) - (x^4 + x^3 + x^2 + x + 1) = x^7 - x^2$, which factors as

$$x^7 - x^2 = x^2(x^5 - 1) = x^2(x - 1)(x^4 + x^3 + x^2 + x + 1).$$

Therefore,

$$
\begin{aligned}
x^7 + x^4 + x^3 + x + 1 &= (x^4 + x^3 + x^2 + x + 1) + (x^7 - x^2) \\
&= (x^4 + x^3 + x^2 + x + 1) + x^2(x - 1)(x^4 + x^3 + x^2 + x + 1) \\
&= (x^4 + x^3 + x^2 + x + 1) + (x^3 - x^2)(x^4 + x^3 + x^2 + x + 1) \\
&= \boxed{(x^3 - x^2 + 1)(x^4 + x^3 + x^2 + x + 1)}.
\end{aligned}
$$

7.4.6   Since $1 + x + x^2 + \cdots + x^{17} = \dfrac{x^{18} - 1}{x - 1}$, we can write

$$
P(x) = (1 + x + x^2 + \cdots + x^{17})^2 - x^{17} = \left(\frac{x^{18} - 1}{x - 1}\right)^2 - x^{17} = \frac{(x^{18} - 1)^2}{(x - 1)^2} - x^{17} = \frac{x^{36} - 2x^{18} + 1}{x^2 - 2x + 1} - x^{17}
$$

$$
= \frac{x^{36} - 2x^{18} + 1 - x^{19} + 2x^{18} - x^{17}}{x^2 - 2x + 1} = \frac{x^{36} - x^{19} - x^{17} + 1}{x^2 - 2x + 1}.
$$

The numerator factors as $x^{36} - x^{19} - x^{17} + 1 = x^{19}(x^{17} - 1) - (x^{17} - 1) = (x^{19} - 1)(x^{17} - 1)$, so

$$
P(x) = \frac{x^{36} - x^{19} - x^{17} + 1}{x^2 - 2x + 1} = \frac{x^{19} - 1}{x - 1} \cdot \frac{x^{17} - 1}{x - 1}.
$$

The roots of $(x^{19} - 1)/(x - 1) = 0$ are the 19th roots of unity other than 1, and the roots of $(x^{17} - 1)/(x - 1) = 0$ are the 17th roots of unity other than 1. Hence, the α_i, in some order, are equal to

$$\frac{1}{17}, \frac{2}{17}, \dots, \frac{16}{17}, \frac{1}{19}, \frac{2}{19}, \dots, \frac{18}{19}.$$

The smallest five α_i are $\alpha_1 = \frac{1}{19}$, $\alpha_2 = \frac{1}{17}$, $\alpha_3 = \frac{2}{19}$, $\alpha_4 = \frac{2}{17}$, and $\alpha_5 = \frac{3}{19}$, so

$$\alpha_1 + \alpha_2 + \alpha_3 + \alpha_4 + \alpha_5 = \frac{1}{19} + \frac{1}{17} + \frac{2}{19} + \frac{2}{17} + \frac{3}{19} = \boxed{\frac{159}{323}}.$$

Review Problems

7.33   In each part, we are asked to convert a rectangular form $x + yi$ to a polar form $r(\cos\theta + i\sin\theta)$. As described in the text, this is equivalent to converting the rectangular coordinates (x, y) to the polar coordinates (r, θ).

(a)   We have $r = \sqrt{(-9)^2 + 0^2} = \sqrt{81} = 9$. We seek an angle θ such that $9\cos\theta = -9$ and $9\sin\theta = 0$, which means $\cos\theta = -1$ and $\sin\theta = 0$. We can take $\theta = \pi$ and express -9 in polar form as $\boxed{9\operatorname{cis}\pi}$.

(b)   We have $r = \sqrt{(\sqrt{6})^2 + (\sqrt{2})^2} = \sqrt{8} = 2\sqrt{2}$. We seek an angle θ such that $2\sqrt{2}\cos\theta = \sqrt{6}$ and $2\sqrt{2}\sin\theta = \sqrt{2}$, which means $\cos\theta = \frac{\sqrt{3}}{2}$ and $\sin\theta = \frac{1}{2}$. We can take $\theta = \frac{\pi}{6}$ and express $\sqrt{6} + \sqrt{2}i$ in polar form as $\boxed{2\sqrt{2}\operatorname{cis}\dfrac{\pi}{6}}$.

(c) We have $r = \sqrt{8^2 + 8^2} = \sqrt{128} = 8\sqrt{2}$. We seek an angle θ such that $8\sqrt{2}\cos\theta = 8$ and $8\sqrt{2}\sin\theta = 8$, which means $\cos\theta = \frac{\sqrt{2}}{2}$ and $\sin\theta = \frac{\sqrt{2}}{2}$. We can take $\theta = \frac{\pi}{4}$ and express $8 + 8i$ in polar form as $\boxed{8\sqrt{2}\operatorname{cis}\dfrac{\pi}{4}}$.

(d) We have $r = \sqrt{\left(-\frac{1}{2}\right)^2 + \left(-\frac{1}{2}\right)^2} = \sqrt{\frac{1}{2}} = \frac{\sqrt{2}}{2}$. We seek an angle θ such that $\frac{\sqrt{2}}{2}\cos\theta = -\frac{1}{2}$ and $\frac{\sqrt{2}}{2}\sin\theta = -\frac{1}{2}$, which means $\cos\theta = -\frac{\sqrt{2}}{2}$ and $\sin\theta = -\frac{\sqrt{2}}{2}$. We can take $\theta = \frac{5\pi}{4}$ and express $-\frac{1}{2} - \frac{1}{2}i$ in polar form as $\boxed{\dfrac{\sqrt{2}}{2}\operatorname{cis}\dfrac{5\pi}{4}}$.

7.34

(a) We have $4\operatorname{cis}120° = 4\cos 120° + 4i\sin 120° = 4\cdot\left(-\frac{1}{2}\right) + 4i\cdot\frac{\sqrt{3}}{2} = \boxed{-2 + 2i\sqrt{3}}$.

(b) We have $\frac{1}{2}\operatorname{cis}315° = \frac{1}{2}\cos 315° + \frac{1}{2}i\sin 315° = \frac{1}{2}\cdot\frac{\sqrt{2}}{2} + \frac{1}{2}i\cdot\left(-\frac{\sqrt{2}}{2}\right) = \boxed{\dfrac{\sqrt{2}}{4} - \dfrac{\sqrt{2}}{4}i}$.

(c) We have $8\operatorname{cis}\frac{\pi}{6} = 8\cos\frac{\pi}{6} + 8i\sin\frac{\pi}{6} = \boxed{4\sqrt{3} + 4i}$.

(d) We have $20\operatorname{cis}97\pi = 20\operatorname{cis}\pi = 20\cos\pi + 20i\sin\pi = \boxed{-20}$.

7.35 We have $\operatorname{cis}\theta = e^{i\theta}$ and $\operatorname{cis}2\theta = e^{2i\theta}$, so the given equation becomes $e^{i\theta} = e^{2i\theta}$. Dividing both sides by $e^{i\theta}$ (which is never 0), we get $e^{i\theta} = 1$. Then $\cos\theta + i\sin\theta = 1$, so $\cos\theta = 1$ and $\sin\theta = 0$. Hence, the solutions in θ are $\boxed{\text{integer multiples of } 2\pi}$.

We also can tackle this problem by noting that $\cos\theta + i\sin\theta = \cos 2\theta + i\sin 2\theta$ if and only if $\cos\theta = \cos 2\theta$ and $\sin\theta = \sin 2\theta$. From the latter equation, we have $\sin\theta = 2\sin\theta\cos\theta$, from which we find that either $\sin\theta = 0$ or $\cos\theta = \frac{1}{2}$.

If $\cos\theta = \frac{1}{2}$, then $\cos 2\theta = 2\cos^2\theta - 1 = -\frac{1}{2}$, so $\cos\theta = \cos 2\theta$ is not satisfied.

If $\sin\theta = 0$, then $\theta = k\pi$ for some integer k. If k is odd, then $\cos\theta = -1$ and $\cos 2\theta = 1$, so we cannot have $\cos\theta = \cos 2\theta$. If k is even, then $\cos\theta = \cos 2\theta = 1$. So, we have $\operatorname{cis}\theta = \operatorname{cis}2\theta$ for even multiples of π, as before.

7.36 By de Moivre's Theorem, we have

$$(\cos\theta + i\sin\theta)^{75} = \operatorname{cis}^{75}\theta = \operatorname{cis}75\theta.$$

Since $|\operatorname{cis}75\theta| = 1$, if $\operatorname{cis}75\theta$ is a real number, then it must be equal to 1 or -1.

But $\operatorname{cis}75\theta$ is equal to 1 or -1 if and only if 75θ is a multiple of $180°$, i.e. $75\theta = 180°k$ for some integer k. Then

$$\theta = \frac{180°k}{75} = \frac{12°k}{5}.$$

Thus, for θ to be an integer (in degrees), k must be a multiple of 5. Let $k = 5t$. Then

$$\theta = \frac{12°\cdot 5t}{5} = 12°t.$$

The angles θ of this form in the range $0° \le \theta \le 90°$ are $\boxed{0°, 12°, 24°, 36°, 48°, 60°, 72°, \text{ and } 84°}$.

7.37

(a) We have $r = \sqrt{24^2 + (-24)^2} = 24\sqrt{2}$. We seek an angle θ such that $24\sqrt{2}\cos\theta = 24$ and $24\sqrt{2}\sin\theta = -24$, which means $\cos\theta = \frac{\sqrt{2}}{2}$ and $\sin\theta = -\frac{\sqrt{2}}{2}$. We can take $\theta = \frac{7\pi}{4}$, so $24 - 24i = \boxed{24\sqrt{2}e^{7\pi i/4}}$.

(b)   We have $r = \sqrt{\left(\frac{4\sqrt{3}}{3}\right)^2 + 4^2} = \sqrt{\frac{64}{3}} = \frac{8\sqrt{3}}{3}$. We seek an angle θ such that $\frac{8\sqrt{3}}{3}\cos\theta = \frac{4\sqrt{3}}{3}$ and $\frac{8\sqrt{3}}{3}\sin\theta = 4$, which means $\cos\theta = \frac{1}{2}$ and $\sin\theta = \frac{\sqrt{3}}{2}$. We can take $\theta = \frac{\pi}{3}$, so $\frac{4\sqrt{3}}{3} + 4i = \boxed{\frac{8\sqrt{3}}{3}e^{\pi i/3}}$.

7.38

(a) $e^{3\pi i/2} = \cos\frac{3\pi}{2} + i\sin\frac{3\pi}{2} = \boxed{-i}$.

(b) $6e^{7\pi i/6} = 6\left(\cos\frac{7\pi}{6} + i\sin\frac{7\pi}{6}\right) = 6\left(-\frac{\sqrt{3}}{2} - \frac{1}{2}i\right) = \boxed{-3\sqrt{3} - 3i}$.

7.39   We see that $e^{-i\theta} = \cos(-\theta) + i\sin(-\theta)$. But

$$e^{-i\theta} = \frac{1}{e^{i\theta}} = \frac{1}{\cos\theta + i\sin\theta} = \frac{\cos\theta - i\sin\theta}{(\cos\theta + i\sin\theta)(\cos\theta - i\sin\theta)} = \frac{\cos\theta - i\sin\theta}{\cos^2\theta + \sin^2\theta} = \cos\theta - i\sin\theta.$$

Equating the real parts and equating the imaginary parts, we find $\cos(-\theta) = \cos\theta$ and $\sin(-\theta) = -\sin\theta$.

7.40   Let $z = re^{i\theta} = r\cos\theta + ir\sin\theta$. Then $\bar{z} = \overline{re^{i\theta}} = \bar{r}\cdot\overline{e^{i\theta}} = re^{-i\theta}$ so

$$\left|\frac{z}{\bar{z}} + \frac{\bar{z}}{z}\right| = \left|\frac{re^{i\theta}}{re^{-i\theta}} + \frac{re^{-i\theta}}{re^{i\theta}}\right| = |e^{2i\theta} + e^{-2i\theta}| = |\cos 2\theta + i\sin 2\theta + \cos(-2\theta) + i\sin(-2\theta)|$$
$$= |\cos 2\theta + i\sin 2\theta + \cos 2\theta - i\sin 2\theta| = 2|\cos 2\theta|.$$

Thus, the given equation becomes $2|\cos 2\theta| = \sqrt{3}$, which means $\cos 2\theta = \frac{\sqrt{3}}{2}$ or $\cos 2\theta = -\frac{\sqrt{3}}{2}$.

If $\cos 2\theta = \frac{\sqrt{3}}{2}$, then $2\theta = 2\pi n + \frac{\pi}{6} = \frac{(12n+1)\pi}{6}$ or $2\theta = 2\pi n + \frac{11\pi}{6} = \frac{(12n+11)\pi}{6}$ for some integer n, so $\theta = \frac{(12n+1)\pi}{12}$ or $\frac{(12n+11)\pi}{12}$.

If $\cos 2\theta = -\frac{\sqrt{3}}{2}$, then $2\theta = 2\pi n + \frac{5\pi}{6} = \frac{(12n+5)\pi}{6}$ or $2\theta = 2\pi n + \frac{7\pi}{6} = \frac{(12n+7)\pi}{6}$ for some integer n, so $\theta = \frac{(12n+5)\pi}{12}$ or $\frac{(12n+7)\pi}{12}$.

Therefore, the solutions are of the form $\boxed{re^{(12n+1)\pi i/12}, re^{(12n+5)\pi i/12}, re^{(12n+7)\pi i/12}, \text{and } re^{(12n+11)\pi i/12}}$, where r is a nonnegative real number and n is an integer. If we relax the restriction that r be nonnegative, then all solutions are of the form $re^{(12n+1)\pi i/12}$ or $re^{(12n+5)\pi i/12}$, since we have

$$re^{(12n+7)\pi i/12} = r\left(e^{(12n+1)\pi i/12}e^{6\pi i/12}\right) = -re^{(12n+1)\pi i/12} \qquad \text{and} \qquad re^{(12n+11)\pi i/12} = r\left(e^{(12n+5)\pi i/12}e^{6\pi i/12}\right) = -re^{(12n+5)\pi i/12}.$$

7.41

(a)   First, we express $\frac{\sqrt{3}}{4} - \frac{1}{4}i$ in exponential form. We have $r = \sqrt{\left(\frac{\sqrt{3}}{4}\right)^2 + \left(-\frac{1}{4}\right)^2} = \sqrt{\frac{1}{4}} = \frac{1}{2}$. We seek an angle θ such that $\frac{1}{2}\cos\theta = \frac{\sqrt{3}}{4}$ and $\frac{1}{2}\sin\theta = -\frac{1}{4}$, which means $\cos\theta = \frac{\sqrt{3}}{2}$ and $\sin\theta = -\frac{1}{2}$. We can take $\theta = \frac{11\pi}{6}$, so $\frac{\sqrt{3}}{4} - \frac{1}{4}i = \frac{1}{2}e^{11\pi i/6}$. Then

$$\left(\frac{\sqrt{3}}{4} - \frac{1}{4}i\right)^8 = \left(\frac{1}{2}e^{11\pi i/6}\right)^8 = \frac{1}{2^8}e^{44\pi i/3} = \frac{1}{256}\left(-\frac{1}{2} + \frac{\sqrt{3}}{2}i\right) = \boxed{-\frac{1}{512} + \frac{\sqrt{3}}{512}i}.$$

We also could have tackled this part by taking $\theta = -\frac{\pi}{6}$ in the exponential form of $\frac{\sqrt{3}}{4} - \frac{1}{4}i$. Then, we have

$$\left(\frac{\sqrt{3}}{4} - \frac{1}{4}i\right)^8 = \left(\frac{1}{2}e^{-\pi i/6}\right)^8 = \frac{1}{2^8}e^{-4\pi i/3} = \frac{1}{2^8}e^{2\pi i/3} = \frac{1}{256}\left(-\frac{1}{2} + \frac{\sqrt{3}}{2}i\right) = \boxed{-\frac{1}{512} + \frac{\sqrt{3}}{512}i}.$$

(b) First, we express $-3\sqrt{3} - 3i$ and $1 + \sqrt{3}i$ in exponential form. For $-3\sqrt{3} - 3i$, we have $r = \sqrt{(-3\sqrt{3})^2 + (-3)^2} = \sqrt{36} = 6$. We seek an angle θ such that $6\cos\theta = -3\sqrt{3}$ and $6\sin\theta = -3$, which means $\cos\theta = -\frac{\sqrt{3}}{2}$ and $\sin\theta = -\frac{1}{2}$. We can take $\theta = \frac{7\pi}{6}$, so $-3\sqrt{3} - 3i = 6e^{7\pi i/6}$.

For $1 + \sqrt{3}i$, we have $r = \sqrt{1^2 + (\sqrt{3})^2} = 2$. We seek an angle θ such that $2\cos\theta = 1$ and $2\sin\theta = \sqrt{3}$, which means $\cos\theta = \frac{1}{2}$ and $\sin\theta = \frac{\sqrt{3}}{2}$. We can take $\theta = \frac{\pi}{3}$, so $1 + \sqrt{3}i = 2e^{\pi i/3}$. Therefore,

$$\frac{(-3\sqrt{3} - 3i)^8}{(1 + \sqrt{3}i)^6} = \frac{(6e^{7\pi i/6})^8}{(2e^{\pi i/3})^6} = \frac{6^8 e^{28\pi i/3}}{2^6 e^{2\pi i}} = 26244\left(-\frac{1}{2} - \frac{\sqrt{3}}{2}i\right) = \boxed{-13122 - 13122\sqrt{3}i}.$$

7.42

(a) By de Moivre's Theorem, we have $\cos 5\theta + i\sin 5\theta = (\cos\theta + i\sin\theta)^5$, so

$$\cos 5\theta + i\sin 5\theta = \cos^5\theta + 5i\cos^4\theta\sin\theta - 10\cos^3\theta\sin^2\theta - 10i\cos^2\theta\sin^3\theta + 5\cos\theta\sin^4\theta + i\sin^5\theta.$$

This time, we equate the imaginary parts, to find

$$\sin 5\theta = 5\cos^4\theta\sin\theta - 10\cos^2\theta\sin^3\theta + \sin^5\theta.$$

Substituting $\cos^2\theta = 1 - \sin^2\theta$, we get

$$5\cos^4\theta\sin\theta - 10\cos^2\theta\sin^3\theta + \sin^5\theta = 5(1 - \sin^2\theta)^2\sin\theta - 10(1 - \sin^2\theta)\sin^3\theta + \sin^5\theta$$
$$= 5(1 - 2\sin^2\theta + \sin^4\theta)\sin\theta - 10(1 - \sin^2\theta)\sin^3\theta + \sin^5\theta$$
$$= 5\sin\theta - 10\sin^3\theta + 5\sin^5\theta - 10\sin^3\theta + 10\sin^5\theta + \sin^5\theta$$
$$= \boxed{16\sin^5\theta - 20\sin^3\theta + 5\sin\theta}.$$

(b) Suppose that there is a function f such that $f(\sin\theta) = \sin 4\theta$ for all θ. Taking $\theta = \frac{\pi}{3}$, we get $f(\sin\frac{\pi}{3}) = \sin\frac{4\pi}{3}$, or $f(\frac{\sqrt{3}}{2}) = -\frac{\sqrt{3}}{2}$. Taking $\theta = \frac{2\pi}{3}$, we get $f(\sin\frac{2\pi}{3}) = \sin\frac{8\pi}{3}$, or $f(\frac{\sqrt{3}}{2}) = \frac{\sqrt{3}}{2}$. But $f(\frac{\sqrt{3}}{2})$ cannot be two different values, so no such function f exists.

7.43 Let $w = e^{\pi i/3}$ and $z = e^{2\pi i/3}$, so $wz = e^{\pi i} = -1$ and $|w| = |z| = 1$. Also,

$$w + z = e^{\pi i/3} + e^{2\pi i/3} = \left(\frac{1}{2} + \frac{\sqrt{3}}{2}i\right) + \left(-\frac{1}{2} + \frac{\sqrt{3}}{2}i\right) = \sqrt{3}i,$$

which is not real, so part (a) is not necessarily true. We claim that $w + z$ is either real or pure imaginary.

To show that (b) is true, we write w and z in exponential form. They have the same magnitude, so we can express them as $w = re^{i\theta}$ and $z = re^{i\phi}$. Therefore, $wz = r^2 e^{i(\theta + \phi)} = r^2(\cos(\theta + \phi) + i\sin(\theta + \phi))$. This is a real number if and only if $\sin(\theta + \phi) = 0$, which is true if and only if $\theta + \phi = k\pi$ for some integer k. If $\theta + \phi = k\pi$ for some integer k, then we have

$$w + z = r(\cos\theta + i\sin\theta) + r(\cos\phi + i\sin\phi) = r(\cos\theta + \cos(k\pi - \theta)) + r(\sin\theta + \sin(k\pi - \theta))i.$$

This means that

$$\text{Re}(w + z) = r(\cos\theta + \cos(k\pi - \theta)) = r(\cos\theta + \cos(k\pi)\cos\theta + \sin(k\pi)\sin\theta) = r(\cos\theta + \cos(k\pi)\cos\theta),$$
$$\text{Im}(w + z) = r(\sin\theta + \sin(k\pi - \theta)) = r(\sin\theta + \sin(k\pi)\cos\theta - \cos(k\pi)\sin\theta) = r(\sin\theta - \cos(k\pi)\sin\theta),$$

where the last step in each line uses the fact that $\sin k\pi = 0$ for all integers k. Therefore, if k is odd, we have $\cos(k\pi) = -1$, so the real part of $w + z$ shown above is 0. If k is even, we have $\cos(k\pi) = 1$, which means the imaginary part of $w + z$ shown above is 0. In either case, we see that $\boxed{w + z \text{ is either real or pure imaginary}}$, as claimed.

7.44 First, we express $-4\sqrt{2} + 4\sqrt{2}i$ in exponential form. We have $r = \sqrt{(-4\sqrt{2})^2 + (4\sqrt{2})^2} = 8$. We seek an angle θ such that $8\cos\theta = -4\sqrt{2}$ and $8\sin\theta = 4\sqrt{2}$, which means $\cos\theta = -\frac{\sqrt{2}}{2}$ and $\sin\theta = \frac{\sqrt{2}}{2}$. We can take $\theta = \frac{3\pi}{4}$, so $-4\sqrt{2} + 4\sqrt{2}i = 8e^{3\pi i/4} = 2^3 e^{3\pi i/4}$.

We can now write the equation as $z^3 = 2^3 e^{3\pi i/4}$. Letting $z = re^{i\theta}$, we have $r^3 e^{3i\theta} = 2^3 e^{3\pi i/4}$, so $r = 2$. Solving $3i\theta = 3\pi i/4$ gives us $z = \boxed{2e^{\pi i/4}}$ as one solution. To find the other two solutions, we note that for any cube root of unity ω, we have $(2\omega e^{\pi i/4})^3 = 2^3 \omega^3 e^{3\pi i/4} = 2^3 e^{3\pi i/4}$, because $\omega^3 = 1$. Therefore, $2\omega e^{\pi i/4}$ is a solution to the original equation for each cube root of unity ω. We've already found the solution corresponding to $\omega = 1$. The ones corresponding to $\omega = e^{2\pi i/3}$ and $\omega = e^{4\pi i/3}$ are $2e^{\pi i/4} \cdot e^{2\pi i/3} = \boxed{2e^{11\pi i/12}}$ and $2e^{\pi i/4} \cdot \left(e^{2\pi i/3}\right)^2 = \boxed{2e^{19\pi i/12}}$.

7.45 *Solution 1.* We have $\frac{1+z}{1-z} = \frac{1 + \cos\theta + i\sin\theta}{1 - \cos\theta - i\sin\theta}$. We rationalize the denominator, to get

$$\frac{1 + \cos\theta + i\sin\theta}{1 - \cos\theta - i\sin\theta} = \frac{(1 + \cos\theta) + i\sin\theta}{(1 - \cos\theta) - i\sin\theta} \cdot \frac{(1 - \cos\theta) + i\sin\theta}{(1 - \cos\theta) + i\sin\theta}$$

$$= \frac{(1 + \cos\theta)(1 - \cos\theta) - \sin^2\theta + i(\sin\theta + \cos\theta\sin\theta + \sin\theta - \cos\theta\sin\theta)}{(1 - \cos\theta)^2 + \sin^2\theta}$$

$$= \frac{1 - \cos^2\theta - \sin^2\theta + 2i\sin\theta}{1 - 2\cos\theta + \cos^2\theta + \sin^2\theta} = \frac{2i\sin\theta}{2 - 2\cos\theta} = \frac{i\sin\theta}{1 - \cos\theta}.$$

From the half-angle formula for tangent, we have $\tan\frac{\theta}{2} = \frac{1 - \cos\theta}{\sin\theta}$, so $\frac{i\sin\theta}{1 - \cos\theta} = i\cot\frac{\theta}{2}$.

Solution 2. We have $\frac{1+z}{1-z} = \frac{1 + e^{i\theta}}{1 - e^{i\theta}}$. Dividing both the numerator and denominator by $e^{i\theta/2}$, we get $\frac{1 + e^{i\theta}}{1 - e^{i\theta}} = \frac{e^{i\theta/2} + e^{-i\theta/2}}{e^{-i\theta/2} - e^{i\theta/2}}$.

Since $\cos\frac{\theta}{2} = \frac{e^{i\theta/2} + e^{-i\theta/2}}{2}$ and $\sin\frac{\theta}{2} = \frac{e^{i\theta/2} - e^{-i\theta/2}}{2i}$, we have $\frac{e^{i\theta/2} + e^{-i\theta/2}}{e^{-i\theta/2} - e^{i\theta/2}} = \frac{2\cos\frac{\theta}{2}}{-2i\sin\frac{\theta}{2}} = i\cot\frac{\theta}{2}$.

7.46

(a) *Solution 1.* The nth roots of unity are $e^0, e^{2\pi i/n}, e^{4\pi i/n}, \ldots, e^{2(n-1)\pi i/n}$, so their product is e^S, where

$$S = 0 + \frac{2\pi i}{n} + \frac{4\pi i}{n} + \frac{6\pi i}{n} + \cdots + \frac{2(n-1)\pi i}{n} = \frac{2\pi i}{n}[1 + 2 + \cdots + (n-1)] = \frac{2\pi i}{n} \cdot \frac{(n-1)n}{2} = (n-1)\pi i.$$

If n is even, then $(n-1)\pi$ is an odd multiple of π, so $e^S = e^{(n-1)\pi i} = -1$, and if n is odd, then $(n-1)\pi$ is an even multiple of π, so $e^S = e^{(n-1)\pi i} = 1$. So to summarize, the product of the nth roots of unity is -1 if n is even, and 1 if n is odd.

Solution 2. In the text, we proved that if ω is an nth root of unity, then so is $\frac{1}{\omega}$. Thus, the nth roots of unity can be paired so that the product of the numbers in each pair is equal to 1. The only roots that cannot be so paired are 1 and -1, because they are their own reciprocals.

If n is odd, then -1 does not appear among the nth roots of unity, so in this case, the product of the roots is 1. However, if n is even, then -1 does appear among the nth roots of unity, so in this case, the product of the roots is -1.

Solution 3. The nth roots of unity are the roots of the equation $x^n - 1 = 0$. Letting $r_1, r_2, \ldots, r_n$ be the roots, we can write the equation as $(x - r_1)(x - r_2) \cdots (x - r_n) = 0$. In the expansion of the left side, the only constant term appears when we take the $-r_i$ from each factor, giving us the product $(-1)^n r_1 r_2 \cdots r_n$. Since the constant term in $x^n - 1$ is -1, we have $(-1)^n r_1 r_2 \cdots r_n = -1$. Therefore, the product of the roots is $(-1)^{1-n}$. Hence, the product of the nth roots of unity is -1 if n is even, and 1 if n is odd. (This approach is essentially equivalent to letting $x = 0$ in the factorization $x^n - 1 = (x - r_1)(x - r_2) \cdots (x - r_n)$.)

(b) We take a cue from Solution 2 of part (a). We know that the nth root of unity $e^{2k\pi i/n}$ is primitive if and only if $\gcd(k, n) = 1$. But

$$\frac{1}{\omega^{2k\pi i/n}} = \frac{\omega^{2nn\pi i/n}}{\omega^{2k\pi i/n}} = \omega^{2(n-k)\pi i/n},$$

and if $\gcd(k, n) = 1$, then $\gcd(n - k, n) = 1$, so $\frac{1}{e^{2k\pi i/n}}$ is also a primitive n^{th} root of unity. Thus, the primitive n^{th} roots of unity can be paired so that the product of the numbers in each pair is equal to 1, and again, the only exceptions are 1 and −1.

   We see that 1 is a primitive n^{th} root unity only for $n = 1$, and −1 is a primitive n^{th} root of unity only for $n = 2$. Hence, the product of the primitive n^{th} roots of unity is 1 for all $n \neq 2$, and −1 for $n = 2$.

7.47

(a) Let x_1 be a root of $x^{2n+1} + 1$, so $x_1^{2n+1} + 1 = 0$. Let $x_2 = -x_1$. Then

$$x_2^{2n+1} - 1 = (-x_1)^{2n+1} - 1 = -x_1^{2n+1} - 1 = -(x_1^{2n+1} + 1) = 0,$$

so x_2 is a root of $x^{2n+1} - 1$. Hence, the roots of $x^{2n+1} + 1$ are the negatives of the roots of $x^{2n+1} - 1$.

(b) If $x_2 = -x_1$, then geometrically, x_2 is the reflection of x_1 through the origin in the complex plane. (Alternatively, we can say that x_2 is a 180° rotation of x_1 about the origin.) In addition, the roots of $x^{2n+1} - 1$ form the vertices of a regular $(2n + 1)$-gon, as do the roots of $x^{2n+1} + 1$. Since $(x^{2n+1} - 1)(x^{2n+1} + 1) = x^{4n+2} - 1$, the two sets of roots together form the vertices of a regular $(4n + 2)$-gon.

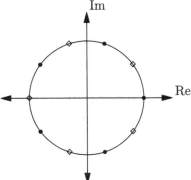

   As an example, we plot the roots for $n = 2$. The roots of $x^5 - 1$ are marked by a dot, and the roots of $x^5 + 1$ are marked by a diamond.

7.48 Let $w = re^{i\theta}$ in exponential form. Let $z_k = r^{1/n}e^{(\theta+2k\pi)i/n}$, for $1 \le k \le n$. Then

$$z_k^n = [r^{1/n}e^{(\theta+2k\pi)i/n}]^n = re^{(\theta+2k\pi)i} = re^{i\theta} \cdot e^{2k\pi i} = w.$$

Hence, $z_1, z_2, \ldots, z_n$ are the n roots of $z^n = w$. These roots all have the same magnitude, and their arguments are $\frac{\theta+2\pi}{n}, \frac{\theta+4\pi}{n}, \ldots, \frac{\theta+2n\pi}{n}$, which form an arithmetic sequence with common difference $\frac{2\pi}{n}$. Therefore, the roots form the vertices of a regular n-gon.

7.49 First, we express $1 + \sqrt{3}i$ in exponential form. We have $r = \sqrt{1^2 + (\sqrt{3})^2} = 2$. We seek an angle θ such that $2\cos\theta = 1$ and $2\sin\theta = \sqrt{3}$, which means $\cos\theta = \frac{1}{2}$ and $\sin\theta = \frac{\sqrt{3}}{2}$. We can take $\theta = \frac{\pi}{3}$, so $1 + \sqrt{3}i = 2e^{\pi i/3}$.

   Then $x_{19} + iy_{19} = (2e^{\pi i/3})^{19}$ and $x_{91} + iy_{91} = (2e^{\pi i/3})^{91}$. We could use these equations to find x_{19}, y_{19}, x_{91}, and y_{91} individually. But instead, we'll show a slicker solution. We don't need to find each of these values; we are only asked for $x_{19}y_{91} + x_{91}y_{19}$. This expression will appear if we take the product of the equations $x_{19} + iy_{19} = (2e^{\pi i/3})^{19}$ and $x_{91} + iy_{91} = (2e^{\pi i/3})^{91}$. This product gives us

$$(x_{19} + iy_{19})(x_{91} + iy_{91}) = (2e^{\pi i/3})^{110}.$$

Expanding the left side gives us

$$(x_{19} + iy_{19})(x_{91} + iy_{91}) = x_{19}x_{91} + ix_{19}y_{91} + ix_{91}y_{19} + i^2y_{19}y_{91} = (x_{19}x_{91} - y_{19}y_{91}) + i(x_{19}y_{91} + x_{91}y_{19}).$$

Thus, $x_{19}y_{91} + x_{91}y_{19}$ is the imaginary part of $(2e^{\pi i/3})^{110} = 2^{110}e^{110\pi i/3} = 2^{110}\left(-\frac{1}{2} + \frac{\sqrt{3}}{2}i\right) = -2^{109} + 2^{109}\sqrt{3}i$, so the answer is $k = \boxed{109}$.

7.50 Let $z = x + yi$. Then $e^z = e^{x+yi} = e^x \cdot e^{iy} = e^x(\cos y + i\sin y)$, so $\overline{e^z} = e^x(\cos y - i\sin y)$. Also, $e^{\bar{z}} = e^{x-yi} = e^x e^{-yi} = e^x(\cos(-y) + i\sin(-y)) = e^x(\cos y - i\sin y)$. Hence, $\overline{e^z} = e^{\bar{z}}$.

7.51

(a) $\cosh^2 x - \sinh^2 x = \left(\dfrac{e^x + e^{-x}}{2}\right)^2 - \left(\dfrac{e^x - e^{-x}}{2}\right)^2 = \dfrac{e^{2x} + 2 + e^{-2x}}{4} - \dfrac{e^{2x} - 2 + e^{-2x}}{4} = 1.$

(b) We have $\sinh(x + y) = \frac{e^{x+y} - e^{-x-y}}{2}$, and

$$\sinh x \cosh y + \cosh x \sinh y = \frac{e^x - e^{-x}}{2} \cdot \frac{e^y + e^{-y}}{2} + \frac{e^x + e^{-x}}{2} \cdot \frac{e^y - e^{-y}}{2}$$
$$= \frac{e^{x+y} + e^{x-y} - e^{-x+y} - e^{-x-y}}{4} + \frac{e^{x+y} - e^{x-y} + e^{-x+y} - e^{-x-y}}{4}$$
$$= \frac{2e^{x+y} - 2e^{-x-y}}{4} = \frac{e^{x+y} - e^{-x-y}}{2}.$$

Hence, $\sinh(x + y) = \sinh x \cosh y + \cosh x \sinh y.$

(c) In general, we have $\tanh x = \frac{e^x - e^{-x}}{e^x + e^{-x}} = \frac{\frac{e^x - e^{-x}}{2}}{\frac{e^x + e^{-x}}{2}} = \frac{\sinh x}{\cosh x}$, so $\tanh(x + y) = \frac{\sinh(x+y)}{\cosh(x+y)}$. We have an identity for $\sinh(x + y)$; let's see if we can make one for $\cosh(x + y)$. We note that the identity for $\sinh(x + y)$ looks a lot like the identity for $\sin(x + y)$. Maybe the identity for $\cosh(x + y)$ will look a lot like the identity for $\cos(x + y)$. Let's write $\cosh(x + y)$, $\cosh x \cosh y$, and $\sinh x \sinh y$ and see if we can relate them:

$$\cosh(x + y) = \frac{e^{x+y} + e^{-(x+y)}}{2},$$
$$\cosh x \cosh y = \frac{e^x + e^{-x}}{2} \cdot \frac{e^y + e^{-y}}{2} = \frac{e^{x+y} + e^{x-y} + e^{y-x} + e^{-(x+y)}}{4},$$
$$\sinh x \sinh y = \frac{e^x - e^{-x}}{2} \cdot \frac{e^y - e^{-y}}{2} = \frac{e^{x+y} - e^{x-y} - e^{y-x} + e^{-(x+y)}}{4}.$$

Sure enough, when we add $\cosh x \cosh y$ and $\sinh x \sinh y$, we get $\cosh(x + y)$. So, we have

$$\tanh(x + y) = \frac{\sinh(x + y)}{\cosh(x + y)} = \frac{\sinh x \cosh y + \cosh x \sinh y}{\cosh x \cosh y + \sinh x \sinh y} = \frac{\frac{\sinh x}{\cosh x} + \frac{\sinh y}{\cosh y}}{1 + \frac{\sinh x \sinh y}{\cosh x \cosh y}} = \frac{\tanh x + \tanh y}{1 + \tanh x \tanh y}.$$

(d) We have $\sinh x + \sinh y = \frac{e^x - e^{-x} + e^y - e^{-y}}{2}$, and

$$2 \sinh \frac{x + y}{2} \cosh \frac{x - y}{2} = 2 \cdot \frac{e^{(x+y)/2} - e^{-(x+y)/2}}{2} \cdot \frac{e^{(x-y)/2} + e^{-(x-y)/2}}{2} = \frac{e^x + e^y - e^{-y} - e^{-x}}{2}.$$

Hence, $\sinh x + \sinh y = 2 \sinh \frac{x+y}{2} \cosh \frac{x-y}{2}.$

7.52 Let $f(n) = x^n + y^n$. First, we express x and y in exponential form. We see that $x = -\frac{1}{2} + \frac{\sqrt{3}}{2}i = \cos \frac{2\pi}{3} + i \sin \frac{2\pi}{3} = e^{2\pi i/3}$, and $y = -\frac{1}{2} - \frac{\sqrt{3}}{2}i = \cos \frac{4\pi}{3} + i \sin \frac{4\pi}{3} = e^{4\pi i/3}$. Then $x^3 = e^{2\pi i} = 1$ and $y^3 = e^{4\pi i} = 1$. Hence, $f(n + 3) = x^{n+3} + y^{n+3} = x^n \cdot x^3 + y^n \cdot y^3 = x^n + y^n = f(n)$. This tells us that $f(n) = f(0)$ if n is divisible by 3, $f(n) = f(1)$ if n is one more than a multiple of 3, and $f(n) = f(2)$ if n is two more than a multiple of 3. Therefore, we only need to check $f(0)$, $f(1)$, and $f(2)$.

We see that $f(0) = x^0 + y^0 = 2$, and $f(1) = x + y = -1$. To compute $f(2)$, note that $x^2 = e^{4\pi i/3} = y$. Therefore, $f(2) = x^2 + y^2 = y + x^4 = y + x = f(1) = -1$. Hence, $f(n) = -1$ if and only if $\boxed{n \text{ is not a multiple of 3}}$.

Challenge Problems

7.53 The polynomial $z^9 - 1$ factors as $z^9 - 1 = (z^3 - 1)(z^6 + z^3 + 1)$. Hence, the roots of $z^6 + z^3 + 1$ are the ninth roots of unity that are not cube roots of unity.

The ninth roots of unity are $\operatorname{cis} 0°$, $\operatorname{cis} 40°$, $\operatorname{cis} 80°$, ..., $\operatorname{cis} 320°$. The cube roots of unity are $\operatorname{cis} 0°$, $\operatorname{cis} 120°$, and $\operatorname{cis} 240°$. The only ninth root of unity that is not a cube root of unity, and whose argument is between $90°$ and $180°$, is $\operatorname{cis} 160°$, so the answer is $\boxed{160°}$.

7.54 Since $x^5 - 1 = (x - 1)(x^4 + x^3 + x^2 + x + 1)$, and $z \neq 1$, z must satisfy the equation $z^4 + z^3 + z^2 + z + 1 = 0$. Then

$$z^{15} + z^{16} + z^{17} + z^{18} + z^{19} = z^{15}(1 + z + z^2 + z^3 + z^4) = 1 \cdot 0 = 0.$$

Similarly, each 5 consecutive terms in the sum add to 0, leaving only $z^{50} = \boxed{1}$.

7.55 We factor the first two terms and the second two terms to write the equation as $z^4(z^3 + 8) - 2i(z^3 + 8) = 0$. Aha! We can factor more, to get $(z^4 - 2i)(z^3 + 8) = 0$. Therefore, the solutions to the equation are the solutions of $z^4 = 2i$ and $z^3 = -8$.

Since $2i = 2e^{\pi i/2}$, the solutions to $z^4 = 2i$ are the four numbers of the form $2^{1/4}\omega e^{\pi i/8}$, where ω is a fourth root of unity. The fourth roots of unity are $1, e^{\pi i/2}, e^{\pi i}, e^{3\pi i/2}$. So, the four solutions to $z^4 = 2i$ are given by the four complex numbers $\boxed{\sqrt[4]{2}e^{\pi i/8}, \sqrt[4]{2}e^{5\pi i/8}, \sqrt[4]{2}e^{9\pi i/8}, \sqrt[4]{2}e^{13\pi i/8}}$.

Since $-8 = 8e^{\pi i}$, the solutions to $z^3 = -8$ are the three numbers of the form $2\omega e^{\pi i/3}$, where ω is a cube root of unity. The cube roots of unity are $1, e^{2\pi i/3}, e^{4\pi i/3}$, so the solutions to $z^3 = -8$ are $\boxed{2e^{\pi i/3}, 2e^{\pi i}, 2e^{5\pi i/3}}$.

7.56 The argument of the complex number $z = 1 + 2i$ is θ. Then

$$z^2 = (1 + 2i)^2 = 1 + 4i - 4 = -3 + 4i,$$
$$z^4 = (z^2)^2 = (-3 + 4i)^2 = 9 - 24i - 16 = -7 - 24i,$$
$$z^8 = (z^4)^2 = (-7 - 24i)^2 = 49 + 336i - 576 = -527 + 336i.$$

Since the real part of z^8 is negative and the imaginary part is positive, 8θ lies in the $\boxed{\text{second quadrant}}$.

7.57 We can rewrite the given equation as $z^9 + z^6 + z^3 + 1 = 0$. Since $z^{12} - 1$ factors as

$$z^{12} - 1 = (z^3 - 1)(z^9 + z^6 + z^3 + 1),$$

the roots of $z^9 + z^6 + z^3 + 1 = 0$ are the 12th roots of unity that are not cube roots of unity.

The 12th roots of unity are e^0, $e^{2\pi i/12}$, $e^{4\pi i/12}$, ..., $e^{22\pi i/12}$. The cube roots of unity are e^0, $e^{2\pi i/3} = e^{8\pi i/12}$, and $e^{4\pi i/3} = e^{16\pi i/12}$. Therefore, the roots of $z^9 + z^6 + z^3 + 1 = 0$ are

$$\boxed{e^{2\pi i/12}, e^{4\pi i/12}, e^{6\pi i/12}, e^{10\pi i/12}, e^{12\pi i/12}, e^{14\pi i/12}, e^{18\pi i/12}, e^{20\pi i/12}, \text{ and } e^{22\pi i/12}}.$$

7.58 Since $(x - 1)(x^2 + x + 1) = x^3 - 1$, the roots of $x^2 + x + 1 = 0$ are the cube roots of unity other than 1. Let these roots be α and β.

Let $f_n(x) = x^{2n} + 1 + (x + 1)^{2n}$. Then by the Factor Theorem, $f_n(x)$ is divisible by $x^2 + x + 1 = (x - \alpha)(x - \beta)$ if and only if $f_n(\alpha) = f_n(\beta) = 0$. We see that $f_n(\alpha) = \alpha^{2n} + 1 + (\alpha + 1)^{2n}$. Since $\alpha^2 + \alpha + 1 = 0$, we have $\alpha^2 = -\alpha - 1$, so

$$(\alpha + 1)^{2n} = ((\alpha + 1)^2)^n = (\alpha^2 + 2\alpha + 1)^n = (-\alpha - 1 + 2\alpha + 1)^n = \alpha^n,$$

so

$$f_n(\alpha) = \alpha^{2n} + 1 + (\alpha + 1)^{2n} = \alpha^{2n} + 1 + \alpha^n = \alpha^{2n} + \alpha^n + 1.$$

This is a familiar form; if we let $y = \alpha^n$, we have $y^2 + y + 1$, which equals $\frac{y^3 - 1}{y - 1}$ if $y \neq 1$. So, if $\alpha^n \neq 1$, we have

$$f_n(\alpha) = \frac{\alpha^{3n} - 1}{\alpha^n - 1}.$$

If $\alpha^n = 1$, we have $f_n(\alpha) = \alpha^{2n} + \alpha^n + 1 = 1 + 1 + 1 = 3$, so α is not a root of $f_n(\alpha)$ if $\alpha^n = 1$. Since α is a primitive cube root of unity, we have $\alpha^n = 1$ if and only if n is a multiple of 3.

If n is not a multiple of 3, then $\alpha^n \neq 1$, and $f_n(\alpha) = \frac{\alpha^{3n}-1}{\alpha^n-1}$. However, because α is a cube root of unity, we have $\alpha^{3n} = (\alpha^3)^n = 1^n = 1$, so $f_n(\alpha) = 0$, which means α is a root of $f_n(\alpha)$. Notice that nothing in this argument depends on which primitive cube root of unity α is, so we have shown that both roots of $x^2 + x + 1$ are roots of $x^{2n} + 1 + (x+1)^{2n}$ if and only if $\boxed{n \text{ is not a multiple of } 3}$.

7.59 Let $w = 2z$, so $z = w/2$. Then

$$32z^5 + 16z^4 + 8z^3 + 4z^2 + 2z + 1 = (2z)^5 + (2z)^4 + (2z)^3 + (2z)^2 + 2z + 1 = w^5 + w^4 + w^3 + w^2 + w + 1.$$

Since $w^6 - 1 = (w - 1)(w^5 + w^4 + w^3 + w^2 + w + 1)$, the roots of $w^5 + w^4 + w^3 + w^2 + w + 1$ are the sixth roots of unity other than 1. Hence, the solutions in z are $\boxed{\frac{1}{2}e^{2\pi i/6}, \frac{1}{2}e^{4\pi i/6}, \frac{1}{2}e^{6\pi i/6}, \frac{1}{2}e^{8\pi i/6}, \text{ and } \frac{1}{2}e^{10\pi i/6}}$.

7.60 A complex number z is real if and only if $z = \bar{z}$. Since x is a root of unity, we have $|x| = 1$, so $\bar{x} = 1/x$. Similarly, $\bar{y} = 1/y$. Then

$$\overline{(x+y)^k} = (\bar{x} + \bar{y})^k = \left(\frac{1}{x} + \frac{1}{y}\right)^k = \left(\frac{x+y}{xy}\right)^k = \frac{(x+y)^k}{x^k y^k} = (x+y)^k,$$

where $x^k = 1$ and $y^k = 1$ because x and y are kth roots of unity. Since $\overline{(x+y)^k} = (x+y)^k$, we know that $(x+y)^k$ is real. (See if you can find a geometric approach to solve the problem!)

7.61 To evaluate the given sum, we pair the terms. Let $1 \leq k \leq 1996$. Then

$$\frac{1}{1+\omega^k} + \frac{1}{1+\omega^{1997-k}} = \frac{1 + \omega^{1997-k} + 1 + \omega^k}{(1+\omega^k)(1+\omega^{1997-k})} = \frac{2 + \omega^k + \omega^{1997-k}}{1 + \omega^k + \omega^{1997-k} + \omega^{1997}} = \frac{2 + \omega^k + \omega^{1997-k}}{2 + \omega^k + \omega^{1997-k}} = 1.$$

Hence,

$$\frac{1}{1+\omega} + \frac{1}{1+\omega^{1996}} = 1,$$
$$\frac{1}{1+\omega^2} + \frac{1}{1+\omega^{1995}} = 1,$$
$$\frac{1}{1+\omega^3} + \frac{1}{1+\omega^{1994}} = 1,$$
$$\vdots$$
$$\frac{1}{1+\omega^{998}} + \frac{1}{1+\omega^{999}} = 1,$$

and

$$\frac{1}{1+\omega^{1997}} = \frac{1}{1+1} = \frac{1}{2}.$$

Therefore,

$$\frac{1}{1+\omega} + \frac{1}{1+\omega^2} + \cdots + \frac{1}{1+\omega^{1997}} = 998 + \frac{1}{2} = \boxed{\frac{1997}{2}}.$$

7.62 Note that $|z_{n+1}| = \left|\frac{iz_n}{\bar{z}_n}\right| = \frac{|z_n|}{|\bar{z}_n|} = 1$ for all $n \geq 0$. We are given that $|z_0| = 1$, so $|z_n| = 1$ for all $n \geq 0$. Then $z_n \bar{z}_n = |z_n|^2 = 1$, so $\bar{z}_n = 1/z_n$. Hence,

$$z_{n+1} = \frac{iz_n}{\bar{z}_n} = \frac{iz_n}{1/z_n} = iz_n^2.$$

for all $n \geq 0$.

Then

$$z_1 = iz_0^2,$$
$$z_2 = iz_1^2 = i(iz_0^2)^2 = -iz_0^4,$$
$$z_3 = iz_2^2 = i(-iz_0^4)^2 = -iz_0^8,$$
$$z_4 = iz_3^2 = i(-iz_0^8)^2 = -iz_0^{16},$$

and so forth. In general, we see that $z_n = -iz_0^{2^n}$ for all $n \geq 2$. Hence, the equation $z_{2005} = 1$ becomes

$$-iz_0^{2^{2005}} = 1,$$

or $z_0^{2^{2005}} = i$. This equation has 2^{2005} roots, so there are $\boxed{2^{2005}}$ possible values for z_0.

7.63 The recursive definitions we are given for a_{n+1} and b_{n+1} have the form of the angle sum and angle difference identities for sine and cosine. We consider taking advantage of this by letting α and β be a sine or cosine, but then we note that α and β can be greater than 1 or less than -1. But thinking of sines and cosines leads us to complex numbers, which also produce expressions of these forms.

We let $z = \alpha + \beta i$, and let $z_n = a_n + b_n i$. Then

$$zz_n = (\alpha + \beta i)(a_n + b_n i) = \alpha a_n + \alpha b_n i + \beta a_n i - \beta b_n = \alpha a_n - \beta b_n + (\beta a_n + \alpha b_n)i = a_{n+1} + b_{n+1}i = z_{n+1}.$$

Also, $z_1 = \alpha + \beta i = z$, so $z_n = z^n$ for all positive integers n.

If $a_{1997} = a_1$ and $b_{1997} = b_1$, then $z_{1997} = z_1$. Since $z_n = z^n$ for all integers n, we must have $z^{1997} = z$. Rearranging gives $z^{1997} - z = 0$, which factors as $z(z^{1996} - 1) = 0$. Hence, there are 1997 possible values of $z = \alpha + \beta i$, namely 0 and the 1996th roots of unity, which means there are $\boxed{1997}$ possible pairs (α, β).

7.64 By de Moivre's Theorem, we have $\cos n\theta + i \sin n\theta = (\cos \theta + i \sin \theta)^n$. But by the Binomial Theorem,

$$(\cos \theta + i \sin \theta)^n = \cos^n \theta + i \binom{n}{1} \cos^{n-1} \theta \sin \theta - \binom{n}{2} \cos^{n-2} \theta \sin^2 \theta - \binom{n}{3} i \cos^{n-3} \theta \sin^3 \theta + \cdots.$$

Equating the real parts and equating the imaginary parts, we get

$$\cos n\theta = \cos^n \theta - \binom{n}{2} \cos^{n-2} \theta \sin^2 \theta + \cdots,$$

$$\sin n\theta = \binom{n}{1} \cos^{n-1} \theta \sin \theta - \binom{n}{3} \cos^{n-3} \theta \sin^3 \theta + \cdots.$$

Now, we can tackle $\tan n\theta$:

$$\tan n\theta = \frac{\sin n\theta}{\cos n\theta} = \frac{\binom{n}{1} \cos^{n-1} \theta \sin \theta - \binom{n}{3} \cos^{n-3} \theta \sin^3 \theta + \cdots}{\cos^n \theta - \binom{n}{2} \cos^{n-2} \theta \sin^2 \theta + \cdots}.$$

Dividing the numerator and denominator by $\cos^n \theta$, we find

$$\frac{\binom{n}{1} \cos^{n-1} \theta \sin \theta - \binom{n}{3} \cos^{n-3} \theta \sin^3 \theta + \cdots}{\cos^n \theta - \binom{n}{2} \cos^{n-2} \theta \sin^2 \theta + \cdots} = \frac{\binom{n}{1} \frac{\sin \theta}{\cos \theta} - \binom{n}{3} \frac{\sin^3 \theta}{\cos^3 \theta} + \cdots}{1 - \binom{n}{2} \frac{\sin^2 \theta}{\cos^2 \theta} + \cdots} = \frac{\binom{n}{1} \tan \theta - \binom{n}{3} \tan^3 \theta + \cdots}{1 - \binom{n}{2} \tan^2 \theta + \cdots}.$$

7.65 From the given equation, we get $z^2 - (2\cos 3°)z + 1 = 0$. Then by the quadratic formula, we have

$$z = \frac{2\cos 3° \pm \sqrt{4\cos^2 3° - 4}}{2} = \frac{2\cos 3° \pm \sqrt{-4(1 - \cos^2 3°)}}{2} = \frac{2\cos 3° \pm 2i\sqrt{\sin^2 3°}}{2} = \cos 3° \pm i \sin 3°.$$

By de Moivre's Theorem,

$$(\cos 3° \pm i \sin 3°)^{2000} = \cos 6000° \pm i \sin 6000° = \cos 240° \pm i \sin 240° = -\frac{1}{2} \mp \frac{\sqrt{3}}{2}i,$$

so $z^{2000} + \frac{1}{z^{2000}} = -\frac{1}{2} - \frac{\sqrt{3}}{2}i - \frac{1}{2} + \frac{\sqrt{3}}{2}i = -1$. The least integer that is greater than -1 is $\boxed{0}$.

7.66

(a) Neither $x^n - 1$ nor any of the $\Phi_{d_i}(x)$ has repeated roots, and all of these polynomials are monic. Also, no complex number ω can be the root of two different $\Phi_{d_i}(x)$, since, by definition, ω is a root of $\Phi_{d_i}(x)$ if and only if d_i is the smallest positive integer such that $\omega^{d_i} = 1$. Therefore, if we show that each root of $x^n - 1$ is a root of exactly one of the $\Phi_{d_i}(x)$, and that each root of each $\Phi_{d_i}(x)$ is a root of $x^n - 1$, then we can conclude that $x^n - 1$ equals the product of the $\Phi_{d_i}(x)$.

First, let ω be a root of $\Phi_{d_i}(x)$, so $\omega^{d_i} = 1$. Since d_i is a divisor of n, we have $k = \frac{n}{d_i}$ for some integer k. Therefore, we have $\omega^n = \left(\omega^{d_i}\right)^{n/d_i} = \left(\omega^{d_i}\right)^k = 1^k = 1$, so ω is a root of $x^n - 1$. This tells us that every root of each $\Phi_{d_i}(x)$ is a root of $x^n - 1$.

Next, we let ω be a root of $x^n - 1$, so $\omega^n = 1$. The set of positive integers k such that $\omega^k = 1$ is non-empty. Let d be the smallest such positive integer. Then ω is a primitive d^{th} root of unity, so $\omega^d = 1$.

Let q and r be the quotient and remainder when n is divided by d, respectively, so $n = qd + r$ and $0 \le r \le d - 1$. Then $\omega^n = \omega^{qd+r} = 1$. But $\omega^{qd+r} = (\omega^d)^q \cdot \omega^r = \omega^r$, so $\omega^r = 1$. But $r \le d - 1$, so r cannot be positive. (By definition, d is the smallest positive integer such that $\omega^d = 1$.) Therefore, $r = 0$, which means $n = qd$. Hence, d must be a divisor of n. This tells us that every root of $x^n - 1$ is a root of of some $\Phi_{d_i}(x)$.

We conclude that $x^n - 1$ can be factored as $\Phi_{d_1}(x)\Phi_{d_2}(x)\Phi_{d_3}(x)\cdots\Phi_{d_k}(x)$, where $d_1, d_2, d_3, \ldots, d_k$ are all of the positive divisors of n.

(b) Taking $n = 1$ in the identity in part (a), we get $\Phi_1(x) = \boxed{x - 1}$.

Taking $n = 2$, we get $\Phi_1(x)\Phi_2(x) = x^2 - 1$, so

$$\Phi_2(x) = \frac{x^2 - 1}{\Phi_1(x)} = \frac{(x-1)(x+1)}{x-1} = \boxed{x+1}.$$

Taking $n = 3$, we get $\Phi_1(x)\Phi_3(x) = x^3 - 1$, so

$$\Phi_3(x) = \frac{x^3 - 1}{\Phi_1(x)} = \frac{(x-1)(x^2+x+1)}{x-1} = \boxed{x^2 + x + 1}.$$

Taking $n = 4$, we get $\Phi_1(x)\Phi_2(x)\Phi_4(x) = x^4 - 1$, so

$$\Phi_4(x) = \frac{x^4 - 1}{\Phi_1(x)\Phi_2(x)}.$$

But $\Phi_1(x)\Phi_2(x) = x^2 - 1$, so

$$\Phi_4(x) = \frac{x^4 - 1}{x^2 - 1} = \frac{(x^2-1)(x^2+1)}{x^2-1} = \boxed{x^2 + 1}.$$

Taking $n = 5$, we get $\Phi_1(x)\Phi_5(x) = x^5 - 1$, so

$$\Phi_5(x) = \frac{x^5 - 1}{\Phi_1(x)} = \frac{(x-1)(x^4+x^3+x^2+x+1)}{x-1} = \boxed{x^4 + x^3 + x^2 + x + 1}.$$

Taking $n = 6$, we get $\Phi_1(x)\Phi_2(x)\Phi_3(x)\Phi_6(x) = x^6 - 1$, so

$$\Phi_6(x) = \frac{x^6 - 1}{\Phi_1(x)\Phi_2(x)\Phi_3(x)}.$$

But $\Phi_1(x)\Phi_3(x) = x^3 - 1$, $\Phi_2(x) = x + 1$, and the numerator factors as

$$x^6 - 1 = (x^3 - 1)(x^3 + 1) = (x^3 - 1)(x + 1)(x^2 - x + 1),$$

so

$$\Phi_6(x) = \frac{(x^3 - 1)(x + 1)(x^2 - x + 1)}{(x^3 - 1)(x + 1)} = \boxed{x^2 - x + 1}.$$

Taking $n = 7$, we get $\Phi_1(x)\Phi_7(x) = x^7 - 1$, so

$$\Phi_7(x) = \frac{x^7 - 1}{\Phi_1(x)} = \frac{(x - 1)(x^6 + x^5 + x^4 + x^3 + x^2 + x + 1)}{x - 1} = \boxed{x^6 + x^5 + x^4 + x^3 + x^2 + x + 1}.$$

Taking $n = 8$, we get $\Phi_1(x)\Phi_2(x)\Phi_4(x)\Phi_8(x) = x^8 - 1$, so

$$\Phi_8(x) = \frac{x^8 - 1}{\Phi_1(x)\Phi_2(x)\Phi_4(x)}.$$

But $\Phi_1(x)\Phi_2(x)\Phi_4(x) = x^4 - 1$, so

$$\Phi_8(x) = \frac{x^8 - 1}{x^4 - 1} = \frac{(x^4 - 1)(x^4 + 1)}{x^4 - 1} = \boxed{x^4 + 1}.$$

Taking $n = 9$, we get $\Phi_1(x)\Phi_3(x)\Phi_9(x) = x^9 - 1$, so

$$\Phi_9(x) = \frac{x^9 - 1}{\Phi_1(x)\Phi_3(x)}.$$

But $\Phi_1(x)\Phi_3(x) = x^3 - 1$, so

$$\Phi_9(x) = \frac{x^9 - 1}{x^3 - 1} = \frac{(x^3 - 1)(x^6 + x^3 + 1)}{x^3 - 1} = \boxed{x^6 + x^3 + 1}.$$

Taking $n = 10$, we get $\Phi_1(x)\Phi_2(x)\Phi_5(x)\Phi_{10}(x) = x^{10} - 1$, so

$$\Phi_{10}(x) = \frac{x^{10} - 1}{\Phi_1(x)\Phi_2(x)\Phi_5(x)}.$$

But $\Phi_1(x)\Phi_5(x) = x^5 - 1$, $\Phi_2(x) = x + 1$, and the numerator factors as

$$x^{10} - 1 = (x^5 - 1)(x^5 + 1) = (x^5 - 1)(x + 1)(x^4 - x^3 + x^2 - x + 1),$$

so

$$\Phi_{10}(x) = \frac{(x^5 - 1)(x + 1)(x^4 - x^3 + x^2 - x + 1)}{(x^5 - 1)(x + 1)} = \boxed{x^4 - x^3 + x^2 - x + 1}.$$

(c) Taking $n = p$, we get $\Phi_1(x)\Phi_p(x) = x^p - 1$, so

$$\Phi_p(x) = \frac{x^p - 1}{\Phi_1(x)} = \frac{x^p - 1}{x - 1} = \boxed{x^{p-1} + x^{p-2} + \cdots + x + 1}.$$

(d) Let $\omega = e^{2\pi i/n}$, and let ω^k be a primitive n^{th} root of unity. Then, we have $\gcd(n, k) = 1$, so $\gcd(n, n-k) = 1$ as well, which means ω^{n-k} is a primitive root of unity. Since $n > 2$, we know that we do not have $k = \frac{n}{2}$, so ω^k and ω^{n-k} are distinct. Therefore, the roots of $\Phi_n(x)$ come in pairs (and none is repeated), so the degree of $\Phi_n(x)$ is even.

(e) Since the degree of $\Phi_n(x)$ is even, the leading coefficient of $\Phi_n(-x)$ is 1. Also, neither $\Phi_{2n}(x)$ nor $\Phi_n(-x)$ has any repeated roots. Therefore, to show that $\Phi_{2n}(x) = \Phi_n(-x)$, it suffices to show that the roots of $\Phi_{2n}(x)$ and $\Phi_n(-x)$ are the same. So, we will show that every root of $\Phi_n(-x)$ is a root of $\Phi_{2n}(x)$, and vice versa.

Every root of $\Phi_n(-x)$ is the negative of a root of $\Phi_n(x)$. The roots of $\Phi_n(x)$ are the primitive n^{th} roots of unity, so each is of the form $e^{2\pi k i/n}$, where k is an integer such that $\gcd(k, n) = 1$. We must therefore show that $-e^{2\pi k i/n}$ is a primitive $(2n)^{\text{th}}$ root of unity. Since $-1 = e^{\pi i}$, we have $-e^{2\pi k i/n} = e^{\pi i}e^{2\pi k i/n} = e^{(2k+n)\pi i/n}$. We then have

$$\left(-e^{2\pi k i/n}\right)^{2n} = \left(e^{(2k+n)\pi i/n}\right)^{2n} = e^{2(2k+n)\pi i} = 1,$$

so $e^{(2k+n)\pi i/n}$ is a $(2n)^{\text{th}}$ root of unity. But is it a primitive $(2n)^{\text{th}}$ root of unity? To see that it is, we note that $e^{(2k+n)\pi i/n} = e^{2(2k+n)\pi i/(2n)} = (e^{2\pi i/(2n)})^{2k+n}$. So, we must show that $\gcd(2k+n, 2n) = 1$. Since n is odd, we know that $2k + n$ is odd. Since $\gcd(k, n) = 1$, there are no divisors of n (besides 1) that are also divisors of $2k + n$. So, we have $\gcd(2k + n, 2n) = 1$, which means that $(e^{2\pi i/(2n)})^{2k+n}$ is indeed a primitive $(2n)^{\text{th}}$ root of unity.

Going the other direction, we must show that every root of $\Phi_{2n}(x)$ is also a root of $\Phi_n(-x)$. To do so, we must show that the negative of any primitive $(2n)^{\text{th}}$ root of unity is a primitive n^{th} root of unity. Let $\omega = e^{2\pi k i/(2n)}$ be a primitive $(2n)^{\text{th}}$ root of unity, so $\gcd(k, 2n) = 1$. Specifically, k is odd. We then have

$$-\omega = -e^{2\pi k i/(2n)} = e^{\pi i}e^{2\pi k i/(2n)} = e^{(2k+2n)\pi i/(2n)} = e^{(k+n)\pi i/n}.$$

This gives us $(-\omega)^n = (e^{(k+n)\pi i/n})^n = e^{(k+n)\pi i}$. Since k and n are odd, the sum $k + n$ is even, which means $(-\omega)^n = e^{(k+n)\pi i} = 1$, so $-\omega$ is an nth root of unity. To show that it is a primitive n^{th} root of unity, we note that $e^{(k+n)\pi i/n} = (e^{2\pi i/n})^{(k+n)/2}$, so we must show that $\gcd\left(\frac{k+n}{2}, n\right) = 1$. (Remember, $k + n$ is even, so $\frac{k+n}{2}$ is an integer.) Since $\gcd(k, 2n) = 1$, we know that $\gcd(k, n) = 1$ as well. Therefore, we have $\gcd(k + n, n) = 1$, so $\gcd\left(\frac{k+n}{2}, n\right) = 1$, as desired.

We have therefore shown that every root $\Phi_n(-x)$ is a root of $\Phi_{2n}(x)$, and vice versa, which means $\Phi_{2n}(x)$ and $\Phi_n(-x)$ are the same.

7.67 The expressions on the left sides of the desired equations resemble applications of the Binomial Theorem, which states that

$$(x + y)^n = \binom{n}{0}x^n y^0 + \binom{n}{1}x^{n-1}y^1 + \binom{n}{2}x^{n-2}y^2 + \cdots + \binom{n}{n-1}x^1 y^{n-1} + \binom{n}{n}x^0 y^n.$$

We can substitute any numbers for x and y. We use the desired equations to guide us. We'd like the $\binom{n}{2}$ term to have a coefficient of -1, but the powers of x and y for this term in the Binomial Theorem are both even. So, we use complex numbers! We let $x = 1$ and $y = i$ We then have

$$(1 + i)^n = \binom{n}{0} + \binom{n}{1}i + \binom{n}{2}i^2 + \binom{n}{3}i^3 + \binom{n}{4}i^4 + \binom{n}{5}i^5 + \cdots$$

$$= \binom{n}{0} + \binom{n}{1}i - \binom{n}{2} - \binom{n}{3}i + \binom{n}{4} + \binom{n}{5}i - \cdots$$

$$= \left[\binom{n}{0} - \binom{n}{2} + \binom{n}{4} - \cdots\right] + \left[\binom{n}{1} - \binom{n}{3} + \binom{n}{5} - \cdots\right]i.$$

Thus, the sums in the problem represent the real and imaginary parts of $(1 + i)^n$.

In exponential notation, we have $1 + i = \sqrt{2}e^{\pi i/4}$, so

$$(1 + i)^n = (\sqrt{2}e^{\pi i/4})^n = (\sqrt{2})^n e^{n\pi i/4} = 2^{n/2}\left(\cos\frac{n\pi}{4} + i\sin\frac{n\pi}{4}\right) = 2^{n/2}\cos\frac{n\pi}{4} + i2^{n/2}\sin\frac{n\pi}{4}.$$

Equating the real parts and equating the imaginary parts gives us

$$\binom{n}{0} - \binom{n}{2} + \binom{n}{4} - \cdots = 2^{n/2} \cos \frac{n\pi}{4},$$

$$\binom{n}{1} - \binom{n}{3} + \binom{n}{5} - \cdots = 2^{n/2} \sin \frac{n\pi}{4}$$

7.68 Let $S = w + 2w^2 + 3w^3 + \cdots + 9w^9$. We can produce another series with many of the same powers of w by multiplying both sides of this equation by w. (We might be inspired to do this by the standard proof of the formula for sum of a geometric series.) We find $wS = w^2 + 2w^3 + 3w^4 + \cdots + 9w^{10}$. Subtracting this from the equation for S gives us a series we know how to handle:

$$(1 - w)S = w + w^2 + w^3 + \cdots + w^9 - 9w^{10} = w(1 + w + w^2 + \ldots w^8) - 9w^{10}.$$

Since $w^9 = (\cos 40° + i \sin 40°)^9 = \cos 360° + i \sin 360° = 1$, we know that w is a ninth root of unity besides 1. Therefore, $1 + w + w^2 + \cdots + w^8 = 0$ and $w^{10} = w^9 \cdot w = w$, which means we have $S(1 - w) = -9w$ from the equation above. Therefore, $S = -9w/(1 - w)$, and

$$|S|^{-1} = \left| \frac{1}{S} \right| = \left| -\frac{1 - w}{9w} \right| = \frac{|w - 1|}{|9w|} = \frac{|w - 1|}{9}, \tag{7.1}$$

since $|w| = 1$. Writing w as $\cos 40° + i \sin 40°$, we have

$$|w - 1| = \sqrt{(\cos 40° - 1)^2 + (\sin 40°)^2} = \sqrt{(\cos^2 40° + \sin^2 40°) + 1 - 2 \cos 40°} = \sqrt{2 - 2 \cos 40°}.$$

However, we have to answer in the form $a \sin b$. To do so, we recognize that $\sqrt{2 - 2 \cos 40°}$ looks a lot like the half-angle formula for sine:

$$\sin \frac{\theta}{2} = \sqrt{\frac{1 - \cos \theta}{2}}.$$

Letting $\theta = 40°$, we have $\sin 20° = \frac{1}{\sqrt{2}} \sqrt{1 - \cos 40°}$, so $\sqrt{2 - 2 \cos 40°} = 2 \sin 20°$, which means that $|S|^{-1} = $

$$\boxed{\frac{2}{9} \sin 20°}.$$

After solving Problem 7.75, come back to this one and see if you can find a faster way to finish starting from line (7.1).

7.69 We prove the result by induction on n. The result clearly holds for $n = 0$ and $n = 1$, so assume it holds for $n = k$ and $n = k + 1$, for some nonnegative integer k, which means $T_k(\cos \theta) = \cos k\theta$ and $T_{k+1}(\cos \theta) = \cos((k + 1)\theta)$. Then

$$T_{k+2}(\cos \theta) = 2 \cos \theta \cdot T_{k+1}(\cos \theta) - T_k(\cos \theta)$$
$$= 2 \cos \theta \cos((k + 1)\theta) - \cos k\theta.$$

We want to show that this is equal to $\cos((k + 2)\theta)$.

By the sum-to-product formula for cosine,

$$\cos((k + 2)\theta) + \cos k\theta = 2 \cos((k + 1)\theta) \cos \theta,$$

so $2 \cos((k + 1)\theta) \cos \theta - \cos k\theta = \cos((k + 2)\theta)$. Therefore, $T_{k+2}(\cos \theta) = \cos((k + 2)\theta)$.

Hence, the result holds for $n = k + 1$ and $n = k + 2$, and by induction, it holds for all nonnegative integer n.

7.70 Consider the polynomial with $e^{i\alpha}$, $e^{i\beta}$, and $e^{i\gamma}$ as roots:

$$(z - e^{i\alpha})(z - e^{i\beta})(z - e^{i\gamma}) = 0.$$

We think of this polynomial for a couple reasons. First, it's a cubic, which we might be able to use to make terms of the form $e^{3i\alpha}$. Second, we know that when we expand the product, the coefficient of the z^2 term will be 0 because the sum of the roots is 0. Specifically, expanding the product gives

$$z^3 - (e^{i\alpha} + e^{i\beta} + e^{i\gamma})z^2 + (e^{i\alpha}e^{i\beta} + e^{i\alpha}e^{i\gamma} + e^{i\beta}e^{i\gamma})z - e^{i\alpha}e^{i\beta}e^{i\gamma} = 0.$$

Now, we just need to figure out what the coefficient of the z term equals. We can get something close by multiplying $e^{i\alpha} + e^{i\beta} + e^{i\gamma} = 0$ by $e^{-i\alpha}e^{-i\beta}e^{-i\gamma}$, which gives us $e^{-i\alpha}e^{-i\beta} + e^{-i\alpha}e^{-i\gamma} + e^{-i\beta}e^{-i\gamma} = 0$. If only we could turn those negative exponents positive... But we can! We take the conjugate of both sides of this equation. On the right, we have 0, and on the left, we have

$$\overline{e^{-i\alpha}e^{-i\beta} + e^{-i\alpha}e^{-i\gamma} + e^{-i\beta}e^{-i\gamma}} = \overline{e^{-i\alpha}e^{-i\beta}} + \overline{e^{-i\alpha}e^{-i\gamma}} + \overline{e^{-i\beta}e^{-i\gamma}}$$

$$= e^{i\alpha}e^{i\beta} + e^{i\alpha}e^{i\gamma} + e^{i\beta}e^{i\gamma}.$$

So, we have $e^{i\alpha}e^{i\beta} + e^{i\alpha}e^{i\gamma} + e^{i\beta}e^{i\gamma} = 0$. (We could have gotten here a little faster with a little more insight. Taking the conjugate of the equation $e^{i\alpha} + e^{i\beta} + e^{i\gamma} = 0$, we get $e^{-i\alpha} + e^{-i\beta} + e^{-i\gamma} = 0$. Multiplying both sides by $e^{i\alpha}e^{i\beta}e^{i\gamma}$, we get $e^{i\alpha}e^{i\beta} + e^{i\alpha}e^{i\gamma} + e^{i\beta}e^{i\gamma} = 0$.)

We now know that the coefficients of z^2 and z are 0, so our polynomial is $z^3 - e^{i\alpha}e^{i\beta}e^{i\gamma} = 0$. Taking $z = e^{i\alpha}$, we find $e^{3i\alpha} = e^{i\alpha}e^{i\beta}e^{i\gamma}$. Similarly, taking $z = e^{i\beta}$ and $z = e^{i\gamma}$, we find $e^{3i\beta} = e^{i\alpha}e^{i\beta}e^{i\gamma}$ and $e^{3i\gamma} = e^{i\alpha}e^{i\beta}e^{i\gamma}$. It follows that $e^{3i\alpha} = e^{3i\beta} = e^{3i\gamma}$.

7.71 We have the identity $\cos\theta = \frac{e^{i\theta}+e^{-i\theta}}{2}$. Let $a = e^{ix/2}$, $b = e^{iy/2}$, and $c = e^{iz/2}$. Then $\cos x = \frac{e^{ix}+e^{-ix}}{2} = \frac{a^2+1/a^2}{2}$. Similarly,

$$\cos y = \frac{b^2 + 1/b^2}{2} \qquad \cos z = \frac{c^2 + 1/c^2}{2}, \qquad \cos(x+y+z) = \frac{a^2b^2c^2 + 1/(a^2b^2c^2)}{2},$$

$$\cos\left(\frac{x+y}{2}\right) = \frac{ab + 1/(ab)}{2}, \qquad \cos\left(\frac{x+z}{2}\right) = \frac{ac + 1/(ac)}{2}, \qquad \cos\left(\frac{y+z}{2}\right) = \frac{bc + 1/(bc)}{2}.$$

Therefore,

$$\cos x + \cos y + \cos z + \cos(x+y+z) = \frac{a^2 + 1/a^2}{2} + \frac{b^2 + 1/b^2}{2} + \frac{c^2 + 1/c^2}{2} + \frac{a^2b^2c^2 + 1/(a^2b^2c^2)}{2}$$

$$= \frac{1}{2}\left(a^2b^2c^2 + a^2 + b^2 + c^2 + \frac{1}{a^2} + \frac{1}{b^2} + \frac{1}{c^2} + \frac{1}{a^2b^2c^2}\right),$$

and

$$4\cos\left(\frac{x+y}{2}\right)\cos\left(\frac{x+z}{2}\right)\cos\left(\frac{y+z}{2}\right) = 4 \cdot \frac{ab + 1/(ab)}{2} \cdot \frac{ac + 1/(ac)}{2} \cdot \frac{bc + 1/(bc)}{2}$$

$$= \frac{1}{2}\left(ab + \frac{1}{ab}\right)\left(ac + \frac{1}{ac}\right)\left(bc + \frac{1}{bc}\right)$$

$$= \frac{1}{2}\left(a^2b^2c^2 + a^2 + b^2 + c^2 + \frac{1}{a^2} + \frac{1}{b^2} + \frac{1}{c^2} + \frac{1}{a^2b^2c^2}\right).$$

Hence,

$$\cos x + \cos y + \cos z + \cos(x+y+z) = 4\cos\left(\frac{x+y}{2}\right)\cos\left(\frac{x+z}{2}\right)\cos\left(\frac{y+z}{2}\right).$$

We also have the identity $\sin\theta = \frac{e^{i\theta}-e^{-i\theta}}{2i}$, so

$$\sin x = \frac{a^2 - 1/a^2}{2i}, \qquad \sin y = \frac{b^2 - 1/b^2}{2i}, \qquad \sin z = \frac{c^2 - 1/c^2}{2i}, \qquad \sin(x+y+z) = \frac{a^2b^2c^2 - 1/(a^2b^2c^2)}{2i},$$

$$\sin\left(\frac{x+y}{2}\right) = \frac{ab - 1/(ab)}{2i}, \qquad \sin\left(\frac{x+z}{2}\right) = \frac{ac - 1/(ac)}{2i}, \qquad \sin\left(\frac{y+z}{2}\right) = \frac{bc - 1/(bc)}{2i}.$$

Therefore,

$$\sin x + \sin y + \sin z - \sin(x+y+z) = \frac{a^2 - 1/a^2}{2i} + \frac{b^2 - 1/b^2}{2i} + \frac{c^2 - 1/c^2}{2i} - \frac{a^2 b^2 c^2 - 1/(a^2 b^2 c^2)}{2i}$$

$$= \frac{1}{2i}\left(-a^2 b^2 c^2 + a^2 + b^2 + c^2 - \frac{1}{a^2} - \frac{1}{b^2} - \frac{1}{c^2} + \frac{1}{a^2 b^2 c^2}\right),$$

and

$$4\sin\left(\frac{x+y}{2}\right)\sin\left(\frac{x+z}{2}\right)\sin\left(\frac{y+z}{2}\right) = 4 \cdot \frac{ab - 1/(ab)}{2i} \cdot \frac{ac - 1/(ac)}{2i} \cdot \frac{bc - 1/(bc)}{2i}$$

$$= \frac{1}{2i^3}\left(ab - \frac{1}{ab}\right)\left(ac - \frac{1}{ac}\right)\left(bc - \frac{1}{bc}\right)$$

$$= -\frac{1}{2i}\left(a^2 b^2 c^2 - a^2 - b^2 - c^2 + \frac{1}{a^2} + \frac{1}{b^2} + \frac{1}{c^2} - \frac{1}{a^2 b^2 c^2}\right)$$

$$= \frac{1}{2i}\left(-a^2 b^2 c^2 + a^2 + b^2 + c^2 - \frac{1}{a^2} - \frac{1}{b^2} - \frac{1}{c^2} + \frac{1}{a^2 b^2 c^2}\right).$$

Hence, $\sin x + \sin y + \sin z - \sin(x+y+z) = 4\sin\left(\frac{x+y}{2}\right)\sin\left(\frac{x+z}{2}\right)\sin\left(\frac{y+z}{2}\right)$.

7.72 If $\theta = \frac{2\pi}{7}$, then $\cos 7\theta = \cos 2\pi = 1$. In an Exercise in Section 7.1, we proved the identity

$$64\cos^7\theta - 112\cos^5\theta + 56\cos^3\theta - 7\cos\theta = \cos 7\theta,$$

so for $\theta = \frac{2\pi}{7}$, we get

$$64\cos^7\frac{2\pi}{7} - 112\cos^5\frac{2\pi}{7} + 56\cos^3\frac{2\pi}{7} - 7\cos\frac{2\pi}{7} = 1.$$

Thus, $x = \cos\frac{2\pi}{7}$ is a root of the equation $64x^7 - 112x^5 + 56x^3 - 7x - 1 = 0$. But $\theta = \frac{4\pi}{7}$ and $\theta = \frac{6\pi}{7}$ also satisfy $\cos 7\theta = 1$, so $x = \cos\frac{4\pi}{7}$ and $x = \cos\frac{6\pi}{7}$ are also roots of the equation $64x^7 - 112x^5 + 56x^3 - 7x - 1 = 0$. We have a polynomial of degree 7, and we know three of the roots, so we need to think about what the other four roots are.

We go back to the equation

$$\cos 7\theta = 1.$$

In addition to $\theta = \frac{2\pi}{7}, \frac{4\pi}{7}$, and $\frac{6\pi}{7}$, we see that the other solutions in the interval $[0, 2\pi)$ are $0, \frac{8\pi}{7}, \frac{10\pi}{7}$, and $\frac{12\pi}{7}$, and that there are no other values of θ in the interval $[0, 2\pi)$ that satisfy the equation. We can quickly see that $x = \cos 0 = 1$ is a root of $64x^7 - 112x^5 + 56x^3 - 7x - 1 = 0$. So, we know that $\cos 0, \cos\frac{2\pi}{7}, \cos\frac{4\pi}{7}$, and $\cos\frac{6\pi}{7}$ are roots of $64x^7 - 112x^5 + 56x^3 - 7x - 1 = 0$, but what are the other three roots?

We found the polynomial $64x^7 - 112x^5 + 56x^3 - 7x - 1 = 0$ by starting with the equation $\cos 7\theta = 1$, which has 7 solutions for $\theta \in [0, 2\pi)$, namely, $0, \frac{2\pi}{7}, \frac{4\pi}{7}, \frac{6\pi}{7}, \frac{8\pi}{7}, \frac{10\pi}{7}$, and $\frac{12\pi}{7}$. We've seen that the cosines of the first four are roots of $64x^7 - 112x^5 + 56x^3 - 7x - 1 = 0$. Noting that $\cos\frac{8\pi}{7} = \cos\frac{6\pi}{7}, \cos\frac{10\pi}{7} = \cos\frac{4\pi}{7}$, and $\cos\frac{12\pi}{7} = \cos\frac{2\pi}{7}$, we conjecture that the polynomial has three double roots corresponding to these equal cosines.

Let's see if that guess is correct. We know that one of the factors of the polynomial is $x - 1$, so we take out a factor of $x - 1$ to get

$$64x^7 - 112x^5 + 56x^3 - 7x - 1 = (x-1)(64x^6 + 64x^5 - 48x^4 - 48x^3 + 8x^2 + 8x + 1).$$

If our guess that the polynomial of degree 6 is the square of a cubic is correct, we must have

$$64x^6 + 64x^5 - 48x^4 - 48x^3 + 8x^2 + 8x + 1 = (ax^3 + bx^2 + cx + d)^2.$$

If we were to expand the righthand side, we would get

$$64x^6 + 64x^5 - 48x^4 - 48x^3 + 8x^2 + 8x + 1 = (ax^3 + bx^2 + cx + d)^2 = a^2 x^6 + 2abx^5 + \cdots.$$

Equating the corresponding coefficients of x^6 and x^5, we get $a^2 = 64$ and $2ab = 64$. Without loss of generality, we can take $a = 8$. (If $a = -8$, then we can flip the sign of all the coefficients of the cubic.) Then $16b = 64$, so $b = 4$. Hence,

$$64x^6 + 64x^5 - 48x^4 - 48x^3 + 8x^2 + 8x + 1 = (8x^3 + 4x^2 + cx + d)^2.$$

Again, expanding the right-hand side, we get

$$64x^6 + 64x^5 - 48x^4 - 48x^3 + 8x^2 + 8x + 1 = 64x^6 + 64x^5 + (16c + 16)x^4 + (8c + 16d)x^3 + \cdots.$$

Then equating the corresponding coefficients of x^4 and of x^3, we get $16c + 16 = -48$ and $8c + 16d = -48$, respectively. Solving for c and d, we find $c = -4$ and $d = -1$. Hence,

$$64x^6 + 64x^5 - 48x^4 - 48x^3 + 8x^2 + 8x + 1 = (8x^3 + 4x^2 - 4x - 1)^2,$$

which we can verify by expanding.

Therefore, a cubic whose roots are $\cos \frac{2\pi}{7}$, $\cos \frac{4\pi}{7}$, and $\cos \frac{6\pi}{7}$ is $\boxed{8x^3 + 4x^2 - 4x - 1}$.

7.73 We see that $z, z^2, \ldots, z^{2005}$ are the 2006th roots of unity other than 1.

Let ω be a 2006th root of unity other than 1. Then $\omega^{2006} - 1 = 0$, which factors as

$$(\omega - 1)(\omega^{2005} + \omega^{2004} + \omega^{2003} + \cdots + \omega + 1) = 0.$$

Since $\omega \neq 1$, ω satisfies the equation $\omega^{2005} + \omega^{2004} + \omega^{2003} + \cdots + \omega + 1 = 0$.

We have

$$f(\omega) = \omega^{2004} + 2\omega^{2003} + 3\omega^{2002} + \cdots + 2004\omega + 2005.$$

This is similar to a geometric series, so we try using the same strategy to evaluate it. We multiply this equation by ω to get

$$\omega f(\omega) = \omega^{2005} + 2\omega^{2004} + 3\omega^{2003} + \cdots + 2004\omega^2 + 2005\omega.$$

Subtracting the equation for $f(\omega)$ from this equation gives

$$
\begin{aligned}
(\omega - 1)f(\omega) &= \omega^{2005} + 2\omega^{2004} + 3\omega^{2003} + \cdots + 2004\omega^2 + 2005\omega - \omega^{2004} - 2\omega^{2003} - 3\omega^{2002} - \cdots - 2004\omega - 2005 \\
&= \omega^{2005} + \omega^{2004} + \omega^{2003} + \cdots + \omega - 2005 \\
&= (\omega^{2005} + \omega^{2004} + \omega^{2003} + \cdots + \omega + 1) - 2006 \\
&= -2006,
\end{aligned}
$$

so $f(\omega) = -\frac{2006}{\omega - 1} = \frac{2006}{1 - \omega}$. Since this holds for any 2006th root of unity other than 1, we have

$$N = f(z)f(z^2) \cdots f(z^{2005}) = \frac{2006}{1 - z} \cdot \frac{2006}{1 - z^2} \cdots \frac{2006}{1 - z^{2005}} = \frac{2006^{2005}}{(1 - z)(1 - z^2) \cdots (1 - z^{2005})}.$$

Since $1, z, z^2, \ldots, z^{2005}$ are the 2006th roots of unity, we have

$$(x - 1)(x - z)(x - z^2) \cdots (x - z^{2005}) = x^{2006} - 1.$$

But $x^{2006} - 1$ factors as $(x - 1)(x^{2005} + x^{2004} + \cdots + x + 1)$, so

$$(x - z)(x - z^2) \cdots (x - z^{2005}) = x^{2005} + x^{2004} + \cdots + x + 1.$$

Taking $x = 1$, we get $(1 - z)(1 - z^2) \cdots (1 - z^{2005}) = 2006$. Therefore,

$$N = \frac{2006^{2005}}{(1 - z)(1 - z^2) \cdots (1 - z^{2005})} = \frac{2006^{2005}}{2006} = \boxed{2006^{2004}}.$$

7.74 The roots of $z^{1997} - 1 = 0$ are the 1997th roots of unity, which are of the form $e^{2k\pi i/1997}$, where $0 \le k \le 1996$. Suppose $v = e^{2j\pi i/1997}$ and $w = e^{2k\pi i/1997}$. We then have

$$|v + w| = \left| e^{2j\pi i/1997} + e^{2k\pi i/1997} \right| = \left| \left(e^{2j\pi i/1997} \right) \left(1 + e^{2(k-j)\pi i/1997} \right) \right| = \left| e^{2j\pi i/1997} \right| \left| 1 + e^{2(k-j)\pi i/1997} \right| = \left| 1 + e^{2(k-j)\pi i/1997} \right|.$$

Since $e^{2(k-j)\pi i/1997}$ is also a solution to $z^{1997} - 1 = 0$, the probability that $|v + w| \ge \sqrt{2 + \sqrt{3}}$ equals the probability that $|1 + t| \ge \sqrt{2 + \sqrt{3}}$, where t is also a nonreal solution to $z^{1997} - 1 = 0$. (We exclude $t = 1$ because v and w must be distinct, so $k \ne j$.)

We let $t = e^{2a\pi i/1997}$ for some a, where $1 \le a \le 1996$. (The argument of t cannot be 0, because v and w are distinct. Thus, there are 1996 possible values of a.) Let $\alpha = \frac{2a\pi}{1997}$. Then $t = e^{2a\pi i/1997} = e^{i\alpha}$, and

$$|1 + t| = |1 + e^{i\alpha}| = |1 + \cos\alpha + i\sin\alpha| = \sqrt{(1 + \cos\alpha)^2 + \sin^2\alpha} = \sqrt{1 + 2\cos\alpha + \cos^2\alpha + \sin^2\alpha} = \sqrt{2 + 2\cos\alpha}.$$

Hence,

$$|1 + t| \ge \sqrt{2 + \sqrt{3}} \quad \Leftrightarrow \quad \sqrt{2 + 2\cos\alpha} \ge \sqrt{2 + \sqrt{3}} \quad \Leftrightarrow \quad 2 + 2\cos\alpha \ge 2 + \sqrt{3} \quad \Leftrightarrow \quad \cos\alpha \ge \frac{\sqrt{3}}{2}.$$

Since $0 < \alpha < 2\pi$, the inequality $\cos\alpha \ge \frac{\sqrt{3}}{2}$ holds if and only if $0 < \alpha \le \frac{\pi}{6}$ or $\frac{11\pi}{6} \le \alpha < 2\pi$.

Since $\alpha = \frac{2a\pi}{1997}$,

$$0 < \alpha \le \frac{\pi}{6} \quad \Leftrightarrow \quad 0 < \frac{2\pi a}{1997} \le \frac{\pi}{6} \quad \Leftrightarrow \quad 0 < a \le \frac{1997}{12}.$$

Since a is an integer, we have $1 \le a \le 166$, for 166 possible values of a. Similarly,

$$\frac{11\pi}{6} \le \alpha < 2\pi \quad \Leftrightarrow \quad \frac{11\pi}{6} \le \frac{2\pi a}{1997} < 2\pi \quad \Leftrightarrow \quad \frac{21967}{12} \le a < 1997.$$

Since a is an integer, we have $1831 \le a \le 1996$, for another 166 possible values of a.

Therefore, the probability that $|v + w| \ge \sqrt{2 + \sqrt{3}}$ is $(166 + 166)/1996 = 332/1996 = \boxed{83/499}$.

7.75 We turn this into a problem involving complex numbers by recalling that $\sin\theta = \frac{e^{i\theta} - e^{-i\theta}}{2i}$, so

$$\sin\frac{\pi}{n} \sin\frac{2\pi}{n} \cdots \sin\frac{(n-1)\pi}{n} = \left(\frac{e^{\pi i/n} - e^{-\pi i/n}}{2i} \right) \left(\frac{e^{2\pi i/n} - e^{-2\pi i/n}}{2i} \right) \left(\frac{e^{3\pi i/n} - e^{-3\pi i/n}}{2i} \right) \cdots \left(\frac{e^{(n-1)\pi i/n} - e^{-(n-1)\pi i/n}}{2i} \right).$$

That's still pretty scary. But, we can do a little factoring to make the numerators expressions that we're more comfortable with:

$$\left(\frac{e^{\pi i/n} - e^{-\pi i/n}}{2i} \right) \left(\frac{e^{2\pi i/n} - e^{-2\pi i/n}}{2i} \right) \left(\frac{e^{3\pi i/n} - e^{-3\pi i/n}}{2i} \right) \cdots \left(\frac{e^{(n-1)\pi i/n} - e^{-(n-1)\pi i/n}}{2i} \right)$$

$$= \left(\frac{e^{-\pi i/n}(e^{2\pi i/n} - 1)}{2i} \right) \left(\frac{e^{-2\pi i/n}(e^{4\pi i/n} - 1)}{2i} \right) \left(\frac{e^{-3\pi i/n}(e^{6\pi i/n} - 1)}{2i} \right) \cdots \left(\frac{e^{-(n-1)\pi i/n}(e^{2(n-1)\pi i/n} - 1)}{2i} \right). \qquad (\spadesuit)$$

Aha! The factors in the numerator look suspiciously like factors of $x^n - 1$. Let's investigate. Let $\omega = e^{2\pi i/n}$. The roots of $x^n - 1 = 0$ are the nth roots of unity, namely $1, \omega, \omega^2, \ldots, \omega^{n-1}$. The polynomial $x^n - 1$ factors as

$$x^n - 1 = (x - 1)(x^{n-1} + x^{n-2} + \cdots + x + 1).$$

Hence, the roots of $x^{n-1} + x^{n-2} + \cdots + x + 1 = 0$ are $\omega, \omega^2, \ldots, \omega^{n-1}$, which means

$$(x - \omega)(x - \omega^2) \cdots (x - \omega^{n-1}) = x^{n-1} + x^{n-2} + \cdots + x + 1.$$

Taking $x = 1$, we get

$$(1 - \omega)(1 - \omega^2) \cdots (1 - \omega^{n-1}) = n.$$

We tie this to the product of sines by reversing the factoring we did on line (♠). We do so by noting that $1 - e^{i\theta} = e^{i\theta/2}(e^{-i\theta/2} - e^{i\theta/2}) = -e^{i\theta/2}(e^{i\theta/2} - e^{-i\theta/2})$, so

$$\left|1 - e^{i\theta}\right| = \left|-e^{i\theta/2}(e^{i\theta/2} - e^{-i\theta/2})\right| = \left|-e^{i\theta/2}\right|\left|e^{i\theta/2} - e^{-i\theta/2}\right| = \left|e^{i\theta/2} - e^{-i\theta/2}\right|.$$

Since $\sin\theta = \frac{e^{i\theta} - e^{-i\theta}}{2i}$, we have

$$\left|e^{i\theta/2} - e^{-i\theta/2}\right| = \left|2i\sin\frac{\theta}{2}\right| = 2\left|\sin\frac{\theta}{2}\right|.$$

Since $0 \le \frac{\theta}{2} \le \pi$, we have $\sin\frac{\theta}{2} \ge 0$, so $|1 - e^{i\theta}| = \sin\frac{\theta}{2}$. In particular, for $\theta = \frac{2k\pi}{n}$, where $1 \le k \le n - 1$, we have $\omega^k = e^{2k\pi i/n} = e^{i\theta}$, so

$$|1 - \omega^k| = 2\sin\frac{k\pi}{n}.$$

Therefore, taking the magnitude of both sides of the equation $(1 - \omega)(1 - \omega^2) \cdots (1 - \omega^{n-1}) = n$, we get

$$2\sin\frac{\pi}{n} \cdot 2\sin\frac{2\pi}{n} \cdots 2\sin\frac{(n-1)\pi}{n} = n.$$

Finally, dividing both sides by 2^{n-1}, we get $\sin\frac{\pi}{n}\sin\frac{2\pi}{n}\cdots\sin\frac{(n-1)\pi}{n} = \frac{n}{2^{n-1}}$, as desired.

CHAPTER

Geometry of Complex Numbers

Exercises for Section 8.1

8.1.1 Since w is a counterclockwise rotation of z by π, we have $w = e^{i\pi}(z - 0) + 0 = \boxed{-z}$. Note that this means that $w + z = 0$, so the origin is the midpoint of the segment connecting w and z. We therefore say that w is the "reflection of z through the origin."

8.1.2

(a) The image of $3 + i$ upon a 30° (or $\frac{\pi}{6}$) counterclockwise rotation about the origin is

$$e^{\pi i/6}(3 + i) = \left(\frac{\sqrt{3}}{2} + \frac{1}{2}i \right)(3 + i) = \frac{3\sqrt{3}}{2} + \frac{\sqrt{3}}{2}i + \frac{3}{2}i - \frac{1}{2} = \boxed{\frac{3\sqrt{3} - 1}{2} + \frac{3 + \sqrt{3}}{2}i}.$$

(b) The image of $3 + i$ upon a 120° (or $\frac{2\pi}{3}$) clockwise rotation about $1 + 2i$ is

$$e^{-2\pi i/3}[(3+i)-(1+2i)]+(1+2i) = \left(-\frac{1}{2} - \frac{\sqrt{3}}{2}i \right)(2-i)+1+2i = -1+\frac{1}{2}i-\sqrt{3}i-\frac{\sqrt{3}}{2}+1+2i = \boxed{-\frac{\sqrt{3}}{2} + \frac{5 - 2\sqrt{3}}{2}i}.$$

8.1.3 It is not true that the argument of $w + z$ is always the average of the argument of w and the argument of z. For example, take $w = 2$ and $z = i$. The argument of $w = 2$ is 0, and the argument of $z = i$ is $\frac{\pi}{2}$, but the argument of $w + z = 2 + i$ is clearly not $\frac{\pi}{4}$.

However, it is true that if $|w| = |z|$, then the argument of $w + z$ is the average of the argument of w and the argument of z. We can see this geometrically.

The origin, w, $w + z$, and z always form a parallelogram. But when $|w| = |z|$, this parallelogram is a rhombus. Hence, the line joining the origin and $w + z$ bisects the angle of the parallelogram at the origin. In other words, the argument of $w + z$ is the average of the argument of w and the argument of z.

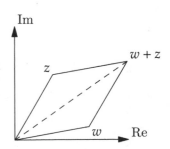

We can also use algebra. We let $w = re^{i\alpha}$ and $z = re^{i\beta}$. We then have

$$w + z = r(e^{i\alpha} + e^{i\beta})$$
$$= r\left(\cos\alpha + \cos\beta + i(\sin\alpha + \sin\beta)\right)$$
$$= 2r\left(\cos\left(\frac{\alpha-\beta}{2}\right)\cos\left(\frac{\alpha+\beta}{2}\right) + i\left(\cos\left(\frac{\alpha-\beta}{2}\right)\sin\left(\frac{\alpha+\beta}{2}\right)\right)\right)$$
$$= 2r\cos\left(\frac{\alpha-\beta}{2}\right)\left(\cos\left(\frac{\alpha+\beta}{2}\right) + i\sin\left(\frac{\alpha+\beta}{2}\right)\right)$$
$$= 2r\cos\left(\frac{\alpha-\beta}{2}\right)e^{i(\alpha+\beta)/2}.$$

(Note that we used the sum-to-product identities for sine and cosine as an intermediate step.) Since $2r\cos(\frac{\alpha-\beta}{2})$ is real, the argument of $w + z$ is $\frac{\alpha+\beta}{2}$.

8.1.4 Since

$$\frac{1}{re^{i\theta}} = \frac{1}{r}e^{-i\theta},$$

the transformation that maps z to $z/(re^{i\theta})$ is a combination of $\boxed{\text{dilating about the origin with a scale factor of } 1/r}$ and $\boxed{\text{rotating by an angle of } \theta \text{ clockwise about the origin}}$.

8.1.5

(a) The function f corresponds to a translation of -5 units horizontally and 2 units vertically.

(b) The function f corresponds to a dilation about the origin with scale factor 3, followed by a translation of -1 unit horizontally. (Note that $f(z) - \frac{1}{2} = 3z - \frac{3}{2} = 3\left(z - \frac{1}{2}\right)$, so we can also describe the transformation as a dilation about $\frac{1}{2}$ with scale factor 3.)

(c) Since $\frac{1}{2} + \frac{\sqrt{3}}{2}i = e^{\pi i/3}$, the function f corresponds to a rotation about the origin counterclockwise by an angle of $\frac{\pi}{3}$.

(d) Since $-1 + i = \sqrt{2}e^{3\pi i/4}$, the function f corresponds to a dilation about the origin with a scale factor of $\sqrt{2}$, followed by a rotation about the origin counterclockwise by an angle of $\frac{3\pi}{4}$ (or vice versa).

(e) Since $-i = e^{3\pi i/2}$, the function f corresponds to a rotation about the origin counterclockwise by an angle of $\frac{3\pi}{2}$, followed by a translation of -1 unit horizontally and 3 units vertically.

We can also describe f as a rotation about a point. We see this by noting that if we rotate z about w by an angle of θ counterclockwise, then the image is $e^{i\theta}(z - w) + w$. Comparing this to $f(z) = -iz - 1 + 3i$, we see that we can find values of w and θ such that $e^{i\theta}(z - w) + w = -iz - 1 + 3i$. Expanding the product on the left, we have

$$e^{i\theta}z - e^{i\theta}w + w = -iz - 1 + 3i.$$

Since this must be true for all z, we must have $e^{i\theta} = -i$, so we can choose $\theta = \frac{3\pi}{2}$, and our equation is

$$-iz + iw + w = -iz - 1 + 3i.$$

Solving for w gives

$$w = \frac{-1 + 3i}{1 + i} = \frac{(-1 + 3i)(1 - i)}{(1 + i)(1 - i)} = \frac{-1 + 3i + i + 3}{2} = \frac{2 + 4i}{2} = 1 + 2i.$$

Hence, f also corresponds to a rotation about $1 + 2i$ counterclockwise by an angle of $\frac{3\pi}{2}$.

8.1.6 We translate the points W, U, and Z to W', U', and Z', so that W' coincides with the origin. Then W', U', and Z' are collinear, and $U'W'/U'Z' = a/b$. Then $U'W'/W'Z' = a/(a+b)$. The corresponding complex numbers are $w' = w - w = 0$, $u' = u - w$, and $z' = z - w$.

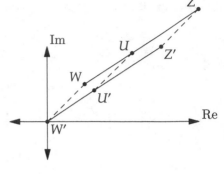

We can think of u' as the result of a dilating z' through the origin with scale factor $a/(a + b)$, so

$$u' = \frac{a}{a + b} z'.$$

Substituting $u' = u - w$ and $z' = z - w$, we get

$$u - w = \frac{a}{a + b}(z - w),$$

so

$$u = \frac{a}{a + b}(z - w) + w = \frac{az - aw + aw + bw}{a + b} = \frac{az + bw}{a + b}.$$

8.1.7 We have $z_2 = e^{\pi i/2}z_1 = iz_1$, and

$$z_3 = e^{\pi i/2}(z_2 - 4 - 3i) + (4 + 3i) = i(iz_1 - 4 - 3i) + (4 + 3i) = -z_1 - 4i + 3 + 4 + 3i = -z_1 + 7 - i.$$

If z_3 is the point obtained when z_1 is rotated counterclockwise about w by an angle θ, then

$$z_3 = e^{i\theta}(z_1 - w) + w = e^{i\theta}z_1 - e^{i\theta}w + w.$$

Since $z_3 = -z_1 + 7 - i$ from above, we seek θ and w such that $e^{i\theta}z_1 - e^{i\theta}w + w = -z_1 + 7 - i$. This means we must have $e^{i\theta} = -1$ and $-e^{i\theta}w + w = 7 - i$. We can take $\theta = \boxed{\pi}$, so $-e^{i\theta}w + w = 2w$, which gives $w = \frac{7-i}{2} = \boxed{\frac{7}{2} - \frac{1}{2}i}$.

Exercises for Section 8.2

8.2.1 The points in the complex plane that represent $1 + 3i$, $5 - 2i$, and z are collinear if and only if $\frac{z-(1+3i)}{z-(5-2i)}$ is real, so we check this value for each the given complex numbers.

For $z = 3 - 2i$, we have $\frac{(3-2i)-(1+3i)}{(3-2i)-(5-2i)} = \frac{2-5i}{-2} = -1 + \frac{5}{2}i$, which is not real.

For $z = 9 - 7i$, we have $\frac{(9-7i)-(1+3i)}{(9-7i)-(5-2i)} = \frac{8-10i}{4-5i} = \frac{2(4-5i)}{4-5i} = 2$, which is real.

For $z = -2 + 5i$, we have $\frac{(-2+5i)-(1+3i)}{(-2+5i)-(5-2i)} = \frac{-3+2i}{-7+7i} = \frac{-3+2i}{7(-1+i)} = \frac{(-3+2i)(-1-i)}{7(-1+i)(-1-i)} = \frac{5+i}{14}$, which is not real.

Therefore, of the given complex numbers, only $\boxed{9 - 7i}$ is collinear with $1 + 3i$ and $5 - 2i$.

8.2.2 We showed in the text that complex numbers z, a, and b are collinear if and only if $\frac{z-a}{b-a}$ is real. But a complex number is real if and only if it is equal to its own conjugate. The conjugate of $\frac{z-a}{b-a}$ is $\frac{\bar{z}-\bar{a}}{\bar{b}-\bar{a}}$. Hence, z, a, and b are collinear if and only if

$$\frac{z - a}{b - a} = \frac{\bar{z} - \bar{a}}{\bar{b} - \bar{a}}.$$

8.2.3 Let the vertices of the quadrilateral be a, b, c, and d. Then the midpoints of the sides are $\frac{a+b}{2}$, $\frac{b+c}{2}$, $\frac{c+d}{2}$, and $\frac{d+a}{2}$. The quadrilateral with these points as vertices is dashed at right. In this quadrilateral, the midpoint of the diagonal joining $\frac{a+b}{2}$ and $\frac{c+d}{2}$ is

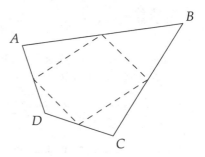

$$\frac{\frac{a+b}{2} + \frac{c+d}{2}}{2} = \frac{a+b+c+d}{4},$$

and the midpoint of the diagonal joining $\frac{b+c}{2}$ and $\frac{d+a}{2}$ is

$$\frac{\frac{b+c}{2} + \frac{d+a}{2}}{2} = \frac{a+b+c+d}{4}.$$

Thus, the midpoints of the two diagonals coincide. In other words, the two diagonals bisect each other. Therefore, the quadrilateral formed by the midpoints of the sides is a parallelogram.

8.2.4 If $\overline{OB'} \perp \overline{OD'}$, then we can rotate D' by 90° to produce a point that is on the ray from the origin through B'. (That is, the complex number corresponding to the image of the rotation has the same argument as b'.) The complex number corresponding to the image of this rotation then is either id' or $-id'$, depending on the orientation of the 90° rotation. Since the image is on the ray from the origin through B', there is a dilation that maps this point to B'. Therefore, we have $b' = k(id')$ or $b' = k(-id')$ for some real constant k. In either case, b'/d' is imaginary.

8.2.5

(a) The complex number corresponding to the midpoint of $\overline{BC}$ is $(b+c)/2$. So, to show that the point corresponding to $(a+b+c)/3$ is on the median from A to $\overline{BC}$, we must show that a, $(a+b+c)/3$, and $(b+c)/2$ are collinear. We have

$$\frac{\frac{a+b+c}{3} - a}{\frac{b+c}{2} - a} = \frac{\frac{b+c-2a}{3}}{\frac{b+c-2a}{2}} = \frac{2}{3},$$

which is real. Therefore, a, $(a+b+c)/3$, and $(b+c)/2$ are collinear. (Note that we cannot have $b+c-2a = 0$ because this gives us $a = (b+c)/2$, which would mean that A is the midpoint of $\overline{BC}$. Since ABC is a triangle, A cannot be the midpoint of $\overline{BC}$.)

Similarly, we can show that G is on each of the other two medians, so the medians of $\triangle ABC$ meet at the point corresponding to $(a+b+c)/3$. (This matches our intuition: the complex number corresponding to the midpoint of a segment is the arithmetic mean of the numbers corresponding to its endpoints, and the complex number corresponding to the centroid of a triangle is the arithmetic mean of the numbers corresponding to the triangle's vertices.)

(b) Our work in the first part pretty much covers this part, too. Letting M be the midpoint of $\overline{BC}$, we have

$$\frac{AG}{AM} = \frac{\left|\frac{a+b+c}{3} - a\right|}{\left|\frac{b+c}{2} - a\right|} = \frac{\left|\frac{b+c-2a}{3}\right|}{\left|\frac{b+c-2a}{2}\right|} = \frac{2}{3} \cdot \frac{|b+c-2a|}{|b+c-2a|} = \frac{2}{3},$$

so AG is $\frac{2}{3}$ of AM, as desired. There's nothing special about the median from A, so the distances from B and C to the centroid are also equal to $\frac{2}{3}$ of the lengths of the respective medians from B and C.

8.2.6 As shown in the text, a, b, and c are collinear if and only if $(a - b)(\overline{b} - \overline{c}) - (\overline{a} - \overline{b})(b - c) = 0$. Expanding the products on the left gives

$$a\overline{b} - a\overline{c} - b\overline{b} + b\overline{c} - \overline{a}b + \overline{a}c + \overline{b}b - \overline{b}c = 0.$$

Cancelling the $b\overline{b}$ terms and rearranging gives the desired

$$a\overline{b} + b\overline{c} + c\overline{a} = \overline{a}b + \overline{b}c + \overline{c}a.$$

8.2.7 Let w, x, y, and z be complex numbers corresponding to W, X, Y, and Z. Then the midpoints of $\overline{WX}$, $\overline{WY}$, and $\overline{XZ}$ are $\frac{w+x}{2}$, $\frac{w+y}{2}$, and $\frac{x+z}{2}$. In the text, we showed that complex numbers a, b, and c are collinear in the complex plane if and only if

$$(b-a)(\overline{b}-\overline{c}) - (\overline{b}-\overline{a})(b-c) = 0.$$

We take $(x+w)/2$ to be b, since it has terms in common with both other points, and we'll therefore have cancellation in each term in parentheses above. Because the three given midpoints are collinear, we have

$$\left(\frac{w+x}{2} - \frac{w+y}{2}\right)\left(\frac{\overline{w+x}}{2} - \frac{\overline{x+z}}{2}\right) - \left(\frac{\overline{w+x}}{2} - \frac{\overline{w+y}}{2}\right)\left(\frac{w+x}{2} - \frac{x+z}{2}\right) = 0.$$

Simplifying each of the differences gives us

$$\frac{1}{4}\left((x-y)(\overline{w}-\overline{z}) - (\overline{x}-\overline{y})(w-z)\right) = 0, \tag{8.1}$$

so $(x-y)(\overline{w}-\overline{z}) - (\overline{x}-\overline{y})(w-z) = 0$.

The midpoint of $\overline{YZ}$ is $\frac{y+z}{2}$. To check if it is on the line through the other three midpoints, we use our test for collinearity on $b = \frac{y+z}{2}$, $a = \frac{w+y}{2}$, and $c = \frac{x+z}{2}$. (Notice that we choose our points strategically to produce cancellation in the test for collinearity.) We have

$$(b-a)(\overline{b}-\overline{c}) - (\overline{b}-\overline{a})(b-c) = \left(\frac{y+z}{2} - \frac{w+y}{2}\right)\left(\frac{\overline{y+z}}{2} - \frac{\overline{x+z}}{2}\right) - \left(\frac{\overline{y+z}}{2} - \frac{\overline{w+y}}{2}\right)\left(\frac{y+z}{2} - \frac{x+z}{2}\right)$$

$$= \frac{1}{4}\left((z-w)(\overline{y}-\overline{x}) - (\overline{z}-\overline{w})(y-x)\right)$$

$$= \frac{1}{4}\left((w-z)(\overline{x}-\overline{y}) - (\overline{w}-\overline{z})(x-y)\right)$$

$$= 0,$$

where we apply the result of Equation (8.1) to provide the last step. Therefore, the midpoint of $\overline{YZ}$ is indeed on the line through the other three midpoints.

8.2.8 Let a, b, c, and d be the complex numbers corresponding to A, B, C, and D, respectively. First, we translate A and B to A' and B', so that A' coincides with the origin O. The corresponding complex numbers are $a' = a - a = 0$ and $b' = b - a = (3+2i) - (1-5i) = 2+7i$.

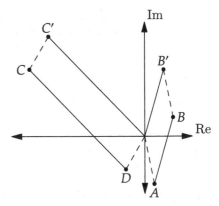

Next, we translate C and D to C' and D', so that D' coincides with the origin O. The corresponding complex numbers are $d' = d - d = 0$ and $c' = c - d = (-7\sqrt{3}+7i) - (-2-2i\sqrt{3}) = (2-7\sqrt{3}) + (7+2\sqrt{3})i$. Note that the angle between $\overline{AB}$ and $\overline{CD}$ is equal to $\angle B'OC'$.

Multiplying a complex number by $re^{i\theta}$ corresponds to the combination of a dilation (about the origin) by a factor of r and a rotation of θ counterclockwise about the origin. Hence, $\angle B'OC'$ is the argument of c'/b'.

We have

$$\frac{c'}{b'} = \frac{(2-7\sqrt{3}) + (7+2\sqrt{3})i}{2+7i} = \frac{[(2-7\sqrt{3}) + (7+2\sqrt{3})i](2-7i)}{(2+7i)(2-7i)}$$

$$= \frac{(4-14\sqrt{3}) + (-14+49\sqrt{3})i + (14+4\sqrt{3})i + (49+14\sqrt{3})}{53} = \frac{53+53\sqrt{3}i}{53} = 1 + \sqrt{3}i.$$

In exponential form, we have $1 + \sqrt{3}i = 2\left(\frac{1}{2} + \frac{\sqrt{3}}{2}i\right) = 2e^{\pi i/3}$, so the acute angle between $\overleftrightarrow{AB}$ and $\overleftrightarrow{CD}$ is $\boxed{\frac{\pi}{3}}$.

Exercises for Section 8.3

8.3.1 Any circle in the complex plane is the graph of an equation of the form $|z - c| = r$, where c is a complex number and r is a positive constant. As described in the text, we can square both sides and write this equation in the form $(z - c)(\bar{z} - \bar{c}) = r^2$, or

$$z\bar{z} - \bar{c}z - c\bar{z} + c\bar{c} = r^2.$$

Comparing this to the given equation, we have $-c = -2 - i$ from the coefficient of $\bar{z}$, so $c = 2 + i$. Then, we have $c\bar{c} = 5$, so the equation for the circle above is $z\bar{z} - (2 - i)z - (2 + i)\bar{z} + 5 = r^2$, or

$$z\bar{z} + (-2 + i)z + (-2 - i)\bar{z} = r^2 - 5.$$

Comparing this to the given equation, we have $r^2 - 5 = 11$, so $r^2 = 16$, which gives $r = \boxed{4}$.

8.3.2 Let a, b, c, and p be complex numbers corresponding to points A, B, C, and P. Then, we wish to show that $|a - c|^2 - |c - p|^2 = |a - b|^2 - |b - p|^2$, which we can also write as

$$|a - c|^2 - |c - p|^2 - |a - b|^2 + |b - p|^2 = 0.$$

We can make life easier on ourselves by choosing one of these points to be the origin. We'll choose p to be the origin, so we wish to show that

$$|a - c|^2 - |c|^2 - |a - b|^2 + |b|^2 = 0.$$

Since $|z|^2 = z\bar{z}$ for any complex number, we have

$$
\begin{aligned}
|a - c|^2 - |c|^2 - |a - b|^2 + |b|^2 &= (a - c)(\overline{a - c}) - c\bar{c} - (a - b)(\overline{a - b}) + b\bar{b} \\
&= (a - c)(\bar{a} - \bar{c}) - c\bar{c} - (a - b)(\bar{a} - \bar{b}) + b\bar{b} \\
&= a\bar{a} - a\bar{c} - c\bar{a} + c\bar{c} - c\bar{c} - a\bar{a} + a\bar{b} + b\bar{a} - b\bar{b} + b\bar{b} \\
&= -a\bar{c} - c\bar{a} + a\bar{b} + b\bar{a} \\
&= a(\bar{b} - \bar{c}) + \bar{a}(b - c).
\end{aligned}
$$

Because $\overleftrightarrow{AP} \perp \overleftrightarrow{BC}$, we have $(a-p)(\bar{b}-\bar{c})+(\bar{a}-\bar{p})(b-c) = 0$. Because we let p be the origin, we have $a(\bar{b}-\bar{c})+\bar{a}(b-c) = 0$, so our expression for $|a - c|^2 - |c|^2 - |a - b|^2 + |b|^2$ equals 0, as desired. Therefore, $AC^2 - CP^2 = AB^2 - BP^2$.

8.3.3

(a) By the Triangle Inequality, we have $|x + y - z| + |x - y + z| \geq |(x + y - z) + (x - y + z)| = |2x| = 2|x|$.

(b) Similarly, $|x + y - z| + |-x + y + z| \geq |(x + y - z) + (-x + y + z)| = |2y| = 2|y|$, and $|x - y + z| + |-x + y + z| \geq |(x - y + z) + (-x + y + z)| = |2z| = 2|z|$. Adding these two and the result from part (a), we get

$$2|x + y - z| + 2|x - y + z| + 2|-x + y + z| \geq 2|x| + 2|y| + 2|z|.$$

Dividing both sides by 2 yields $|x + y - z| + |x - y + z| + |-x + y + z| \geq |x| + |y| + |z|$.

8.3.4 Because $|w| = 1$, the point corresponding to w in the complex plane is on the unit circle. Similarly, $w + a$ and $w + bi$ are on the unit circle. Let the points corresponding to w, $w + a$, and $w + bi$ be W, A, and B. Since $w + a$ is a horizontal translation of w, point A is directly to the left or right of W. Similarly, since $w + bi$ is a vertical translation of W, point B is directly above or below W. Therefore, $\angle AWB$ is a right angle. Because $\angle AWB$ is a right angle that is inscribed in a circle, the arc of this circle connecting A to B is a semicircle. Therefore, $\overline{AB}$ is a diameter of the unit circle, which means its length is 2. From the Pythagorean Theorem, we also have $AB^2 = WA^2 + WB^2 = a^2 + b^2$, so $a^2 + b^2 = AB^2 = 4$.

8.3.5 Let the quadrilateral be $ABCD$ and let a, b, c, and d correspond to A, B, C, and D. Then, we wish to show that

$$|a - c| + |b - d| < |a - b| + |b - c| + |c - d| + |d - a|.$$

We see two different pairs on the right side to which we can apply the Triangle Inequality in order to produce an $|a - c|$ term on the lesser side. We have

$$|a - b| + |b - c| > |(a - b) + (b - c)| = |a - c|,$$
$$|c - d| + |d - a| > |(c - d) + (d - a)| = |c - a| = |a - c|,$$

both by the Triangle Inequality. Note that we can write both inequalities as strict inequalities because ABC and ACD are triangles, so A cannot be on $\overline{BC}$ or $\overline{CD}$. Adding these two inequalities gives us

$$|a - b| + |b - c| + |c - d| + |d - a| > 2|a - c|. \qquad (\spadesuit)$$

We can do the same thing with $|b - d|$. The Triangle Inequality gives

$$|b - c| + |c - d| > |(b - c) + (c - d)| = |b - d|,$$
$$|d - a| + |a - b| > |(d - a) + (a - b)| = |d - b| = |b - d|,$$

where again we have strict inequality because we cannot have any three vertices of the quadrilateral be collinear. Adding these inequalities gives

$$|a - b| + |b - c| + |c - d| + |d - a| > 2|b - d|,$$

and adding this inequality to inequality $(\spadesuit)$, and then dividing the resulting inequality by 2, gives the desired

$$|a - b| + |b - c| + |c - d| + |d - a| > |a - c| + |b - d|.$$

8.3.6 Since both $|w + z|$ and $|w| + |z|$ are positive, we can compare $|w + z|^2$ and $(|w| + |z|)^2$ in order to compare $|w + z|$ and $|w| + |z|$. We introduce the squares because they are easier to work with. We have

$$(|w| + |z|)^2 = |w|^2 + 2|w||z| + |z|^2 = w\overline{w} + 2|w||z| + z\overline{z},$$
$$|w + z|^2 = (w + z)(\overline{w} + \overline{z}) = w\overline{w} + w\overline{z} + z\overline{w} + z\overline{z}.$$

Therefore, we have

$$(|w| + |z|)^2 - |w + z|^2 = 2|w||z| - (w\overline{z} + z\overline{w}).$$

We'd like to simplify $w\overline{z} + z\overline{w}$. Where have we seen an expression like this before? We first note that $w\overline{z}$ and $z\overline{w}$ are conjugates:

$$\overline{w\overline{z}} = \overline{w} \cdot \overline{\overline{z}} = \overline{w}z.$$

Then, we recall that the sum of a complex number and its conjugate is twice its real part: $a + bi + \overline{a + bi} = 2a$. So, we have

$$(|w| + |z|)^2 - |w + z|^2 = 2|w||z| - 2\text{Re}(w\overline{z}).$$

We want to show that the expression on the right is nonnegative, so we wish to show that $|w||z| \geq \text{Re}(w\overline{z})$. So that we'll be working with like expressions, we note that $|z| = |\overline{z}|$, so $|w||z| = |w||\overline{z}| = |w\overline{z}|$, which means we'd now like to prove that $|w\overline{z}| \geq \text{Re}(w\overline{z})$. Now that we're comparing apples to apples, the result is obvious. The magnitude of any complex number is at least as large as its real part:

$$|a + bi| = \sqrt{a^2 + b^2} \geq \sqrt{a^2} = |a| \geq a.$$

Note that equality holds here if and only if $b = 0$ and $|a| = a$; that is, if the complex number is a positive real number.

We therefore have $|w\overline{z}| \geq \text{Re}(w\overline{z})$, which means $|w||z| \geq \text{Re}(w\overline{z})$, so

$$(|w| + |z|)^2 - |w + z|^2 = 2|w||z| - 2\text{Re}(w\overline{z}) \geq 0,$$

and we can conclude that $|w| + |z| \geq |w + z|$.

Equality holds in $|w| + |z| \geq |w + z|$ if and only if it also holds in our key inequality step, $|w\bar{z}| \geq \mathrm{Re}(w\bar{z})$. As explained above, this inequality only holds if $w\bar{z}$ is a positive real number. Multiplying $w\bar{z} = c$ by z gives $wz\bar{z} = cz$, so $\frac{w}{z} = \frac{c}{|z|^2}$. Since c is positive, this tells us that w and z are on the same ray from the origin.

Putting this all together, we have $|w| + |z| \geq |w + z|$, where equality holds if an only if w/z is a nonnegative real number or $z = 0$.

Exercises for Section 8.4

8.4.1 Let r_1 and r_2 be the complex numbers corresponding to the two possible locations of R, which are points R_1 and R_2 shown at right.

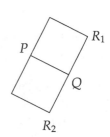

Then R_1 is obtained by rotating P clockwise around Q by $\frac{\pi}{2}$, so

$$r_1 = e^{-\pi i/2}(p - q) + q = -i(p - q) + q = \boxed{-ip + (1 + i)q}.$$

Also, R_2 is obtained by rotating P counterclockwise around Q by $\frac{\pi}{2}$, so

$$r_2 = e^{\pi i/2}(p - q) + q = i(p - q) + q = \boxed{ip + (1 - i)q}.$$

8.4.2 Let a be the complex number corresponding to A, and similarly for the other points. Let ζ be a primitive sixth root of unity, so $\zeta^2 = \zeta - 1$, and let O be the origin.

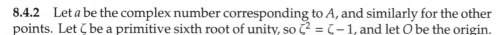

Then $b = \zeta a$, $d = \zeta c$, and $f = \zeta e$, so

$$l = \frac{b + c}{2} = \frac{\zeta a + c}{2}, \quad m = \frac{d + e}{2} = \frac{\zeta c + e}{2}, \quad n = \frac{f + a}{2} = \frac{\zeta e + a}{2}.$$

To show that triangle LMN is equilateral, it suffices to show that $n - l = \zeta(m - l)$. Substituting, we find

$$n - l = \frac{\zeta e + a}{2} - \frac{\zeta a + c}{2} = \frac{(1 - \zeta)a - c + \zeta e}{2},$$

and

$$m - l = \frac{\zeta c + e}{2} - \frac{\zeta a + c}{2} = \frac{-\zeta a + (-1 + \zeta)c + e}{2},$$

so

$$\zeta(m - l) = \frac{-\zeta^2 a + (-\zeta + \zeta^2)c + \zeta e}{2} = \frac{(1 - \zeta)a - c + \zeta e}{2},$$

where we use the fact that $\zeta^2 = \zeta - 1$ in the final step. Hence, $n - l = \zeta(m - l)$, so triangle LMN is equilateral.

8.4.3 Let a be the complex number corresponding to A, and similarly for the other points. We let b be the origin, so that $d = a + c$. Next, we find f and g in terms of a and c.

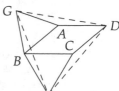

Because $\triangle ABG$ is equilateral and B is the origin, we have $g = \zeta a$ for some sixth root of unity ζ. From equilateral triangle CFB, we have $c = \zeta f$. We are careful to choose the same orientation of the rotation as we chose when considering $\triangle ABG$, so that the sixth root of unity in $g = \zeta a$ is the same as the sixth root of unity in $c = \zeta f$. Solving for f gives $f = c/\zeta$.

Now, in order to check if $\triangle DGF$ is equilateral, we must check if $d - g = \zeta(f - g)$, where we again strategically choose to check the rotation of F counterclockwise about G because doing so will allow us to cancel the ζ in the denominator of $f = c/\zeta$.

We have

$$d - g = (a + c) - \zeta a = (1 - \zeta)a + c,$$

$$\zeta(f - g) = \zeta\left(\frac{c}{\zeta} - \zeta a\right) = c - \zeta^2 a.$$

As described in the text, we have $\zeta^2 - \zeta + 1 = 0$, so $\zeta^2 = \zeta - 1$, and we have

$$\zeta(f - g) = c - \zeta^2 a = c - (\zeta - 1)a = (1 - \zeta)a + c = d - g,$$

which matches our expression for $d - g$ from above. So, triangle DFG is equilateral.

8.4.4

(a) The relationship we must prove looks a lot like the condition we proved in the text under which a triangle is equilateral. There, we showed that $\triangle ABC$ is equilateral if and only if $a - b = \zeta(c - b)$ for some primitive sixth root of unity ζ. We make this look even more like $a - b = \omega(b - c)$ by writing it as $a - b = -\zeta(b - c)$.

Now, we just have to show that $-\zeta$ is a primitive cube root of unity if and only if ζ is a primitive sixth root of unity. If $\zeta = e^{\pi i/3}$, we have

$$-\zeta = (-1)e^{\pi i/3} = e^{\pi i} \cdot e^{\pi i/3} = e^{4\pi i/3},$$

which is a primitive cube root of unity. Similarly, if $\zeta = e^{5\pi i/3}$, then $-\zeta = e^{2\pi i/3}$, which is also a primitive cube root of unity. We conclude that $-\zeta$ is a primitive cube root of unity if and only if ζ is a primitive sixth root of unity. Combining this with the fact that $a - b = -\zeta(b - c)$ if and only if $\triangle ABC$ is equilateral, we see that $\triangle ABC$ is equilateral if and only if $a - b = \omega(b - c)$ for some primitive cube root of unity ω.

The diagram at right gives us some geometric intuition for the relationship between the primitive cube roots and the primitive sixth roots of unity. The vertices of regular hexagon $ABCDEF$ are the sixth roots of unity. Points B and F correspond to the primitive sixth roots. Points C and E correspond to sixth roots of unity that are also primitive cube roots of unity. Each primitive sixth root of unity is the image of rotating a primitive cube root of unity by $180°$ about the origin. Therefore, each primitive sixth root of unity is the negative of some primitive cube root of unity.

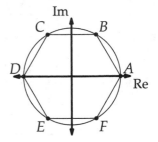

(b) We'd like to use the previous part to knock off this part, since part (a) already has a primitive cube root of unity in it. However, we need a way to introduce ω^2. We could multiply both sides of $a - b = \omega(b - c)$ by ω, but that gives us $\omega(a - b) = \omega^2(b - c)$, which doesn't appear to be much help. If we rearrange this equation, we can group the b terms to give

$$\omega a - (\omega + \omega^2)b + \omega^2 c = 0.$$

Where have we seen $\omega + \omega^2$ before when thinking about cube roots of unity? In the text, we showed that the primitive cube roots of unity are the roots of $z^2 + z + 1$, which means that $\omega^2 + \omega + 1 = 0$, so $\omega^2 + \omega = -1$. Therefore, our equation $\omega a - (\omega + \omega^2)b + \omega^2 c = 0$ becomes $\omega a + b + \omega^2 c = 0$. So, $\triangle ABC$ is equilateral if and only if $b + \omega a + \omega^2 c = 0$. Uh-oh. That's not quite what we wanted. We wanted $a + \omega b + \omega^2 c = 0$. Is that a problem?

No, it's not. We just showed that if $(a - b) = \omega(b - c)$, then $b + \omega a + \omega^2 c$. From part (a), we also know that we must have $(b - a) = \omega'(a - c)$, where all we have done is swapped a and b, for some primitive cube root of unity ω'. If we follow the same steps we used above to show $b + \omega a + \omega^2 c = 0$, but start with $(b - a) = \omega'(a - c)$, we will end at $a + \omega' b + (\omega')^2 c = 0$. Since ω' is a cube root of unity, we have proved the desired relationship. (What's really going on here is that ω and ω' are not the same primitive cube root of unity, which is why starting with $(a - b) = \omega(b - c)$ doesn't get you exactly what you want, but starting with $(b - a) = \omega'(a - c)$ does.)

8.4.5 Let a be the complex number corresponding to A, and similarly for the other points. We place the square $ABCD$ in the complex plane so that $a = 1$, $b = i$, $c = -1$, and $d = -i$. Then the center of the square O coincides with the origin.

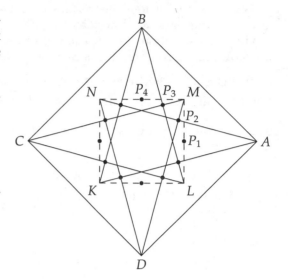

Then K is obtained by rotating B by $\frac{\pi}{3}$ counterclockwise around A, so

$$k = e^{\pi i/3}(b - a) + a = \left(\frac{1}{2} + \frac{\sqrt{3}}{2}i \right)(i - 1) + 1 = \frac{1 - \sqrt{3}}{2} + \frac{1 - \sqrt{3}}{2}i.$$

By symmetry, L can be obtained by rotating K by $\frac{\pi}{2}$ counterclockwise around the origin, so

$$l = ik = i\left(\frac{1 - \sqrt{3}}{2} + \frac{1 - \sqrt{3}}{2}i \right) = \frac{\sqrt{3} - 1}{2} + \frac{1 - \sqrt{3}}{2}i.$$

Since M and N are rotations by π of K and L about the origin, we have $m = -k = \frac{\sqrt{3}-1}{2} + \frac{\sqrt{3}-1}{2}i$ and $n = -l = \frac{1-\sqrt{3}}{2} + \frac{\sqrt{3}-1}{2}i$.

Let P_1 be the midpoint of $\overline{ML}$, so $p_1 = \frac{l+m}{2} = \frac{\sqrt{3}-1}{2}$. Let P_2 be the midpoint of $\overline{AN}$, so

$$p_2 = \frac{a + n}{2} = \frac{1 + \frac{1-\sqrt{3}}{2} + \frac{\sqrt{3}-1}{2}i}{2} = \frac{3 - \sqrt{3}}{4} + \frac{\sqrt{3} - 1}{4}i.$$

Then

$$\frac{p_2}{p_1} = \frac{\frac{3-\sqrt{3}}{4} + \frac{\sqrt{3}-1}{4}i}{\frac{\sqrt{3}-1}{2}} = \frac{(3 - \sqrt{3}) + (\sqrt{3} - 1)i}{2(\sqrt{3} - 1)} = \frac{\sqrt{3}(\sqrt{3} - 1) + (\sqrt{3} - 1)i}{2(\sqrt{3} - 1)} = \frac{\sqrt{3} + i}{2},$$

which is equal to

$$\frac{\sqrt{3}}{2} + \frac{1}{2}i = \cos\frac{\pi}{6} + i\sin\frac{\pi}{6} = e^{\pi i/6}.$$

Hence, P_2 can be obtained by rotating P_1 by $\frac{\pi}{6}$ counterclockwise about the origin.

Now, let P_3 be the midpoint of $\overline{BL}$, and let P_4 be the midpoint of $\overline{MN}$. Then by symmetry, P_4 can be obtained by rotating P_3 by $\frac{\pi}{6}$ counterclockwise about the origin. Furthermore, $OP_1 = OP_4$, and $\angle P_1 O P_4 = \frac{\pi}{2}$, so $OP_2 = OP_3$ and $\angle P_2 O P_3 = \frac{\pi}{2} - \frac{\pi}{6} - \frac{\pi}{6} = \frac{\pi}{6}$. Hence, P_1, P_2, P_3, and P_4 are four of the vertices of a regular dodecagon, centered at O.

By symmetry, all 12 midpoints form the vertices of a regular dodecagon.

Review Problems

8.32 If $z = x + yi$, then $\bar{z} = x - yi$. Thus, $\bar{z}$ has the same real part as z, but the imaginary parts of z and $\bar{z}$ are opposites. Hence, $\boxed{\text{reflection over the real axis}}$ maps z to $\bar{z}$.

8.33 We see that $f_1(z) = z + 2 - 3i$ and $f_2(z) = e^{\pi i/2}(z - 4i) + 4i = (i)(z - 4i) + 4i = iz + 4 + 4i$

(a) We have $(f_2 \circ f_1)(z) = f_2(f_1(z)) = f_2(z + 2 - 3i) = i(z + 2 - 3i) + 4 + 4i = \boxed{iz + 7 + 6i}$.

(b) We have $(f_1 \circ f_2)(z) = f_1(f_2(z)) = f_1(iz + 4 + 4i) = iz + 4 + 4i + 2 - 3i = \boxed{iz + 6 + i}$.

Note that our results for parts (a) and (b) are not the same! We did not make a mistake; this is an example of the fact that a translation followed by a rotation is not necessarily the same as doing the rotation first, and following it with the translation. (See if you can prove that the only time these two are the same is when one of the two transformations does nothing, such as a rotation of 2π. Here's a hint: think about what happens to the origin.)

8.34 Let $z = x + yi$. Then $f(z) = i\bar{z} = i(x - yi) = y + xi$. Thus, f swaps the real and imaginary parts of z. The transformation corresponding to f is $\boxed{\text{reflection over the line } x = y}$.

8.35 In both parts, let points A, B, and C correspond to a, b, and c, respectively. Then, we have $|a - b| = AB$, $|a - c| = AC$, and $|b - c| = BC$.

(a) We have $AB + AC = BC$. By the Triangle Inequality, this equation holds if and only if A is on $\overline{BC}$.

(b) We have $AB^2 + AC^2 = BC^2$. By the Pythagorean Theorem, this equation holds if and only if $\angle BAC$ is a right angle.

8.36

(a) First, note that the magnitude of $\frac{z}{|z|}$ is $\left| \frac{z}{|z|} \right| = \frac{|z|}{|z|} = 1$, so $\frac{z}{|z|}$ lies on the unit circle. Furthermore, the complex number $\frac{z}{|z|}$ can be obtained by dilating z through the origin by a factor of $\frac{1}{|z|}$. Hence, $\frac{z}{|z|}$ is the intersection of the unit circle and the ray joining the origin to z.

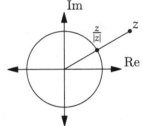

(b) Let $z = x + yi$, where x and y are real numbers. Then $\bar{z} = x - yi$, so $z + \bar{z} = 2x$. Hence, the foot of the altitude from z to the x-axis, which we'll call point M, is the midpoint of the segment from the origin to $z + \bar{z}$. After locating M, we draw the circle with center M and radius OM, where O is the origin. The other point at which this circle intersects the real axis corresponds to $z + \bar{z}$.

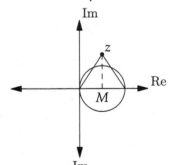

(c) By definition, z lies on the circle centered at the origin with radius $|z|$. Furthermore, $|z|$ is a real number, i.e. $|z|$ is the intersection of this circle and the nonnegative real axis. The point corresponding to $|z|$ is $|z|$ to the right of the origin. The point corresponding to $z + |z|$ is $|z|$ to the right of z. Therefore, if we build a quadrilateral with vertices 0, z, $z + |z|$, and $|z|$ in the complex plane, the side connecting 0 to $|z|$ and the side connecting z to $z + |z|$ are parallel and equal in length. Therefore, to locate $z + |z|$, we draw the line through $|z|$ that is parallel to the line from the origin to z. The intersection of this line with the line through z parallel to the real axis is the point corresponding to $z + |z|$.

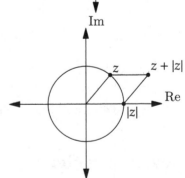

8.37 Let A, B, and C be the vertices of the triangle, and let a, b, and c be the corresponding complex numbers. Let M and N be the midpoints of $\overline{AB}$ and $\overline{AC}$, respectively, so the corresponding complex numbers are $m = \frac{a+b}{2}$ and $n = \frac{a+c}{2}$, respectively.

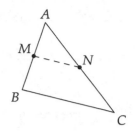

We therefore have $m - n = \frac{a+b}{2} - \frac{a+c}{2} = \frac{b-c}{2}$. This means that $MN = |m - n| = \left|\frac{b-c}{2}\right| = \frac{1}{2}|b - c| = \frac{BC}{2}$, and that $\frac{m-n}{b-c} = \frac{1}{2}$. Since $\frac{m-n}{b-c}$ is real we know that $\overline{MN} \parallel \overline{BC}$. Therefore, the segment connecting the midpoints of $\overline{AB}$ and $\overline{AC}$ is parallel to $\overline{BC}$ and has half the length of $\overline{BC}$.

8.38

(a) Let $a = 3 - 2i$ and $b = 1 - 5i$. As shown in the text, the line joining a and b is perpendicular to the line joining z and b if and only if

$$\frac{z - b}{a - b} = \frac{z - (1 - 5i)}{(3 - 2i) - (1 - 5i)} = \frac{z - (1 - 5i)}{2 + 3i}$$

is imaginary, so we check this value for each the given complex numbers.

For $z = 2 - 7i$, we have

$$\frac{(2 - 7i) - (1 - 5i)}{2 + 3i} = \frac{1 - 2i}{2 + 3i} = \frac{(1 - 2i)(2 - 3i)}{(2 + 3i)(2 - 3i)} = \frac{2 - 3i - 4i - 6}{13} = \frac{-4 - 7i}{13},$$

which is not imaginary.

For $z = -6 + 2i$, we have

$$\frac{(-6 + 2i) - (1 - 5i)}{2 + 3i} = \frac{-7 + 7i}{2 + 3i} = \frac{(-7 + 7i)(2 - 3i)}{(2 + 3i)(2 - 3i)} = \frac{-14 + 21i + 14i + 21}{13} = \frac{7 + 35i}{13},$$

which is not imaginary.

For $z = -5 - i$, we have

$$\frac{(-5 - i) - (1 - 5i)}{2 + 3i} = \frac{-6 + 4i}{2 + 3i} = \frac{(-6 + 4i)(2 - 3i)}{(2 + 3i)(2 - 3i)} = \frac{-12 + 18i + 8i + 12}{13} = \frac{26i}{13} = 2i,$$

which is imaginary.

Therefore, of the given complex numbers, only $\boxed{-5 - i}$ has the desired property.

(b) If z is on the line through b such that the line is perpendicular to the line through a and b, then we must have

$$(z - b)(\bar{a} - \bar{b}) + (\bar{z} - \bar{b})(a - b) = 0,$$

as proved in the text. Therefore, the desired equation is

$$(z - (1 - 5i))((3 + 2i) - (1 + 5i)) + (\bar{z} - (1 + 5i))((3 - 2i) - (1 - 5i)) = 0.$$

Simplifying the left side gives

$$(z - 1 + 5i)(2 - 3i) + (\bar{z} - 1 - 5i)(2 + 3i) = 0.$$

Expanding and simplifying the left side gives $\boxed{(2 - 3i)z + (2 + 3i)\bar{z} + 26 = 0}$.

8.39 Let a, b, and d be complex numbers corresponding to points A, B, and D in the complex plane. We know that any circle in the complex plane can be described as the graph of some equation $|z - c| = r$, where c is the center and r is the radius. We can make our work a lot easier if we choose the midpoint of $\overline{AB}$ to be the origin, so the circle through A and B in the complex plane has center 0 and radius $|a|$. Therefore, the circle with diameter $\overline{AB}$ is the graph of the equation $|z| = |a|$. So, to show that D is on the circle, we must prove that $|d| = |a|$.

If $\overline{AD} \perp \overline{DB}$, then we have

$$(a - d)(\overline{b} - \overline{d}) + (\overline{a} - \overline{d})(b - d) = 0.$$

Since we chose the midpoint of $\overline{AB}$ to be the origin, we have $b = -a$, and the equation above becomes

$$(a - d)(-\overline{a} - \overline{d}) + (\overline{a} - \overline{d})(-a - d) = 0.$$

Expanding the left side, we have

$$-a\overline{a} - a\overline{d} + d\overline{a} + d\overline{d} - a\overline{a} - d\overline{a} + a\overline{d} + d\overline{d} = 0.$$

Massive cancellation ensues, and we're left with

$$2d\overline{d} - 2a\overline{a} = 0.$$

Dividing by 2 and rearranging gives $d\overline{d} = a\overline{a}$, from which we conclude that $|d|^2 = |a|^2$ (because $z\overline{z} = |z|^2$ for any $|z|$). Finally, since $|d|$ and $|a|$ are nonnegative, we can take the square root to find $|d| = |a|$, as desired. Therefore, we conclude that D is on the circle with diameter $\overline{AB}$.

8.40 The center of the hexagon is given by $\frac{(2+i)+(8-3i)}{2} = 5 - i$. Two vertices are the results of counterclockwise rotations by $\frac{\pi}{3}$ and by $\frac{2\pi}{3}$ of $2 + i$ about the center of the hexagon, and two vertices are the results of counterclockwise rotations by $\frac{\pi}{3}$ and by $\frac{2\pi}{3}$ of $8 - 3i$ about the center of the hexagon. Therefore, the vertices are:

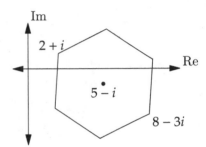

Then the other vertices of the hexagon are given by

$$e^{\pi i/3}[(2 + i) - (5 - i)] + (5 - i) = \left(\frac{1}{2} + \frac{\sqrt{3}}{2}i\right)(-3 + 2i) + (5 - i)$$

$$= \boxed{\frac{7 - 2\sqrt{3}}{2} - \frac{3\sqrt{3}}{2}i},$$

$$e^{2\pi i/3}[(2 + i) - (5 - i)] + (5 - i) = \left(-\frac{1}{2} + \frac{\sqrt{3}}{2}i\right)(-3 + 2i) + (5 - i) = \boxed{\frac{13 - 2\sqrt{3}}{2} - \frac{4 + 3\sqrt{3}}{2}i},$$

$$e^{\pi i/3}[(8 - 3i) - (5 - i)] + (5 - i) = \left(\frac{1}{2} + \frac{\sqrt{3}}{2}i\right)(3 - 2i) + (5 - i) = \boxed{\frac{13 + 2\sqrt{3}}{2} + \frac{-4 + 3\sqrt{3}}{2}i},$$

$$e^{2\pi i/3}[(8 - 3i) - (5 - i)] + (5 - i) = \left(-\frac{1}{2} + \frac{\sqrt{3}}{2}i\right)(3 - 2i) + (5 - i) = \boxed{\frac{7 + 2\sqrt{3}}{2} + \frac{3\sqrt{3}}{2}i}.$$

8.41 Let the vertices of the trapezoid be $A, B, C,$ and D, where the parallel sides are $\overline{AB}$ and $\overline{CD}$, and let the corresponding complex numbers be $a, b, c,$ and d, respectively. Without loss of generality, we can orient trapezoid $ABCD$ so that both a and b are real and $b > a$. Then $AB = b - a$.

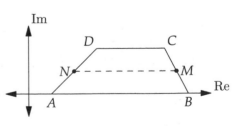

Since $\overline{AB} \parallel \overline{CD}$, c and d have the same imaginary part, say h. Let $c = p + hi$ and $d = q + hi$, where p and q are real numbers. Note that $c - d = (q + hi) - (p + hi) = p - q$, so $CD = p - q$.

Let M and N be the midpoints of $\overline{BC}$ and $\overline{AD}$, so the corresponding complex numbers are

$$m = \frac{b + c}{2} = \frac{b + p + hi}{2} \quad \text{and} \quad n = \frac{a + d}{2} = \frac{a + q + hi}{2}.$$

Then

$$m - n = \frac{b + p + hi}{2} - \frac{a + q + hi}{2} = \frac{(b-a)+(p-q)}{2},$$

which is a real number, so $\overline{MN}$ is parallel to $\overline{AB}$ and $\overline{CD}$. Furthermore, $MN = \frac{AB+CD}{2}$, as desired.

8.42   Without loss of generality, we can assume that one of the vertices of the parallelogram is the origin. Then the other vertices of the parallelogram are of the form a, b, and $a + b$.

The diagonals are perpendicular if and only if the line through 0 and $a + b$ is perpendicular to the line through a and b. This occurs if and only if

$$((a+b)-0)(\overline{a}-\overline{b}) + (\overline{a+b}-0)(a-b) = 0,$$

which is equivalent to $(a+b)(\overline{a}-\overline{b}) + (\overline{a}+\overline{b})(a-b) = 0$. Expanding the left side gives

$$a\overline{a} - a\overline{b} + \overline{a}b - b\overline{b} + a\overline{a} + a\overline{b} - \overline{a}b - b\overline{b} = 0,$$

which simplifies as $a\overline{a} = b\overline{b}$, or $|a|^2 = |b|^2$, or $|a| = |b|$. But two sides of the parallelogram are equal in length if and only if it is a rhombus.

8.43   We know that z lies on the perpendicular bisector if and only if the line joining z and $\frac{a+b}{2}$ is perpendicular to the line joining a and b, which is true if and only if

$$(\overline{a}-\overline{b})\left(z - \frac{a+b}{2}\right) + (a-b)\left(\overline{z} - \frac{\overline{a}+\overline{b}}{2}\right) = 0.$$

Expanding the products gives

$$(\overline{a}-\overline{b})z - \frac{(\overline{a}-\overline{b})(a+b)}{2} + (a-b)\overline{z} - \frac{(a-b)(\overline{a}+\overline{b})}{2} = 0,$$

and rearranging this gives

$$(\overline{a}-\overline{b})z + (a-b)\overline{z} = \frac{(a+b)(\overline{a}-\overline{b}) + (a-b)(\overline{a}+\overline{b})}{2} = \frac{a\overline{a} - a\overline{b} + \overline{a}b - b\overline{b} + a\overline{a} + a\overline{b} - \overline{a}b - b\overline{b}}{2} = a\overline{a} - b\overline{b}.$$

8.44   We consider the case in which $WXYZ$ is oriented as shown at right. (We can tackle the other possible orientation of $WXYZ$ in essentially the same way.) If $\triangle TSZ$ is an isosceles triangle, it looks like we must have $\angle TSZ = 90°$ and $TS = SZ$. We have $\angle ZST = 90°$ and $TS = SZ$ if and only if T is a $90°$ rotation of Z about S. Letting s, t, and z be the complex numbers corresponding to S, T, and Z, we can write our condition for $\triangle TSZ$ to be an isosceles triangle as simply $t - s = (z - s)i$ (for the case that t is a $90°$ counterclockwise rotation of z about s). That's a pretty simple target, so let's plow ahead with complex numbers.

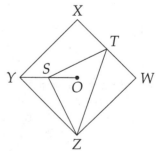

We can make life somewhat simpler by choosing convenient coordinates. We choose the center of the square to be the origin, and orient the axes so that W corresponds to 1, X to i, Y to -1, and Z to $-i$. Since T is the midpoint of $\overline{WX}$, we have $t = \frac{1+i}{2}$. Since S is the midpoint of $\overline{OY}$, we have $s = -\frac{1}{2}$.

Now, we're all set to finish off the proof. We have $z = -i$, $s = -\frac{1}{2}$, and $t = \frac{1+i}{2}$. Therefore, we have $(z-s)i = \left(-i - \left(-\frac{1}{2}\right)\right)i = 1 + \frac{i}{2}$ and $t - s = \frac{1+i}{2} - \left(-\frac{1}{2}\right) = 1 + \frac{i}{2}$, so $(z-s)i = t-s$. This tells us that t is a $90°$ counterclockwise rotation of z about s, as desired. Therefore, $\triangle TSZ$ is an isosceles right triangle.

8.45 We could tackle this problem with the regular tools from Euclidean geometry, but we'd have to be very careful to make sure that our proof covered all possible configurations. Instead, we'll use complex numbers, and tackle all possible configurations at once.

We place the problem on the complex plane such that X is the origin and W is at 1. Since $\overline{XW}$ is horizontal, point Y is the image of a vertical translation of X and Z is the image of a vertical translation of W. Therefore, Y is at ai for some real constant a, and Z is at $1 + bi$ for some real constant b. The midpoint of $\overline{YZ}$ then is $\frac{ai+1+bi}{2} = \frac{1}{2} + \frac{a+b}{2}i$.

Let this midpoint be M. We wish to show that $MW = MX$. We have $MW = \left|\left(\frac{1}{2} + \frac{a+b}{2}i\right) - 1\right| = \left|-\frac{1}{2} + \frac{a+b}{2}i\right|$ and $MX = \left|\frac{1}{2} + \frac{a+b}{2}i\right|$. Since $|z| = |\overline{z}|$ for any complex number z, we have

$$MW = \left|-\frac{1}{2} + \frac{a+b}{2}i\right| = \left|\overline{-\frac{1}{2} + \frac{a+b}{2}i}\right| = \left|-\frac{1}{2} - \frac{a+b}{2}i\right| = |-1|\left|\frac{1}{2} + \frac{a+b}{2}i\right| = MX.$$

We could also have noted that M is the result of a vertical translation of the point corresponding to $\frac{1}{2}$. Therefore, M is on the perpendicular bisector of $\overline{WX}$, which means that it is equidistant from W and Z.

We have no diagram for this problem because there are two possible configurations: Y and Z can be on opposite sides of $\overleftrightarrow{WX}$ or on the same side of $\overleftrightarrow{WX}$. Using complex numbers allows us to tackle both at once.

8.46 Let $ABCD$ be a quadrilateral and let a, b, c, and d be complex numbers corresponding to the respective vertices of the quadrilateral. Then, the midpoints of $\overline{AB}$ and $\overline{CD}$ are $\frac{a+b}{2}$ and $\frac{c+d}{2}$. The midpoint of the segment connecting these two points is $\frac{(a+b)/2+(c+d)/2}{2} = \frac{a+b+c+d}{4}$. Since this equals $\frac{(a+d)/2+(b+c)/2}{2}$, the complex number $\frac{a+b+c+d}{4}$ also corresponds to the midpoint of the segment connecting the midpoints of $\overline{AD}$ and $\overline{BC}$.

The midpoints of the diagonals $\overline{AC}$ and $\overline{BD}$ are $\frac{a+c}{2}$ and $\frac{b+d}{2}$. The midpoint of the segment connecting these two points is $\frac{(a+c)/2+(b+d)/2}{2} = \frac{a+b+c+d}{4}$. This coincides with the midpoint of the segment whose endpoints are the midpoints of $\overline{AB}$ and $\overline{CD}$, so the two segments described in the problem bisect each other. (We can go through the same steps with sides $\overline{BC}$ and $\overline{AD}$, as well.)

8.47 *Solution 1: Use Translation.* The translation that maps A to B also maps D to C. The translation that maps A to B maps any complex number z to $z + (b - a)$, since $a + (b - a) = b$. Therefore, this translation maps d to $d + (b - a)$. Since this translation also maps D to C, we must have $d + (b - a) = c$, so $d = c + a - b$.

Solution 2: Use Diagonals. Since $ABCD$ is a parallelogram, the diagonals must bisect each other. That is, their midpoints must be the same, so $\frac{a+c}{2} = \frac{b+d}{2}$, from which we find $d = a + c - b$, as before.

8.48 From the Triangle Inequality, we have $|w - z| + |z| \geq |(w - z) + z| = |w|$, so $|w - z| \geq |w| - |z|$.

8.49 Let p, a, b, c, and d be complex numbers corresponding to P, A, B, C, and D, respectively. So, we must show that
$$|a - p|^2 + |c - p|^2 = |b - p|^2 + |d - p|^2.$$

We can get rid of the magnitudes by noting that $|z|^2 = z\overline{z}$ for all complex numbers z, but first we simplify the problem by choosing our origin wisely. We let the center of the rectangle (the intersection of the diagonals) be the origin. This point is equidistant from all 4 vertices and is the midpoint of $\overline{AC}$ and $\overline{BD}$. Therefore, we have $c = -a$, $d = -b$, and $|a| = |b|$. So, we have

$$|a - p|^2 + |c - p|^2 = |a - p|^2 + |-a - p|^2 = (a - p)(\overline{a} - \overline{p}) + (-a - p)(-\overline{a} - \overline{p})$$
$$= a\overline{a} - a\overline{p} - p\overline{a} + p\overline{p} + a\overline{a} + a\overline{p} + p\overline{a} + p\overline{p}$$
$$= 2(a\overline{a} + p\overline{p}).$$

Similarly, we have $|b - p|^2 + |d - p|^2 = 2(b\overline{b} + p\overline{p})$. Since $|a| = |b|$, we have $|a|^2 = |b|^2$, so $a\overline{a} = b\overline{b}$ and $a\overline{a} + p\overline{p} = b\overline{b} + p\overline{p}$. Therefore, we have $|a - p|^2 + |c - p|^2 = |b - p|^2 + |d - p|^2$, so $AP^2 + CP^2 = BP^2 + DP^2$, as desired.

Challenge Problems

8.50   We save ourselves a lot of work by choosing the origin and real axis wisely. We place the problem on the complex plane by letting O be the origin and letting k be the real axis. Then the distance from each vertex to k is the real part of the complex number that corresponds to the vertex. We let $x + yi$ correspond to one vertex of the square. We find another vertex by rotating $x + yi$ by $\frac{\pi}{2}$ counterclockwise about the origin, to get $(x + yi)i = -y + xi$. The next vertex is a $\frac{\pi}{2}$ counterclockwise rotation of $-y + xi$ about the origin, so it is at $(-y + xi)i = -x - yi$. The last vertex is a $\frac{\pi}{2}$ rotation of $-x - yi$ about the origin, or $(-x - yi)i = y - xi$.

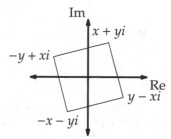

The sum of the squares of the distances from each vertex to the real axis is $y^2 + x^2 + y^2 + x^2 = 2(x^2 + y^2)$. The quantity $x^2 + y^2$ equals the square of the distance from any vertex to the origin. This distance equals half the length of a diagonal of the square, and this length is the same no matter how the square is oriented about the origin. Therefore, the quantity $x^2 + y^2$ is the same no matter how the square is oriented about the origin, which means that the sum of the squares of the distances from the vertices of $ABCD$ to k is the same for any orientation of the square about the line k. Equivalently, this sum is the same for any location of k through the center of a fixed square $ABCD$.

8.51   From the diagram, we expect that if $\triangle PAQ$ is an isosceles right triangle, then $\angle PAQ = \frac{\pi}{2}$ and $PA = AQ$, so we want to show that P is the image of a $\frac{\pi}{2}$ rotation of Q about A. We place the problem on the complex plane, letting the lowercase version of each letter be the complex number corresponding to the respective uppercase point. With the diagram oriented as shown, showing that P is the image of a $\frac{\pi}{2}$ counterclockwise rotation of Q about A is the equivalent of showing that $p - a = e^{\pi i/2}(q - a) = i(q - a)$.

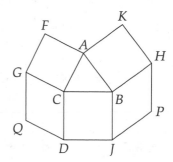

We simplify the problem by letting A be the origin, so $a = 0$ and we wish to prove that $p = iq$. Now, we must find p and q in terms of b and c, and check if we have $p = iq$.

In square $CAFG$, point G is a $\frac{\pi}{2}$ counterclockwise rotation of A around C, so we have

$$g - c = e^{\pi i/2}(a - c) = i(a - c).$$

Since $a = 0$, we have $g = c - ic$.

Similarly, D is a $\frac{\pi}{2}$ *clockwise* rotation of B about C, so

$$d - c = e^{-\pi i/2}(b - c) = (-i)(b - c) = -ib + ic,$$

which means $d = -ib + c + ic$.

Finally, we find q from parallelogram $CDQG$. The diagonals of the parallelogram bisect each other, so $\frac{q+c}{2} = \frac{d+g}{2}$, which means

$$q = d + g - c = (-ib + c + ic) + (c - ic) - c = -ib + c.$$

Similarly, H is a $\frac{\pi}{2}$ clockwise rotation of A about B, so $h - b = e^{-\pi i/2}(a - b) = (-i)(0 - b) = ib$, which means $h = b + ib$. Since J is a $\frac{\pi}{2}$ counterclockwise rotation of C about B, we have $j - b = e^{\pi i/2}(c - b) = i(c - b)$, so $j = b + i(c - b) = (1 - i)b + ic$. Parallelogram $BHPJ$ then gives

$$p = j + h - b = ((1 - i)b + ic) + (b + ib) - b = b + ic.$$

We therefore have $iq = i(-ib + c) = b + ic = p$, so p is a $\frac{\pi}{2}$ counterclockwise rotation of q about the origin. Since A is the origin, we have proved that P is a $\frac{\pi}{2}$ rotation of Q about A, which means that $\triangle PAQ$ is an isosceles right triangle, as desired.

8.52 We find p first. Since a, p, and b are collinear, we have

$$(p - b)(\bar{a} - \bar{b}) - (\bar{p} - \bar{b})(a - b) = 0.$$

Since the line joining p and z is perpendicular to the line joining a and b, we have

$$(p - z)(\bar{a} - \bar{b}) + (\bar{p} - \bar{z})(a - b) = 0.$$

We can cancel $\bar{p}$ by adding these two equations. When we add the equations, we group the $(\bar{a} - \bar{b})$ terms and group the $(a - b)$ terms. This gives us

$$(2p - b - z)(\bar{a} - \bar{b}) + (\bar{b} - \bar{z})(a - b) = 0.$$

Rearranging this equation gives

$$(2p - b - z)(\bar{a} - \bar{b}) = (\bar{z} - \bar{b})(a - b).$$

Since $a \neq b$, we can divide by $(\bar{a} - \bar{b})$, and then isolate p to find

$$p = \frac{1}{2}\left(\frac{(\bar{z} - \bar{b})(a - b)}{\bar{a} - \bar{b}} + b + z \right) = \frac{\bar{z}(a - b) - a\bar{b} + b\bar{b} + b\bar{a} - b\bar{b} + z(\bar{a} - \bar{b})}{2(\bar{a} - \bar{b})} = \boxed{\frac{(\bar{a} - \bar{b})z + (a - b)\bar{z} - a\bar{b} + \bar{a}b}{2(\bar{a} - \bar{b})}}.$$

Since p is the midpoint of z and w, we have $p = \frac{z+w}{2}$. Solving for w gives $w = 2p - z = \boxed{\dfrac{(a - b)\bar{z} - a\bar{b} + \bar{a}b}{\bar{a} - \bar{b}}}$.

8.53 The expressions $\sqrt{x^2 + (x - 1)^2}$ and $\sqrt{x^2 + (x + 1)^2}$ resemble the distance formula. We are further inspired to think of them in this way by being asked to minimize their sum, because minimizing a sum of distances reminds us of the Triangle Inequality, $AB + BC \geq AC$. But what points should we take to be A, B, and C to give the expressions in the problem? Taking these to be points in the complex plane, we can let B represent $x + xi$, A represent i, and C represent $-i$. Then, we have

$$AB + BC = \sqrt{x^2 + (x - 1)^2} + \sqrt{x^2 + (x + 1)^2}.$$

By the Triangle Inequality, this sum is minimized when B is on $\overline{AC}$, which is a fixed segment that does not depend on x. Since $\overline{AC}$ is the segment from i to $-i$, point B must be on the imaginary axis to minimize $AB + BC$. The only point of the form $x + xi$ on the imaginary axis is 0, so the minimum value of f is $f(0) = \boxed{2}$.

Looking back over our geometric argument, we can now construct a purely algebraic argument to show that the minimum value of f is 2. Let $z = x + xi$, $a = i$, and $b = -i$. Then

$$\begin{aligned}
|z - a| + |z - b| &= |(x + xi) - i| + |(x + xi) + i| \\
&= |x + (x - 1)i| + |x + (x + 1)i| \\
&= \sqrt{x^2 + (x - 1)^2} + \sqrt{x^2 + (x + 1)^2},
\end{aligned}$$

which is $f(x)$. By the Triangle Inequality, we have $|z - a| + |z - b| = |a - z| + |z - b| \geq |(a - z) + (z - b)| = |a - b| = |i - (-i)| = |2i| = 2$, so $f(x) \geq 2$ for all x. Furthermore, $f(0) = \sqrt{0^2 + 1^2} + \sqrt{0^2 + 1^2} = 1 + 1 = 2$, so the minimum value of $f(x)$ is $\boxed{2}$.

8.54 We present two solutions—one that is largely geometric and the other largely algebraic.

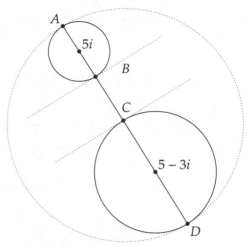

Solution 1: Geometry. The inequality $|z_1 - 5 + 3i| \le 4$ means that z_1 lies inside or on the circle with center $5 - 3i$ and radius 4, and the inequality $|z_1 - 5i| \le 2$ means that z_2 lies inside or on the circle with center $5i$ and radius 2. The distance between the centers of the circles is $|(5i) - (5 - 3i)| = \sqrt{89}$, and the sum of their radii is 6. Since $6 < \sqrt{89}$, the circles do not intersect.

Let the line joining $5 - 3i$ and $5i$ intersect the two circles at A, B, C, and D, as shown at right. The expression $|z_1 - z_2|$ equals the distance between z_1, which is in or on the larger circle, and z_2, which is in or on the smaller circle. We claim that the shortest distance between two points with one point on each circle is BC and the longest distance is AD.

To see why BC is the shortest possible distance, consider the line through B tangent to the circle centered at $5i$ and the line through C tangent to the circle centered at $5 - 3i$. (These lines are dotted in the diagram.) These two tangents are perpendicular to $\overleftrightarrow{BC}$, so they are parallel to each other. Neither circle passes through a point that lies between the two tangents, so $|z_2 - z_1|$ is at least as great as the distance between the two tangents, which is BC.

To see why AD is the largest possible distance, consider the circle with diameter $\overline{AD}$. (This circle is dotted in the diagram.) The circle with diameters $\overline{AB}$ and $\overline{CD}$ are internally tangent to the circle with diameter $\overline{AD}$, so, other than points A and D, both circles are entirely inside the large circle. Therefore, $|z_2 - z_1|$ can be no larger than the length of the diameter of the large circle, which is AD.

To find the lengths of BC and AD, we recall that the distance between the centers of the smaller circles is $\sqrt{89}$. Then, BC is this distance minus each of the radii of the circles, and AD is this distance plus each of the radii of the circles. That gives us $BC = \sqrt{89} - 2 - 4 = \boxed{\sqrt{89} - 6}$ as the minimum value of $|z_1 - z_2|$ and $AD = \sqrt{89} + 2 + 4 = \boxed{6 + \sqrt{89}}$ as the maximum value of $|z_1 - z_2|$.

Solution 2: The Triangle Inequality. This solution is a little less intuitive, but our first solution helps guide us. First, we'll tackle the maximum value. We wish to show that $|z_1 - z_2| \le k$, where k is some constant. To use the Triangle Inequality, we want to break $z_1 - z_2$ into a sum of expressions that we know something about. We know that $|z_1 - 5 + 3i| \le 4$ and $|z_2 - 5i| \le 2$, and we can write the second inequality as $|5i - z_2| \le 2$. But $z_1 - z_2$ is not exactly the sum of $z_1 - 5 + 3i$ and $5i - z_2$, so we'll need one more term, $(5 - 3i) - 5i$, and we have

$$z_1 - z_2 = (z_1 - 5 + 3i) + ((5 - 3i) - 5i) + (5i - z_2).$$

We have a hint that we're barking at the right tree when we notice that the middle term on the right is the difference of the complex numbers corresponding to the centers of the circles in our first solution.

Apply the Triangle Inequality gives

$$|z_1 - z_2| = |[z_1 - 5 + 3i] + [(5 - 3i) - 5i] + [5i - z_2]| \le |z_1 - 5 + 3i| + |(5 - 3i) - 5i| + |5i - z_2|.$$

We know that $|z_1 - 5 + 3i| \le 4$, $|(5 - 3i) - 5i| = |5 - 8i| = \sqrt{89}$, and $|z_2 - 5i| \le 2$, so we have

$$|z_1 - z_2| \le |z_1 - 5 + 3i| + |(5 - 3i) - 5i| + |5i - z_2| \le 4 + \sqrt{89} + 2 = 6 + \sqrt{89}.$$

We can show that the maximum can be achieved in the same way we did in the first solution. We consider the circles that are the graphs of $|z - 5 + 3i| = 4$ and $|z_2 - 5i| = 2$, and take point D on the first circle and A on the second such that the centers of both circles are on $\overline{AD}$. The distance between the centers is $\sqrt{89}$, and the sum of the radii is 6, so $AD = \boxed{6 + \sqrt{89}}$.

Finding the minimum is a little trickier, because here, we want $|z_1 - z_2| \ge k$, where k is some constant. To apply the Triangle Inequality to produce $|z_1 - z_2|$ on the greater side, we'll have to add more expressions to it. But we have some candidates to try. We know that $|z_1 - 5 + 3i| \le 4$ and $|z_2 - 5i| \le 2$. This time, we "reverse" the terms in the first inequality and write it as $|5 - 3i - z_1| \le 4$, where we do this so that z_1 will cancel when we apply the Triangle Inequality:

$$|5 - 3i - z_1| + |z_1 - z_2| + |z_2 - 5i| \ge |(5 - 3i - z_1) + (z_1 - z_2) + (z_2 - 5i)| = |5 - 3i - 5i| = \sqrt{89}.$$

Therefore, to minimize $|z_1 - z_2|$, we wish to maximize the sum of the other two terms on the left above, which are $|5 - 3i - z_1|$ and $|z_2 - 5i|$. We can use the given inequalities to write

$$|5 - 3i - z_1| + |z_1 - z_2| + |z_2 - 5i| \le 4 + |z_1 - z_2| + 2 = 6 + |z_1 - z_2|.$$

Putting this together with the Triangle Inequality result, we have

$$6 + |z_1 - z_2| \ge |5 - 3i - z_1| + |z_1 - z_2| + |z_2 - 5i| \ge \sqrt{89}.$$

Taking the two ends of this, we find $|z_1 - z_2| \ge -6 + \sqrt{89}$. We can show that the minimum can be achieved by considering the circles that are the graphs of $|z - 5 + 3i| = 4$ and $|z_2 - 5i| = 2$. We take point C on the first and B on the second such that the segment connecting the centers of both circles contains B and C. The distance between the centers is $\sqrt{89}$, and the sum of the radii is 6, so $BC = \boxed{-6 + \sqrt{89}}$.

8.55 Let $a = 2$, $b = 1 + 2i$, $c = -1 + i$, and $d = -1 - i$, so $|z - 2| + |z - (1 + 2i)| + |z - (-1 + i)| + |z - (-1 - i)| = |z - a| + |z - b| + |z - c| + |z - d|$.

By the Triangle Inequality, we have

$$|z - a| + |z - c| = |a - z| + |z - c| \ge |(a - z) + (z - c)| = |a - c| = |3 - i| = \sqrt{10},$$

and

$$|z - b| + |z - d| = |b - z| + |z - d| \ge |(b - z) + (z - d)| = |b - d| = |2 + 3i| = \sqrt{13}.$$

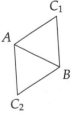

Furthermore, equality occurs in both inequalities when z is on the segment from a to c and on the segment from b to d. Therefore, we will have equality at the point at which these segments intersect, if they do intersect. The diagram above shows that they do indeed intersect (we could also show this algebraically), so the minimum value is $\boxed{\sqrt{10} + \sqrt{13}}$.

8.56

(a) *Solution 1: Apply Vieta's Formulas for quadratics.* Let C_1 and C_2 be the two possible locations of C such that A, B, and C form an equilateral triangle, as shown at right, and let c_1 and c_2 be the complex numbers corresponding to C_1 and C_2.

Let ζ be a primitive sixth root of unity. Since C_1 is the image of a $\frac{\pi}{3}$ counterclockwise rotation of B around A, we have

$$c_1 = \zeta(b - a) + a = (1 - \zeta)a + \zeta b.$$

Also, C_2 is the image of a $\frac{\pi}{3}$ counterclockwise rotation of A around B, so

$$c_2 = \zeta(a - b) + b = \zeta a + (1 - \zeta)b.$$

The equation we wish to prove true can be viewed as a quadratic in c. This equation has the roots c_1 and c_2 given above. We know that if a quadratic has the form $x^2 + px + q$, then the sum of the roots is $-p$ and the product of the roots is q. So, we can find the quadratic in c with roots c_1 and c_2 by finding $c_1 + c_2$ and $c_1 c_2$.

First, we have

$$c_1 + c_2 = (1 - \zeta)a + \zeta b + \zeta a + (1 - \zeta)b = a + b,$$

Then, as explained in the text, ζ is a root of $x^2 - x + 1$, so $\zeta^2 - \zeta + 1 = 0$, which means $\zeta^2 = \zeta - 1$, and we have

$$
\begin{aligned}
c_1 c_2 &= [(1 - \zeta)a + \zeta b][\zeta a + (1 - \zeta)b] \\
&= (\zeta - \zeta^2)a^2 + (1 - \zeta)^2 ab + \zeta^2 ab + (\zeta - \zeta^2)b^2 \\
&= (\zeta - \zeta^2)a^2 + (1 - 2\zeta + \zeta^2)ab + \zeta^2 ab + (\zeta - \zeta^2)b^2 \\
&= (\zeta - \zeta^2)a^2 + (1 - 2\zeta + 2\zeta^2)ab + (\zeta - \zeta^2)b^2 \\
&= (\zeta - \zeta + 1)a^2 + (1 - 2\zeta + 2\zeta - 2)ab + (\zeta - \zeta + 1)b^2 \\
&= a^2 - ab + b^2.
\end{aligned}
$$

Therefore, as explained above, c_1 and c_2 are the roots of

$$c^2 - (a + b)c + a^2 - ab + b^2 = 0,$$

which rearranges as $a^2 + b^2 + c^2 = ab + ac + bc$.

Solution 2: Apply Vieta's Formulas for cubics. Let $f(x) = (x - r)(x - s)(x - t)$. Expanding the right side gives

$$f(x) = x^3 - (r + s + t)x^2 + (rs + st + tr)x - rst.$$

Therefore, the sum of the roots of a cubic of the form $x^3 + a_2 x^2 + a_1 x + a_0$ is $-a_2$, the sum of all three products of pairs of roots is a_1, and the product of all three roots is $-a_0$. These are Vieta's formulas for a cubic.

To apply them to this problem, consider the cubic $f(x) = (x - a)(x - b)(x - c)$, where a, b, and c are complex numbers. The roots of $f(x)$ correspond to the vertices of the triangle formed by connecting a, b, and c in the complex plane. Let this be $\triangle ABC$. We'd like to relate f to a cubic we know a lot about, and preferably a cubic whose roots are the vertices of an equilateral triangle in the complex plane. A natural choice is $g(z) = z^3 - 1$, whose roots are the cube roots of unity, which are the vertices of an equilateral triangle in the complex plane. Let this triangle be $\triangle PQR$.

Triangle ABC is equilateral if an only if we can map it to $\triangle PQR$ by a translation followed by a dilation, followed by a rotation. First, we translate $\triangle ABC$ such that its centroid is the origin. Then, we dilate the triangle about its center by a factor of $\frac{1}{r}$, where r is the circumradius, and finally, we rotate the triangle about the origin such that one of the vertices is at 1 in the complex plane. In each transformation, the original triangle and its image are similar, so $\triangle ABC$ is equilateral if and only if the result of these three transformations is. Let's see what the effect of these transformations is in complex numbers.

The translation maps x to $x - t$, where t is a complex number. The dilation maps $x - t$ to $k(x - t)$, where k is a nonzero real number. The rotation maps $k(x - t)$ to $\omega k(x - t)$, where ω is complex number such that $|\omega| = 1$.

Therefore, we can map the roots of $f(x) = (x - a)(x - b)(x - c)$ to the roots of $g(z) = z^3 - 1$ with the transformation $z = \omega k(x - t) = \omega k x - \omega k t$, where z is a root of $g(z)$ and x is a root of $f(x)$. Letting $u = \omega k$ and $v = -\omega k t$, we have $z = ux + v$ for some complex numbers u and v. Substituting this into our expression for g gives

$$g(ux + v) = (ux + v)^3 - 1 = u^3 x^3 + 3u^2 v x^2 + 3uv^2 x + v^3 - 1 = u^3 \left(x^3 + \frac{3v}{u}x^2 + \frac{3v^2}{u^2}x + \frac{v^3}{u^3} - \frac{1}{u^3} \right).$$

Since this transformation maps $\triangle ABC$ to $\triangle PQR$, the polynomials $f(x)$ and $g(ux + v)$ have the same roots. Furthermore, since f is monic, we have

$$f(x) = \frac{1}{u^3}g(ux + v) = x^3 + \frac{3v}{u}x^2 + \frac{3v^2}{u^2}x + \frac{v^3}{u^3} - \frac{1}{u^3}.$$

Applying Vieta's formulas for a cubic, the sum of the roots of the cubic in x must be $-\frac{3v}{u}$ and the sum of pairs of these roots must be $\frac{3v^2}{u^2}$. Since the roots of f are a, b, and c, we have

$$a + b + c = -\frac{3v}{u},$$
$$ab + bc + ca = \frac{3v^2}{u^2}.$$

Squaring the first equation, and noting that writing $\frac{9v^2}{u^2} = 3 \cdot \frac{3v^2}{u^2}$ allows us to substitute $ab + bc + ca$ for $\frac{3v^2}{u^2}$, gives

$$(a + b + c)^2 = \frac{9v^2}{u^2} = 3 \cdot \frac{3v^2}{u^2} = 3(ab + bc + ca).$$

Expanding $(a + b + c)^2$ gives $a^2 + b^2 + c^2 + 2(ab + bc + ca) = 3(ab + bc + ca)$, so $a^2 + b^2 + c^2 = ab + bc + ca$, as desired. Therefore, the triangle is equilateral if and only if $a^2 + b^2 + c^2 = ab + bc + ca$.

(b) Multiplying both sides by $(a - b)(b - c)(c - a)$, the desired equation becomes

$$(b - c)(c - a) + (c - a)(a - b) + (a - b)(b - c) = 0$$
$$\Leftrightarrow \quad bc - ab - c^2 + ac + ac - bc - a^2 + ab + ab - ac - b^2 + bc = 0$$
$$\Leftrightarrow \quad ab + ac + bc - a^2 - b^2 - c^2 = 0$$
$$\Leftrightarrow \quad a^2 + b^2 + c^2 = ab + ac + bc,$$

which is the same equation as in part (a).

(c) Expanding, the left-hand side becomes

$$(b - c)^2 + (c - a)^2 + (a - b)^2 = b^2 - 2bc + c^2 + c^2 - 2ac + a^2 + a^2 - 2ab + b^2 = 2a^2 + 2b^2 + 2c^2 - 2ab - 2ac - 2bc = 0,$$

which simplifies as $a^2 + b^2 + c^2 = ab + ac + bc$. This is the same equation as in part (a).

8.57 Since S_n is the minimum value of the sum, we have

$$\sqrt{(2 \cdot 1 - 1)^2 + a_1^2} + \sqrt{(2 \cdot 2 - 1)^2 + a_2^2} + \cdots + \sqrt{(2 \cdot n - 1)^2 + a_n^2} \geq S_n$$

for all n. An inequality with a sum on the greater side—that looks like the Triangle Inequality. If we write each z_j in

$$|z_1| + |z_2| + \cdots + |z_n| \geq |z_1 + z_2 + \cdots + z_n|$$

as $x_j + y_j i$, so $|z_j| = \sqrt{x_j^2 + y_j^2}$, we have

$$\sqrt{x_1^2 + y_1^2} + \sqrt{x_2^2 + y_2^2} + \cdots + \sqrt{x_n^2 + y_n^2} \geq \sqrt{(x_1 + x_2 + \cdots + x_n)^2 + (y_1 + y_2 + \cdots + y_n)^2}.$$

Applying this to the sum of square roots in our problem, we have

$$\sqrt{(2 \cdot 1 - 1)^2 + a_1^2} + \sqrt{(2 \cdot 2 - 1)^2 + a_2^2} + \cdots + \sqrt{(2 \cdot n - 1)^2 + a_n^2} \geq$$
$$\sqrt{[(2 \cdot 1 - 1) + (2 \cdot 2 - 1) + \cdots + (2 \cdot n - 1)]^2 + (a_1 + a_2 + \cdots + a_n)^2}.$$

The sum $a_1 + a_2 + \cdots + a_n$ is given as 17, and

$$(2 \cdot 1 - 1) + (2 \cdot 2 - 1) + \cdots + (2 \cdot n - 1) = 2(1 + 2 + \cdots + n) - n = 2 \cdot \frac{n(n + 1)}{2} - n = n^2,$$

so the inequality now is simply

$$\sqrt{(2 \cdot 1 - 1)^2 + a_1^2} + \sqrt{(2 \cdot 2 - 1)^2 + a_2^2} + \cdots + \sqrt{(2 \cdot n - 1)^2 + a_n^2} \geq \sqrt{n^4 + 17^2}.$$

Now, our problem is to find an integer n such that $S_n = \sqrt{n^4 + 17^2}$ is an integer. We square both sides to get rid of the square root and find $S_n^2 = n^4 + 17^2$. Subtracting n^4 from both sides and factoring $S_n^2 - n^4$ as a difference of squares gives $(S_n - n^2)(S_n + n^2) = 17^2$. We know that $S_n + n^2$ is positive, so $S_n - n^2$ must be positive as well. Both of these expressions must be integers, so we have only two options: either $S_n + n^2 = S_n - n^2 = 17$, or $S_n + n^2 = 17^2$ and $S_n - n^2 = 1$. The former possibility gives $S_n = 17$ and $n = 0$, but n must be positive. The latter gives $S_n + n^2 = 289$ and $S_n - n^2 = 1$, which means $S_n = 145$ and $n = \boxed{12}$.

8.58   We let the origin be the point at which the four given lines meet, and we let a, b, c, d, e be complex numbers corresponding to the respective vertices in the diagram. Because the origin is on the line through A and the midpoint of $\overline{EB}$, we have

$$\left(0 - \frac{b+e}{2}\right)(\bar{0} - \bar{a}) - \left(\bar{0} - \frac{\bar{b}+\bar{e}}{2}\right)(0 - a) = 0.$$

Simplifying this equation gives

$$b\bar{a} + e\bar{a} - a\bar{b} - a\bar{e} = 0.$$

Similarly, the other three given lines pass through the origin and yield

$$c\bar{b} + a\bar{b} - b\bar{c} - b\bar{a} = 0,$$
$$d\bar{c} + b\bar{c} - c\bar{d} - c\bar{b} = 0,$$
$$e\bar{d} + c\bar{d} - d\bar{e} - d\bar{c} = 0.$$

Adding all four of these equations gives a whole lot of cancellation, and leaves

$$e\bar{a} + e\bar{d} - a\bar{e} - d\bar{e} = 0. \tag{8.2}$$

We'd like to show that the line through E and the midpoint of $\overline{AD}$ passes through the origin, which is true if and only if

$$\left(0 - \frac{a+d}{2}\right)(\bar{0} - \bar{e}) - \left(\bar{0} - \frac{\bar{a}+\bar{d}}{2}\right)(0 - e) = 0.$$

Simplifying this equation gives

$$a\bar{e} + d\bar{e} - e\bar{a} - e\bar{d} = 0.$$

Multiplying Equation (8.2) by -1 gives us this desired equation, so the line through E and the midpoint of $\overline{AD}$ passes through the origin.

8.59

(a) Let $\omega = e^{2\pi i/n}$ be a primitive n^{th} root of unity, so $\omega^n = 1$. Without loss of generality, we can assume that the point P_k corresponds to the complex number ω^k in the complex plane, $0 \leq k \leq n - 1$. Let z be the complex number represented by P, so $|z| = 1$. Because ω is an nth root of unity, we have $\overline{\omega}^k = \omega^{-k} = \omega^{n-k}$. Then

$$PP_k^2 = |z - \omega^k|^2 = (z - \omega^k)(\bar{z} - \overline{\omega}^k) = (z - \omega^k)(\bar{z} - \omega^{-k}) = (z - \omega^k)(\bar{z} - \omega^{n-k})$$
$$= z\bar{z} - \omega^{n-k}z - \omega^k\bar{z} + \omega^n = |z|^2 - \omega^{n-k}z - \omega^k\bar{z} + 1 = 2 - \omega^{n-k}z - \omega^k\bar{z},$$

so

$$PP_0^2 + PP_1^2 + PP_2^2 + \cdots + PP_{n-1}^2$$

$$= (2 - \omega^n z - \overline{z}) + (2 - \omega^{n-1} z - \omega\overline{z}) + (2 - \omega^{n-2} z - \omega^2\overline{z}) + \cdots + (2 - \omega z - \omega^{n-1}\overline{z})$$

$$= 2n - (\omega + \omega^2 + \omega^3 + \cdots + \omega^n)z - (1 + \omega + \omega^2 + \cdots + \omega^{n-1})\overline{z}$$

$$= 2n - (\omega + \omega^2 + \cdots + \omega^{n-1} + 1)z - (1 + \omega + \omega^2 + \cdots + \omega^{n-1})\overline{z}$$

$$= 2n - (1 + \omega + \omega^2 + \cdots + \omega^{n-1})z - (1 + \omega + \omega^2 + \cdots + \omega^{n-1})\overline{z}$$

By the formula for a geometric series, we have $1 + \omega + \omega^2 + \cdots + \omega^{n-1} = \dfrac{1 - \omega^n}{1 - \omega} = 0$, so

$$PP_0^2 + PP_1^2 + PP_2^2 + \cdots + PP_{n-1}^2 = 2n.$$

(b) From the equation $PP_k^2 = 2 - \omega^{n-k} z - \omega^k \overline{z}$, we have

$$PP_k^4 = (2 - \omega^{n-k} z - \omega^k \overline{z})^2 = 4 + \omega^{2n-2k} z^2 + \omega^{2k} \overline{z}^2 - 4\omega^{n-k} z - 4\omega^k \overline{z} + 2\omega^n z\overline{z}$$

$$= 4 + \omega^{2n-2k} z^2 + \omega^{2k} \overline{z}^2 - 4\omega^{n-k} z - 4\omega^k \overline{z} + 2 = 6 + \omega^{2n-2k} z^2 + \omega^{2k} \overline{z}^2 - 4\omega^{n-k} z - 4\omega^k \overline{z}.$$

Then

$$PP_0^4 + PP_1^4 + PP_2^4 + \cdots + PP_{n-1}^4 = (6 + \omega^{2n} z^2 + \overline{z}^2 - 4\omega^n z - 4\overline{z})$$

$$+ (6 + \omega^{2n-2} z^2 + \omega^2 \overline{z}^2 - 4\omega^{n-1} z - 4\omega\overline{z})$$

$$+ (6 + \omega^{2n-4} z^2 + \omega^4 \overline{z}^2 - 4\omega^{n-2} z - 4\omega^2\overline{z})$$

$$+ \cdots + (6 + \omega^2 z^2 + \omega^{2n-2} \overline{z}^2 - 4\omega z - 4\omega^{n-1}\overline{z})$$

$$= 6n + (\omega^2 + \omega^4 + \omega^6 + \cdots + \omega^{2n}) z^2 + (1 + \omega^2 + \omega^4 + \cdots + \omega^{2n-2}) \overline{z}^2$$

$$- 4(\omega + \omega^2 + \omega^3 + \cdots + \omega^n) z - 4(1 + \omega + \omega^2 + \cdots + \omega^{n-1})\overline{z}$$

$$= 6n + \omega^2(1 + \omega^2 + \omega^4 + \cdots + \omega^{2n-2}) z^2 + (1 + \omega^2 + \omega^4 + \cdots + \omega^{2n-2}) \overline{z}^2$$

$$- 4\omega(1 + \omega + \omega^2 + \cdots + \omega^{n-1}) z - 4(1 + \omega + \omega^2 + \cdots + \omega^{n-1})\overline{z},$$

where we used $\omega^n = 1$ and $\omega^{2n} = (\omega^n)^2 = 1$ in the last step. We know that $1 + \omega + \omega^2 + \cdots + \omega^{n-1} = 0$, and

$$1 + \omega^2 + \omega^4 + \cdots + \omega^{2n-2} = \frac{1 - \omega^{2n}}{1 - \omega^2} = \frac{1 - (\omega^n)^2}{1 - \omega^2} = 0,$$

so $PP_0^4 + PP_1^2 + PP_2^4 + \cdots + PP_{n-1}^4 = 6n.$

8.60 We could let $z^8 = x + yi$ and go through some clever algebraic manipulations to find the possible values of z^{28} and z^8, but a little geometric insight offers a significant shortcut. Rearranging the given equation as $z^{28} = z^8 + 1$, we see that z^{28} is the image of a translation of z^8 by 1 unit to the right. We also know that $|z| = 1$, so $|z^{28}| = |z^8| = 1$ as well, which means that z^{28} and z^8 are on the unit circle. There are only two pairs of points on the unit circle such that one point is 1 unit to the right of the other. These are shown at right.

To see that these are the only two possibilities, we note that there are only at most two points on the unit circle with a given y-coordinate. The difference in the x-coordinates of these two points is 2 when the y-coordinate is 0, and the difference strictly decreases as y moves away from 0. Therefore, this difference can only equal 1 once above the x-axis and once below the x-axis, and these are at the values shown in the diagram.

Since z^{28} is the result of translating z^8 by 1 unit to the right, the two possibilities are $z^8 = e^{2\pi i/3}$ and $z^{28} = e^{\pi i/3}$, and $z^8 = e^{4\pi i/3}$ and $z^{28} = e^{5\pi i/3}$.

If $z^8 = e^{2\pi i/3}$ and $z^{28} = e^{\pi i/3}$, then $z^{24} = (z^8)^3 = (e^{2\pi i/3})^3 = e^{2\pi i} = 1$, and

$$z^4 = \frac{z^{28}}{z^{24}} = e^{\pi i/3}.$$

The solutions to this are of the form $z = e^{\pi i/12 + k\pi/2}$, where $k = 0$, 1, 2, and 3. These arguments correspond to the angles $15°$, $105°$, $195°$, and $285°$. Furthermore, if $z^4 = e^{\pi i/3}$, then $z^8 = (z^4)^2 = (e^{\pi i/3})^2 = e^{2\pi i/3}$, and $z^{28} = (z^4)^7 = (e^{\pi i/3})^7 = e^{7\pi i/3} = e^{\pi i/3}$, so all these solutions work.

If $z^8 = e^{4\pi i/3}$ and $z^{28} = e^{5\pi i/3}$, then $z^{24} = (z^8)^3 = (e^{4\pi i/3})^3 = e^{12\pi i} = 1$, and

$$z^4 = \frac{z^{28}}{z^{24}} = e^{5\pi i/3}.$$

The solutions to this are of the form $z = e^{5\pi i/12 + k\pi/2}$, where $k = 0$, 1, 2, and 3. These arguments correspond to the angles $75°$, $165°$, $255°$, and $345°$. Furthermore, if $z^4 = e^{5\pi i/3}$, then $z^8 = (z^4)^2 = (e^{5\pi i/3})^2 = e^{10\pi i/3} = e^{4\pi i/3}$, and $z^{28} = (z^4)^7 = (e^{5\pi i/3})^7 = e^{35\pi i/3} = e^{5\pi i/3}$, so all these solutions work.

Therefore the angles θ_i in increasing order are $15°$, $75°$, $105°$, $165°$, $195°$, $255°$, $285°$, and $345°$, and the sum we seek is $75 + 165 + 255 + 345 = \boxed{840}$.

8.61 The center of the decagon is 2 (as a complex number), in the complex plane. Moreover, the decagon is a rightward translation by 2 units of the decagon whose vertices are the tenth roots of unity. Hence, each complex number $x_k + y_k i$ is of the form $2 + z$, where z is a 10th root of unity. Let $\omega = e^{2\pi i/10}$. Then

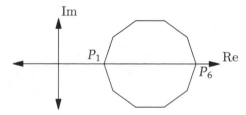

$$(x_1 + y_1 i)(x_2 + y_2 i)(x_3 + y_3 i) \cdots (x_{10} + y_{10} i) = (2+1)(2+\omega)(2+\omega^2) \cdots (2+\omega^9).$$

The 10th roots of unity are the roots of $x^{10} - 1 = 0$, so

$$(x-1)(x-\omega)(x-\omega^2) \cdots (x-\omega^9) = x^{10} - 1.$$

Taking $x = -2$, we get $(-2-1)(-2-\omega)(-2-\omega^2) \cdots (-2-\omega^9) = (-2)^{10} - 1$, which simplifies as

$$(2+1)(2+\omega)(2+\omega^2) \cdots (2+\omega^9) = \boxed{1023}.$$

8.62 The area of the equilateral triangle is maximized when one vertex of the triangle coincides with a vertex of the rectangle, and the other two vertices of the triangle lie on the sides of the rectangle, so let the vertices of the equilateral triangle be A, P, and Q, where P lies on $\overline{BC}$ and Q lies on $\overline{CD}$.

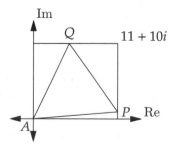

We place the rectangle in the complex plane so that A corresponds to the origin and C corresponds to $11 + 10i$. Let p and q be the complex numbers corresponding to P and Q, respectively, so that the real part of p is 11 and the imaginary part of q is 10. Let $p = 11 + xi$ and $q = y + 10i$.

Then q can be obtained by rotating p by $\frac{\pi}{3}$ counterclockwise about the origin, so $q = e^{\pi i/3} p$, or

$$y + 10i = \left(\cos\frac{\pi}{3} + i\sin\frac{\pi}{3}\right)(11 + xi) \quad \Rightarrow \quad y + 10i = \left(\frac{1}{2} + \frac{\sqrt{3}}{2}i\right)(11 + xi) \quad \Rightarrow \quad y + 10i = \frac{11}{2} + \frac{x}{2}i + \frac{11\sqrt{3}}{2}i - \frac{x\sqrt{3}}{2}.$$

Equating the real parts and equating the imaginary parts, we obtain the system of equations

$$y = \frac{11}{2} - \frac{\sqrt{3}}{2}x,$$

$$10 = \frac{1}{2}x + \frac{11\sqrt{3}}{2}.$$

From the second equation, we have $x = 20 - 11\sqrt{3}$. Then

$$|p|^2 = |11 + xi|^2 = 11^2 + x^2 = 121 + (20 - 11\sqrt{3})^2 = 884 - 440\sqrt{3},$$

so the area of triangle APQ is $\frac{\sqrt{3}}{4}|p|^2 = \frac{\sqrt{3}}{4}(884 - 440\sqrt{3}) = \boxed{-330 + 221\sqrt{3}}$.

8.63 Let $\omega = e^{2\pi i/3}$, a primitive cube root of unity, so $\omega^2 + \omega + 1 = 0$. Without loss of generality, let the vertices of the equilateral triangle in the complex plane be 1, ω, and ω^2, so $s = \sqrt{3}$.

Let p be the complex number corresponding to the point P. Then in some order, PA^2, PB^2, PC^2 are

$$|p - 1|^2 = (p - 1)(\bar{p} - 1) = |p|^2 - p - \bar{p} + 1,$$
$$|p - \omega|^2 = (p - \omega)(\bar{p} - \bar{\omega}) = (p - \omega)(\bar{p} - \omega^2) = |p|^2 - \omega^2 p - \omega\bar{p} + 1,$$
$$|p - \omega^2|^2 = (p - \omega^2)(\bar{p} - \bar{\omega}^2) = (p - \omega^2)(\bar{p} - \omega) = |p|^2 - \omega p - \omega^2\bar{p} + 1.$$

Then

$$\begin{aligned}
PA^2 + PB^2 + PC^2 + s^2 &= (|p|^2 - p - \bar{p} + 1) + (|p|^2 - \omega^2 p - \omega\bar{p} + 1) + (|p|^2 - \omega p - \omega^2\bar{p} + 1) + s^2 \\
&= 3|p|^2 - (1 + \omega + \omega^2)p - (1 + \omega + \omega^2)\bar{p} + 6 \\
&= 3|p|^2 + 6,
\end{aligned}$$

so $(PA^2 + PB^2 + PC^2 + s^2)^2 = 9|p|^4 + 36|p|^2 + 36$.

Also, we have

$$\begin{aligned}
PA^4 + PB^4 + PC^4 &= (|p|^2 - p - \bar{p} + 1)^2 + (|p|^2 - \omega^2 p - \omega\bar{p} + 1)^2 + (|p|^2 - \omega p - \omega^2\bar{p} + 1)^2 \\
&= |p|^4 + p^2 + \bar{p}^2 + 1 - 2|p|^2 p - 2|p|^2\bar{p} + 2|p|^2 + 2|p|^2 - 2p - 2\bar{p} \\
&\quad + |p|^4 + \omega p^2 + \omega^2\bar{p}^2 + 1 - 2\omega^2|p|^2 p - 2\omega|p|^2\bar{p} + 2|p|^2 + 2|p|^2 - 2\omega^2 p - 2\omega\bar{p} \\
&\quad + |p|^4 + \omega^2 p^2 + \omega\bar{p}^2 + 1 - 2\omega|p|^2 p - 2\omega^2|p|^2\bar{p} + 2|p|^2 + 2|p|^2 - 2\omega p - 2\omega^2\bar{p} \\
&= 3|p|^4 + (1 + \omega + \omega^2)p^2 + (1 + \omega + \omega^2)\bar{p}^2 + 3 - 2(1 + \omega + \omega^2)p - 2(1 + \omega + \omega^2)|p|^2 p \\
&\quad - 2(1 + \omega + \omega^2)|p|^2\bar{p} + 12|p|^2 - 2(1 + \omega + \omega^2)p - 2(1 + \omega + \omega^2)\bar{p} \\
&= 3|p|^4 + 12|p|^2 + 3,
\end{aligned}$$

where we used the fact that $1 + \omega + \omega^2 = 0$ to eliminate many terms in the last step. We therefore have

$$3(PA^4 + PB^4 + PC^4 + s^4) = 3(3|p|^4 + 12|p|^2 + 3 + 9) = 9|p|^4 + 36|p|^2 + 36,$$

so $3(PA^4 + PB^4 + PC^4 + s^4) = (PA^2 + PB^2 + PC^2 + s^2)^2$.

8.64 We know how to handle equilateral triangles and testing for perpendicularity on the complex plane, so we place the problem on the complex plane. Let ζ be a primitive sixth root of unity and let the lowercase version of each letter be the complex number corresponding to the respective uppercase point.

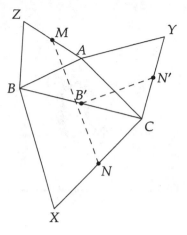

From equilateral triangles BCX, CAY, and ABZ, we have

$$x = \zeta(b - c) + c = \zeta b + (1 - \zeta)c,$$
$$y = \zeta(c - a) + a = \zeta c + (1 - \zeta)a,$$
$$z = \zeta(a - b) + b = \zeta a + (1 - \zeta)b,$$

so

$$m = \frac{a + z}{2} = \frac{(1 + \zeta)a + (1 - \zeta)b}{2},$$
$$n = \frac{c + x}{2} = \frac{\zeta b + (2 - \zeta)c}{2},$$
$$b' = \frac{b + c}{2},$$
$$n' = \frac{c + y}{2} = \frac{(1 - \zeta)a + (1 + \zeta)c}{2},$$

so

$$m - n = \frac{(1 + \zeta)a + (1 - 2\zeta)b + (-2 + \zeta)c}{2},$$

and

$$n' - b' = \frac{(1 - \zeta)a - b + \zeta c}{2}.$$

We have $\overline{B'N'} \perp \overline{MN}$ if and only if $(m - n)/(n' - b')$ is imaginary. We can make our work a little easier by choosing A to be the origin, so $a = 0$, and we have

$$\frac{m - n}{n' - b'} = \frac{(1 - 2\zeta)b + (-2 + \zeta)c}{-b + \zeta c}.$$

Uh-oh. That expression on the right doesn't look imaginary. The fact that it still has b and c in it is a problem, since b and c can be anything. So, we wonder if this expression is equivalent to an expression that doesn't have b or c. This will be the case if the ratio of the coefficients of b equals the ratio of the coefficients of c, which will give us some convenient cancellation. Let's check if that's the case by comparing $\frac{1-2\zeta}{-1}$ to $\frac{-2+\zeta}{\zeta}$. We have

$$\frac{\frac{1-2\zeta}{-1}}{\frac{-2+\zeta}{\zeta}} = \frac{\zeta - 2\zeta^2}{2 - \zeta} = \frac{\zeta - 2(\zeta - 1)}{2 - \zeta} = \frac{2 - \zeta}{2 - \zeta} = 1.$$

Sure enough, they're the same. So, there's a constant k such that $\frac{1-2\zeta}{-1} = \frac{-2+\zeta}{\zeta} = k$, and we have

$$\frac{m - n}{n' - b'} = \frac{(1 - 2\zeta)b + (-2 + \zeta)c}{-b + \zeta c} = \frac{k(-1)b + k(\zeta)c}{-b + \zeta c} = k\frac{-b + \zeta c}{-b + \zeta c} = k.$$

Now, we just have to show that k is imaginary. We have $k = -1 + 2\zeta$, and the only possibilities for ζ are the primitive sixth roots of unity, which are $\frac{1}{2} + \frac{\sqrt{3}}{2}i$ and $\frac{1}{2} - \frac{\sqrt{3}}{2}i$. For each of these values of ζ, the value of $k = -1 + 2\zeta$ is indeed imaginary. We have therefore proved that $\frac{m-n}{n'-b'}$ is imaginary, so $\overline{B'N'} \perp \overline{MN}$.

CHAPTER **9**

_____ **Vectors in Two Dimensions**

Exercises for Section 9.1

9.1.1 Let $\mathbf{v} = \begin{pmatrix} v_1 \\ v_2 \end{pmatrix}$. Then, we have

$$\mathbf{0} + \mathbf{v} = \begin{pmatrix} 0 \\ 0 \end{pmatrix} + \begin{pmatrix} v_1 \\ v_2 \end{pmatrix} = \begin{pmatrix} 0 + v_1 \\ 0 + v_2 \end{pmatrix} = \begin{pmatrix} v_1 \\ v_2 \end{pmatrix} = \mathbf{v},$$

$$0\mathbf{v} = 0 \begin{pmatrix} v_1 \\ v_2 \end{pmatrix} = \begin{pmatrix} 0v_1 \\ 0v_2 \end{pmatrix} = \begin{pmatrix} 0 \\ 0 \end{pmatrix} = \mathbf{0}.$$

9.1.2

(a) $\left\| \begin{pmatrix} 4 \\ -5 \end{pmatrix} \right\| = \sqrt{4^2 + (-5)^2} = \boxed{\sqrt{41}}.$

(b) $\|7\mathbf{i} - 24\mathbf{j}\| = \sqrt{7^2 + (-24)^2} = \sqrt{625} = \boxed{25}.$

9.1.3

(a) $\begin{pmatrix} 4 \\ -3 \end{pmatrix} + 2 \begin{pmatrix} 5 \\ -3 \end{pmatrix} - \begin{pmatrix} -1 \\ 7 \end{pmatrix} = \begin{pmatrix} 4 \\ -3 \end{pmatrix} + \begin{pmatrix} 10 \\ -6 \end{pmatrix} - \begin{pmatrix} -1 \\ 7 \end{pmatrix} = \boxed{\begin{pmatrix} 15 \\ -16 \end{pmatrix}}.$

(b) $\begin{pmatrix} -1 \\ 6 \end{pmatrix} - 2 \begin{pmatrix} -6 \\ 2 \end{pmatrix} - 3 \begin{pmatrix} 0 \\ 4 \end{pmatrix} = \begin{pmatrix} -1 \\ 6 \end{pmatrix} - \begin{pmatrix} -12 \\ 4 \end{pmatrix} - \begin{pmatrix} 0 \\ 12 \end{pmatrix} = \boxed{\begin{pmatrix} 11 \\ -10 \end{pmatrix}}.$

9.1.4 The vector $\overrightarrow{AB}$ represents the vector pointing from A to B, and the vector $\overrightarrow{CD}$ represents the vector pointing from C to D. If these two vectors are equal, then $\overline{AB}$ and $\overline{CD}$ are equal in length and parallel, so quadrilateral $ABDC$ is a parallelogram.

9.1.5 Let $\mathbf{v} = \begin{pmatrix} x \\ y \end{pmatrix}$. Then $c\mathbf{v} = \begin{pmatrix} cx \\ cy \end{pmatrix}$, so

$$\|c\mathbf{v}\| = \left\| \begin{pmatrix} cx \\ cy \end{pmatrix} \right\| = \sqrt{(cx)^2 + (cy)^2} = \sqrt{c^2x^2 + c^2y^2} = \sqrt{c^2(x^2 + y^2)} = \sqrt{c^2}\sqrt{x^2 + y^2} = |c|\|\mathbf{v}\|.$$

Exercises for Section 9.2

9.2.1

(a) $\begin{pmatrix} 5 \\ -3 \end{pmatrix} \cdot \begin{pmatrix} -2 \\ 4 \end{pmatrix} = 5 \cdot (-2) + (-3) \cdot 4 = -10 + (-12) = \boxed{-22}$.

(b) $\begin{pmatrix} -6 \\ -1 \end{pmatrix} \cdot \begin{pmatrix} 0 \\ -3 \end{pmatrix} = (-6) \cdot 0 + (-1) \cdot (-3) = 0 + 3 = \boxed{3}$.

9.2.2 We have $\mathbf{u} - \mathbf{v} = (5\mathbf{i} - 12\mathbf{j}) - (3\mathbf{i} + 4\mathbf{j}) = 2\mathbf{i} - 16\mathbf{j}$, and $\mathbf{u} + \mathbf{v} = (5\mathbf{i} - 12\mathbf{j}) + (3\mathbf{i} + 4\mathbf{j}) = 8\mathbf{i} - 8\mathbf{j}$, so

$$(\mathbf{u} - \mathbf{v}) \cdot (\mathbf{u} + \mathbf{v}) = (2\mathbf{i} - 16\mathbf{j}) \cdot (8\mathbf{i} - 8\mathbf{j}) = 2 \cdot 8 + (-16) \cdot (-8) = 16 + 128 = \boxed{144}.$$

We can also distribute:

$$(\mathbf{u} - \mathbf{v}) \cdot (\mathbf{u} + \mathbf{v}) = \mathbf{u} \cdot (\mathbf{u} + \mathbf{v}) - \mathbf{v} \cdot (\mathbf{u} + \mathbf{v}) = \mathbf{u} \cdot \mathbf{u} + \mathbf{u} \cdot \mathbf{v} - (\mathbf{v} \cdot \mathbf{u} + \mathbf{v} \cdot \mathbf{v}) = \mathbf{u} \cdot \mathbf{u} + \mathbf{u} \cdot \mathbf{v} - \mathbf{u} \cdot \mathbf{v} - \mathbf{v} \cdot \mathbf{v}$$

$$= \mathbf{u} \cdot \mathbf{u} - \mathbf{v} \cdot \mathbf{v} = [5^2 + (-12)^2] - (3^2 + 4^2) = 169 - 25 = \boxed{144}.$$

9.2.3 If the vector $\begin{pmatrix} x \\ y \end{pmatrix}$ is orthogonal to $\begin{pmatrix} -1 \\ 2 \end{pmatrix}$, then $\begin{pmatrix} x \\ y \end{pmatrix} \cdot \begin{pmatrix} -1 \\ 2 \end{pmatrix} = 0$, which gives $-x + 2y = 0$. If $x = 5$, then $y = x/2 = 5/2$. If $y = 5$, then $x = 2y = 10$. Hence, possible vectors are $\boxed{\begin{pmatrix} 5 \\ 5/2 \end{pmatrix}}$ and $\boxed{\begin{pmatrix} 10 \\ 5 \end{pmatrix}}$.

9.2.4 The vectors $\mathbf{p} = \begin{pmatrix} s \\ t \end{pmatrix}$ and $\mathbf{q} = \begin{pmatrix} u \\ v \end{pmatrix}$ are orthogonal if and only if $\mathbf{p} \cdot \mathbf{q} = \begin{pmatrix} s \\ t \end{pmatrix} \cdot \begin{pmatrix} u \\ v \end{pmatrix} = su + tv = 0$. Computing this value for each pair of vectors, we find that the only pairs of orthogonal vectors are $\boxed{\left\{ \begin{pmatrix} -3 \\ 4 \end{pmatrix}, \begin{pmatrix} 8 \\ 6 \end{pmatrix} \right\}}$, and $\boxed{\left\{ \begin{pmatrix} 4 \\ -6 \end{pmatrix}, \begin{pmatrix} 9 \\ 6 \end{pmatrix} \right\}}$.

9.2.5 We have $\vec{AB} = \begin{pmatrix} 1 \\ -3 \end{pmatrix} - \begin{pmatrix} 0 \\ 5 \end{pmatrix} = \begin{pmatrix} 1 \\ -8 \end{pmatrix}$, and $\vec{AC} = \begin{pmatrix} -5 \\ 4 \end{pmatrix} - \begin{pmatrix} 0 \\ 5 \end{pmatrix} = \begin{pmatrix} -5 \\ -1 \end{pmatrix}$. Then

$$\cos A = \frac{\vec{AB} \cdot \vec{AC}}{\|\vec{AB}\| \|\vec{AC}\|} = \frac{1 \cdot (-5) + (-8) \cdot (-1)}{\sqrt{1^2 + (-8)^2} \sqrt{(-5)^2 + (-1)^2}} = \frac{-5 + 8}{\sqrt{65} \sqrt{26}} = \frac{3}{13\sqrt{10}},$$

so $A = \arccos\left(\frac{3}{13\sqrt{10}}\right) \approx \boxed{86°}$.

Similarly, we have $\vec{BC} = \begin{pmatrix} -5 \\ 4 \end{pmatrix} - \begin{pmatrix} 1 \\ -3 \end{pmatrix} = \begin{pmatrix} -6 \\ 7 \end{pmatrix}$ and $\vec{BA} = \begin{pmatrix} 0 \\ 5 \end{pmatrix} - \begin{pmatrix} 1 \\ -3 \end{pmatrix} = \begin{pmatrix} -1 \\ 8 \end{pmatrix}$, so

$$\cos B = \frac{(-6) \cdot (-1) + 7 \cdot 8}{\sqrt{(-6)^2 + 7^2} \sqrt{(-1)^2 + 8^2}} = \frac{6 + 56}{\sqrt{85} \sqrt{65}} = \frac{62}{5\sqrt{221}},$$

and $B = \arccos\left(\frac{62}{5\sqrt{221}}\right) \approx \boxed{33°}$.

The vectors emanating from C are $\vec{CA} = \begin{pmatrix} 0 \\ 5 \end{pmatrix} - \begin{pmatrix} -5 \\ 4 \end{pmatrix} = \begin{pmatrix} 5 \\ 1 \end{pmatrix}$ and $\vec{CB} = \begin{pmatrix} 1 \\ -3 \end{pmatrix} - \begin{pmatrix} -5 \\ 4 \end{pmatrix} = \begin{pmatrix} 6 \\ -7 \end{pmatrix}$, so

$$\cos C = \frac{5 \cdot 6 + 1 \cdot (-7)}{\sqrt{5^2 + 1^2} \sqrt{6^2 + (-7)^2}} = \frac{30 - 7}{\sqrt{26} \sqrt{85}} = \frac{23}{\sqrt{2210}},$$

and $C = \arccos\left(\frac{23}{\sqrt{2210}}\right) \approx \boxed{61°}$.

9.2.6

(a) We have $v_1 = r_v \cos\theta_v$ and $v_2 = r_v \sin\theta_v$, so

$$\|\mathbf{v}\| = \sqrt{v_1^2 + v_2^2} = \sqrt{(r_v\cos\theta_v)^2 + (r_v\sin\theta_v)^2} = \sqrt{r_v^2\cos^2\theta_v + r_v^2\sin^2\theta_v} = \sqrt{r_v^2(\cos^2\theta_v + \sin^2\theta_v)} = \sqrt{r_v^2} = r_v.$$

Similarly, $\|\mathbf{w}\| = r_w$. Therefore, $\|\mathbf{v}\|\|\mathbf{w}\|\cos\phi = r_v r_w \cos\phi$. Since ϕ is the angle between $\mathbf{v}$ and $\mathbf{w}$, angles θ_v and θ_w differ by ϕ, so $\cos(\theta_v - \theta_w) = \cos\phi$, and we have $\|\mathbf{v}\|\|\mathbf{w}\|\cos\phi = r_v r_w \cos(\theta_v - \theta_w)$.

(b) We also have $w_1 = r_w \cos\theta_w$ and $w_2 = r_w \sin\theta_w$, so by the angle difference formula for cosine,

$$\begin{aligned}
v_1 w_1 + v_2 w_2 &= (r_v \cos\theta_v)(r_w \cos\theta_w) + (r_v \sin\theta_v)(r_w \sin\theta_w) \\
&= r_v r_w (\cos\theta_v \cos\theta_w + \sin\theta_v \sin\theta_w) \\
&= r_v r_w \cos(\theta_v - \theta_w).
\end{aligned}$$

Combining this with part (a), we have

$$\mathbf{v} \cdot \mathbf{w} = v_1 w_1 + v_2 w_2 = \|\mathbf{v}\|\|\mathbf{w}\|\cos(\theta_v - \theta_w).$$

Exercises for Section 9.3

9.3.1 We have $x = 3 - 4s$, $y = -1 + 5s$. Multiplying the first equation by 5 and the second by 4 gives $5x = 15 - 20s$, $4y = -4 + 20s$. Adding these two equations gives $\boxed{5x + 4y = 11}$.

9.3.2 Solving for y gives $y = 3x - 4$. Letting $x = 0$ gives $y = -4$. So, taking $t = x$ as a parameter, the graph of $3x - y = 4$ is the same as the graph of $\boxed{\begin{pmatrix} x \\ y \end{pmatrix} = \begin{pmatrix} 0 \\ -4 \end{pmatrix} + t\begin{pmatrix} 1 \\ 3 \end{pmatrix}}$. There are infinitely many other possible solutions. To test your solution of the form $\begin{pmatrix} x \\ y \end{pmatrix} = \mathbf{v} + t\mathbf{w}$, make sure the point corresponding to your $\mathbf{v}$ satisfies $3x - y = 4$, and make sure your $\mathbf{w}$ is some multiple of $\begin{pmatrix} 1 \\ 3 \end{pmatrix}$.

9.3.3 As explained in the text, two vectors in are linearly dependent if and only if one is a multiple of the other. Using this condition, we find that the only pairs of linearly dependent vectors are $\boxed{\left\{\begin{pmatrix} 4 \\ -6 \end{pmatrix}, \begin{pmatrix} -6 \\ 9 \end{pmatrix}\right\}}$ and $\boxed{\left\{\begin{pmatrix} 12 \\ 8 \end{pmatrix}, \begin{pmatrix} 9 \\ 6 \end{pmatrix}\right\}}$.

9.3.4

(a) We know that at least one such pair (a, b) exists, since every point (x, y) is in the graph of $\begin{pmatrix} x \\ y \end{pmatrix} = a\mathbf{u} + b\mathbf{v}$. We will show that it is impossible for there to be two such pairs.

Let $a_1\mathbf{u} + b_1\mathbf{v} = \mathbf{w}$ and $a_2\mathbf{u} + b_2\mathbf{v} = \mathbf{w}$, where (a_1, b_1) and (a_2, b_2) are distinct pairs of nonzero scalars, so $a_1\mathbf{u} + b_1\mathbf{v} = a_2\mathbf{u} + b_2\mathbf{v}$. Hence,

$$(a_1 - a_2)\mathbf{u} + (b_1 - b_2)\mathbf{v} = 0.$$

Since $\mathbf{u}$ and $\mathbf{v}$ are linearly independent, we must have $a_1 - a_2 = b_1 - b_2 = 0$, so $a_1 = a_2$ and $b_1 = b_2$. Therefore, the pair (a, b) such that $\mathbf{w} = a\mathbf{u} + b\mathbf{v}$ is unique.

(b) This part follows immediately from the previous part, which tells us that it is impossible for there to be two pairs (a, b) if **u** and **v** are linearly independent. Therefore, if two different such pairs exist, then **u** and **v** must be linearly dependent.

Exercises for Section 9.4

9.4.1

(a) We have $\text{proj}_\mathbf{v}\mathbf{u} = \dfrac{\mathbf{u} \cdot \mathbf{v}}{\|\mathbf{v}\|^2}\mathbf{v} = \dfrac{(2)(3) + (0)(4)}{\left(\sqrt{3^2 + 4^2}\right)^2}\begin{pmatrix}3\\4\end{pmatrix} = \dfrac{6}{25}\begin{pmatrix}3\\4\end{pmatrix} = \boxed{\begin{pmatrix}\frac{18}{25}\\\frac{24}{25}\end{pmatrix}}$.

(b) We have $\text{proj}_\mathbf{v}\mathbf{u} = \dfrac{\mathbf{u} \cdot \mathbf{v}}{\|\mathbf{v}\|^2}\mathbf{v} = \dfrac{(-3)(2) + (1)(-2)}{\left(\sqrt{(2)^2 + (-2)^2}\right)^2}\begin{pmatrix}2\\-2\end{pmatrix} = \dfrac{-8}{8}\begin{pmatrix}2\\-2\end{pmatrix} = \boxed{\begin{pmatrix}-2\\2\end{pmatrix}}$.

9.4.2

(a) As explained in the text, $\text{proj}_\mathbf{v}\mathbf{u}$ depends on the direction of **v**, but not on the magnitude of **v**. Since **w** has the same direction as **v**, the projection of **u** onto **w** is the same as the projection of **u** onto **v**. The fact that **w** has three times the magnitude of **v** is irrelevant. Therefore, $\text{proj}_\mathbf{w}\mathbf{u} = \boxed{9\mathbf{i} - 3\mathbf{j}}$.

(b) We have $\text{proj}_\mathbf{v}\mathbf{x} = \dfrac{\mathbf{x} \cdot \mathbf{v}}{\|\mathbf{v}\|^2}\mathbf{v} = \dfrac{(3\mathbf{u}) \cdot \mathbf{v}}{\|\mathbf{v}\|^2}\mathbf{v} = 3\left(\dfrac{\mathbf{u} \cdot \mathbf{v}}{\|\mathbf{v}\|^2}\mathbf{v}\right) = 3\text{proj}_\mathbf{v}\mathbf{u} = \boxed{27\mathbf{i} - 9\mathbf{j}}$. Geometrically speaking, this shows that tripling the magnitude of a vector triples the magnitude of the vector's projection onto any other vector.

9.4.3

(a) Let (x, y) be a point on the line. Since $5\mathbf{i} - 3\mathbf{j}$ is normal to the line through (x, y) and $(3, 4)$, it must be orthogonal to the vector $(x - 3)\mathbf{i} + (y - 4)\mathbf{j}$. Therefore, we must have $(5\mathbf{i} - 3\mathbf{j}) \cdot ((x - 3)\mathbf{i} + (y - 4)\mathbf{j}) = 0$, so $(5)(x - 3) + (-3)(y - 4) = 0$. Expanding and rearranging gives $\boxed{5x - 3y - 3 = 0}$.

Alternatively, because $5\mathbf{i} - 3\mathbf{j}$ is normal to the line, we know the line is the graph of the equation $5x - 3y + c = 0$ for some constant c. Since $(3, 4)$ is on the line, we must have $5(3) - 3(4) + c = 0$, so $c = -3$. Therefore, the line is the graph of $\boxed{5x - 3y - 3 = 0}$.

(b) As explained in the text, the vector $2\mathbf{i} + 4\mathbf{j}$ is normal to the graph of $2x + 4y = 7$. But this vector has magnitude $\sqrt{2^2 + 4^2} = 2\sqrt{5}$. We seek a vector with magnitude 5, but with the same direction as $2\mathbf{i} + 4\mathbf{j}$. Since $2\mathbf{i} + 4\mathbf{j}$ has magnitude $2\sqrt{5}$, we can multiply this vector by $\sqrt{5}/2$ to get the desired vector, $\boxed{\sqrt{5}\mathbf{i} + 2\sqrt{5}\mathbf{j}}$. We also could have multiplied by $-\sqrt{5}/2$ to produce $\boxed{-\sqrt{5}\mathbf{i} - 2\sqrt{5}\mathbf{j}}$, which also satisfies the problem.

9.4.4 For any vector $\mathbf{w} = w_1\mathbf{i} + w_2\mathbf{j}$, we have

$$\text{proj}_\mathbf{i}\mathbf{w} = \dfrac{\mathbf{w} \cdot \mathbf{i}}{\|\mathbf{i}\|^2}\mathbf{i} = \dfrac{(w_1)(1) + (w_2)(0)}{1^2}\mathbf{i} = w_1\mathbf{i}.$$

Similarly, we have $\text{proj}_\mathbf{j}\mathbf{w} = w_2\mathbf{j}$. Since $\text{proj}_\mathbf{x}\mathbf{u} = \text{proj}_\mathbf{x}\mathbf{v}$ for all **x**, it must hold for $\mathbf{x} = \mathbf{i}$ and $\mathbf{x} = \mathbf{j}$. The equation $\text{proj}_\mathbf{i}\mathbf{u} = \text{proj}_\mathbf{i}\mathbf{v}$ tells us that the first components of **u** and **v** are the same, and $\text{proj}_\mathbf{j}\mathbf{u} = \text{proj}_\mathbf{j}\mathbf{v}$ tells us that the second components of **u** and **j** are same. Combining these observations gives us $\mathbf{u} = \mathbf{v}$.

9.4.5

(a) No. We have

$$\text{proj}_{\mathbf{v}}(-\mathbf{w}) = \frac{(-\mathbf{w}) \cdot \mathbf{v}}{\|\mathbf{v}\|^2}\mathbf{v} = -\frac{\mathbf{w} \cdot \mathbf{v}}{\|\mathbf{v}\|^2}\mathbf{v} = -\text{proj}_{\mathbf{v}}\mathbf{w}.$$

If we also have $\text{proj}_{\mathbf{v}}(-\mathbf{w}) = \text{proj}_{\mathbf{v}}\mathbf{w}$, then $\text{proj}_{\mathbf{v}}\mathbf{w} = -\text{proj}_{\mathbf{v}}\mathbf{w}$, which means that $\text{proj}_{\mathbf{v}}\mathbf{w} = \mathbf{0}$. All this tells us is that $\mathbf{w}$ is orthogonal to $\mathbf{v}$. If $\mathbf{w}$ is orthogonal to $\mathbf{v}$, then both $\mathbf{w} \cdot \mathbf{v}$ and $(-\mathbf{w}) \cdot \mathbf{v} = 0$, so the projections of $\mathbf{w}$ and $-\mathbf{w}$ onto $\mathbf{v}$ are both $\mathbf{0}$.

(b) No. Geometrically speaking, the projection of $\mathbf{w}$ onto a vector is determined by finding the foot of the perpendicular from the head of $\mathbf{w}$ to the line through the origin that contains the vector from the origin onto which we are projecting $\mathbf{w}$. This line is the same when we project $\mathbf{w}$ onto $\mathbf{v}$ as it is when we project $\mathbf{w}$ onto $-\mathbf{v}$. So, the projection is the same in both cases.

Algebraically speaking, we have

$$\text{proj}_{(-\mathbf{v})}\mathbf{w} = \frac{\mathbf{w} \cdot (-\mathbf{v})}{\|-\mathbf{v}\|^2}(-\mathbf{v}) = -\frac{\mathbf{w} \cdot \mathbf{v}}{\|\mathbf{v}\|^2}(-\mathbf{v}) = \frac{\mathbf{w} \cdot \mathbf{v}}{\|\mathbf{v}\|^2}(\mathbf{v}) = \text{proj}_{\mathbf{v}}\mathbf{w}$$

for any vectors $\mathbf{w}$ and $\mathbf{v}$.

9.4.6

(a) We have $\mathbf{u} \cdot \mathbf{u}^{\perp} = (x)(-y) + (y)(x) = 0$, so the two vectors are orthogonal. Therefore, the angle between them is $\pi/2$.

(b) Let $\vec{OX}$ and $\vec{OY}$ be the projections of $\mathbf{a}$ onto $\mathbf{u}$ and $\mathbf{u}^{\perp}$. Since $\mathbf{u}$ and $\mathbf{u}^{\perp}$ are orthogonal, we have $\angle OYA = \angle OXA = \angle XOY = 90°$, which means that $OYAX$ is a rectangle. Every rectangle is also a parallelogram, so we have $\vec{A} = \vec{X} + \vec{Y}$. This gives us $\mathbf{a} = \text{proj}_{\mathbf{u}}\mathbf{a} + \text{proj}_{(\mathbf{u}^{\perp})}\mathbf{a}$.

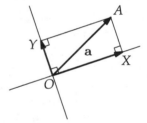

Algebraically, we note that $\text{proj}_{\mathbf{u}}\mathbf{a} = \dfrac{\mathbf{u} \cdot \mathbf{a}}{\|\mathbf{u}\|^2}\mathbf{u} = (\mathbf{u} \cdot \mathbf{a})\mathbf{u}$ because $\mathbf{u}$ is a unit vector. Letting $\mathbf{a} = a_1\mathbf{i} + a_2\mathbf{j}$, we have $\mathbf{a} \cdot \mathbf{u} = a_1 x + a_2 y$, so

$$\text{proj}_{\mathbf{u}}\mathbf{a} = (a_1 x + a_2 y)(x\mathbf{i} + y\mathbf{j}) = (a_1 x^2 + a_2 xy)\mathbf{i} + (a_1 xy + a_2 y^2)\mathbf{j}.$$

Similarly, we have

$$\text{proj}_{(\mathbf{u}^{\perp})}\mathbf{a} = (\mathbf{a} \cdot \mathbf{u}^{\perp})\mathbf{u}^{\perp} = (-a_1 y + a_2 x)(-y\mathbf{i} + x\mathbf{j}) = (a_1 y^2 - a_2 xy)\mathbf{i} + (-a_1 xy + a_2 x^2)\mathbf{j}.$$

This gives us

$$\text{proj}_{\mathbf{u}}\mathbf{a} + \text{proj}_{(\mathbf{u}^{\perp})}\mathbf{a} = (a_1 x^2 + a_2 xy)\mathbf{i} + (a_1 xy + a_2 y^2)\mathbf{j} + (a_1 y^2 - a_2 xy)\mathbf{i} + (-a_1 xy + a_2 x^2)\mathbf{j}$$
$$= a_1(x^2 + y^2)\mathbf{i} + a_2(x^2 + y^2)\mathbf{j}.$$

Since $\mathbf{u}$ is a unit vector, we have $x^2 + y^2 = 1$, so $\text{proj}_{\mathbf{u}}\mathbf{a} + \text{proj}_{(\mathbf{u}^{\perp})}\mathbf{a} = a_1\mathbf{i} + a_2\mathbf{j} = \mathbf{a}$, as desired.

(c) There are many ways to see why this equation holds. First, you might see that it is just the Pythagorean Theorem! Let's see why. We saw above that $\text{proj}_{\mathbf{u}}\mathbf{a} = (\mathbf{a} \cdot \mathbf{u})\mathbf{u}$. Since $\mathbf{u}$ is a unit vector, the magnitude of $\text{proj}_{\mathbf{u}}\mathbf{a}$ is $\mathbf{a} \cdot \mathbf{u}$. Similarly, the magnitude of $\text{proj}_{(\mathbf{u}^{\perp})}\mathbf{a}$ is $\mathbf{a} \cdot \mathbf{u}^{\perp}$. In the diagram we used for our geometric solution to part (b), triangle OXA is a right triangle with hypotenuse $\|\mathbf{a}\|$ and legs of lengths $\|\text{proj}_{\mathbf{u}}\mathbf{a}\|$ and $\|\text{proj}_{(\mathbf{u}^{\perp})}\mathbf{a}\|$. Using our expressions above for $\|\text{proj}_{\mathbf{u}}\mathbf{a}\|$ and $\|\text{proj}_{(\mathbf{u}^{\perp})}\mathbf{a}\|$, we have

$$\|\mathbf{a}\|^2 = (\mathbf{a} \cdot \mathbf{u})^2 + (\mathbf{a} \cdot \mathbf{u}^{\perp})^2$$

from the Pythagorean Theorem.

We also could have used the relationship we proved in part (b), rather than the Pythagorean Theorem. Since $\mathbf{u}$ and $\mathbf{u}^\perp$ are orthogonal, the projections of $\mathbf{a}$ onto these vectors are orthogonal:

$$\text{proj}_{\mathbf{u}}\mathbf{a} \cdot \text{proj}_{(\mathbf{u}^\perp)}\mathbf{a} = ((\mathbf{a}\cdot\mathbf{u})\mathbf{u})\cdot((\mathbf{a}\cdot\mathbf{u}^\perp)\mathbf{u}^\perp) = (\mathbf{a}\cdot\mathbf{u})(\mathbf{a}\cdot\mathbf{u}^\perp)(\mathbf{u}\cdot\mathbf{u}^\perp) = 0.$$

(There are no denominators in the expressions for the projections because $\mathbf{u}$ and $\mathbf{u}^\perp$ are unit vectors.) We therefore have

$$\begin{aligned}\|\mathbf{a}\|^2 = \mathbf{a}\cdot\mathbf{a} &= (\text{proj}_{\mathbf{u}}\mathbf{a} + \text{proj}_{\mathbf{u}^\perp}\mathbf{a})\cdot(\text{proj}_{\mathbf{u}}\mathbf{a} + \text{proj}_{\mathbf{u}^\perp}\mathbf{a})\\ &= (\text{proj}_{\mathbf{u}}\mathbf{a})\cdot(\text{proj}_{\mathbf{u}}\mathbf{a}) + (\text{proj}_{\mathbf{u}^\perp}\mathbf{a})\cdot(\text{proj}_{\mathbf{u}^\perp}\mathbf{a}) + 2(\text{proj}_{\mathbf{u}}\mathbf{a})\cdot(\text{proj}_{\mathbf{u}^\perp}\mathbf{a})\\ &= \|\text{proj}_{\mathbf{u}}\mathbf{a}\|^2 + \|\text{proj}_{\mathbf{u}^\perp}\mathbf{a}\|^2 + 2(0)\\ &= (\mathbf{a}\cdot\mathbf{u})^2 + (\mathbf{a}\cdot\mathbf{u}^\perp)^2.\end{aligned}$$

9.4.7 We have

$$\frac{\mathbf{u}\cdot\mathbf{v}}{(\text{proj}_{\mathbf{v}}\mathbf{u})\cdot(\text{proj}_{\mathbf{u}}\mathbf{v})} = \frac{\mathbf{u}\cdot\mathbf{v}}{\left(\dfrac{\mathbf{u}\cdot\mathbf{v}}{\|\mathbf{v}\|^2}\mathbf{v}\right)\cdot\left(\dfrac{\mathbf{v}\cdot\mathbf{u}}{\|\mathbf{u}\|^2}\mathbf{u}\right)} = \frac{\mathbf{u}\cdot\mathbf{v}}{\dfrac{(\mathbf{u}\cdot\mathbf{v})^3}{\|\mathbf{u}\|^2\|\mathbf{v}\|^2}}$$

$$= \frac{\|\mathbf{u}\|^2\|\mathbf{v}\|^2}{(\mathbf{u}\cdot\mathbf{v})^2} = \frac{\|\mathbf{u}\|^2\|\mathbf{v}\|^2}{\|\mathbf{u}\|^2\|\mathbf{v}\|^2\cos^2\theta} = \boxed{\sec^2\theta}.$$

Review Problems

9.29 We have $\left\|\begin{pmatrix}-6\\2\end{pmatrix}\right\| = \sqrt{(-6)^2+2^2} = \boxed{2\sqrt{10}}$, and $\|15\mathbf{i}-8\mathbf{j}\| = \sqrt{15^2+(-8)^2} = \boxed{17}$.

9.30 A set of two vectors is linearly dependent if and only if one of the vectors is a multiple of the other vector. Let $4\mathbf{i}-5\mathbf{j} = t(-3\mathbf{i}+k\mathbf{j}) = -3t\mathbf{i}+tk\mathbf{j}$. Then $4 = -3t$ and $-5 = tk$, so $t = -4/3$, and $k = -5/t = -5/(-4/3) = \boxed{15/4}$.

9.31 First, we find the unit vector in the direction of $4\mathbf{i}-6\mathbf{j}$, which is

$$\frac{4\mathbf{i}-6\mathbf{j}}{\|4\mathbf{i}-6\mathbf{j}\|} = \frac{4\mathbf{i}-6\mathbf{j}}{\sqrt{4^2+(-6)^2}} = \frac{4\mathbf{i}-6\mathbf{j}}{\sqrt{16+36}} = \frac{4\mathbf{i}-6\mathbf{j}}{\sqrt{52}} = \frac{4\mathbf{i}-6\mathbf{j}}{2\sqrt{13}} = \frac{2\sqrt{13}\mathbf{i}-3\sqrt{13}\mathbf{j}}{13}.$$

Then the vector $\mathbf{v}$ with magnitude 10 in the direction of $4\mathbf{i}-6\mathbf{j}$ is given by

$$\mathbf{v} = 10\cdot\frac{2\sqrt{13}\mathbf{i}-3\sqrt{13}\mathbf{j}}{13} = \boxed{\frac{20\sqrt{13}\mathbf{i}-30\sqrt{13}\mathbf{j}}{13}}.$$

9.32 We have $\vec{AB}+\vec{BC}+\vec{CA} = (\vec{B}-\vec{A})+(\vec{C}-\vec{B})+(\vec{A}-\vec{C}) = \boxed{0}$.

9.33 Let $\vec{D} = \vec{A}+\vec{B}$ and let O be the origin. We have $\vec{C} = \frac{1}{2}\vec{D}$, so C is the midpoint of $\vec{D}$. Since $OADB$ is a parallelogram, diagonals $\overline{OD}$ and $\overline{AB}$ bisect each other. Therefore, the midpoint of $\overline{OD}$ is also the midpoint of $\overline{AB}$, which means C is the midpoint of $\overline{AB}$.

9.34 Squaring the given inequality, we find that $\|\mathbf{v}\|+\|\mathbf{w}\| \ge \|\mathbf{v}+\mathbf{w}\|$ is equivalent to

$$\|\mathbf{v}\|^2 + 2\|\mathbf{v}\|\|\mathbf{w}\| + \|\mathbf{w}\|^2 \ge \|\mathbf{v}+\mathbf{w}\|^2.$$

Let θ be the angle between $\mathbf{v}$ and $\mathbf{w}$. We have $\|\mathbf{v}\|^2 = \mathbf{v}\cdot\mathbf{v}$, $\|\mathbf{w}\|^2 = \mathbf{w}\cdot\mathbf{w}$, and

$$\|\mathbf{v}+\mathbf{w}\|^2 = (\mathbf{v}+\mathbf{w})\cdot(\mathbf{v}+\mathbf{w}) = \mathbf{v}\cdot(\mathbf{v}+\mathbf{w}) + \mathbf{w}\cdot(\mathbf{v}+\mathbf{w}) = \mathbf{v}\cdot\mathbf{v} + \mathbf{v}\cdot\mathbf{w} + \mathbf{w}\cdot\mathbf{v} + \mathbf{w}\cdot\mathbf{w} = \mathbf{v}\cdot\mathbf{v} + 2\mathbf{v}\cdot\mathbf{w} + \mathbf{w}\cdot\mathbf{w}$$

$$= \|\mathbf{v}\|^2 + \|\mathbf{w}\|^2 + 2\|\mathbf{v}\|\|\mathbf{w}\|\cos\theta.$$

Since $\cos\theta \leq 1$, we have

$$\|\mathbf{v} + \mathbf{w}\|^2 = \|\mathbf{v}\|^2 + \|\mathbf{w}\|^2 + 2\|\mathbf{v}\|\,\|\mathbf{w}\|\cos\theta$$
$$\leq \|\mathbf{v}\|^2 + \|\mathbf{w}\|^2 + 2\|\mathbf{v}\|\,\|\mathbf{w}\| = (\|\mathbf{v}\| + \|\mathbf{w}\|)^2.$$

Since the magnitude of any vector is nonnegative, we can take the square root of both ends to find that we have $\|\mathbf{v} + \mathbf{w}\| \leq \|\mathbf{v}\| + \|\mathbf{w}\|$, as desired.

As we explained in the text, we can find the vector $\mathbf{v} + \mathbf{w}$ by drawing $\mathbf{w}$ starting from the head of $\mathbf{v}$. The vector from the tail of $\mathbf{v}$ to the head of this $\mathbf{w}$ is $\mathbf{v} + \mathbf{w}$, as shown. The expressions $\|\mathbf{v} + \mathbf{w}\|$, $\|\mathbf{v}\|$, and $\|\mathbf{w}\|$ are the side lengths of this triangle. So, the relationship we just proved is equivalent to the Triangle Inequality.

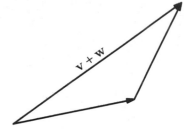

9.35

(a) $(3\mathbf{i} - 2\mathbf{j}) \cdot 8\mathbf{i} = 3 \cdot 8 + (-2) \cdot 0 = 24 + 0 = \boxed{24}$.

(b) $\begin{pmatrix} -2 \\ 5 \end{pmatrix} \cdot \begin{pmatrix} -4 \\ 3 \end{pmatrix} = (-2) \cdot (-4) + 5 \cdot 3 = 8 + 15 = \boxed{23}$.

9.36 In general, $\mathbf{u} \cdot \mathbf{v} = \|\mathbf{u}\|\|\mathbf{v}\|\cos\theta$, where θ is the angle formed by the vectors $\mathbf{u}$ and $\mathbf{v}$. In particular, if $\mathbf{u}$ and $\mathbf{v}$ have the same direction, then $\theta = 0$, so $\cos\theta = 1$, and $\boxed{\mathbf{u} \cdot \mathbf{v} = \|\mathbf{u}\|\|\mathbf{v}\|}$.

If $\mathbf{u}$ and $\mathbf{v}$ have opposite directions, then $\theta = \pi$, so $\cos\theta = -1$, and $\boxed{\mathbf{u} \cdot \mathbf{v} = -\|\mathbf{u}\|\|\mathbf{v}\|}$.

9.37

(a) Let $\mathbf{a} = \begin{pmatrix} a_1 \\ a_2 \end{pmatrix}$, $\mathbf{b} = \begin{pmatrix} b_1 \\ b_2 \end{pmatrix}$, and $\mathbf{c} = \begin{pmatrix} c_1 \\ c_2 \end{pmatrix}$. Then $\mathbf{a} \cdot \mathbf{b}$ and $\mathbf{a} \cdot \mathbf{c} = 0$ gives us $a_1 b_1 + a_2 b_2 = 0$ and $a_1 c_1 + a_2 c_2 = 0$. Since the vector $\mathbf{a}$ is nonzero, at least one of the components a_1 and a_2 is nonzero. Without loss of generality, assume that a_1 is nonzero. Then from these equations, we have $b_1 = -a_2 b_2 / a_1$ and $c_1 = -a_2 c_2 / a_1$, so

$$c_2 \mathbf{b} - b_2 \mathbf{c} = c_2 \begin{pmatrix} b_1 \\ b_2 \end{pmatrix} - b_2 \begin{pmatrix} c_1 \\ c_2 \end{pmatrix} = c_2 \begin{pmatrix} -a_2 b_2 / a_1 \\ b_2 \end{pmatrix} - b_2 \begin{pmatrix} -a_2 c_2 / a_1 \\ c_2 \end{pmatrix} = \begin{pmatrix} -a_2 b_2 c_2 / a_1 \\ b_2 c_2 \end{pmatrix} - \begin{pmatrix} -a_2 b_2 c_2 / a_1 \\ b_2 c_2 \end{pmatrix} = \begin{pmatrix} 0 \\ 0 \end{pmatrix}.$$

If $c_2 = 0$, then $c_1 = -a_2 c_2 / a_1 = 0$, so $\mathbf{c} = \mathbf{0}$, in which case $\mathbf{b}$ and $\mathbf{c}$ are linearly dependent. If c_2 is nonzero, then the linear combination above shows that $\mathbf{b}$ and $\mathbf{c}$ are linearly dependent. Hence, in either case, $\mathbf{b}$ and $\mathbf{c}$ are linearly dependent.

We can also reason geometrically. The equations $\mathbf{a} \cdot \mathbf{b} = \mathbf{a} \cdot \mathbf{c} = 0$ imply that $\mathbf{a}$ is perpendicular to both $\mathbf{b}$ and $\mathbf{c}$, so $\mathbf{b}$ and $\mathbf{c}$ are parallel. Hence, one vector is a multiple of the other vector, i.e. the set $\{\mathbf{b}, \mathbf{c}\}$ is linearly dependent.

(b) Yes. If $\mathbf{c}$ and $\mathbf{b}$ are linearly dependent and are both nonzero, then there is a nonzero constant k such that $\mathbf{c} = k\mathbf{b}$. So, we have $\mathbf{a} \cdot \mathbf{c} = \mathbf{a} \cdot (k\mathbf{b}) = k\mathbf{a} \cdot \mathbf{b} = 0$. However, we are given that $\mathbf{a} \cdot \mathbf{c} \neq 0$, so we conclude that $\mathbf{b}$ and $\mathbf{c}$ cannot be linearly dependent after all. Therefore, $\mathbf{b}$ and $\mathbf{c}$ are linearly independent.

9.38 Let $\mathbf{v} = \begin{pmatrix} x \\ y \end{pmatrix}$. Then $\mathbf{u} \cdot \mathbf{v} = \begin{pmatrix} 3 \\ 2 \end{pmatrix} \cdot \begin{pmatrix} x \\ y \end{pmatrix} = 3x + 2y$, so $3x + 2y = 9$. Thus, the heads of all the vectors $\mathbf{v}$ form a $\boxed{\text{line}}$.

9.39 Since both sides of the given equation are nonnegative, squaring both sides gives the equivalent equation $\|\mathbf{v} + \mathbf{w}\|^2 = \|\mathbf{v} - \mathbf{w}\|^2$. But

$$\|\mathbf{v} + \mathbf{w}\|^2 = (\mathbf{v} + \mathbf{w}) \cdot (\mathbf{v} + \mathbf{w}) = \mathbf{v} \cdot (\mathbf{v} + \mathbf{w}) + \mathbf{w} \cdot (\mathbf{v} + \mathbf{w}) = \mathbf{v} \cdot \mathbf{v} + \mathbf{v} \cdot \mathbf{w} + \mathbf{w} \cdot \mathbf{v} + \mathbf{w} \cdot \mathbf{w} = \mathbf{v} \cdot \mathbf{v} + 2\mathbf{v} \cdot \mathbf{w} + \mathbf{w} \cdot \mathbf{w},$$

and

$$\|\mathbf{v} - \mathbf{w}\|^2 = \mathbf{v} \cdot \mathbf{v} + 2\mathbf{v} \cdot (-\mathbf{w}) + (-\mathbf{w}) \cdot (-\mathbf{w}) = \mathbf{v} \cdot \mathbf{v} - 2\mathbf{v} \cdot \mathbf{w} + \mathbf{w} \cdot \mathbf{w}.$$

Hence,

$$\|\mathbf{v} + \mathbf{w}\|^2 = \|\mathbf{v} - \mathbf{w}\|^2 \quad \Leftrightarrow \quad \mathbf{v} \cdot \mathbf{v} + 2\mathbf{v} \cdot \mathbf{w} + \mathbf{w} \cdot \mathbf{w} = \mathbf{v} \cdot \mathbf{v} - 2\mathbf{v} \cdot \mathbf{w} + \mathbf{w} \cdot \mathbf{w} \quad \Leftrightarrow \quad 4\mathbf{v} \cdot \mathbf{w} = 0 \quad \Leftrightarrow \quad \mathbf{v} \cdot \mathbf{w} = 0.$$

Thus, $\|\mathbf{v} + \mathbf{w}\| = \|\mathbf{v} - \mathbf{w}\|$ if and only if $\mathbf{v}$ and $\mathbf{w}$ are orthogonal.

9.40

(a) Let $\mathbf{v}$ and $\mathbf{w}$ be two vectors that have the same direction, and let $\mathbf{u}$ be the unit vector in the common direction. Then both $\mathbf{v}$ and $\mathbf{w}$ are scalar multiples of $\mathbf{u}$, because by definition, $\mathbf{v} = \|\mathbf{v}\|\mathbf{u}$ and $\mathbf{w} = \|\mathbf{w}\|\mathbf{u}$. Therefore, one of $\mathbf{v}$ and $\mathbf{w}$ is a multiple of the other.

(b) No. Let $\mathbf{v}$ be a nonzero vector, and let $\mathbf{w} = -\mathbf{v}$. Then $\mathbf{w}$ is a scalar multiple of $\mathbf{v}$, but they do not have the same direction, because they have opposite directions.

9.41 Yes. Let θ be the angle between $\mathbf{u}$ and $\mathbf{v}$. Then, we have $\mathbf{u} \cdot \mathbf{v} = \|\mathbf{u}\|\,\|\mathbf{v}\|\cos\theta$. Since $\|\mathbf{u}\|$ and $\|\mathbf{v}\|$ are nonnegative, the sign of $\mathbf{u} \cdot \mathbf{v}$ matches the sign of $\cos\theta$. Since $\cos\theta$ is positive for acute angles θ and negative for obtuse angles θ, then a positive dot product $\mathbf{u} \cdot \mathbf{v}$ means the angle between $\mathbf{u}$ and $\mathbf{v}$ is acute, and a negative dot product means the angle between the vectors is obtuse.

9.42

(a) $\operatorname{proj}_{\mathbf{u}} \mathbf{v} = \dfrac{\mathbf{v} \cdot \mathbf{u}}{\|\mathbf{u}\|^2} \mathbf{u} = \dfrac{(0)(3) + (-3)(5)}{\left(\sqrt{3^2 + 5^2}\right)^2} \begin{pmatrix} 3 \\ 5 \end{pmatrix} = \dfrac{-15}{34} \begin{pmatrix} 3 \\ 5 \end{pmatrix} = \boxed{\begin{pmatrix} -45/34 \\ -75/34 \end{pmatrix}}.$

(b) $\operatorname{proj}_{\mathbf{u}} \mathbf{v} = \dfrac{\mathbf{v} \cdot \mathbf{u}}{\|\mathbf{u}\|^2} \mathbf{u} = \dfrac{(2)(5) + (-4)(7)}{\left(\sqrt{5^2 + 7^2}\right)^2} \begin{pmatrix} 5 \\ 7 \end{pmatrix} = \dfrac{-18}{74} \begin{pmatrix} 5 \\ 7 \end{pmatrix} = \boxed{\begin{pmatrix} -45/37 \\ -63/37 \end{pmatrix}}.$

9.43 The vector $\operatorname{proj}_{\mathbf{v}} \mathbf{u}$ must be a multiple of $\mathbf{v}$. Letting $\operatorname{proj}_{\mathbf{v}} \mathbf{u} = k\mathbf{v}$, the equation $\operatorname{proj}_{\mathbf{v}} \mathbf{u} = \mathbf{u}$ gives us $\mathbf{u} = k\mathbf{v}$. Therefore, we have $\mathbf{u} + (-k)\mathbf{v} = \mathbf{0}$, which means that $\mathbf{u}$ and $\mathbf{v}$ are linearly dependent.

9.44 Algebraically, letting θ be the angle between $\mathbf{a}$ and $\mathbf{b}$, we have

$$\left\|\operatorname{proj}_{\mathbf{b}} \mathbf{a}\right\| = \left\|\dfrac{\mathbf{a} \cdot \mathbf{b}}{\|\mathbf{b}\|^2} \mathbf{b}\right\| = \dfrac{|\mathbf{a} \cdot \mathbf{b}|}{\|\mathbf{b}\|^2} \|\mathbf{b}\| = \dfrac{\|\mathbf{a}\|\,\|\mathbf{b}\|\,|\cos\theta|}{\|\mathbf{b}\|^2} \|\mathbf{b}\| = \|\mathbf{a}\|\,|\cos\theta|.$$

Since $-1 \le \cos\theta \le 1$, we have $|\cos\theta| \le 1$, so $\|\mathbf{a}\|\,|\cos\theta| \le \|\mathbf{a}\|$. Therefore, we have $\left\|\operatorname{proj}_{\mathbf{b}} \mathbf{a}\right\| \le \|\mathbf{a}\|$.

Geometrically, we showed in the text that if $\mathbf{a}$ is not in the same or opposite direction as $\mathbf{b}$, then the arrow from the origin representing $\mathbf{a}$ is the hypotenuse of a right triangle with the arrow representing $\operatorname{proj}_{\mathbf{a}} \mathbf{b}$ as one of the legs. The hypotenuse of any right triangle is longer than each leg of the triangle, so $\|\mathbf{a}\| > \left\|\operatorname{proj}_{\mathbf{b}} \mathbf{a}\right\|$. If $\mathbf{a}$ is in the same or opposite direction as $\mathbf{b}$, then $\operatorname{proj}_{\mathbf{b}} \mathbf{a} = \mathbf{a}$, so $\|\mathbf{a}\| = \left\|\operatorname{proj}_{\mathbf{b}} \mathbf{a}\right\|$. Combining these, we have $\|\mathbf{a}\| \ge \left\|\operatorname{proj}_{\mathbf{b}} \mathbf{a}\right\|$.

Challenge Problems

9.45 If the set of vectors $\{\mathbf{u}, \mathbf{v}\}$ is linearly dependent, then the set of vectors $\{\mathbf{u}, \mathbf{v}, \mathbf{w}\}$ is linearly dependent.

Otherwise, the set of vectors $\{\mathbf{u}, \mathbf{v}\}$ is linearly independent. Hence, there exist constants a and b such that $a\mathbf{u} + b\mathbf{v} = \mathbf{w}$. Rewriting this equation as $a\mathbf{u} + b\mathbf{v} + (-1)\mathbf{w} = \mathbf{0}$, we see that there exist constants c_1, c_2, c_3 such that not all three are 0 and $c_1\mathbf{u} + c_2\mathbf{v} + c_3\mathbf{w} = \mathbf{0}$. Therefore, the set of vectors $\{\mathbf{u}, \mathbf{v}, \mathbf{w}\}$ is linearly dependent.

9.46 If the vector $\begin{pmatrix} x \\ 2 \end{pmatrix}$ has magnitude 6, then $x^2 + 4 = 36$, then $x^2 = 32$, or $x = \pm 4\sqrt{2}$. If the vector $\begin{pmatrix} 2 \\ y \end{pmatrix}$ has magnitude 6, then $4 + y^2 = 36$, so $y^2 = 32$, or $y = \pm 4\sqrt{2}$. Therefore, there are $\boxed{4}$ such vectors:

$$\begin{pmatrix} 2 \\ 4\sqrt{2} \end{pmatrix}, \begin{pmatrix} 2 \\ -4\sqrt{2} \end{pmatrix}, \begin{pmatrix} 4\sqrt{2} \\ 2 \end{pmatrix}, \text{ and } \begin{pmatrix} -4\sqrt{2} \\ 2 \end{pmatrix}.$$

Another way to think about this problem is to note that if $\begin{pmatrix} x \\ y \end{pmatrix}$ fits the problem, then (x, y) is on the circle with radius 6 centered at the origin, and either $x = 2$ or $y = 2$. Therefore each such (x, y) is the intersection of the circle and either the line $x = 2$ or $y = 2$. Since each of these lines is less than 6 from the origin, each line intersects the circle twice. Moreover, the lines meet at $(2, 2)$, which is not 6 units from the origin, so the lines meet the circle at $\boxed{4}$ distinct points altogether.

9.47 Let $\mathbf{u} = \begin{pmatrix} u_1 \\ 2 \end{pmatrix}$, $\mathbf{v} = \begin{pmatrix} 0 \\ 1 \end{pmatrix}$, and $\mathbf{w} = \begin{pmatrix} w_1 \\ -3 \end{pmatrix}$. Then $\mathbf{u} \cdot \mathbf{v} = u_1 \cdot 0 + 2 \cdot 1 = 2$, and $\mathbf{v} \cdot \mathbf{w} = 0 \cdot w_1 + 1 \cdot (-3) = -3$. Also, $\mathbf{u} \cdot \mathbf{w} = u_1 w_1 + 2 \cdot (-3) = u_1 w_1 - 6$. Since u_1 and w_1 can be any real numbers, $\mathbf{u} \cdot \mathbf{w}$ can also be any $\boxed{\text{any real number}}$.

9.48

(a) Geometrically speaking, let O be the origin, let X be $(6, -2)$, so $\vec{X} = \text{proj}_\mathbf{a} \mathbf{b} = 2\mathbf{a}$. Let B be a point such that $\mathbf{b} = \vec{B}$ and $\text{proj}_\mathbf{a} \mathbf{b} = 2\mathbf{a} = \vec{X}$. Therefore, we must have $\overline{BX} \perp \overleftrightarrow{OX}$. So, point B must be on the line through X perpendicular to $\overleftrightarrow{OX}$. Moreover, any point on this line can be B, including X. So, we can take $\mathbf{b} = \vec{X} = 2\mathbf{a} = 6\mathbf{i} - 2\mathbf{j}$. To find other possible values of $\mathbf{b}$, we note that $\overleftrightarrow{OX}$ is the graph of $y = -x/3$. Rearranging this gives $x + 3y = 0$, so the vector $\mathbf{i} + 3\mathbf{j}$ is normal to $\overleftrightarrow{OX}$. Letting $\mathbf{n} = \mathbf{i} + 3\mathbf{j}$, we can then let $\mathbf{b} = \boxed{2\mathbf{a} + t\mathbf{n} \text{ for any constant } t}$.

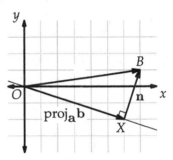

Checking, we see that if $\mathbf{b} = 2\mathbf{a} + t\mathbf{n}$, we have

$$\mathbf{b} \cdot \mathbf{a} = (2\mathbf{a} + t\mathbf{n}) \cdot \mathbf{a} = 2\mathbf{a} \cdot \mathbf{a} + t\mathbf{n} \cdot \mathbf{a} = 2\|\mathbf{a}\|^2 + t(0) = 2\|\mathbf{a}\|^2,$$

since $\mathbf{n}$ and $\mathbf{a}$ are orthogonal. Therefore,

$$\text{proj}_\mathbf{a} \mathbf{b} = \frac{\mathbf{b} \cdot \mathbf{a}}{\|\mathbf{a}\|^2} \mathbf{a} = \frac{2\|\mathbf{a}\|^2}{\|\mathbf{a}\|^2} \mathbf{a} = 2\mathbf{a},$$

as desired.

(b) Continuing with the geometric set-up of part (a), we know that B is on the line through X that is perpendicular to $\overleftrightarrow{OX}$. Since $\|\mathbf{b}\| = 10$, we know that $OB = 10$. Therefore, B must also be on the circle with radius 10 centered at the origin. Since $OX = \sqrt{6^2 + (-2)^2} = \sqrt{40} < 100$, point X is inside this circle. This means that the line that must contain X passes inside this circle, so this line intersects the circle at $\boxed{2}$ points B such that $\mathbf{b} = \vec{B}$ satisfies both $\|\mathbf{b}\| = 10$ and $\text{proj}_\mathbf{a} \mathbf{b} = 6\mathbf{i} - 2\mathbf{j}$.

(c) If $\text{proj}_\mathbf{j} \mathbf{b} = -\mathbf{j}$, then B is on the horizontal line through $(0, -1)$. This line intersects the line through X perpendicular to $\overleftrightarrow{OX}$ at exactly $\boxed{1}$ point, and the vector from the origin to this point is therefore the only vector $\mathbf{b}$ that satisfies $\text{proj}_\mathbf{j} \mathbf{b} = -\mathbf{j}$ and $\text{proj}_\mathbf{a} \mathbf{b} = 6\mathbf{i} - 2\mathbf{j}$.

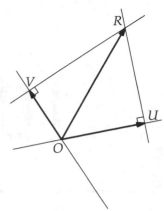

9.49 Let U and V be points such that $\vec{U} = \text{proj}_{\mathbf{u}}\mathbf{r}$ and $\vec{V} = \text{proj}_{\mathbf{v}}\mathbf{r}$, and let O be the origin. $\vec{U}$ is in the same or opposite direction of $\mathbf{u}$, and $\vec{V}$ is in the same or opposite direction of $\mathbf{v}$. Since $\mathbf{u}$ and $\mathbf{v}$ are linearly independent, so are $\vec{U}$ and $\vec{V}$. Therefore, neither U nor V is the origin, and $\vec{U}$ and $\vec{V}$ do not have the same direction or opposite directions. This means that $\overleftrightarrow{OU}$ and $\overleftrightarrow{OV}$ are two different lines. Let R be a point such that $\mathbf{r} = \vec{R}$, and $\text{proj}_{\mathbf{u}}\mathbf{r}$ and $\text{proj}_{\mathbf{v}}\mathbf{r}$ have the given values. Then R must be on the line through U perpendicular to $\overleftrightarrow{OU}$, as well as on the line through V perpendicular to $\overleftrightarrow{OV}$. Since $\overleftrightarrow{OU}$ and $\overleftrightarrow{OV}$ are different lines through the same point, they have different directions. Therefore, the two perpendiculars that pass through R cannot be the same line, and cannot be parallel, which means they must intersect at exactly one point. So, there is exactly one point R such that $\text{proj}_{\mathbf{u}}\mathbf{r}$ and $\text{proj}_{\mathbf{v}}\mathbf{r}$ have the given values.

9.50 First, suppose $\text{proj}_{\mathbf{v}}\mathbf{u} = 0$. Then, we have $\mathbf{v} \cdot \mathbf{u} = 0$, and from $\mathbf{u} = \text{proj}_{\mathbf{v}}\mathbf{u} + \text{proj}_{\mathbf{w}}\mathbf{u} = 0 + \text{proj}_{\mathbf{w}}\mathbf{u} = \text{proj}_{\mathbf{w}}\mathbf{u}$, we know that $\mathbf{u} = k\mathbf{w}$ for some nonzero constant k. Therefore, we have $\mathbf{v} \cdot (k\mathbf{w}) = 0$, so $\mathbf{v} \cdot \mathbf{w} = 0$.

Next suppose that both projections are nonzero. Let O be the origin and let U, V, and W be points such that $\vec{U} = \mathbf{u}$, $\vec{V} = \text{proj}_{\mathbf{v}}\mathbf{u}$, and $\vec{W} = \text{proj}_{\mathbf{w}}\mathbf{u}$. Since $\vec{U} = \vec{V} + \vec{W}$, we know that $OVUW$ is a parallelogram. Since $\vec{OV}$ is a projection of $\vec{U}$ onto $\mathbf{v}$, we know that $\vec{VU} \perp \vec{OV}$, which means $\angle OVU = 90°$. Since $OVUW$ is a parallelogram, we have $\angle OWU = \angle OVU = 90°$ and $\angle VUW = \angle VOW = 180° - \angle OVU = 90°$. So, $OVUW$ is a rectangle, which means $\overline{OV} \perp \overline{OW}$, so $\vec{V} \cdot \vec{W} = 0$. We have $\vec{V} = k_1\mathbf{v}$ and $\vec{W} = k_2\mathbf{w}$ for some nonzero constants k_1, k_2, since $\vec{V}$ and $\vec{W}$ are projections of $\vec{U}$ onto $\mathbf{v}$ and $\mathbf{w}$, respectively. Substituting these into $\vec{V} \cdot \vec{W} = 0$ gives $\mathbf{v} \cdot \mathbf{w} = 0$.

9.51 Let (x, y) be a point on one of the angle bisectors. Any point on the bisector of an angle must be equidistant from the sides of the angle. Therefore, (x, y) must be equidistant from $4x - 3y = 5$ and $x - 2y + 7 = 0$. Applying the formula for the distance between a point and line that we proved in the text, we have

$$\frac{|4x - 3y - 5|}{\sqrt{4^2 + (-3)^2}} = \frac{|x - 2y + 7|}{\sqrt{1^2 + (-2)^2}}.$$

We have two cases to consider:

Case 1: The signs of $4x - 3y - 5$ and $x - 2y + 7$ are the same. Then, we must have

$$\frac{4x - 3y - 5}{\sqrt{4^2 + (-3)^2}} = \frac{x - 2y + 7}{\sqrt{1^2 + (-2)^2}}.$$

Multiplying both sides by 5 gives $4x - 3y - 5 = \sqrt{5}(x - 2y + 7)$, and rearranging gives

$$\boxed{(4 - \sqrt{5})x + (-3 + 2\sqrt{5})y - 5 - 7\sqrt{5} = 0}.$$

Case 2: The signs of $4x - 3y - 5$ and $x - 2y + 7$ are opposite. Then, we must have

$$\frac{4x - 3y - 5}{\sqrt{4^2 + (-3)^2}} = -\left(\frac{x - 2y + 7}{\sqrt{1^2 + (-2)^2}}\right).$$

Multiplying both sides by 5 gives $4x - 3y - 5 = -\sqrt{5}(x - 2y + 7)$, and rearranging gives

$$\boxed{(4 + \sqrt{5})x + (-3 - 2\sqrt{5})y - 5 + 7\sqrt{5} = 0}.$$

CHAPTER 10

Matrices in Two Dimensions

Exercises for Section 10.1

10.1.1

(a) $\begin{pmatrix} 3 & -4 \\ 5 & -2 \end{pmatrix} \begin{pmatrix} 3 \\ -5 \end{pmatrix} = \begin{pmatrix} 3 \cdot 3 + (-4) \cdot (-5) \\ 5 \cdot 3 + (-2) \cdot (-5) \end{pmatrix} = \begin{pmatrix} 9 + 20 \\ 15 + 10 \end{pmatrix} = \boxed{\begin{pmatrix} 29 \\ 25 \end{pmatrix}}$.

(b) $\begin{pmatrix} -1 & 0 \\ 4 & -2 \end{pmatrix} \begin{pmatrix} -1 \\ 3 \end{pmatrix} = \begin{pmatrix} (-1) \cdot (-1) + 0 \cdot 3 \\ 4 \cdot (-1) + (-2) \cdot 3 \end{pmatrix} = \begin{pmatrix} 1 + 0 \\ (-4) + (-6) \end{pmatrix} = \boxed{\begin{pmatrix} 1 \\ -10 \end{pmatrix}}$.

10.1.2 Let $D = \begin{pmatrix} d_{11} & d_{12} \\ d_{21} & d_{22} \end{pmatrix}$. Since $D = A - B$, matrix D satisfies $D + B = A$, which means that we must have

$$\begin{pmatrix} d_{11} & d_{12} \\ d_{21} & d_{22} \end{pmatrix} + \begin{pmatrix} b_{11} & b_{12} \\ b_{21} & b_{22} \end{pmatrix} = \begin{pmatrix} a_{11} & a_{12} \\ a_{21} & a_{22} \end{pmatrix}.$$

Adding the two matrices on the left gives

$$\begin{pmatrix} d_{11} + b_{11} & d_{12} + b_{12} \\ d_{21} + b_{21} & d_{22} + b_{22} \end{pmatrix} = \begin{pmatrix} a_{11} & a_{12} \\ a_{21} & a_{22} \end{pmatrix}.$$

Therefore, we must have $d_{11} + b_{11} = a_{11}$, so $d_{11} = a_{11} - b_{11}$. Similarly, each entry in D is the difference of the corresponding entries in A and B, so we have

$$A - B = D = \begin{pmatrix} a_{11} - b_{11} & a_{12} - b_{12} \\ a_{21} - b_{21} & a_{22} - b_{22} \end{pmatrix}.$$

10.1.3

(a) We see that

$$A \begin{pmatrix} 2 \\ -3 \end{pmatrix} = \begin{pmatrix} 6 & -4 \\ 3 & -2 \end{pmatrix} \begin{pmatrix} 2 \\ -3 \end{pmatrix} = \begin{pmatrix} 6 \cdot 2 + (-4) \cdot (-3) \\ 3 \cdot 2 + (-2) \cdot (-3) \end{pmatrix} = \begin{pmatrix} 12 + 12 \\ 6 + 6 \end{pmatrix} = \boxed{\begin{pmatrix} 24 \\ 12 \end{pmatrix}},$$

and

$$A \begin{pmatrix} 5 \\ -1 \end{pmatrix} = \begin{pmatrix} 6 & -4 \\ 3 & -2 \end{pmatrix} \begin{pmatrix} 5 \\ -1 \end{pmatrix} = \begin{pmatrix} 6 \cdot 5 + (-4) \cdot (-1) \\ 3 \cdot 5 + (-2) \cdot (-1) \end{pmatrix} = \begin{pmatrix} 30 + 4 \\ 15 + 2 \end{pmatrix} = \boxed{\begin{pmatrix} 34 \\ 17 \end{pmatrix}}.$$

(b) We showed in the text that a pair of vectors is linearly dependent if and only if one of the vectors is a multiple of the other. Neither of the given vectors is a multiple of the other vector, so the set of vectors $\left\{ \begin{pmatrix} 2 \\ -3 \end{pmatrix}, \begin{pmatrix} 5 \\ -1 \end{pmatrix} \right\}$ is linearly independent.

(c) The set of vectors $\left\{\binom{24}{12}, \binom{34}{17}\right\}$ is linearly dependent, because $\binom{34}{17} = \frac{17}{12} \cdot \binom{24}{12}$.

(d) For any vector $\binom{x}{y}$, we have

$$\mathbf{A}\binom{x}{y} = \begin{pmatrix} 6 & -4 \\ 3 & -2 \end{pmatrix}\binom{x}{y} = \binom{6x - 4y}{3x - 2y} = (3x - 2y)\binom{2}{1},$$

so $\mathbf{A}\binom{x}{y}$ is always a scalar multiple of the vector $\binom{2}{1}$. Therefore, for any vectors $\mathbf{v}$ and $\mathbf{w}$, the set of vectors $\{\mathbf{Av}, \mathbf{Aw}\}$ is linearly dependent.

Exercises for Section 10.2

10.2.1

(a) $\begin{pmatrix} 3 & 4 \\ -2 & -6 \end{pmatrix} \begin{pmatrix} -1 & 0 \\ -5 & -7 \end{pmatrix} = \begin{pmatrix} 3 \cdot (-1) + 4 \cdot (-5) & 3 \cdot 0 + 4 \cdot (-7) \\ (-2) \cdot (-1) + (-6) \cdot (-5) & (-2) \cdot 0 + (-6) \cdot (-7) \end{pmatrix} = \boxed{\begin{pmatrix} -23 & -28 \\ 32 & 42 \end{pmatrix}}.$

(b) $\begin{pmatrix} -4 & -8 \\ 0 & 9 \end{pmatrix} \begin{pmatrix} -\frac{1}{2} & 2 \\ -3 & 3 \end{pmatrix} = \begin{pmatrix} (-4) \cdot (-\frac{1}{2}) + (-8) \cdot (-3) & (-4) \cdot 2 + (-8) \cdot 3 \\ 0 \cdot (-\frac{1}{2}) + 9 \cdot (-3) & 0 \cdot 2 + 9 \cdot 3 \end{pmatrix} = \boxed{\begin{pmatrix} 26 & -32 \\ -27 & 27 \end{pmatrix}}.$

10.2.2 Let $\begin{pmatrix} a_1 & 0 \\ 0 & a_2 \end{pmatrix}$ and $\begin{pmatrix} b_1 & 0 \\ 0 & b_2 \end{pmatrix}$ be two diagonal matrices. Then their product is

$$\begin{pmatrix} a_1 & 0 \\ 0 & a_2 \end{pmatrix} \begin{pmatrix} b_1 & 0 \\ 0 & b_2 \end{pmatrix} = \begin{pmatrix} a_1 b_1 & 0 \\ 0 & a_2 b_2 \end{pmatrix},$$

which is also diagonal.

10.2.3 Let $\mathbf{A} = \begin{pmatrix} a & b \\ c & d \end{pmatrix}$ and $\mathbf{F} = \begin{pmatrix} 0 & 1 \\ 1 & 0 \end{pmatrix}$. Then

$$\mathbf{FA} = \begin{pmatrix} 0 & 1 \\ 1 & 0 \end{pmatrix} \begin{pmatrix} a & b \\ c & d \end{pmatrix} = \begin{pmatrix} 0 \cdot a + 1 \cdot c & 0 \cdot b + 1 \cdot d \\ 1 \cdot a + 0 \cdot c & 1 \cdot b + 0 \cdot d \end{pmatrix} = \begin{pmatrix} c & d \\ a & b \end{pmatrix}.$$

Thus, the matrix $\mathbf{FA}$ is the same as the matrix $\mathbf{A}$ when the rows are swapped.

Furthermore,

$$\mathbf{AF} = \begin{pmatrix} a & b \\ c & d \end{pmatrix} \begin{pmatrix} 0 & 1 \\ 1 & 0 \end{pmatrix} = \begin{pmatrix} a \cdot 0 + b \cdot 1 & a \cdot 1 + b \cdot 0 \\ c \cdot 0 + d \cdot 1 & c \cdot 1 + d \cdot 0 \end{pmatrix} = \begin{pmatrix} b & a \\ d & c \end{pmatrix}.$$

Thus, the matrix $\mathbf{AF}$ is the same as the matrix $\mathbf{A}$ when the columns are swapped.

10.2.4 The answer is $\boxed{\text{no}}$. For example, let $\mathbf{A} = \mathbf{B} = \begin{pmatrix} 0 & 1 \\ 0 & 0 \end{pmatrix}$. Then

$$\mathbf{AB} = \begin{pmatrix} 0 & 1 \\ 0 & 0 \end{pmatrix} \begin{pmatrix} 0 & 1 \\ 0 & 0 \end{pmatrix} = \begin{pmatrix} 0 \cdot 0 + 1 \cdot 0 & 0 \cdot 1 + 1 \cdot 0 \\ 0 \cdot 0 + 0 \cdot 0 & 0 \cdot 1 + 0 \cdot 0 \end{pmatrix} = \begin{pmatrix} 0 & 0 \\ 0 & 0 \end{pmatrix} = \mathbf{0}.$$

10.2.5 Let $\mathbf{T} = \begin{pmatrix} a & b \\ c & d \end{pmatrix}$. Then

$$\mathbf{T}\binom{4}{-3} = \begin{pmatrix} a & b \\ c & d \end{pmatrix} \binom{4}{-3} = \binom{4a - 3b}{4c - 3d},$$

so from the given equation $\mathbf{T}\begin{pmatrix}4\\-3\end{pmatrix} = \begin{pmatrix}-3\\4\end{pmatrix}$, we have $4a - 3b = -3$ and $4c - 3d = 4$. Also,

$$\mathbf{T}\begin{pmatrix}-3\\1\end{pmatrix} = \begin{pmatrix}a & b\\c & d\end{pmatrix}\begin{pmatrix}-3\\1\end{pmatrix} = \begin{pmatrix}-3a + b\\-3c + d\end{pmatrix},$$

so from the given equation $\mathbf{T}\begin{pmatrix}-3\\1\end{pmatrix} = \begin{pmatrix}0\\5\end{pmatrix}$, we have $-3a + b = 0$ and $-3c + d = 5$. Hence,

$$\mathbf{T}\begin{pmatrix}4 & -3\\-3 & 1\end{pmatrix} = \begin{pmatrix}a & b\\c & d\end{pmatrix}\begin{pmatrix}4 & -3\\-3 & 1\end{pmatrix} = \begin{pmatrix}4a - 3b & -3a + b\\4c - 3d & -3c + d\end{pmatrix} = \boxed{\begin{pmatrix}-3 & 0\\4 & 5\end{pmatrix}}.$$

We also could have jumped straight to the answer by noting that the first column of $\mathbf{T}\begin{pmatrix}4 & -3\\-3 & 1\end{pmatrix}$ equals the product of $\mathbf{T}$ and the first column of $\begin{pmatrix}4 & -3\\-3 & 1\end{pmatrix}$, and the second column of the product is the product of $\mathbf{T}$ and the second column of $\begin{pmatrix}4 & -3\\-3 & 1\end{pmatrix}$. We are given that the first product is $\begin{pmatrix}-3\\4\end{pmatrix}$ and the second is $\begin{pmatrix}0\\5\end{pmatrix}$, so $\mathbf{T}\begin{pmatrix}4 & -3\\-3 & 1\end{pmatrix} = \begin{pmatrix}-3 & 0\\4 & 5\end{pmatrix}$.

10.2.6 If we multiply both sides of the equation $\mathbf{A}\begin{pmatrix}1\\2\end{pmatrix} = \begin{pmatrix}17\\0\end{pmatrix}$ by -2, then the left side becomes

$$-2\mathbf{A}\begin{pmatrix}1\\2\end{pmatrix} = \mathbf{A}\left(-2\begin{pmatrix}1\\2\end{pmatrix}\right) = \mathbf{A}\begin{pmatrix}-2\\-4\end{pmatrix},$$

and the right side becomes $-2\begin{pmatrix}17\\0\end{pmatrix} = \begin{pmatrix}-34\\0\end{pmatrix}$, so $\mathbf{A}\begin{pmatrix}-2\\-4\end{pmatrix} = \begin{pmatrix}-34\\0\end{pmatrix}$. Thus, the two equations are the same.

Exercises for Section 10.3

10.3.1 Let $\mathbf{A} = \begin{pmatrix}3 & 3\\0 & 1\end{pmatrix}$. Then $\mathbf{A}\begin{pmatrix}3\\-2\end{pmatrix} = \begin{pmatrix}3 & 3\\0 & 1\end{pmatrix}\begin{pmatrix}3\\-2\end{pmatrix} = \begin{pmatrix}3\cdot3 + 3\cdot(-2)\\0\cdot3 + 1\cdot(-2)\end{pmatrix} = \begin{pmatrix}3\\-2\end{pmatrix}$. Hence, $\mathbf{A}$ is not necessarily the identity matrix.

10.3.2 The reflection of the vector $\begin{pmatrix}x\\y\end{pmatrix}$ over the x-axis is $\begin{pmatrix}x\\-y\end{pmatrix}$. Hence, we seek a matrix $\mathbf{A} = \begin{pmatrix}a & b\\c & d\end{pmatrix}$ such that

$$\begin{pmatrix}a & b\\c & d\end{pmatrix}\begin{pmatrix}x\\y\end{pmatrix} = \begin{pmatrix}x\\-y\end{pmatrix}$$

for all x and y. We see that $\begin{pmatrix}a & b\\c & d\end{pmatrix}\begin{pmatrix}x\\y\end{pmatrix} = \begin{pmatrix}ax + by\\cx + dy\end{pmatrix}$, so we can take $a = 1$, $b = 0$, $c = 0$, and $d = -1$, so the matrix we seek is $\mathbf{A} = \boxed{\begin{pmatrix}1 & 0\\0 & -1\end{pmatrix}}$.

10.3.3 Any linear function of the form $f(\mathbf{v}) = \mathbf{A}\mathbf{v}$ satisfies $f(\mathbf{v} + \mathbf{w}) = f(\mathbf{v}) + f(\mathbf{w})$ for all vectors $\mathbf{v}$ and $\mathbf{w}$. But if we take $\mathbf{v} = \mathbf{w} = \begin{pmatrix}1\\1\end{pmatrix}$, then

$$f(\mathbf{v} + \mathbf{w}) = f\left(\begin{pmatrix}1\\1\end{pmatrix} + \begin{pmatrix}1\\1\end{pmatrix}\right) = f\left(\begin{pmatrix}2\\2\end{pmatrix}\right) = \begin{pmatrix}2^2\\2^2\end{pmatrix} = \begin{pmatrix}4\\4\end{pmatrix},$$

and

$$f(\mathbf{v}) + f(\mathbf{w}) = f\left(\begin{pmatrix} 1 \\ 1 \end{pmatrix}\right) + f\left(\begin{pmatrix} 1 \\ 1 \end{pmatrix}\right) = \begin{pmatrix} 1^2 \\ 1^2 \end{pmatrix} + \begin{pmatrix} 1^2 \\ 1^2 \end{pmatrix} = \begin{pmatrix} 1 \\ 1 \end{pmatrix} + \begin{pmatrix} 1 \\ 1 \end{pmatrix} = \begin{pmatrix} 2 \\ 2 \end{pmatrix},$$

so $f(\mathbf{v} + \mathbf{w}) \neq f(\mathbf{v}) + f(\mathbf{w})$. We showed in the text that any function of the form $g(\mathbf{v}) = \mathbf{A}\mathbf{v}$ is a linear function, so $f(\mathbf{v})$ cannot be written in the form $\mathbf{A}\mathbf{v}$.

10.3.4 The matrix corresponding to a rotation of $\alpha + \beta$ counterclockwise about the origin is

$$\begin{pmatrix} \cos(\alpha + \beta) & -\sin(\alpha + \beta) \\ \sin(\alpha + \beta) & \cos(\alpha + \beta) \end{pmatrix}.$$

But such a rotation is also the composition of two counterclockwise rotations, with angles α and β, respectively, which corresponds to the product

$$\begin{pmatrix} \cos\beta & -\sin\beta \\ \sin\beta & \cos\beta \end{pmatrix} \begin{pmatrix} \cos\alpha & -\sin\alpha \\ \sin\alpha & \cos\alpha \end{pmatrix} = \begin{pmatrix} \cos\alpha\cos\beta - \sin\alpha\sin\beta & -\sin\alpha\cos\beta - \cos\alpha\sin\beta \\ \sin\alpha\cos\beta + \cos\alpha\sin\beta & \cos\alpha\cos\beta - \sin\alpha\sin\beta \end{pmatrix}.$$

Therefore, $\cos(\alpha + \beta) = \cos\alpha\cos\beta - \sin\alpha\sin\beta$ and $\sin(\alpha + \beta) = \sin\alpha\cos\beta + \cos\alpha\sin\beta$.

10.3.5 As we found in Exercise 10.3.2, the matrix corresponding to reflection over the x-axis is $\mathbf{A} = \begin{pmatrix} 1 & 0 \\ 0 & -1 \end{pmatrix}$, and the matrix corresponding to a $90°$ counterclockwise about the origin is $\mathbf{B} = \begin{pmatrix} 0 & -1 \\ 1 & 0 \end{pmatrix}$. Then the problem becomes determining whether $\mathbf{BA}$ is equal to $\mathbf{AB}$.

We see that

$$\mathbf{BA} = \begin{pmatrix} 0 & -1 \\ 1 & 0 \end{pmatrix} \begin{pmatrix} 1 & 0 \\ 0 & -1 \end{pmatrix} = \begin{pmatrix} 0 \cdot 1 + (-1) \cdot 0 & 0 \cdot 0 + (-1) \cdot (-1) \\ 1 \cdot 1 + 0 \cdot 0 & 1 \cdot 0 + 0 \cdot (-1) \end{pmatrix} = \begin{pmatrix} 0 & 1 \\ 1 & 0 \end{pmatrix},$$

and

$$\mathbf{AB} = \begin{pmatrix} 1 & 0 \\ 0 & -1 \end{pmatrix} \begin{pmatrix} 0 & -1 \\ 1 & 0 \end{pmatrix} = \begin{pmatrix} 1 \cdot 0 + 0 \cdot 1 & 1 \cdot (-1) + 0 \cdot 0 \\ 0 \cdot 0 + (-1) \cdot 1 & 0 \cdot (-1) + (-1) \cdot 0 \end{pmatrix} = \begin{pmatrix} 0 & -1 \\ -1 & 0 \end{pmatrix}.$$

These two matrices are not the same, so the two operations of reflecting over the x-axis and rotating $90°$ counterclockwise about the origin do not commute, i.e. the order in which you perform them makes a difference.

Exercises for Section 10.4

10.4.1 First, we rewrite the given equations as

$$6x + 4y = 7,$$
$$2x + 5y = 9.$$

By Cramer's rule, we have

$$x = \frac{\begin{vmatrix} 7 & 4 \\ 9 & 5 \end{vmatrix}}{\begin{vmatrix} 6 & 4 \\ 2 & 5 \end{vmatrix}} = \frac{7 \cdot 5 - 4 \cdot 9}{6 \cdot 5 - 4 \cdot 2} = \boxed{-\frac{1}{22}}, \quad \text{and} \quad y = \frac{\begin{vmatrix} 6 & 7 \\ 2 & 9 \end{vmatrix}}{\begin{vmatrix} 6 & 4 \\ 2 & 5 \end{vmatrix}} = \frac{6 \cdot 9 - 7 \cdot 2}{6 \cdot 5 - 4 \cdot 2} = \frac{40}{22} = \boxed{\frac{20}{11}}.$$

10.4.2

(a) $\begin{vmatrix} -3 & -5 \\ 6 & 18 \end{vmatrix} = (-3) \cdot 18 - (-5) \cdot 6 = -54 - (-30) = \boxed{-24}$.

(b) To make the determinant easier to evaluate, we take a factor of 800 from the first row and a factor of 1/3 from the second row:

$$\begin{vmatrix} 2400 & 3200 \\ 25/3 & 31/3 \end{vmatrix} = 800 \begin{vmatrix} 3 & 4 \\ 25/3 & 31/3 \end{vmatrix} = \frac{800}{3} \begin{vmatrix} 3 & 4 \\ 25 & 31 \end{vmatrix} = \frac{800}{3}(3 \cdot 31 - 4 \cdot 25) = \frac{800}{3} \cdot (-7) = \boxed{-\frac{5600}{3}}.$$

10.4.3 For all matrices $\mathbf{A}$ and $\mathbf{B}$, $\det(\mathbf{AB}) = \det(\mathbf{A})\det(\mathbf{B}) = \det(\mathbf{B})\det(\mathbf{A}) = \det(\mathbf{BA})$.

10.4.4 Let $\mathbf{A} = \begin{pmatrix} a & b \\ c & d \end{pmatrix}$, so $\mathbf{B} = \begin{pmatrix} a & b+ka \\ c & d+kc \end{pmatrix}$, and

$$\det(\mathbf{B}) = \begin{vmatrix} a & b+ka \\ c & d+kc \end{vmatrix} = a(d+kc) - (b+ka)c = ad + kac - bc - kac = ad - bc = \det(\mathbf{A}).$$

10.4.5

(a) Let $\mathbf{A} = \begin{pmatrix} 1 & 0 \\ 0 & 0 \end{pmatrix}$ and $\mathbf{B} = \begin{pmatrix} 0 & 0 \\ 0 & 1 \end{pmatrix}$, so $\mathbf{A} + \mathbf{B} = \begin{pmatrix} 1 & 0 \\ 0 & 0 \end{pmatrix} + \begin{pmatrix} 0 & 0 \\ 0 & 1 \end{pmatrix} = \begin{pmatrix} 1 & 0 \\ 0 & 1 \end{pmatrix}$. Then

$$\det(\mathbf{A}) = \begin{vmatrix} 1 & 0 \\ 0 & 0 \end{vmatrix} = 1 \cdot 0 - 0 \cdot 0 = 0, \quad \det(\mathbf{B}) = \begin{vmatrix} 0 & 0 \\ 0 & 1 \end{vmatrix} = 0 \cdot 1 - 0 \cdot 0 = 0,$$

and

$$\det(\mathbf{A} + \mathbf{B}) = \begin{vmatrix} 1 & 0 \\ 0 & 1 \end{vmatrix} = 1 \cdot 1 - 0 \cdot 0 = 1.$$

So the conditions $\det(\mathbf{A}) = 0$ and $\det(\mathbf{B}) = 0$ are satisfied, which means $\det(\mathbf{A}) + \det(\mathbf{B}) = 0 + 0 = 0$, but $\det(\mathbf{A} + \mathbf{B}) = 1$, so $\det(\mathbf{A}) + \det(\mathbf{B})$ and $\det(\mathbf{A} + \mathbf{B})$ are not necessarily equal.

(b) Let $\mathbf{A} = \begin{pmatrix} 1 & 0 \\ 0 & 0 \end{pmatrix}$ and $\mathbf{B} = \begin{pmatrix} 1 & 0 \\ 0 & 1 \end{pmatrix}$, so $\mathbf{A} + \mathbf{B} = \begin{pmatrix} 1 & 0 \\ 0 & 0 \end{pmatrix} + \begin{pmatrix} 1 & 0 \\ 0 & 1 \end{pmatrix} = \begin{pmatrix} 2 & 0 \\ 0 & 1 \end{pmatrix}$. Then

$$\det(\mathbf{A}) = \begin{vmatrix} 1 & 0 \\ 0 & 0 \end{vmatrix} = 1 \cdot 0 - 0 \cdot 0 = 0, \quad \det(\mathbf{B}) = \begin{vmatrix} 1 & 0 \\ 0 & 1 \end{vmatrix} = 1 \cdot 1 - 0 \cdot 0 = 1,$$

and

$$\det(\mathbf{A} + \mathbf{B}) = \begin{vmatrix} 2 & 0 \\ 0 & 1 \end{vmatrix} = 2 \cdot 1 - 0 \cdot 0 = 2.$$

So the conditions $\det(\mathbf{A}) = 0$ and $\det(\mathbf{B}) \neq 0$ are satisfied, but $\det(\mathbf{A}) + \det(\mathbf{B}) = 0 + 1 = 1$ and $\det(\mathbf{A} + \mathbf{B}) = 2$, so $\det(\mathbf{A}) + \det(\mathbf{B})$ and $\det(\mathbf{A} + \mathbf{B})$ are not necessarily equal.

Exercises for Section 10.5

10.5.1 Let $\mathbf{R} = \begin{pmatrix} \cos\theta & -\sin\theta \\ \sin\theta & \cos\theta \end{pmatrix}$. The determinant of $\mathbf{R}$ is given by

$$\begin{vmatrix} \cos\theta & -\sin\theta \\ \sin\theta & \cos\theta \end{vmatrix} = \cos\theta \cdot \cos\theta - (-\sin\theta) \cdot \sin\theta = \cos^2\theta + \sin^2\theta = 1.$$

As discussed in the text, multiplying a vector by this matrix rotates the vector. Rotation does not affect area (or signed area), so the parallelogram with adjacent sides $\mathbf{Ri}$ and $\mathbf{Rj}$ is a rotation of the parallelogram with adjacent sides $\mathbf{i}$ and $\mathbf{j}$. This latter parallelogram is a square with side length one, and rotation doesn't affect signed area, so the parallelogram with adjacent sides $\mathbf{Ri}$ and $\mathbf{Rj}$ also has area 1. Therefore, the matrix has determinant 1.

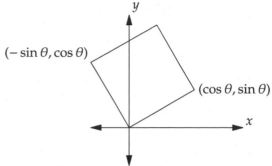

Another way to look at this is that $\mathbf{Ri} = (\cos\theta)\mathbf{i} + (\sin\theta)\mathbf{j}$ and $\mathbf{Rj} = (-\sin\theta)\mathbf{i} + (\cos\theta)\mathbf{j}$. The point $(-\sin\theta, \cos\theta)$ is obtained by rotating the point $(\cos\theta, \sin\theta)$ by $90°$ counterclockwise about the origin, so $\mathbf{Rj}$ is a $90°$ counterclockwise rotation about the origin. Moreover, the magnitudes of $\mathbf{Ri}$ and $\mathbf{Rj}$ are both equal to 1. Hence, the parallelogram with adjacent sides $\mathbf{Ri}$ and $\mathbf{Rj}$ is a square with side length 1, and therefore has area 1.

10.5.2 We know how to find the area of a triangle given its vertices, so we split the quadrilateral into two triangles, as shown. As described in the text area of triangle ABD is given by

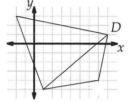

$$\frac{1}{2}\left|\det\begin{pmatrix}1-(-2) & 8-(-2)\\ -5-3 & 1-3\end{pmatrix}\right| = \frac{1}{2}\left|\det\begin{pmatrix}3 & 10\\ -8 & -2\end{pmatrix}\right| = \frac{1}{2}|3\cdot(-2)-(-8)\cdot 10| = \frac{1}{2}|74| = 37,$$

and the area of triangle BCD is given by

$$\frac{1}{2}\left|\det\begin{pmatrix}7-1 & 8-1\\ -4-(-5) & 1-(-5)\end{pmatrix}\right| = \frac{1}{2}\left|\det\begin{pmatrix}6 & 7\\ 1 & 6\end{pmatrix}\right| = \frac{1}{2}|6\cdot 6-1\cdot 7| = \frac{1}{2}|29| = \frac{29}{2},$$

so the area of quadrilateral $ABCD$ is $37 + \frac{29}{2} = \boxed{\frac{103}{2}}$.

10.5.3 Let $\mathbf{v}$ and $\mathbf{w}$ be the column vectors of $\mathbf{A}$, so that $\mathbf{v}$ is the column we multiply by k. Let $\mathbf{A}'$ be the resulting matrix when $\mathbf{v}$ is multiplied by k.

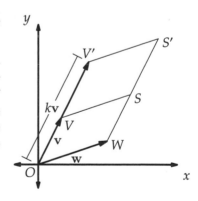

Suppose first that k is positive. Geometrically, $\det(\mathbf{A})$ is the area of the parallelogram formed by $\mathbf{v}$ and $\mathbf{w}$, and $\det(\mathbf{A}')$ is the area of the parallelogram formed by $k\mathbf{v}$ and $\mathbf{w}$. In the diagram at right, $OVSW$ is the first parallelogram and $OV'S'W$ is the second. The area of a parallelogram equals the length of a side times the height to this side from a point on the opposite side. The height from W to side $\overline{OV}$ (extended if necessary) equals the height from W to $\overline{OV'}$. Therefore, the ratio of the area of $OV'S'W$ to the area of $OVSW$ equals OV'/OV, which equals k.

If k is negative, the situation is essentially the same, except now, as shown in the diagram at right, the orientation of parallelogram $OV'S'W$ is opposite the orientation of $OVSW$. Again, we have $OV'/OV = |k|$, and the ratio of the *signed* areas of the parallelogram is k.

Since the area of the parallelogram with sides $k\mathbf{v}$ and $\mathbf{w}$ is always k times the area of the matrix with sides $\mathbf{v}$ and $\mathbf{w}$, the determinant of the matrix with columns $k\mathbf{v}$ and $\mathbf{w}$ is k times the determinant of the matrix with columns $\mathbf{v}$ and $\mathbf{w}$.

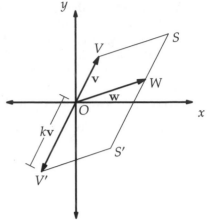

10.5.4 Let the vertices of $\mathcal{T}$ be the lattice points (x_1, y_1), (x_2, y_2), and (x_3, y_3), so all the coordinates are integers. Then the area of triangle $\mathcal{T}$ is

$$\frac{1}{2} \left| \det \begin{pmatrix} x_2 - x_1 & x_3 - x_1 \\ y_2 - y_1 & y_3 - y_1 \end{pmatrix} \right|.$$

When we double this quantity, we get

$$\left| \det \begin{pmatrix} x_2 - x_1 & x_3 - x_1 \\ y_2 - y_1 & y_3 - y_1 \end{pmatrix} \right| = |(x_2 - x_1)(y_3 - y_1) - (x_3 - x_1)(y_2 - y_1)|,$$

which is clearly an integer.

10.5.5 We know that $\det(\mathbf{A})$ is the signed area of the parallelogram formed by the column vectors of $\mathbf{A}$. Since multiplying $\mathbf{A}$ on the right by $\mathbf{F}$ swaps the columns of $\mathbf{A}$, the parallelogram formed by the column vectors of $\mathbf{AF}$ is the same as the original parallelogram, except that the orientation has been reversed, so $\det(\mathbf{AF}) = -\det(\mathbf{A})$.

But $\det(\mathbf{AF}) = \det(\mathbf{A}) \det(\mathbf{F})$, so $\det(\mathbf{A}) \det(\mathbf{F}) = -\det(\mathbf{A})$. For this to hold for all matrices $\mathbf{A}$, $\det(\mathbf{F})$ must be equal to $\boxed{-1}$.

Exercises for Section 10.6

10.6.1 Since $\mathbf{I} \cdot \mathbf{I} = \mathbf{I}$, the inverse of $\mathbf{I}$ is $\boxed{\mathbf{I}}$.

10.6.2

(a) $\begin{pmatrix} 6 & 8 \\ -2 & 4 \end{pmatrix}^{-1} = \dfrac{1}{6 \cdot 4 - 8 \cdot (-2)} \begin{pmatrix} 4 & -8 \\ 2 & 6 \end{pmatrix} = \dfrac{1}{40} \begin{pmatrix} 4 & -8 \\ 2 & 6 \end{pmatrix} = \boxed{\begin{pmatrix} \frac{1}{10} & -\frac{1}{5} \\ \frac{1}{20} & \frac{3}{20} \end{pmatrix}}.$

(b) $\begin{pmatrix} 2 & -1 \\ -0.9 & 0.5 \end{pmatrix}^{-1} = \dfrac{1}{2 \cdot 0.5 - (-1) \cdot (-0.9)} \begin{pmatrix} 0.5 & 1 \\ 0.9 & 2 \end{pmatrix} = \dfrac{1}{0.1} \begin{pmatrix} 0.5 & 1 \\ 0.9 & 2 \end{pmatrix} = \boxed{\begin{pmatrix} 5 & 10 \\ 9 & 20 \end{pmatrix}}.$

(c) Since $\det \begin{pmatrix} -40 & 30 \\ 8 & -6 \end{pmatrix} = (-40) \cdot (-6) - 8 \cdot 30 = 0$, the matrix $\begin{pmatrix} -40 & 30 \\ 8 & -6 \end{pmatrix}$ does not have an inverse.

10.6.3

(a) If $\mathbf{A}$ is nonsingular, then by definition, $\det(\mathbf{A})$ is nonzero, so $\mathbf{A}$ is invertible. Hence, we can multiply both sides of the equation $\mathbf{Aw} = \mathbf{v}$ by $\mathbf{A}^{-1}$ to get $\mathbf{A}^{-1}\mathbf{Aw} = \mathbf{A}^{-1}\mathbf{v}$, which simplifies as $\mathbf{w} = \boxed{\mathbf{A}^{-1}\mathbf{v}}$. So, $\boxed{\text{yes}}$, there must be a vector $\mathbf{w}$ such that $\mathbf{Aw} = \mathbf{v}$.

(b) No, even if $\mathbf{A}$ is singular, it is still possible that there is a vector $\mathbf{w}$ such that $\mathbf{Aw} = \mathbf{v}$. For example, if $\mathbf{A} = \begin{pmatrix} 0 & 0 \\ 0 & 0 \end{pmatrix}$ and $\mathbf{v} = \begin{pmatrix} 0 \\ 0 \end{pmatrix}$, then any vector $\mathbf{w}$ satisfies the equation $\mathbf{Aw} = \mathbf{v}$. If $\mathbf{A}$ is singular, all we can conclude is that the equation $\mathbf{Aw} = \mathbf{v}$ does not necessarily have a solution $\mathbf{w}$.

10.6.4 We have that $\det(\mathbf{M})\det(\mathbf{M}^{-1}) = \det(\mathbf{M}\mathbf{M}^{-1}) = \det(\mathbf{I}) = 1$. However, since all the entries of $\mathbf{M}$ and $\mathbf{M}^{-1}$ are integers, $\det(\mathbf{M})$ and $\det(\mathbf{M}^{-1})$ must be integers as well. The only integers that have 1 as a multiple are $\boxed{1}$ and $\boxed{-1}$.

If $\mathbf{M} = \mathbf{I}$, then $\mathbf{M}^{-1} = \mathbf{I}$ and $\det(\mathbf{M}) = 1$, and if $\mathbf{M} = \begin{pmatrix} 1 & 0 \\ 0 & -1 \end{pmatrix}$, then

$$\mathbf{M}^{-1} = \frac{1}{1 \cdot (-1) - 0 \cdot 0} \begin{pmatrix} -1 & 0 \\ 0 & 1 \end{pmatrix} = \frac{1}{-1} \begin{pmatrix} -1 & 0 \\ 0 & 1 \end{pmatrix} = \begin{pmatrix} 1 & 0 \\ 0 & -1 \end{pmatrix},$$

and $\det(\mathbf{M}) = 1 \cdot (-1) - 0 \cdot 0 = -1$, so both 1 and -1 are possible values of $\det(\mathbf{M})$.

10.6.5 Let $\mathbf{A} = \mathbf{B} = \mathbf{I} = \begin{pmatrix} 1 & 0 \\ 0 & 1 \end{pmatrix}$, so $\mathbf{A} + \mathbf{B} = \begin{pmatrix} 1 & 0 \\ 0 & 1 \end{pmatrix} + \begin{pmatrix} 1 & 0 \\ 0 & 1 \end{pmatrix} = \begin{pmatrix} 2 & 0 \\ 0 & 2 \end{pmatrix}$. Then $\mathbf{A}^{-1} = \mathbf{B}^{-1} = \mathbf{I}^{-1} = \mathbf{I}$, so $\mathbf{B}^{-1} + \mathbf{A}^{-1} = 2\mathbf{I} = \begin{pmatrix} 2 & 0 \\ 0 & 2 \end{pmatrix}$, and

$$(\mathbf{A} + \mathbf{B})^{-1} = \begin{pmatrix} 2 & 0 \\ 0 & 2 \end{pmatrix}^{-1} = \frac{1}{2 \cdot 2 - 0 \cdot 0} \begin{pmatrix} 2 & 0 \\ 0 & 2 \end{pmatrix} = \frac{1}{4} \begin{pmatrix} 2 & 0 \\ 0 & 2 \end{pmatrix} = \begin{pmatrix} \frac{1}{2} & 0 \\ 0 & \frac{1}{2} \end{pmatrix}.$$

Hence, $(\mathbf{A} + \mathbf{B})^{-1}$ need not equal $\mathbf{B}^{-1} + \mathbf{A}^{-1}$. (In fact, it's possible that $\mathbf{A} + \mathbf{B}$ is not even invertible! For example, consider the case $\mathbf{A} = \mathbf{I}$ and $\mathbf{B} = -\mathbf{I}$. Then, both $\mathbf{A}$ and $\mathbf{B}$ are invertible, but $\mathbf{A} + \mathbf{B}$ is not!)

10.6.6 For any complex number z, if we multiply z by $a + bi$, and then divide the product by $a + bi$, the result is z. Correspondingly, if we multiply a vector $\begin{pmatrix} x \\ y \end{pmatrix}$ by the matrix $\begin{pmatrix} a & -b \\ b & a \end{pmatrix}$ and then multiply the result by the matrix $\mathbf{M}$ that corresponds to dividing by $a + bi$, we should get $\begin{pmatrix} x \\ y \end{pmatrix}$ back in the end. In other words, multiplying by $\mathbf{M}$ "undoes" multiplication by $\begin{pmatrix} a & -b \\ b & a \end{pmatrix}$, which means that the desired matrix is the inverse of $\begin{pmatrix} a & -b \\ b & a \end{pmatrix}$. We have

$$\begin{pmatrix} a & -b \\ b & a \end{pmatrix}^{-1} = \frac{1}{a \cdot a - (-b) \cdot b} \begin{pmatrix} a & b \\ -b & a \end{pmatrix} = \frac{1}{a^2 + b^2} \begin{pmatrix} a & b \\ -b & a \end{pmatrix} = \boxed{\begin{pmatrix} \frac{a}{a^2+b^2} & \frac{b}{a^2+b^2} \\ -\frac{b}{a^2+b^2} & \frac{a}{a^2+b^2} \end{pmatrix}}.$$

(Note that the determinant of our matrix for multiplying by $a + bi$ is $a^2 + b^2 = |a + bi|^2$, which is nonzero if $a + bi \neq 0$, so this matrix is indeed invertible.)

Review Problems

10.42 We have

$$\begin{pmatrix} a & 4 \\ 7 & 2 \end{pmatrix} + c \begin{pmatrix} 6 & b \\ -3 & -1 \end{pmatrix} = \begin{pmatrix} a & 4 \\ 7 & 2 \end{pmatrix} + \begin{pmatrix} 6c & bc \\ -3c & -c \end{pmatrix} = \begin{pmatrix} a + 6c & bc + 4 \\ -3c + 7 & -c + 2 \end{pmatrix},$$

so a, b, and c must satisfy the equations $a + 6c = 32$, $bc + 4 = 9$, $-3c + 7 = -8$, and $-c + 2 = -3$.

From the equation $-c + 2 = -3$, we have $c = 5$. This value of c also satisfies the equation $-3c + 7 = -8$. Then from the equation $a + 6c = 32$, we have $a = 32 - 6c = 32 - 6 \cdot 5 = 2$, and from the equation $bc + 4 = 9$, we have $5b + 4 = 9$, so $b = (9 - 4)/5 = 1$. To summarize, all four equations are satisfied by $\boxed{a = 2, b = 1, \text{ and } c = 5}$.

10.43 We have

$$\mathbf{MN} = \begin{pmatrix} 2 & 3 \\ -7 & 0 \end{pmatrix} \begin{pmatrix} -3 & -1 \\ 5 & 2 \end{pmatrix} = \begin{pmatrix} 2 \cdot (-3) + 3 \cdot 5 & 2 \cdot (-1) + 3 \cdot 2 \\ (-7) \cdot (-3) + 0 \cdot 5 & (-7) \cdot (-1) + 0 \cdot 2 \end{pmatrix} = \boxed{\begin{pmatrix} 9 & 4 \\ 21 & 7 \end{pmatrix}},$$

and

$$\mathbf{NM} = \begin{pmatrix} -3 & -1 \\ 5 & 2 \end{pmatrix}\begin{pmatrix} 2 & 3 \\ -7 & 0 \end{pmatrix} = \begin{pmatrix} (-3)\cdot 2 + (-1)\cdot(-7) & (-3)\cdot 3 + (-1)\cdot 0 \\ 5\cdot 2 + 2\cdot(-7) & 5\cdot 3 + 2\cdot 0 \end{pmatrix} = \boxed{\begin{pmatrix} 1 & -9 \\ -4 & 15 \end{pmatrix}}.$$

10.44 *Solution 1: Use the determinant.* Let the matrix be $\begin{pmatrix} a & b \\ c & d \end{pmatrix}$. We know that the rows of the matrix are linearly dependent if and only if the determinant is 0, i.e. $ad - bc = 0$.

Now consider the matrix $\begin{pmatrix} a & c \\ b & d \end{pmatrix}$. The rows of this matrix are linearly dependent if and only if its determinant is 0, i.e. $ad - bc = 0$. Furthermore, the rows of this matrix are the same as the columns of our original matrix. Hence, the rows of a matrix are linearly dependent if and only if the columns of the matrix are linearly dependent.

Solution 2: The longer way. Let the matrix be $\begin{pmatrix} a & b \\ c & d \end{pmatrix}$. If the rows are linearly dependent and no entries are 0, then $a = ct$ and $b = dt$ for some constant t. Then, we have $\frac{a}{c} = \frac{b}{d}$, so $\frac{a}{b} = \frac{c}{d}$. Letting these ratios equal k, we have $a = bk$ and $c = dk$, so the first column is k times the second, which means the column vectors are linearly dependent. If one entry is 0, then we have two cases to consider: either the other entry in the row is 0 or the corresponding entry in the other row is 0, because the rows are linearly dependent. A little more casework shows that in both cases one column is a multiple of the other. (The casework caused by 0 makes this the *much* longer way.)

Similarly, if the columns are linearly dependent and no entries are 0, then $a = bk$ and $c = dk$ for some constant k, from which we find $\frac{a}{b} = \frac{c}{d}$. Rearranging this gives $\frac{a}{c} = \frac{b}{d}$, so one row is a multiple of the other, meaning the rows are linearly dependent. As before, if one entry is 0, we can work through some more simple cases to show that the rows are linearly dependent.

10.45 Suppose that such a matrix $\mathbf{F} = \begin{pmatrix} x & y \\ z & w \end{pmatrix}$ exists. Then

$$\mathbf{FI} = \begin{pmatrix} x & y \\ z & w \end{pmatrix}\begin{pmatrix} 1 & 0 \\ 0 & 1 \end{pmatrix} = \begin{pmatrix} 0 & 1 \\ 1 & 0 \end{pmatrix}.$$

But

$$\mathbf{FI} = \mathbf{F} = \begin{pmatrix} x & y \\ z & w \end{pmatrix},$$

so $x = 0$, $y = 1$, $z = 1$, and $w = 0$. Therefore, if such an $\mathbf{F}$ exists, it must be $\mathbf{F} = \begin{pmatrix} 0 & 1 \\ 1 & 0 \end{pmatrix}$. But for an arbitrary matrix $\mathbf{A} = \begin{pmatrix} a & b \\ c & d \end{pmatrix}$, we have

$$\mathbf{FA} = \begin{pmatrix} 0 & 1 \\ 1 & 0 \end{pmatrix}\begin{pmatrix} a & b \\ c & d \end{pmatrix} = \begin{pmatrix} 0\cdot a + 1\cdot c & 0\cdot b + 1\cdot d \\ 1\cdot a + 0\cdot c & 1\cdot b + 0\cdot d \end{pmatrix} = \begin{pmatrix} c & d \\ a & b \end{pmatrix},$$

but $\begin{pmatrix} c & d \\ a & b \end{pmatrix}$ is not the matrix formed when the columns of $\mathbf{A}$ are swapped. Hence, no such matrix $\mathbf{F}$ exists.

10.46 From the given information, we have

$$\mathbf{A}(3\mathbf{v} - 4\mathbf{w}) = \mathbf{A}(3\mathbf{v}) - \mathbf{A}(4\mathbf{w}) = 3\mathbf{Av} - 4\mathbf{Aw} = 3\begin{pmatrix} 3 \\ -2 \end{pmatrix} - 4\begin{pmatrix} -7 \\ 4 \end{pmatrix} = \begin{pmatrix} 9 \\ -6 \end{pmatrix} - \begin{pmatrix} -28 \\ 16 \end{pmatrix} = \boxed{\begin{pmatrix} 37 \\ -22 \end{pmatrix}}.$$

10.47

(a) Let $\mathbf{A}_1 = \begin{pmatrix} 1 & 0 \\ 0 & 0 \end{pmatrix}$. Then

$$\mathbf{A}_1^2 = \begin{pmatrix} 1 & 0 \\ 0 & 0 \end{pmatrix} \begin{pmatrix} 1 & 0 \\ 0 & 0 \end{pmatrix} = \begin{pmatrix} 1 \cdot 1 + 0 \cdot 0 & 1 \cdot 0 + 0 \cdot 0 \\ 0 \cdot 1 + 0 \cdot 0 & 0 \cdot 0 + 0 \cdot 0 \end{pmatrix} = \begin{pmatrix} 1 & 0 \\ 0 & 0 \end{pmatrix} = \mathbf{A}_1.$$

Also, let $\mathbf{A}_2 = \begin{pmatrix} 0 & 0 \\ 0 & 1 \end{pmatrix}$. Then

$$\mathbf{A}_2^2 = \begin{pmatrix} 0 & 0 \\ 0 & 1 \end{pmatrix} \begin{pmatrix} 0 & 0 \\ 0 & 1 \end{pmatrix} = \begin{pmatrix} 0 \cdot 0 + 0 \cdot 0 & 0 \cdot 0 + 0 \cdot 1 \\ 0 \cdot 0 + 1 \cdot 0 & 0 \cdot 0 + 1 \cdot 1 \end{pmatrix} = \begin{pmatrix} 0 & 0 \\ 0 & 1 \end{pmatrix} = \mathbf{A}_2.$$

(b) Let $\mathbf{A} = \begin{pmatrix} 0 & 1 \\ 1 & 0 \end{pmatrix}$. Then

$$\mathbf{A}^2 = \begin{pmatrix} 0 & 1 \\ 1 & 0 \end{pmatrix} \begin{pmatrix} 0 & 1 \\ 1 & 0 \end{pmatrix} = \begin{pmatrix} 0 \cdot 0 + 1 \cdot 1 & 0 \cdot 1 + 1 \cdot 0 \\ 1 \cdot 0 + 0 \cdot 1 & 1 \cdot 1 + 0 \cdot 0 \end{pmatrix} = \begin{pmatrix} 1 & 0 \\ 0 & 1 \end{pmatrix} = \mathbf{I},$$

so $\mathbf{A}^2 \neq \mathbf{A}$ and $\mathbf{A}^3 = \mathbf{A}^2 \cdot \mathbf{A} = \mathbf{IA} = \mathbf{A}$.

There are many other matrices that satisfy this part. Any matrix such that $\mathbf{A}^2 = \mathbf{I}$ and $\mathbf{A} \neq \mathbf{I}$ will do. For example, a matrix $\mathbf{A}$ will satisfy $\mathbf{A}^2 = \mathbf{I}$ if $\mathbf{A}\mathbf{v}$ is a reflection of $\mathbf{v}$ over some line for any vector $\mathbf{v}$. This is because multiplying $\mathbf{A}\mathbf{v}$ by $\mathbf{A}$ (to give $\mathbf{A}^2\mathbf{v}$) just reflects $\mathbf{A}\mathbf{v}$ back over the line, which results in $\mathbf{v}$.

10.48

(a) The reflection of $\begin{pmatrix} x \\ y \end{pmatrix}$ over the y-axis is $\begin{pmatrix} -x \\ y \end{pmatrix}$, so we seek a matrix $\mathbf{A} = \begin{pmatrix} a & b \\ c & d \end{pmatrix}$ such that

$$\begin{pmatrix} a & b \\ c & d \end{pmatrix} \begin{pmatrix} x \\ y \end{pmatrix} = \begin{pmatrix} -x \\ y \end{pmatrix}$$

for all x, y. We can take $a = -1$, $b = 0$, $c = 0$, and $d = 1$, so we can take $\mathbf{A} = \boxed{\begin{pmatrix} -1 & 0 \\ 0 & 1 \end{pmatrix}}$.

(b) The reflection of $\begin{pmatrix} x \\ y \end{pmatrix}$ over the graph of $x = y$ is $\begin{pmatrix} y \\ x \end{pmatrix}$, so we seek a matrix $\mathbf{B} = \begin{pmatrix} a & b \\ c & d \end{pmatrix}$ such that

$$\begin{pmatrix} a & b \\ c & d \end{pmatrix} \begin{pmatrix} x \\ y \end{pmatrix} = \begin{pmatrix} y \\ x \end{pmatrix}$$

for all x, y. We can take $a = 0$, $b = 1$, $c = 1$, and $d = 0$, so we can take $\mathbf{B} = \boxed{\begin{pmatrix} 0 & 1 \\ 1 & 0 \end{pmatrix}}$.

10.49 A clockwise rotation by θ is equivalent to a counterclockwise rotation by $-\theta$. We can therefore substitute $-\theta$ as the angle into the rotation matrix we found in the text, to find

$$\mathbf{R} = \begin{pmatrix} \cos(-\theta) & -\sin(-\theta) \\ \sin(-\theta) & \cos(-\theta) \end{pmatrix} = \boxed{\begin{pmatrix} \cos\theta & \sin\theta \\ -\sin\theta & \cos\theta \end{pmatrix}}.$$

10.50 We see that

$$\mathbf{T}^2 = \begin{pmatrix} \cos\theta & -\sin\theta \\ \sin\theta & \cos\theta \end{pmatrix} \begin{pmatrix} \cos\theta & -\sin\theta \\ \sin\theta & \cos\theta \end{pmatrix} = \begin{pmatrix} \cos^2\theta - \sin^2\theta & -2\sin\theta\cos\theta \\ 2\sin\theta\cos\theta & \cos^2\theta - \sin^2\theta \end{pmatrix}.$$

But by the double angle formulas for cosine and sine, $\cos 2\theta = \cos^2\theta - \sin^2\theta$ and $\sin 2\theta = 2\sin\theta\cos\theta$, so

$$\mathbf{T}^2 = \begin{pmatrix} \cos 2\theta & -\sin 2\theta \\ \sin 2\theta & \cos 2\theta \end{pmatrix}.$$

Thus, $\mathbf{T}^2$ corresponds to a rotation counterclockwise about the origin by an angle of 2θ.

We also have

$$\mathbf{T}^3 = \mathbf{T}\mathbf{T}^2 = \begin{pmatrix} \cos\theta & -\sin\theta \\ \sin\theta & \cos\theta \end{pmatrix} \begin{pmatrix} \cos 2\theta & -\sin 2\theta \\ \sin 2\theta & \cos 2\theta \end{pmatrix} = \begin{pmatrix} \cos\theta\cos 2\theta - \sin\theta\sin 2\theta & -\cos\theta\sin 2\theta - \sin\theta\cos 2\theta \\ \sin\theta\cos 2\theta + \cos\theta\sin 2\theta & -\sin\theta\sin 2\theta + \cos\theta\cos 2\theta \end{pmatrix}$$

$$= \begin{pmatrix} \cos(\theta + 2\theta) & -\sin(\theta + 2\theta) \\ \sin(\theta + 2\theta) & \cos(\theta + 2\theta) \end{pmatrix} = \begin{pmatrix} \cos 3\theta & -\sin 3\theta \\ \sin 3\theta & \cos 3\theta \end{pmatrix}.$$

10.51 We see that

$$\mathbf{M} = \begin{pmatrix} \sqrt{2} & -\sqrt{2} \\ \sqrt{2} & \sqrt{2} \end{pmatrix} = 2\begin{pmatrix} \frac{\sqrt{2}}{2} & -\frac{\sqrt{2}}{2} \\ \frac{\sqrt{2}}{2} & \frac{\sqrt{2}}{2} \end{pmatrix} = 2\begin{pmatrix} \cos 45^\circ & -\sin 45^\circ \\ \sin 45^\circ & \cos 45^\circ \end{pmatrix}.$$

The matrix $\begin{pmatrix} \cos 45^\circ & -\sin 45^\circ \\ \sin 45^\circ & \cos 45^\circ \end{pmatrix}$ corresponds to a 45° counterclockwise rotation about the origin, so

$$\begin{pmatrix} \cos 45^\circ & -\sin 45^\circ \\ \sin 45^\circ & \cos 45^\circ \end{pmatrix}^{10} = \begin{pmatrix} \cos(10\cdot 45^\circ) & -\sin(10\cdot 45^\circ) \\ \sin(10\cdot 45^\circ) & \cos(10\cdot 45^\circ) \end{pmatrix} = \begin{pmatrix} \cos 450^\circ & -\sin 450^\circ \\ \sin 450^\circ & \cos 450^\circ \end{pmatrix} = \begin{pmatrix} 0 & -1 \\ 1 & 0 \end{pmatrix}.$$

Therefore,

$$\mathbf{M}^{10} = \left(2\begin{pmatrix} \cos 45^\circ & -\sin 45^\circ \\ \sin 45^\circ & \cos 45^\circ \end{pmatrix}\right)^{10} = 2^{10}\begin{pmatrix} \cos 45^\circ & -\sin 45^\circ \\ \sin 45^\circ & \cos 45^\circ \end{pmatrix}^{10} = 1024\begin{pmatrix} 0 & -1 \\ 1 & 0 \end{pmatrix} = \boxed{\begin{pmatrix} 0 & -1024 \\ 1024 & 0 \end{pmatrix}}.$$

10.52 To recover point P, we can take the point $(-2 - \sqrt{3}, 1 - 2\sqrt{3})$, rotate 60° clockwise about the origin, and then reflect over the y-axis.

Rotating 60° clockwise about the origin, we get

$$\begin{pmatrix} \cos(-60^\circ) & -\sin(-60^\circ) \\ \sin(-60^\circ) & \cos(-60^\circ) \end{pmatrix}\begin{pmatrix} -2 - \sqrt{3} \\ 1 - 2\sqrt{3} \end{pmatrix} = \begin{pmatrix} \frac{1}{2} & \frac{\sqrt{3}}{2} \\ -\frac{\sqrt{3}}{2} & \frac{1}{2} \end{pmatrix}\begin{pmatrix} -2 - \sqrt{3} \\ 1 - 2\sqrt{3} \end{pmatrix} = \begin{pmatrix} \frac{1}{2}\cdot(-2 - \sqrt{3}) + \frac{\sqrt{3}}{2}\cdot(1 - 2\sqrt{3}) \\ (-\frac{\sqrt{3}}{2})\cdot(-2 - \sqrt{3}) + \frac{1}{2}\cdot(1 - 2\sqrt{3}) \end{pmatrix} = \begin{pmatrix} -4 \\ 2 \end{pmatrix},$$

and the reflection of the point $(-4, 2)$ over the y-axis is $(4, 2)$. Therefore, $P = \boxed{(4, 2)}$.

10.53 For any linear function f, we have $f(\mathbf{0}) = f(0\cdot\mathbf{0}) = 0\cdot f(\mathbf{0}) = \mathbf{0}$.

10.54

(a) $\begin{vmatrix} 7 & -4 \\ -2 & 7 \end{vmatrix} = 7\cdot 7 - (-4)\cdot(-2) = \boxed{41}$.

(b) $\begin{vmatrix} 0.27 & -175 \\ -0.15 & 95 \end{vmatrix} = 0.27\cdot 95 - (-175)\cdot(-0.15) = \boxed{-0.6}$.

10.55

(a) Let $\mathbf{A}$ and $\mathbf{B}$ be two singular matrices, so $\det(\mathbf{A}) = \det(\mathbf{B}) = 0$. Then $\det(\mathbf{AB}) = \det(\mathbf{A})\det(\mathbf{B}) = 0$, so the matrix $\mathbf{AB}$ is singular. Hence, the product of two singular matrices is also singular.

(b) Let $\mathbf{A}$ and $\mathbf{B}$ be two nonsingular matrices, so $\det(\mathbf{A}) \neq 0$ and $\det(\mathbf{B}) \neq 0$. Then $\det(\mathbf{AB}) = \det(\mathbf{A})\det(\mathbf{B}) \neq 0$, so the matrix $\mathbf{AB}$ is nonsingular. Hence, the product of two nonsingular matrices is also nonsingular.

(c) Let $\mathbf{A}$ be a singular matrix and let $\mathbf{B}$ be a nonsingular matrix, so $\det(\mathbf{A}) = 0$ and $\det(\mathbf{B}) \neq 0$. Then $\det(\mathbf{AB}) = \det(\mathbf{A}) \det(\mathbf{B}) = 0$, so the matrix $\mathbf{AB}$ is singular. Hence, the product of a singular matrix and a nonsingular matrix is always singular.

10.56 Let $\mathbf{A} = \mathbf{I} = \begin{pmatrix} 1 & 0 \\ 0 & 1 \end{pmatrix}$. Then $\mathbf{C} = \begin{pmatrix} 1 & k \\ j & 1 \end{pmatrix}$, so $\det(\mathbf{C}) = \det \begin{pmatrix} 1 & k \\ j & 1 \end{pmatrix} = 1 - jk$. However, $\det(\mathbf{A}) = \det(\mathbf{I}) = 1$, so $\det(\mathbf{A})$ and $\det(\mathbf{C})$ are not necessarily equal.

10.57 Recall that the determinant of a matrix is the signed area of the parallelogram formed by the column vectors of the matrix. Swapping the columns of the matrix reverses the orientation of this parallelogram, and reversing the orientation of the parallelogram changes the sign of the signed area.

10.58

(a) $\begin{pmatrix} -2 & 3 \\ -2 & 8 \end{pmatrix}^{-1} = \dfrac{1}{(-2) \cdot 8 - 3 \cdot (-2)} \begin{pmatrix} 8 & -3 \\ 2 & -2 \end{pmatrix} = \dfrac{1}{-10} \begin{pmatrix} 8 & -3 \\ 2 & -2 \end{pmatrix} = \boxed{\begin{pmatrix} -\frac{4}{5} & \frac{3}{10} \\ -\frac{1}{5} & \frac{1}{5} \end{pmatrix}}$.

(b) $\begin{pmatrix} 0 & -3 \\ -2 & 0 \end{pmatrix}^{-1} = \dfrac{1}{0 \cdot 0 - (-3) \cdot (-2)} \begin{pmatrix} 0 & 3 \\ 2 & 0 \end{pmatrix} = \dfrac{1}{-6} \begin{pmatrix} 0 & 3 \\ 2 & 0 \end{pmatrix} = \boxed{\begin{pmatrix} 0 & -\frac{1}{2} \\ -\frac{1}{3} & 0 \end{pmatrix}}$.

(c) Since $\det \begin{pmatrix} 6 & 12 \\ -2 & -4 \end{pmatrix} = 6 \cdot (-4) - 12 \cdot (-2) = 0$, the matrix $\begin{pmatrix} 6 & 12 \\ -2 & -4 \end{pmatrix}$ does not have an inverse.

10.59 Since $\mathbf{A}^{-1}\mathbf{A} = \mathbf{I}$, the equation $\mathbf{A}^{-1} = \mathbf{A}$ tells us that $\mathbf{A}^2 = \mathbf{I}$, so $\det(\mathbf{A}^2) = \det(\mathbf{I}) = 1$. But $\det(\mathbf{A}^2) = [\det(\mathbf{A})]^2 = 1$, so $\det \mathbf{A} = 1$ or -1.

If $\mathbf{A} = \mathbf{I}$, then $\mathbf{A}^{-1} = \mathbf{I}^{-1} = \mathbf{I} = \mathbf{A}$, and $\det(\mathbf{A}) = \det(\mathbf{I}) = 1$. If $\mathbf{A} = \begin{pmatrix} 1 & 0 \\ 0 & -1 \end{pmatrix}$, then

$$\mathbf{A}^2 = \begin{pmatrix} 1 & 0 \\ 0 & -1 \end{pmatrix} \begin{pmatrix} 1 & 0 \\ 0 & -1 \end{pmatrix} = \begin{pmatrix} 1 \cdot 1 + 0 \cdot 0 & 1 \cdot 0 + 0 \cdot (-1) \\ 0 \cdot 1 + (-1) \cdot 0 & 0 \cdot 0 + (-1) \cdot (-1) \end{pmatrix} = \begin{pmatrix} 1 & 0 \\ 0 & 1 \end{pmatrix} = \mathbf{I},$$

so $\mathbf{A}^{-1} = \mathbf{A}$, and $\det(\mathbf{A}) = \det \begin{pmatrix} 1 & 0 \\ 0 & -1 \end{pmatrix} = 1 \cdot (-1) - 0 \cdot 0 = -1$. Hence, values of $\boxed{1}$ and $\boxed{-1}$ are both possible.

10.60

(a) Let $\mathbf{A} = \begin{pmatrix} a & b \\ c & d \end{pmatrix}$ and $\mathbf{B} = \begin{pmatrix} x & y \\ z & w \end{pmatrix}$, so $\mathbf{A} + \mathbf{B} = \begin{pmatrix} a & b \\ c & d \end{pmatrix} + \begin{pmatrix} x & y \\ z & w \end{pmatrix} = \begin{pmatrix} a + x & b + y \\ c + z & d + w \end{pmatrix}$, and

$$\mathbf{A}^T = \begin{pmatrix} a & b \\ c & d \end{pmatrix}^T = \begin{pmatrix} a & c \\ b & d \end{pmatrix}, \quad \mathbf{B}^T = \begin{pmatrix} x & y \\ z & w \end{pmatrix}^T = \begin{pmatrix} x & z \\ y & w \end{pmatrix},$$

so

$$\mathbf{A}^T + \mathbf{B}^T = \begin{pmatrix} a & c \\ b & d \end{pmatrix} + \begin{pmatrix} x & z \\ y & w \end{pmatrix} = \begin{pmatrix} a + x & c + z \\ b + y & d + w \end{pmatrix},$$

and

$$(\mathbf{A} + \mathbf{B})^T = \begin{pmatrix} a + x & b + y \\ c + z & d + w \end{pmatrix}^T = \begin{pmatrix} a + x & c + z \\ b + y & d + w \end{pmatrix}.$$

Hence, $(\mathbf{A} + \mathbf{B})^T = \mathbf{A}^T + \mathbf{B}^T$.

(b) Let $\mathbf{A} = \begin{pmatrix} a & b \\ c & d \end{pmatrix}$. Then $\det(\mathbf{A}^T) = \det \begin{pmatrix} a & c \\ b & d \end{pmatrix} = ad - bc = \det(\mathbf{A})$. Furthermore, $\mathbf{A}$ is invertible if and only if $\det(\mathbf{A}) \neq 0$. Therefore, if $\mathbf{A}$ is invertible, then $\mathbf{A}^T$ is invertible.

(c)   Let $\mathbf{A} = \begin{pmatrix} a & b \\ c & d \end{pmatrix}$, where $\mathbf{A}^T = \begin{pmatrix} a & c \\ b & d \end{pmatrix}$ is invertible. Then

$$(\mathbf{A}^T)^{-1} = \begin{pmatrix} a & c \\ b & d \end{pmatrix}^{-1} = \frac{1}{ad - bc} \begin{pmatrix} d & -c \\ -b & a \end{pmatrix}.$$

Also,

$$\mathbf{A}^{-1} = \begin{pmatrix} a & b \\ c & d \end{pmatrix}^{-1} = \frac{1}{ad - bc} \begin{pmatrix} d & -b \\ -c & a \end{pmatrix},$$

so

$$(\mathbf{A}^{-1})^T = \frac{1}{ad - bc} \begin{pmatrix} d & -c \\ -b & a \end{pmatrix}.$$

Hence, $(\mathbf{A}^T)^{-1} = (\mathbf{A}^{-1})^T$.

(d)   Given $\mathbf{R} = \begin{pmatrix} 2 & 3 \\ -1 & 4 \end{pmatrix}$ and $\mathbf{S} = \begin{pmatrix} -1 & 5 \\ -6 & 3 \end{pmatrix}$, we have $\mathbf{R}^T = \begin{pmatrix} 2 & -1 \\ 3 & 4 \end{pmatrix}$ and $\mathbf{S}^T = \begin{pmatrix} -1 & -6 \\ 5 & 3 \end{pmatrix}$, so

$$\mathbf{RS} = \begin{pmatrix} 2 & 3 \\ -1 & 4 \end{pmatrix} \begin{pmatrix} -1 & 5 \\ -6 & 3 \end{pmatrix} = \begin{pmatrix} 2 \cdot (-1) + 3 \cdot (-6) & 2 \cdot 5 + 3 \cdot 3 \\ (-1) \cdot (-1) + 4 \cdot (-6) & (-1) \cdot 5 + 4 \cdot 3 \end{pmatrix} = \boxed{\begin{pmatrix} -20 & 19 \\ -23 & 7 \end{pmatrix}},$$

$$(\mathbf{RS})^T = \begin{pmatrix} -20 & 19 \\ -23 & 7 \end{pmatrix}^T = \boxed{\begin{pmatrix} -20 & -23 \\ 19 & 7 \end{pmatrix}},$$

$$\mathbf{R}^T\mathbf{S}^T = \begin{pmatrix} 2 & -1 \\ 3 & 4 \end{pmatrix} \begin{pmatrix} -1 & -6 \\ 5 & 3 \end{pmatrix} = \begin{pmatrix} 2 \cdot (-1) + (-1) \cdot 5 & 2 \cdot (-6) + (-1) \cdot 3 \\ 3 \cdot (-1) + 4 \cdot 5 & 3 \cdot (-6) + 4 \cdot 3 \end{pmatrix} = \boxed{\begin{pmatrix} -7 & -15 \\ 17 & -6 \end{pmatrix}},$$

$$\mathbf{S}^T\mathbf{R}^T = \begin{pmatrix} -1 & -6 \\ 5 & 3 \end{pmatrix} \begin{pmatrix} 2 & -1 \\ 3 & 4 \end{pmatrix} = \begin{pmatrix} (-1) \cdot 2 + (-6) \cdot 3 & (-1) \cdot (-1) + (-6) \cdot 4 \\ 5 \cdot 2 + 3 \cdot 3 & 5 \cdot (-1) + 3 \cdot 4 \end{pmatrix} = \boxed{\begin{pmatrix} -20 & -23 \\ 19 & 7 \end{pmatrix}}.$$

Note that $(\mathbf{RS})^T = \mathbf{S}^T\mathbf{R}^T$. We claim this is true in general.

Let $\mathbf{R} = \begin{pmatrix} a & b \\ c & d \end{pmatrix}$ and $\mathbf{S} = \begin{pmatrix} x & y \\ z & w \end{pmatrix}$, so

$$\mathbf{RS} = \begin{pmatrix} a & b \\ c & d \end{pmatrix} \begin{pmatrix} x & y \\ z & w \end{pmatrix} = \begin{pmatrix} ax + bz & ay + bw \\ cx + dz & cy + dw \end{pmatrix},$$

and $(\mathbf{RS})^T = \begin{pmatrix} ax + bz & cx + dz \\ ay + bw & cy + dw \end{pmatrix}$. Also, $\mathbf{R}^T = \begin{pmatrix} a & c \\ b & d \end{pmatrix}$ and $\mathbf{S}^T = \begin{pmatrix} x & z \\ y & w \end{pmatrix}$, so

$$\mathbf{S}^T\mathbf{R}^T = \begin{pmatrix} x & z \\ y & w \end{pmatrix} \begin{pmatrix} a & c \\ b & d \end{pmatrix} = \begin{pmatrix} ax + bz & cx + dz \\ ay + bw & cy + dw \end{pmatrix}.$$

Hence, $(\mathbf{RS})^T = \mathbf{S}^T\mathbf{R}^T$ for all matrices $\mathbf{R}$ and $\mathbf{S}$.

Challenge Problems

10.61   Let $\mathbf{A} = \begin{pmatrix} 1 & 0 \\ 0 & 2 \end{pmatrix}$ then $\mathbf{B} = \begin{pmatrix} 2 & 0 \\ 0 & 1 \end{pmatrix}$. Then $\mathbf{AB} = \mathbf{BA} = \begin{pmatrix} 2 & 0 \\ 0 & 2 \end{pmatrix}$. To find examples of $\mathbf{A}$ and $\mathbf{B}$ where none of the entries are equal to 0, we can use rotation matrices. Let

$$\mathbf{A} = \begin{pmatrix} \cos 30° & -\sin 30° \\ \sin 30° & \cos 30° \end{pmatrix} = \begin{pmatrix} \frac{\sqrt{3}}{2} & -\frac{1}{2} \\ \frac{1}{2} & \frac{\sqrt{3}}{2} \end{pmatrix} \quad \text{and} \quad \mathbf{B} = \begin{pmatrix} \cos 60° & -\sin 60° \\ \sin 60° & \cos 60° \end{pmatrix} = \begin{pmatrix} \frac{1}{2} & -\frac{\sqrt{3}}{2} \\ \frac{\sqrt{3}}{2} & \frac{1}{2} \end{pmatrix}.$$

Then $\mathbf{AB} = \mathbf{BA} = \begin{pmatrix} \cos 90° & -\sin 90° \\ \sin 90° & \cos 90° \end{pmatrix} = \begin{pmatrix} 0 & -1 \\ 1 & 0 \end{pmatrix}.$

10.62 If $\mathbf{A}^6 = \mathbf{A}$, then $\det(\mathbf{A}^6) = \det(\mathbf{A})$. But $\det(\mathbf{A}^6) = [\det(\mathbf{A})]^6$, so $[\det(\mathbf{A})]^6 = \det(\mathbf{A})$.

Let $x = \det(\mathbf{A})$, so $x^6 = x$. Then $x^6 - x = 0$, so $x(x^5 - 1) = 0$. Therefore, either $x = 0$ or x is a fifth root of unity. The only fifth root of unity that is real is 1, so if all the entries of $\mathbf{A}$ are real, then $\det(\mathbf{A})$ must be equal to $\boxed{0 \text{ or } 1}$.

If $\mathbf{A} = \begin{pmatrix} 0 & 0 \\ 0 & 0 \end{pmatrix}$, then $\mathbf{A}^6 = \mathbf{A}$, and $\det(\mathbf{A}) = 0$, and if $\mathbf{A} = \mathbf{I}$, then $\mathbf{A}^6 = \mathbf{A} = \mathbf{I}$, and $\det(\mathbf{A}) = 1$. Hence, both 0 and 1 are possible values of $\det(\mathbf{A})$.

If the entries of $\mathbf{A}$ can be nonreal, then in addition, $\det(\mathbf{A})$ can be any fifth root of unity. Let $\mathbf{A} = \begin{pmatrix} \omega & 0 \\ 0 & 1 \end{pmatrix}$, where ω is a fifth root of unity. Then

$$\mathbf{A}^6 = \begin{pmatrix} \omega^6 & 0 \\ 0 & 1 \end{pmatrix} = \begin{pmatrix} \omega & 0 \\ 0 & 1 \end{pmatrix} = \mathbf{A},$$

and $\det(\mathbf{A}) = \omega$.

10.63 If $\mathbf{A}^4 = \mathbf{A}$, then $\mathbf{A}^3 = \mathbf{I}$. So, we seek a matrix $\mathbf{A}$ such that multiplying any vector by $\mathbf{A}$ three times leaves the vector unchanged. Rotation by 120° is a transformation that leaves a vector unchanged if applied to the vector three times. We expect that the rotation matrix for a 120° counterclockwise rotation will fit the problem. Let's check. We let

$$\mathbf{A} = \begin{pmatrix} \cos 120° & -\sin 120° \\ \sin 120° & \cos 120° \end{pmatrix} = \begin{pmatrix} -\frac{1}{2} & -\frac{\sqrt{3}}{2} \\ \frac{\sqrt{3}}{2} & -\frac{1}{2} \end{pmatrix}.$$

Then

$$\mathbf{A}^4 = \begin{pmatrix} \cos 480° & -\sin 480° \\ \sin 480° & \cos 480° \end{pmatrix} = \begin{pmatrix} \cos 120° & -\sin 120° \\ \sin 120° & \cos 120° \end{pmatrix} = \mathbf{A},$$

but

$$\mathbf{A}^2 = \begin{pmatrix} \cos 240° & -\sin 240° \\ \sin 240° & \cos 240° \end{pmatrix} = \begin{pmatrix} -\frac{1}{2} & \frac{\sqrt{3}}{2} \\ -\frac{\sqrt{3}}{2} & -\frac{1}{2} \end{pmatrix} \ne \mathbf{A}.$$

So, we have $\mathbf{A}^4 = \mathbf{A}$, but $\mathbf{A}^2 \ne \mathbf{A}$, as desired.

10.64 Let $\mathbf{A} = \begin{pmatrix} a_1 & a_2 \\ a_3 & a_4 \end{pmatrix}$ and $\mathbf{B} = \begin{pmatrix} b_1 & b_2 \\ b_3 & b_4 \end{pmatrix}$. Then

$$\det(\mathbf{A} + \mathbf{B}) = \det \begin{pmatrix} a_1 + b_1 & a_2 + b_2 \\ a_3 + b_3 & a_4 + b_4 \end{pmatrix} = (a_1 + b_1)(a_4 + b_4) - (a_2 + b_2)(a_3 + b_3)$$

$$= a_1 a_4 + a_1 b_4 + a_4 b_1 + b_1 b_4 - a_2 a_3 - a_2 b_3 - a_3 b_2 - b_2 b_3,$$

and

$$\det(\mathbf{A} - \mathbf{B}) = \det \begin{pmatrix} a_1 - b_1 & a_2 - b_2 \\ a_3 - b_3 & a_4 - b_4 \end{pmatrix} = (a_1 - b_1)(a_4 - b_4) - (a_2 - b_2)(a_3 - b_3)$$

$$= a_1 a_4 - a_1 b_4 - a_4 b_1 + b_1 b_4 - a_2 a_3 + a_2 b_3 + a_3 b_2 - b_2 b_3,$$

so

$$\det(\mathbf{A} + \mathbf{B}) + \det(\mathbf{A} - \mathbf{B}) = (a_1 a_4 + a_1 b_4 + a_4 b_1 + b_1 b_4 - a_2 a_3 - a_2 b_3 - a_3 b_2 - b_2 b_3)$$
$$+ (a_1 a_4 - a_1 b_4 - a_4 b_1 + b_1 b_4 - a_2 a_3 + a_2 b_3 + a_3 b_2 - b_2 b_3)$$
$$= 2a_1 a_4 + 2b_1 b_4 - 2a_2 a_3 - 2b_2 b_3 = 2(a_1 a_4 - a_2 a_3) + 2(b_1 b_4 - b_2 b_3) = 2[\det(\mathbf{A}) + \det(\mathbf{B})].$$

10.65   If we compute the determinant, we get

$$(x - x_1)(y - y_2) - (x - x_2)(y - y_1) = 0.$$

When we expand the left side, the xy terms cancel and we are left with an equation of the form $ax + by + c = 0$, where a, b, and c are constants. The graph of such an equation is a line. Furthermore, if $(x, y) = (x_1, y_1)$, then

$$\begin{vmatrix} x - x_1 & y - y_1 \\ x - x_2 & y - y_2 \end{vmatrix} = \begin{vmatrix} 0 & 0 \\ x_1 - x_2 & y_1 - y_2 \end{vmatrix} = 0,$$

and if $(x, y) = (x_2, y_2)$, then

$$\begin{vmatrix} x - x_1 & y - y_1 \\ x - x_2 & y - y_2 \end{vmatrix} = \begin{vmatrix} x_2 - x_1 & y_2 - y_1 \\ 0 & 0 \end{vmatrix} = 0.$$

Thus, both points (x_1, y_1) and (x_2, y_2) satisfy this linear equation. Hence, it represents the equation of the line passing through both points.

10.66

(a)   We prove the result by induction. For $n = 1$, we have

$$\mathbf{A} = \begin{pmatrix} 1 & 1 \\ 1 & 0 \end{pmatrix} = \begin{pmatrix} F_2 & F_1 \\ F_1 & F_0 \end{pmatrix},$$

so the result holds for $n = 1$. Assume that the result holds for some positive integer $n = k$, so

$$\mathbf{A}^k = \begin{pmatrix} F_{k+1} & F_k \\ F_k & F_{k-1} \end{pmatrix}.$$

Then

$$\mathbf{A}^{k+1} = \mathbf{A} \cdot \mathbf{A}^k = \begin{pmatrix} 1 & 1 \\ 1 & 0 \end{pmatrix} \begin{pmatrix} F_{k+1} & F_k \\ F_k & F_{k-1} \end{pmatrix} = \begin{pmatrix} F_{k+1} + F_k & F_k + F_{k-1} \\ F_{k+1} & F_k \end{pmatrix} = \begin{pmatrix} F_{k+2} & F_{k+1} \\ F_{k+1} & F_k \end{pmatrix}.$$

Hence, the result holds for $n = k + 1$, so by induction, it holds for all positive integers n.

(b)   Taking the determinant of both sides of $\mathbf{A}^n = \begin{pmatrix} F_{n+1} & F_n \\ F_n & F_{n-1} \end{pmatrix}$, we get

$$\det(\mathbf{A}^n) = \det \begin{pmatrix} F_{n+1} & F_n \\ F_n & F_{n-1} \end{pmatrix} = F_{n+1}F_{n-1} - F_n^2.$$

But

$$\det(\mathbf{A}^n) = [\det(\mathbf{A})]^n = \left[\det \begin{pmatrix} 1 & 1 \\ 1 & 0 \end{pmatrix} \right]^n = (-1)^n,$$

so $F_{n+1}F_{n-1} - F_n^2 = (-1)^n$.

10.67   We see that

$$\begin{pmatrix} \cos\theta & -\sin\theta \\ \sin\theta & \cos\theta \end{pmatrix} \begin{pmatrix} x \\ y \end{pmatrix} = \begin{pmatrix} x\cos\theta - y\sin\theta \\ x\sin\theta + y\cos\theta \end{pmatrix} = \begin{pmatrix} 3 \\ 2 \end{pmatrix}.$$

In other words, if the vector $\begin{pmatrix} x \\ y \end{pmatrix}$ is rotated by an angle of θ counterclockwise about the origin, then the vector $\begin{pmatrix} 3 \\ 2 \end{pmatrix}$ is obtained. The only such vectors $\begin{pmatrix} x \\ y \end{pmatrix}$ are the vectors such that (x, y) is on the same circle, centered at the origin, as $(3, 2)$. Hence, there exists such a θ if and only if $\boxed{x^2 + y^2 = 13}$.

10.68

(a) Let ℓ be a line through the origin and let $f(\mathbf{v})$ be the reflection of $\mathbf{v}$ over ℓ. To show that f is linear, we must show that $f(\mathbf{v} + \mathbf{w}) = f(\mathbf{v}) + f(\mathbf{w})$ and that $f(k\mathbf{v}) = kf(\mathbf{v})$. The diagram at left below exhibits the former and the diagram at right exhibits the latter.

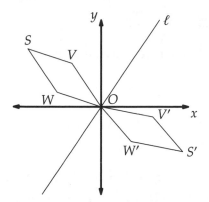 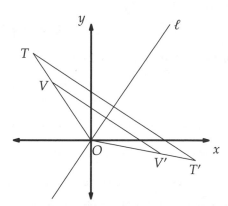

In both diagrams, ℓ is the graph of a line through the origin. In the diagram at left, parallelogram $OVSW$ is constructed, and $OV'S'W'$ is the reflection of $OVSW$ over ℓ. Therefore $f(\vec{V} + \vec{W}) = f(\vec{S}) = \vec{S'}$ and $f(\vec{V}) + f(\vec{W}) = \vec{V'} + \vec{W'} = \vec{S'}$. In the diagram at right, $\vec{T} = k\vec{V}$, and V' and T' are the reflections of V and T, respectively, over ℓ, so $f(\vec{T}) = \vec{T'}$. By SAS Similarity, we have $\triangle VOV' \sim \triangle TOT'$, so $OT'/OV' = OT/OV = k$, which gives $\vec{T'} = k\vec{V'}$. Therefore, we have $f(k\vec{V}) = f(\vec{T}) = \vec{T'} = k\vec{V'} = kf(\vec{V})$. We can go through essentially the same steps if $k < 0$. We can therefore conclude that f is linear.

(b) Since the reflection is a linear transformation, we can represent the action of the transformation on any vector as a product of the matrix and the vector. So, we just have to find the matrix $\mathbf{A}$ such that $\mathbf{Av}$ is the reflection of $\mathbf{v}$ over $y = 2x$ for any vector $\mathbf{v}$. We know the results of two reflections, and can use these to find $\mathbf{A}$. Specifically, the vector $\begin{pmatrix} 1 \\ 2 \end{pmatrix}$ is in the direction of the line, so it is its own reflection. Rewriting the equation as $2x - y = 0$, we see that $\begin{pmatrix} 2 \\ -1 \end{pmatrix}$ is normal to the line, so it is the opposite of its reflection. Therefore, we have

$$\mathbf{A}\begin{pmatrix} 1 \\ 2 \end{pmatrix} = \begin{pmatrix} 1 \\ 2 \end{pmatrix} \qquad \text{and} \qquad \mathbf{A}\begin{pmatrix} 2 \\ -1 \end{pmatrix} = \begin{pmatrix} -2 \\ 1 \end{pmatrix}.$$

Letting $\mathbf{A} = \begin{pmatrix} a & b \\ c & d \end{pmatrix}$, we have

$$\begin{pmatrix} a + 2b \\ c + 2d \end{pmatrix} = \begin{pmatrix} 1 \\ 2 \end{pmatrix}, \qquad \text{and} \qquad \begin{pmatrix} 2a - b \\ 2c - d \end{pmatrix} = \begin{pmatrix} -2 \\ 1 \end{pmatrix}.$$

From the system $a + 2b = 1$, $2a - b = -2$, we have $a = -\frac{3}{5}$, $b = \frac{4}{5}$. From the system $c + 2d = 2$, $2c - d = 1$, we have $c = \frac{4}{5}$, $d = \frac{3}{5}$ so $\mathbf{A} = \begin{pmatrix} -\frac{3}{5} & \frac{4}{5} \\ \frac{4}{5} & \frac{3}{5} \end{pmatrix}$.

To find the reflection of $(4, -6)$, we note that

$$\begin{pmatrix} -\frac{3}{5} & \frac{4}{5} \\ \frac{4}{5} & \frac{3}{5} \end{pmatrix} \begin{pmatrix} 4 \\ -6 \end{pmatrix} = \begin{pmatrix} -\frac{36}{5} \\ -\frac{2}{5} \end{pmatrix},$$

so the desired point is $\boxed{\left(-\frac{36}{5}, -\frac{2}{5}\right)}$.

(c) If f is linear, we must have $f(0 + 0) = f(0) + f(0)$, which gives $f(0) = 2f(0)$, so $f(0) = 0$. However, the reflection of $(0,0)$ over $x + y = 6$ is not the origin, so f cannot be linear.

10.69

(a) Let $\mathbf{v} = \begin{pmatrix} x \\ y \end{pmatrix}$. Then $\mathbf{Av} = \begin{pmatrix} -2 & 6 \\ 2 & -1 \end{pmatrix}\begin{pmatrix} x \\ y \end{pmatrix} = \begin{pmatrix} -2x + 6y \\ 2x - y \end{pmatrix}$, and $\lambda\mathbf{v} = \begin{pmatrix} \lambda x \\ \lambda y \end{pmatrix}$, so $\mathbf{Av} = \lambda\mathbf{v}$ gives the system of equations $-2x + 6y = \lambda x$ and $2x - y = \lambda y$. Then $6y = (\lambda + 2)x$, so $y = (\lambda + 2)x/6$. From the equation $2x - y = \lambda y$, we have

$$2x = (\lambda + 1)y = (\lambda + 1) \cdot \frac{(\lambda + 2)x}{6} = \frac{(\lambda + 1)(\lambda + 2)x}{6}.$$

If $x = 0$, then $y = (\lambda + 2)x/6 = 0$. But we seek nonzero vectors $\mathbf{v}$, so we may divide both sides by x in the equation above to get

$$2 = \frac{(\lambda + 1)(\lambda + 2)}{6} \quad \Rightarrow \quad (\lambda + 1)(\lambda + 2) = 12.$$

This simplifies as $\lambda^2 + 3\lambda - 10 = 0$, so $(\lambda - 2)(\lambda + 5) = 0$, which gives $\lambda = 2$ or $\lambda = -5$ as the only possible values of λ. So, we seek vectors $\mathbf{v} = \begin{pmatrix} x \\ y \end{pmatrix}$ such that either $\mathbf{Av} = 2\mathbf{v}$ or $\mathbf{Av} = -5\mathbf{v}$.

For $\mathbf{Av} = 2\mathbf{v}$, we have $-2x + 6y = 2x$ and $2x - y = 2y$, both of which give $2x = 3y$. So, we can take $x = 1$, and $y = 4 \cdot 1/6 = 2/3$, which gives $\mathbf{v} = \begin{pmatrix} 1 \\ 2/3 \end{pmatrix}$. Any multiple of this vector also satisfies $\mathbf{Av} = 2\mathbf{v}$

For $\mathbf{Av} = -5\mathbf{v}$, we have $-2x + 6y = -5x$ and $2x - y = -5y$, both of which give $x = -2y$. we can take $x = 1$, and $y = -1/2$, so $\mathbf{v} = \begin{pmatrix} 1 \\ -1/2 \end{pmatrix}$. Any multiple of this vector also satisfies $\mathbf{Av} = -5\mathbf{v}$.

(b) For $\lambda = 2$, we have

$$\det(\mathbf{A} - \lambda\mathbf{I}) = \det\left(\begin{pmatrix} -2 & 6 \\ 2 & -1 \end{pmatrix} - 2\begin{pmatrix} 1 & 0 \\ 0 & 1 \end{pmatrix}\right) = \det\begin{pmatrix} -4 & 6 \\ 2 & -3 \end{pmatrix} = (-4) \cdot (-3) - 6 \cdot 2 = 0,$$

and for $\lambda = -5$, we have

$$\det(\mathbf{A} - \lambda\mathbf{I}) = \det\left(\begin{pmatrix} -2 & 6 \\ 2 & -1 \end{pmatrix} + 5\begin{pmatrix} 1 & 0 \\ 0 & 1 \end{pmatrix}\right) = \det\begin{pmatrix} 3 & 6 \\ 2 & 4 \end{pmatrix} = 3 \cdot 4 - 6 \cdot 2 = 0.$$

If $\mathbf{Av} = \lambda\mathbf{v}$, then we have $(\mathbf{A} - \lambda\mathbf{I})\mathbf{v} = \mathbf{Av} - \lambda\mathbf{v} = \mathbf{0}$. Furthermore, if $\det(\mathbf{A} - \lambda\mathbf{I}) \neq 0$, then $\mathbf{A} - \lambda\mathbf{I}$ is an invertible matrix. Multiplying both sides of $(\mathbf{A} - \lambda\mathbf{I})\mathbf{v} = \mathbf{0}$ by this inverse gives us $\mathbf{v} = \mathbf{0}$. So, we cannot have a nonzero solution $\mathbf{v}$ to the equation $\mathbf{Av} = \lambda\mathbf{v}$ if $\det(\mathbf{A} - \lambda\mathbf{I}) \neq 0$. Therefore, if there exists a nonzero vector $\mathbf{v}$ that satisfies $(\mathbf{A} - \lambda\mathbf{I})\mathbf{v} = \mathbf{0}$, then we must have $\det(\mathbf{A} - \lambda\mathbf{I}) = 0$.

10.70 First, we give an algebraic proof. Let $\mathbf{A} = \begin{pmatrix} a & b \\ c & d \end{pmatrix}$, so

$$\det(\mathbf{A}) = \begin{vmatrix} a & b \\ c & d \end{vmatrix} = ad - bc.$$

We are given that the columns of $\mathbf{A}$ are orthogonal, so $\begin{pmatrix} a \\ c \end{pmatrix} \cdot \begin{pmatrix} b \\ d \end{pmatrix} = 0$, which means $ab + cd = 0$. The product of the norms of the columns is $\sqrt{a^2 + c^2}\,\sqrt{b^2 + d^2}$. To make this quantity easier to work with, we square it, to get

$$(a^2 + c^2)(b^2 + d^2) = a^2b^2 + a^2d^2 + b^2c^2 + c^2d^2 = (a^2b^2 + 2abcd + c^2d^2) + (a^2d^2 - 2abcd + b^2c^2)$$
$$= (ab + cd)^2 + (ad - bc)^2 = 0 + (ad - bc)^2 = (ad - bc)^2,$$

so $\sqrt{a^2 + c^2}\,\sqrt{b^2 + d^2} = |ad - bc| = |\det(\mathbf{A})|$.

Now we give a geometric proof. We know that $|\det(\mathbf{A})|$ is the area of the parallelogram formed by the vectors $\mathbf{v} = \begin{pmatrix} a \\ c \end{pmatrix}$ and $\mathbf{w} = \begin{pmatrix} b \\ d \end{pmatrix}$. But we are given that these vectors are orthogonal, so the parallelogram is a rectangle. This means its area is simply the product of the lengths of consecutive sides, which is equal to the product of the norms of $\mathbf{v}$ and $\mathbf{w}$.

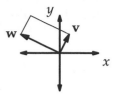

10.71

(a) In the text we showed that the area of a triangle with vertices (x_1, y_1), (x_2, y_2), and (x_3, y_3), in counterclockwise order, is

$$\frac{1}{2} \begin{vmatrix} x_2 - x_1 & x_3 - x_1 \\ y_2 - y_1 & y_3 - y_1 \end{vmatrix} = \frac{1}{2}((x_2 - x_1)(y_3 - y_1) - (y_2 - y_1)(x_3 - x_1)) = \frac{1}{2}(x_1 y_2 + x_2 y_3 + x_3 y_1 - x_2 y_1 - x_3 y_2 - x_1 y_3),$$

which is exactly the expression given by the Shoelace Theorem.

(b) We split the quadrilateral into a triangle with vertices (x_1, y_1), (x_2, y_2), and (x_3, y_3) and a triangle with vertices (x_3, y_3), (x_4, y_4), and (x_1, y_1). We then apply part (a) to both triangles, which tells us that the area of the first triangle is

$$\frac{1}{2}(x_1 y_2 + x_2 y_3 + x_3 y_1 - x_2 y_1 - x_3 y_2 - x_1 y_3)$$

and the second is

$$\frac{1}{2}(x_3 y_4 + x_4 y_1 + x_1 y_3 - x_4 y_3 - x_1 y_4 - x_3 y_1).$$

Adding these gives

$$\frac{1}{2}(x_1 y_2 + x_2 y_3 + x_3 y_4 + x_4 y_1 - x_2 y_1 - x_3 y_2 - x_4 y_3 - x_1 y_4),$$

which is again exactly the expression given by the Shoelace Theorem.

(c) We apply induction. We assume the Shoelace Theorem gives the area for a polygon with k sides, where $k \geq 3$. We let the vertices, in counterclockwise order, of a polygon with $k+1$ sides be (x_1, y_1), (x_2, y_2), ..., (x_{k+1}, y_{k+1}). We then break this polygon into the k-sided polygon with vertices (x_1, y_1), (x_2, y_2), ..., (x_k, y_k) and the triangle with vertices (x_k, y_k), (x_{k+1}, y_{k+1}), (x_1, y_1). Applying the inductive assumption, the area of the k-sided polygon is

$$\frac{1}{2}(x_1 y_2 + x_2 y_3 + \cdots + x_{k-1} y_k + x_k y_1 - x_2 y_1 - x_3 y_2 - \cdots - x_k y_{k-1} - x_1 y_k)$$

and the area of the triangle is

$$\frac{1}{2}(x_k y_{k+1} + x_{k+1} y_1 + x_1 y_k - x_{k+1} y_k - x_1 y_{k+1} - x_k y_1).$$

Adding these gives the total area of the $k + 1$-sided polygon as

$$\frac{1}{2}(x_1 y_2 + x_2 y_3 + \cdots + x_k y_{k+1} + x_{k+1} y_1 - x_2 y_1 - x_3 y_2 - \cdots - x_{k+1} y_k - x_1 y_{k+1}),$$

and the induction is complete.

10.72

(a) Let $\mathbf{r} = \begin{pmatrix} r_1 \\ r_2 \end{pmatrix}$ and $\mathbf{s} = \begin{pmatrix} s_1 \\ s_2 \end{pmatrix}$, and take $\mathbf{B} = \begin{pmatrix} r_1 & s_1 \\ r_2 & s_2 \end{pmatrix}$. Then

$$\mathbf{Bi} = \begin{pmatrix} r_1 & s_1 \\ r_2 & s_2 \end{pmatrix} \begin{pmatrix} 1 \\ 0 \end{pmatrix} = \begin{pmatrix} r_1 \\ r_2 \end{pmatrix} = \mathbf{r},$$

and

$$\mathbf{Bj} = \begin{pmatrix} r_1 & s_1 \\ r_2 & s_2 \end{pmatrix} \begin{pmatrix} 0 \\ 1 \end{pmatrix} = \begin{pmatrix} s_1 \\ s_2 \end{pmatrix} = \mathbf{s}.$$

(b) The signed area of parallelogram with $\mathbf{r}$ and $\mathbf{s}$ as sides is $\det(\mathbf{B})$.

(c) Since $\mathbf{Ar} = \mathbf{A}(\mathbf{B}i) = (\mathbf{AB})i$ and $\mathbf{As} = \mathbf{A}(\mathbf{B}j) = (\mathbf{AB})j$, the signed area of the parallelogram with $\mathbf{Ar}$ and $\mathbf{As}$ as sides is $\det(\mathbf{AB})$.

(d) Since $\det(\mathbf{AB}) = \det(\mathbf{A})\det(\mathbf{B})$, the signed area of the parallelogram with $\mathbf{Ar}$ and $\mathbf{As}$ as sides is $\det(\mathbf{A})$ times the signed area of parallelogram with $\mathbf{r}$ and $\mathbf{s}$ as sides.

10.73 As described in the text, $\det(\mathbf{T})$ is the signed area of the parallelogram with $\mathbf{T}i$ and $\mathbf{T}j$ as sides. Since multiplying by $\mathbf{T}$ is equivalent to a reflection, the area of the parallelogram formed by $\mathbf{T}i$ and $\mathbf{T}j$ has the same magnitude as the area of the parallelogram formed by i and j, which is 1. But since multiplying by $\mathbf{T}$ is a reflection, the orientation of the parallelogram is reversed, so the area of the parallelogram formed by $\mathbf{T}i$ and $\mathbf{T}j$ is -1. Therefore, $\det(\mathbf{T}) = -1$.

10.74 Let $\mathbf{A} = \begin{pmatrix} a & b \\ c & d \end{pmatrix}$. From $\mathbf{A}^2 = \mathbf{I}$, we have $\det(\mathbf{A}^2) = \det(\mathbf{I}) = 1$. But $\det(\mathbf{A}^2) = [\det(\mathbf{A})]^2$, so $\det(\mathbf{A}) = \pm 1$, which means $ad - bc = \pm 1$.

We know that $\mathbf{A}$ is invertible, because $\mathbf{A} \cdot \mathbf{A} = \mathbf{I}$. Hence, $\mathbf{A}^{-1} = \mathbf{A}$. But

$$\mathbf{A}^{-1} = \frac{1}{ad - bc} \begin{pmatrix} d & -b \\ -c & a \end{pmatrix}.$$

If $ad - bc = 1$, then we must have $\begin{pmatrix} a & b \\ c & d \end{pmatrix} = \mathbf{A}^{-1} = \begin{pmatrix} d & -b \\ -c & a \end{pmatrix}$, so $a = d$, $b = -b$, $c = -c$, and $d = a$. Hence, $b = 0$ and $c = 0$. Substituting into $ad - bc = 1$, we get $a^2 - 0 = 1$, so $a = \pm 1$.

If $ad - bc = -1$, then we must have $\begin{pmatrix} a & b \\ c & d \end{pmatrix} = \mathbf{A}^{-1} = \begin{pmatrix} -d & b \\ c & -a \end{pmatrix}$, so $a = -d$, $b = b$, $c = c$, and $d = -a$. Substituting into $ad - bc = -1$, we get $-a^2 - bc = -1$, so $bc = 1 - a^2$. If $b = 0$, then $1 - a^2 = 0$, so $a = \pm 1$ and c can be anything. If $b \neq 0$, then we can solve for c to get $c = (1 - a^2)/b$. Hence, the solutions to $\mathbf{A}^2 = \mathbf{I}$ are of the form

$$\boxed{\begin{pmatrix} 1 & 0 \\ 0 & 1 \end{pmatrix}, \quad \begin{pmatrix} -1 & 0 \\ 0 & -1 \end{pmatrix}, \quad \begin{pmatrix} 1 & 0 \\ c & -1 \end{pmatrix}, \quad \begin{pmatrix} -1 & 0 \\ c & 1 \end{pmatrix}, \quad \begin{pmatrix} a & b \\ \frac{1-a^2}{b} & -a \end{pmatrix},}$$

where a and c are any real numbers, and b is any nonzero real number.

10.75 Expanding, we get $(\mathbf{A} - \mathbf{B})(\mathbf{A} + \mathbf{B}) = \mathbf{A}(\mathbf{A} + \mathbf{B}) - \mathbf{B}(\mathbf{A} + \mathbf{B}) = \mathbf{A}^2 + \mathbf{AB} - \mathbf{BA} - \mathbf{B}^2$, which is equal to $\mathbf{A}^2 - \mathbf{B}^2$ if and only if $\mathbf{AB} = \mathbf{BA}$, i.e. the matrices $\mathbf{A}$ and $\mathbf{B}$ commute. Therefore, we do not have $\mathbf{A}^2 - \mathbf{B}^2 = (\mathbf{A} - \mathbf{B})(\mathbf{A} + \mathbf{B})$ for all matrices $\mathbf{A}$ and $\mathbf{B}$. (See if you can construct a specific example in which $\mathbf{A}^2 - \mathbf{B}^2$ and $(\mathbf{A} - \mathbf{B})(\mathbf{A} + \mathbf{B})$ are not equal.)

10.76 Rearranging $\mathbf{A} + \mathbf{B} = \mathbf{AB}$ gives $\mathbf{AB} - \mathbf{A} - \mathbf{B} = 0$. Factoring the first two terms gives $\mathbf{A}(\mathbf{B} - \mathbf{I}) - \mathbf{B} = 0$. Adding $\mathbf{I}$ to both sides lets us factor more:

$$\mathbf{A}(\mathbf{B} - \mathbf{I}) - \mathbf{B} + \mathbf{I} = \mathbf{I} \quad \Leftrightarrow \quad \mathbf{A}(\mathbf{B} - \mathbf{I}) - \mathbf{I}(\mathbf{B} - \mathbf{I}) = \mathbf{I} \quad \Leftrightarrow \quad (\mathbf{A} - \mathbf{I})(\mathbf{B} - \mathbf{I}) = \mathbf{I}.$$

The equation $(\mathbf{A} - \mathbf{I})(\mathbf{B} - \mathbf{I}) = \mathbf{I}$ means that $\mathbf{A} - \mathbf{I}$ and $\mathbf{B} - \mathbf{I}$ are inverses. We showed in the text that inverses commute (that is, we have $\mathbf{M}^{-1}\mathbf{M} = \mathbf{M}\mathbf{M}^{-1}$ for any invertible matrix $\mathbf{M}$), so $(\mathbf{B} - \mathbf{I})(\mathbf{A} - \mathbf{I}) = \mathbf{I}$. This expands as

$$\mathbf{BA} - \mathbf{B} - \mathbf{A} + \mathbf{I} = \mathbf{I},$$

so $\mathbf{BA} = \mathbf{A} + \mathbf{B} = \mathbf{AB}$.

10.77 Since $\mathbf{A}\begin{pmatrix} 1 \\ 3 \end{pmatrix} = \begin{pmatrix} 4 \\ 5 \end{pmatrix}$, we have $\mathbf{A}^2\begin{pmatrix} 1 \\ 3 \end{pmatrix} = \mathbf{A}\begin{pmatrix} 4 \\ 5 \end{pmatrix}$. But

$$\mathbf{A}^2\begin{pmatrix} 1 \\ 3 \end{pmatrix} = (\mathbf{A} - 5\mathbf{I})\begin{pmatrix} 1 \\ 3 \end{pmatrix} = \mathbf{A}\begin{pmatrix} 1 \\ 3 \end{pmatrix} - 5\begin{pmatrix} 1 \\ 3 \end{pmatrix} = \begin{pmatrix} 4 \\ 5 \end{pmatrix} - \begin{pmatrix} 5 \\ 15 \end{pmatrix} = \begin{pmatrix} -1 \\ -10 \end{pmatrix}.$$

Now, let $\mathbf{A} = \begin{pmatrix} a & b \\ c & d \end{pmatrix}$. Then, we have $\mathbf{A}\begin{pmatrix} 1 \\ 3 \end{pmatrix} = \begin{pmatrix} a & b \\ c & d \end{pmatrix}\begin{pmatrix} 1 \\ 3 \end{pmatrix} = \begin{pmatrix} a + 3b \\ c + 3d \end{pmatrix}$ and $\mathbf{A}\begin{pmatrix} 4 \\ 5 \end{pmatrix} = \begin{pmatrix} a & b \\ c & d \end{pmatrix}\begin{pmatrix} 4 \\ 5 \end{pmatrix} = \begin{pmatrix} 4a + 5b \\ 4c + 5d \end{pmatrix}$, so the equations $\mathbf{A}\begin{pmatrix} 1 \\ 3 \end{pmatrix} = \begin{pmatrix} 4 \\ 5 \end{pmatrix}$ and $\mathbf{A}\begin{pmatrix} 4 \\ 5 \end{pmatrix} = \begin{pmatrix} -1 \\ -10 \end{pmatrix}$ give us

$$\begin{pmatrix} a + 3b \\ c + 3d \end{pmatrix} = \begin{pmatrix} 4 \\ 5 \end{pmatrix}, \qquad \text{and} \qquad \begin{pmatrix} 4a + 5b \\ 4c + 5d \end{pmatrix} = \begin{pmatrix} -1 \\ -10 \end{pmatrix}.$$

Solving the system of equations $a + 3b = 4$ and $4a + 5b = -1$, we get $a = -23/7$ and $b = 17/7$, and from the system of equations $c + 3d = 5$ and $4c + 5d = -10$, we get $c = -55/7$ and $d = 30/7$. Therefore,

$$\mathbf{A} = \boxed{\begin{pmatrix} -\frac{23}{7} & \frac{17}{7} \\ -\frac{55}{7} & \frac{30}{7} \end{pmatrix}}.$$

10.78

(a) Let $\mathbf{A} = \begin{pmatrix} 0 & 1 \\ 0 & 0 \end{pmatrix}$. Then $\mathbf{A}^2 = \begin{pmatrix} 0 & 1 \\ 0 & 0 \end{pmatrix}\begin{pmatrix} 0 & 1 \\ 0 & 0 \end{pmatrix} = \begin{pmatrix} 0 & 0 \\ 0 & 0 \end{pmatrix} = \mathbf{0}$. Hence, if $\mathbf{A}^2 = \mathbf{0}$, $\mathbf{A}$ is not necessarily equal to $\mathbf{0}$.

(b) Let $\mathbf{A} = \begin{pmatrix} a & b \\ c & d \end{pmatrix}$. First, we claim that $\mathbf{A}^2 - (a + d)\mathbf{A} + (ad - bc)\mathbf{I} = \mathbf{0}$. Expanding, we get

$$\begin{aligned} \mathbf{A}^2 - (a + d)\mathbf{A} + (ad - bc)\mathbf{I} &= \begin{pmatrix} a & b \\ c & d \end{pmatrix}\begin{pmatrix} a & b \\ c & d \end{pmatrix} - (a + d)\begin{pmatrix} a & b \\ c & d \end{pmatrix} + (ad - bc)\begin{pmatrix} 1 & 0 \\ 0 & 1 \end{pmatrix} \\ &= \begin{pmatrix} a^2 + bc & ab + bd \\ ac + cd & bc + d^2 \end{pmatrix} - \begin{pmatrix} a^2 + ad & ab + bd \\ ac + cd & ad + d^2 \end{pmatrix} + \begin{pmatrix} ad - bc & 0 \\ 0 & ad - bc \end{pmatrix} \\ &= \begin{pmatrix} a^2 + bc - a^2 - ad + ad - bc & ab + bd - ab - bd \\ ac + cd - ac - cd & bc + d^2 - ad - d^2 + ad - bc \end{pmatrix} \\ &= \begin{pmatrix} 0 & 0 \\ 0 & 0 \end{pmatrix} \\ &= \mathbf{0}. \end{aligned}$$

Given $\mathbf{A}^3 = \mathbf{0}$, we take the determinant of both sides to get $\det(\mathbf{A}^3) = \det(\mathbf{0}) = 0$. But $\det(\mathbf{A}^3) = [\det(\mathbf{A})]^3$, so $[\det(\mathbf{A})]^3 = 0$. Therefore, $\det(\mathbf{A}) = 0$, so $ad - bc = 0$. Hence, the equation $\mathbf{A}^2 - (a + d)\mathbf{A} + (ad - bc)\mathbf{I} = \mathbf{0}$ becomes

$$\mathbf{A}^2 - (a + d)\mathbf{A} = \mathbf{0},$$

which means $\mathbf{A}^2 = (a + d)\mathbf{A}$. Multiplying both sides by $\mathbf{A}$, we get $\mathbf{A}^3 = (a + d)\mathbf{A}^2$, so $(a + d)\mathbf{A}^2 = \mathbf{0}$.

If $a + d \neq 0$, we can divide both sides by $a + d$, to get $\mathbf{A}^2 = \mathbf{0}$. If $a + d = 0$, then $d = -a$, and

$$\mathbf{A}^2 = \begin{pmatrix} a^2 + bc & ab + bd \\ ac + cd & bc + d^2 \end{pmatrix} = \begin{pmatrix} a^2 + bc & b(a + (-a)) \\ c(a + (-a)) & bc + (-a)^2 \end{pmatrix} = \begin{pmatrix} a^2 + bc & 0 \\ 0 & a^2 + bc \end{pmatrix} = (a^2 + bc)\mathbf{I}.$$

Let $x = a^2 + bc$, so $\mathbf{A}^2 = x\mathbf{I}$. Multiplying both sides by $\mathbf{A}$, we get $\mathbf{A}^3 = x\mathbf{A}$, so $x\mathbf{A} = \mathbf{0}$. If $x \neq 0$, we can divide both sides by x, to get $\mathbf{A} = \mathbf{0}$, and so $\mathbf{A}^2 = \mathbf{0}$. If $x = 0$, then $\mathbf{A}^2 = x\mathbf{I} = \mathbf{0}$. In either case, $\mathbf{A}^2 = \mathbf{0}$.

CHAPTER 11

Vectors and Matrices in Three Dimensions, Part 1

Exercises for Section 11.1

11.1.1

(a) If $\mathbf{u} - 3\mathbf{w} = 2\mathbf{v}$, then $\mathbf{w} = \frac{\mathbf{u}-2\mathbf{v}}{3} = \frac{(4\mathbf{i}+2\mathbf{j}-\mathbf{k})-2(-\mathbf{i}+5\mathbf{j}+2\mathbf{k})}{3} = \frac{6\mathbf{i}-8\mathbf{j}-5\mathbf{k}}{3} = \boxed{2\mathbf{i} - \frac{8}{3}\mathbf{j} - \frac{5}{3}\mathbf{k}}$.

(b) We seek constants a and b such that $a(4\mathbf{i} + 2\mathbf{j} - \mathbf{k}) + b(-\mathbf{i} + 5\mathbf{j} + 2\mathbf{k}) = \mathbf{0}$. Rearranging the left side gives $(4a - b)\mathbf{i} + (2a + 5b)\mathbf{j} + (-a + 2b)\mathbf{k} = \mathbf{0}$. Therefore, we must have $4a = b$, $2a = -5b$, and $a = 2b$. Substituting $a = 2b$ into $4a = b$ gives $8b = b$, from which we find $b = 0$ and $a = 0$. Therefore, there are not nonzero scalars a and b such that $a\mathbf{u} + b\mathbf{v} = \mathbf{0}$.

(c) We have
$$x\mathbf{u} + y\mathbf{v} = x(4\mathbf{i} + 2\mathbf{j} - \mathbf{k}) + y(-\mathbf{i} + 5\mathbf{j} + 2\mathbf{k}) = (4x - y)\mathbf{i} + (2x + 5y)\mathbf{j} + (-x + 2y)\mathbf{k}.$$
We claim that this vector cannot be equal to $\mathbf{k}$.

If $(4x - y)\mathbf{i} + (2x + 5y)\mathbf{j} + (-x + 2y)\mathbf{k} = \mathbf{k}$, then $4x - y = 0$, $2x + 5y = 0$, and $-x + 2y = 1$. The first equation gives $y = 4x$, and substituting this into the second equation gives $2x + 20x = 0$. This gives us $x = 0$, and $y = 4x$ then gives $y = 0$. But $x = y = 0$ does not satisfy $-x + 2y = 1$, so there do not exist scalars x and y such that $x\mathbf{u} + y\mathbf{v} = \mathbf{k}$.

11.1.2 If $\begin{pmatrix} 1 \\ x \\ y \end{pmatrix}$ is orthogonal to $\begin{pmatrix} 3 \\ -1 \\ 4 \end{pmatrix}$, then $\begin{pmatrix} 1 \\ x \\ y \end{pmatrix} \cdot \begin{pmatrix} 3 \\ -1 \\ 4 \end{pmatrix} = 0$, which gives $3 - x + 4y = 0$.

If $\begin{pmatrix} 1 \\ x \\ y \end{pmatrix}$ is orthogonal to $\begin{pmatrix} -4 \\ 1 \\ 2 \end{pmatrix}$, then $\begin{pmatrix} 1 \\ x \\ y \end{pmatrix} \cdot \begin{pmatrix} -4 \\ 1 \\ 2 \end{pmatrix} = 0$, which gives $-4 + x + 2y = 0$.

Hence, we have $x - 4y = 3$ and $x + 2y = 4$. Subtracting the first equation from the second, we get $6y = 1$, so $y = 1/6$. Substituting, we get $x - 4/6 = 3$, so $x = 11/3$. The only solution is $(x, y) = \boxed{(11/3, 1/6)}$.

11.1.3 Let θ be the angle between the two vectors. Then
$$\cos\theta = \frac{\begin{pmatrix} 3 \\ -4 \\ 2 \end{pmatrix} \cdot \begin{pmatrix} -1 \\ 4 \\ 7 \end{pmatrix}}{\left\| \begin{pmatrix} 3 \\ -4 \\ 2 \end{pmatrix} \right\| \left\| \begin{pmatrix} -1 \\ 4 \\ 7 \end{pmatrix} \right\|} = \frac{-3 - 16 + 14}{\sqrt{3^2 + (-4)^2 + 2^2}\sqrt{(-1)^2 + 4^2 + 7^2}} = \frac{-5}{\sqrt{29}\sqrt{66}} = \frac{-5}{\sqrt{1914}},$$

so $\theta = \arccos(\frac{-5}{\sqrt{1914}}) \approx \boxed{97°}$.

11.1.4   Since $\|\mathbf{u}\|$ is the magnitude of $\mathbf{u}$ and ρ is the distance from the point to the origin, we have $\|\mathbf{u}\| = \rho$. From here, we can finish either algebraically, or geometrically.

Geometric finish. By the definition of the ϕ-coordinate in spherical coordinates, the angle between $\mathbf{u}$ and $\mathbf{k}$ is ϕ. We therefore have $\mathbf{u} \cdot \mathbf{k} = \|\mathbf{u}\|\,\|\mathbf{k}\| \cos\phi = \|\mathbf{u}\| \cos\phi$, so $\cos\phi = \frac{\mathbf{u}\cdot\mathbf{k}}{\|\mathbf{u}\|}$, which means $\phi = \boxed{\arccos\left(\frac{\mathbf{u}\cdot\mathbf{k}}{\|\mathbf{u}\|}\right)}$.

Algebraic finish. We have $z = \rho\cos\phi = \|\mathbf{u}\| \cos\phi$, so $\cos\phi = \frac{z}{\|\mathbf{u}\|}$. To get z in terms of $\mathbf{u}$ and $\mathbf{k}$, we note that $\mathbf{u} \cdot \mathbf{k} = (x\mathbf{i} + y\mathbf{j} + z\mathbf{k}) \cdot \mathbf{k} = z$. Hence, $\cos\phi = \frac{\mathbf{u}\cdot\mathbf{k}}{\|\mathbf{u}\|}$, so $\phi = \boxed{\arccos\left(\frac{\mathbf{u}\cdot\mathbf{k}}{\|\mathbf{u}\|}\right)}$.

11.1.5

(a)   Geometrically speaking, the equation $\|\mathbf{v}\| = 4$ tells us that the length of $\mathbf{v}$ is 4, so (x, y, z) is 4 units from the origin. Therefore, the graph is a $\boxed{\text{sphere with radius 4 and center } (0,0,0)}$.

Algebraically speaking, we can note that $\|\mathbf{v}\| = \|x\mathbf{i} + y\mathbf{j} + z\mathbf{k}\| = \sqrt{x^2 + y^2 + z^2}$, so we must have $\sqrt{x^2 + y^2 + z^2} = 4$. Squaring both sides gives $x^2 + y^2 + z^2 = 4^2$, which is the equation of a sphere with radius 4 and center $(0,0,0)$.

(b)   Geometrically speaking, the equation $\|\mathbf{v} - \mathbf{w}\| = 6$ tells us that the vector from $(1, -2, 4)$ to (x, y, z) has length 6. In other words, (x, y, z) is 6 units from $(1, -2, 4)$, which means that it is on the

$$\boxed{\text{sphere with radius 6 and center } (1, -2, 4)}.$$

Algebraically speaking, we have

$$\|\mathbf{v} - \mathbf{w}\| = \|(x\mathbf{i} + y\mathbf{j} + z\mathbf{k}) - (\mathbf{i} - 2\mathbf{j} + 4\mathbf{k})\| = \|(x - 1)\mathbf{i} + (y + 2)\mathbf{j} + (z - 4)\mathbf{k}\| = \sqrt{(x-1)^2 + (y+2)^2 + (z-4)^2},$$

so we must have $\sqrt{(x-1)^2 + (y+2)^2 + (z-4)^2} = 6$. This is the equation of a sphere with radius 6 and center $(1, -2, 4)$.

(c)   The origin is on the graph, since both sides of the given equation are 0 when $\mathbf{v} = \mathbf{0}$. Otherwise, let θ between the angle between $\mathbf{v}$ and $\mathbf{w}$. We know that $\mathbf{v} \cdot \mathbf{w} = \|\mathbf{v}\|\|\mathbf{w}\| \cos\theta$, so the given equation becomes $2\|\mathbf{v}\|\|\mathbf{w}\| \cos\theta = \|\mathbf{v}\|\|\mathbf{w}\|$. We know that $\|\mathbf{w}\| \neq 0$, and assuming that $\|\mathbf{v}\| \neq 0$ (i.e. $\mathbf{v} \neq \mathbf{0}$), we can divide both sides by $\|\mathbf{v}\|\|\mathbf{w}\|$ to get $2\cos\theta = 1$, or $\cos\theta = \frac{1}{2}$.

From $\cos\theta = \frac{1}{2}$, we know that θ is $\frac{\pi}{3}$ or $-\frac{\pi}{3}$ (or a multiple of 2π greater or less than these values). But θ is the angle between $\mathbf{v}$ and $\mathbf{w}$. Hence, the graph is a $\boxed{\text{cone}}$, where the apex is the origin, the axis has direction vector $\mathbf{w}$, and each line on the cone passing through the origin makes an angle of $\frac{\pi}{3}$ with $\mathbf{w}$.

Exercises for Section 11.2

11.2.1

(a) $\begin{pmatrix} 4 & -1 & 3 \\ -1 & 0 & 5 \\ -3 & -1 & 6 \end{pmatrix} \begin{pmatrix} -1 \\ 4 \\ -5 \end{pmatrix} = \begin{pmatrix} 4\cdot(-1) + (-1)\cdot 4 + 3\cdot(-5) \\ (-1)\cdot(-1) + 0\cdot 4 + 5\cdot(-5) \\ (-3)\cdot(-1) + (-1)\cdot 4 + 6\cdot(-5) \end{pmatrix} = \boxed{\begin{pmatrix} -23 \\ -24 \\ -31 \end{pmatrix}}$.

(b) $\begin{pmatrix} 0 & -3 & 2 \\ -1 & -2 & 7 \\ -3 & 2 & 1 \end{pmatrix} \begin{pmatrix} 0 \\ -3 \\ 3 \end{pmatrix} = \begin{pmatrix} 0\cdot 0 + (-3)\cdot(-3) + 2\cdot 3 \\ (-1)\cdot 0 + (-2)\cdot(-3) + 7\cdot 3 \\ (-3)\cdot 0 + 2\cdot(-3) + 1\cdot 3 \end{pmatrix} = \boxed{\begin{pmatrix} 15 \\ 27 \\ -3 \end{pmatrix}}$.

11.2.2 $\begin{pmatrix} -1 & -3 & 6 \\ -2 & 4 & 0 \\ -1 & -2 & 0 \end{pmatrix} \begin{pmatrix} 0 & -1 & 3 \\ 10 & -2 & -3 \\ 4 & 6 & 0.5 \end{pmatrix} = \boxed{\begin{pmatrix} -6 & 43 & 9 \\ 40 & -6 & -18 \\ -20 & 5 & 3 \end{pmatrix}}$.

11.2.3 Let $\mathbf{M} = \boxed{k\mathbf{I}}$. Then $\mathbf{MA} = (k\mathbf{I})\mathbf{A} = k(\mathbf{IA}) = k\mathbf{A}$ for any matrix $\mathbf{A}$.

11.2.4 In the text, we found $\mathbf{T}$ by finding a matrix such that the product $\mathbf{TA}$ is formed by moving the first two rows of $\mathbf{A}$ down one row, and moving the last row of $\mathbf{A}$ to the first row of the product. This isn't the only way we can reorder the rows! We could also have moved the bottom two rows of $\mathbf{A}$ up one, and moved the top row to the last row. Performing this operation three times gives us $\mathbf{A}$ back again. So, we seek the matrix $\mathbf{U}$ such that $\mathbf{Ui} = \mathbf{k}$, $\mathbf{Uj} = \mathbf{i}$, and $\mathbf{Uk} = \mathbf{j}$. So, the columns of $\mathbf{U}$, in order, are $\mathbf{k}$, $\mathbf{i}$, and $\mathbf{j}$. Checking, we see that

$$\begin{pmatrix} 0 & 1 & 0 \\ 0 & 0 & 1 \\ 1 & 0 & 0 \end{pmatrix} \begin{pmatrix} a_{11} & a_{12} & a_{13} \\ a_{21} & a_{22} & a_{23} \\ a_{31} & a_{32} & a_{33} \end{pmatrix} = \begin{pmatrix} a_{21} & a_{22} & a_{23} \\ a_{31} & a_{32} & a_{33} \\ a_{11} & a_{12} & a_{13} \end{pmatrix}.$$

Notice that $\mathbf{U} = \mathbf{T}^2$, where $\mathbf{T}$ is the matrix we found in the text such that $\mathbf{T}^3 = \mathbf{I}$. So, we have

$$\mathbf{U}^3 = (\mathbf{T}^2)^3 = \mathbf{T}^6 = (\mathbf{T}^3)^2 = \mathbf{I}^2 = \mathbf{I},$$

and clearly $\mathbf{U} \neq \mathbf{I}$. There are many other matrices that fit the problem; see if you can find others.

Exercises for Section 11.3

11.3.1

(a)
$$\begin{vmatrix} 1 & 5 & 0 \\ -3 & -2 & 6 \\ -1 & 7 & -3 \end{vmatrix} = 1 \begin{vmatrix} -2 & 6 \\ 7 & -3 \end{vmatrix} - 5 \begin{vmatrix} -3 & 6 \\ -1 & -3 \end{vmatrix} + 0 \begin{vmatrix} -3 & -2 \\ -1 & 7 \end{vmatrix}$$

$$= 1[(-2) \cdot (-3) - 6 \cdot 7] - 5[(-3) \cdot (-3) - 6 \cdot (-1)] + 0$$

$$= 1(-36) - 5(15) = \boxed{-111}.$$

(b)
$$\begin{vmatrix} -3 & 4 & 0.5 \\ 0.25 & -6 & -1 \\ 0 & 8 & 2 \end{vmatrix} = -3 \begin{vmatrix} -6 & -1 \\ 8 & 2 \end{vmatrix} - 4 \begin{vmatrix} 0.25 & -1 \\ 0 & 2 \end{vmatrix} + 0.5 \begin{vmatrix} 0.25 & -6 \\ 0 & 8 \end{vmatrix}$$

$$= -3[(-6) \cdot 2 - (-1) \cdot 8] - 4[0.25 \cdot 2 - (-1) \cdot 0] + 0.5[0.25 \cdot 8 - (-6) \cdot 0]$$

$$= -3(-4) - 4(0.5) + 0.5(2) = \boxed{11}.$$

11.3.2 We have

$$\mathbf{A} + \mathbf{B} = \begin{pmatrix} a_{11} & a_{12} & a_{13} \\ a_{21} & a_{22} & a_{23} \\ a_{31} & a_{32} & a_{33} \end{pmatrix} + \begin{pmatrix} b_{11} & b_{12} & b_{13} \\ b_{21} & b_{22} & b_{23} \\ b_{31} & b_{32} & b_{33} \end{pmatrix} = \begin{pmatrix} a_{11} + b_{11} & a_{12} + b_{12} & a_{13} + b_{13} \\ a_{21} + b_{21} & a_{22} + b_{22} & a_{23} + b_{23} \\ a_{31} + b_{31} & a_{32} + b_{32} & a_{33} + b_{33} \end{pmatrix},$$

so

$$(\mathbf{A} + \mathbf{B})^T = \begin{pmatrix} a_{11} + b_{11} & a_{21} + b_{21} & a_{31} + b_{31} \\ a_{12} + b_{12} & a_{22} + b_{22} & a_{32} + b_{32} \\ a_{13} + b_{13} & a_{23} + b_{23} & a_{33} + b_{33} \end{pmatrix} = \begin{pmatrix} a_{11} & a_{21} & a_{31} \\ a_{12} & a_{22} & a_{32} \\ a_{13} & a_{23} & a_{33} \end{pmatrix} + \begin{pmatrix} b_{11} & b_{21} & b_{31} \\ b_{12} & b_{22} & b_{32} \\ b_{13} & b_{23} & b_{33} \end{pmatrix} = \mathbf{A}^T + \mathbf{B}^T.$$

11.3.3 Since $\mathbf{A}$ and $\mathbf{B}$ have the same second row and same third row, we can express these matrices as

$$\mathbf{A} = \begin{pmatrix} a_{11} & a_{12} & a_{13} \\ a_{21} & a_{22} & a_{23} \\ a_{31} & a_{32} & a_{33} \end{pmatrix} \qquad \text{and} \qquad \mathbf{B} = \begin{pmatrix} b_{11} & b_{12} & b_{13} \\ a_{21} & a_{22} & a_{23} \\ a_{31} & a_{32} & a_{33} \end{pmatrix}.$$

As shown in the text, we have

$$\det(\mathbf{A}) + \det(\mathbf{B}) = \begin{vmatrix} a_{11} & a_{12} & a_{13} \\ a_{21} & a_{22} & a_{23} \\ a_{31} & a_{32} & a_{33} \end{vmatrix} + \begin{vmatrix} b_{11} & b_{12} & b_{13} \\ a_{21} & a_{22} & a_{23} \\ a_{31} & a_{32} & a_{33} \end{vmatrix} = \begin{vmatrix} a_{11} + b_{11} & a_{12} + b_{12} & a_{13} + b_{13} \\ a_{21} & a_{22} & a_{23} \\ a_{31} & a_{32} & a_{33} \end{vmatrix}.$$

We then have

$$\det(\mathbf{A} + \mathbf{B}) = \begin{vmatrix} a_{11} + b_{11} & a_{12} + b_{12} & a_{13} + b_{13} \\ 2a_{21} & 2a_{22} & 2a_{23} \\ 2a_{31} & 2a_{32} & 2a_{33} \end{vmatrix} = 2 \cdot 2 \cdot \begin{vmatrix} a_{11} + b_{11} & a_{12} + b_{12} & a_{13} + b_{13} \\ a_{21} & a_{22} & a_{23} \\ a_{31} & a_{32} & a_{33} \end{vmatrix} = 4(\det(\mathbf{A}) + \det(\mathbf{B})).$$

11.3.4 Subtracting twice the second row from the first row, we get

$$\begin{vmatrix} 30 & 51 & 20 \\ 15 & 24 & 10 \\ -46 & 28 & -29 \end{vmatrix} = \begin{vmatrix} 0 & 3 & 0 \\ 15 & 24 & 10 \\ -46 & 28 & -29 \end{vmatrix}.$$

Expanding the determinant along the first row, we get $\begin{vmatrix} 0 & 3 & 0 \\ 15 & 24 & 10 \\ -46 & 28 & -29 \end{vmatrix} = -3 \begin{vmatrix} 15 & 10 \\ -46 & -29 \end{vmatrix}.$

Adding three times the first row to the second row, we get

$$-3 \begin{vmatrix} 15 & 10 \\ -46 & -29 \end{vmatrix} = -3 \begin{vmatrix} 15 & 10 \\ -1 & 1 \end{vmatrix} = -3[15 \cdot 1 - 10 \cdot (-1)] = -3(25) = \boxed{-75}.$$

11.3.5 Expanding the determinant along the first column, we get $\begin{vmatrix} a_{11} & a_{12} & a_{13} \\ 0 & a_{22} & a_{23} \\ 0 & 0 & a_{33} \end{vmatrix} = a_{11} \begin{vmatrix} a_{22} & a_{23} \\ 0 & a_{33} \end{vmatrix} = \boxed{a_{11} a_{22} a_{33}}.$

Exercises for Section 11.4

11.4.1

$$\begin{pmatrix} 3 & 4 \\ -1 & 5 \\ 0 & 2 \end{pmatrix} \begin{pmatrix} -3 & 4 & -2 & -1 \\ 7 & 8 & -1 & 0 \end{pmatrix} = \begin{pmatrix} (3)(-3) + (4)(7) & (3)(4) + (4)(8) & (3)(-2) + (4)(-1) & (3)(-1) + (4)(0) \\ (-1)(-3) + (5)(7) & (-1)(4) + (5)(8) & (-1)(-2) + (5)(-1) & (-1)(-1) + (5)(0) \\ (0)(-3) + (2)(7) & (0)(4) + (2)(8) & (0)(-2) + (2)(-1) & (0)(-1) + (2)(0) \end{pmatrix}$$

$$= \boxed{\begin{pmatrix} 19 & 44 & -10 & -3 \\ 38 & 36 & -3 & 1 \\ 14 & 16 & -2 & 0 \end{pmatrix}}.$$

11.4.2 $\begin{pmatrix} 4 \\ 1 \end{pmatrix} \begin{pmatrix} -1 & 5 & -3 \end{pmatrix} = \begin{pmatrix} (4)(-1) & (4)(5) & (4)(-3) \\ (1)(-1) & (1)(5) & (1)(-3) \end{pmatrix} = \boxed{\begin{pmatrix} -4 & 20 & -12 \\ -1 & 5 & -3 \end{pmatrix}}.$

11.4.3 We have $\mathbf{v} \cdot \mathbf{w} = v_1 w_1 + v_2 w_2 + \cdots + v_n w_n = w_1 v_1 + w_2 v_2 + \cdots + w_n v_n = \mathbf{w} \cdot \mathbf{v}.$

11.4.4

(a) We have

$$\mathbf{v} \cdot (a\mathbf{w}) = v_1(aw_1) + v_2(aw_2) + \cdots + v_n(aw_n) = a(v_1 w_1 + v_2 w_2 + \cdots + v_n w_n) = a\mathbf{v} \cdot \mathbf{w},$$

$$(a\mathbf{v}) \cdot \mathbf{w} = (av_1)w_1 + (av_2)w_2 + \cdots + (av_n)w_n = a(v_1 w_1 + v_2 w_2 + \cdots + v_n w_n) = a\mathbf{v} \cdot \mathbf{w},$$

so $\mathbf{v} \cdot (a\mathbf{w}) = (a\mathbf{v}) \cdot \mathbf{w} = a\mathbf{v} \cdot \mathbf{w}.$

(b)   We have

$$\mathbf{u} \cdot (\mathbf{v} + \mathbf{w}) = u_1(v_1 + w_1) + u_2(v_2 + w_2) + \cdots + u_n(v_n + w_n)$$
$$= (u_1v_1 + u_1w_1) + (u_2v_2 + u_2w_2) + \cdots + (u_nv_n + u_nw_n)$$
$$= u_1v_1 + u_2v_2 + \cdots + u_nv_n + u_1w_1 + u_2w_2 + \cdots + u_nw_n$$
$$= \mathbf{u} \cdot \mathbf{v} + \mathbf{u} \cdot \mathbf{w}.$$

11.4.5   In order for $\mathbf{IA}$ to be defined, $\mathbf{I}$ must have 2 columns, since $\mathbf{A}$ has 2 rows. Since the product $\mathbf{IA}$ must equal $\mathbf{A}$, the product must have 2 rows. Therefore, $\mathbf{I}$ must have 2 rows. This suggests that we try using the 2×2 identity matrix as $\mathbf{I}$. Sure enough, we find that

$$\begin{pmatrix} 1 & 0 \\ 0 & 1 \end{pmatrix} \begin{pmatrix} a_{11} & a_{12} & a_{13} \\ a_{21} & a_{22} & a_{23} \end{pmatrix} = \begin{pmatrix} a_{11} & a_{12} & a_{13} \\ a_{21} & a_{22} & a_{23} \end{pmatrix}.$$

However, is this the only possible $\mathbf{I}$? Letting $\mathbf{I} = \begin{pmatrix} p & q \\ r & s \end{pmatrix}$, we have

$$\begin{pmatrix} p & q \\ r & s \end{pmatrix} \begin{pmatrix} a_{11} & a_{12} & a_{13} \\ a_{21} & a_{22} & a_{23} \end{pmatrix} = \begin{pmatrix} a_{11} & a_{12} & a_{13} \\ a_{21} & a_{22} & a_{23} \end{pmatrix}.$$

From the entry in the first column and first row of the product, we have $pa_{11} + qa_{21}$, which can only equal a_{11} if $p = 1$ and $q = 0$. Similarly, we can show that $r = 0$ and $s = 1$, so the 2×2 identity matrix is the only matrix $\mathbf{I}$ such that $\mathbf{IA} = \mathbf{A}$.

Review Problems

11.29

(a)   We have $\|\mathbf{u}\| = \left\| \begin{pmatrix} 2 \\ 3 \\ -5 \end{pmatrix} \right\| = \sqrt{2^2 + 3^2 + (-5)^2} = \boxed{\sqrt{38}}$, and $\|\mathbf{v}\| = \left\| \begin{pmatrix} 4 \\ 5 \\ -1 \end{pmatrix} \right\| = \sqrt{4^2 + 5^2 + (-1)^2} = \boxed{\sqrt{42}}$.

(b)
$$(2\mathbf{u} + 3\mathbf{v}) \cdot (\mathbf{u} - 3\mathbf{v}) = \begin{pmatrix} 2 \cdot 2 + 3 \cdot 4 \\ 2 \cdot 3 + 3 \cdot 5 \\ 2 \cdot (-5) + 3 \cdot (-1) \end{pmatrix} \cdot \begin{pmatrix} 2 - 3 \cdot 4 \\ 3 - 3 \cdot 5 \\ -5 - 3 \cdot (-1) \end{pmatrix} = \begin{pmatrix} 16 \\ 21 \\ -13 \end{pmatrix} \cdot \begin{pmatrix} -10 \\ -12 \\ -2 \end{pmatrix}$$
$$= 16 \cdot (-10) + 21 \cdot (-12) + (-13) \cdot (-2) = \boxed{-386}.$$

11.30   Checking the dot product of all possible pairs, we see that only the $\boxed{\text{second and fourth}}$ vectors are orthogonal:

$$\begin{pmatrix} -3 \\ 4 \\ 2 \end{pmatrix} \cdot \begin{pmatrix} -2 \\ -3 \\ 3 \end{pmatrix} = (-3) \cdot (-2) + 4 \cdot (-3) + 2 \cdot 3 = 0.$$

11.31   Let $\mathbf{v} = \begin{pmatrix} x_1 \\ x_2 \\ x_3 \end{pmatrix}$ and $\mathbf{w} = \begin{pmatrix} y_1 \\ y_2 \\ y_3 \end{pmatrix}$, and let θ be the angle between $\mathbf{v}$ and $\mathbf{w}$. Then $\mathbf{v} \cdot \mathbf{w} = \|\mathbf{v}\|\|\mathbf{w}\| \cos \theta$. Squaring both sides, we get $(\mathbf{v} \cdot \mathbf{w})^2 = \|\mathbf{v}\|^2\|\mathbf{w}\|^2 \cos^2 \theta$, so

$$(x_1y_1 + x_2y_2 + x_3y_3)^2 = (x_1^2 + x_2^2 + x_3^2)(y_1^2 + y_2^2 + y_3^2) \cos^2 \theta.$$

But $\cos^2 \theta \leq 1$, so $(x_1y_1 + x_2y_2 + x_3y_3)^2 \leq (x_1^2 + x_2^2 + x_3^2)(y_1^2 + y_2^2 + y_3^2)$.

11.32

(a) $\begin{pmatrix} 4 & 5 & -2 \\ 1 & -3 & 4 \\ 0 & 1 & -2 \end{pmatrix} \begin{pmatrix} -1 \\ 1 \\ 4 \end{pmatrix} = \begin{pmatrix} 4 \cdot (-1) + 5 \cdot 1 + (-2) \cdot 4 \\ 1 \cdot (-1) + (-3) \cdot 1 + 4 \cdot 4 \\ 0 \cdot (-1) + 1 \cdot 1 + (-2) \cdot 4 \end{pmatrix} = \boxed{\begin{pmatrix} -7 \\ 12 \\ -7 \end{pmatrix}}.$

(b) There are many 3×3 matrices that have the same property as in part (a). For example,

$$\begin{pmatrix} 0 & -7 & 0 \\ 0 & 12 & 0 \\ 0 & -7 & 0 \end{pmatrix} \begin{pmatrix} -1 \\ 1 \\ 4 \end{pmatrix} = \begin{pmatrix} -7 \\ 12 \\ -7 \end{pmatrix}.$$

(c) Consider the equation

$$\begin{pmatrix} 4 & 5 & -2 \\ 1 & -3 & 4 \\ 0 & 1 & -2 \end{pmatrix} \begin{pmatrix} x \\ y \\ z \end{pmatrix} = \begin{pmatrix} -7 \\ 12 \\ -7 \end{pmatrix}.$$

The determinant of the 3×3 matrix is

$$\begin{vmatrix} 4 & 5 & -2 \\ 1 & -3 & 4 \\ 0 & 1 & -2 \end{vmatrix} = 4 \begin{vmatrix} -3 & 4 \\ 1 & -2 \end{vmatrix} - 5 \begin{vmatrix} 1 & 4 \\ 0 & -2 \end{vmatrix} + (-2) \begin{vmatrix} 1 & -3 \\ 0 & 1 \end{vmatrix}$$

$$= 4[(-3) \cdot (-2) - 4 \cdot 1] - 5[1 \cdot (-2) - 4 \cdot 0] - 2[1 \cdot 1 - (-3) \cdot 0]$$

$$= 4(2) - 5(-2) - 2(1)$$

$$= 16.$$

Because this determinant is nonzero, the system of equations has a unique solution. In other words, the vector $\begin{pmatrix} -1 \\ 1 \\ 4 \end{pmatrix}$ is the only vector that can be multiplied by the matrix in part (a) to get $\begin{pmatrix} -7 \\ 12 \\ -7 \end{pmatrix}$.

11.33

(a) Let $\mathbf{A} = \begin{pmatrix} 1 & 0 & 0 \\ 0 & 0 & 0 \\ 0 & 0 & 0 \end{pmatrix}$ and $\mathbf{v} = \begin{pmatrix} 0 \\ 0 \\ 1 \end{pmatrix}$. Then

$$\mathbf{Av} = \begin{pmatrix} 1 & 0 & 0 \\ 0 & 0 & 0 \\ 0 & 0 & 0 \end{pmatrix} \begin{pmatrix} 0 \\ 0 \\ 1 \end{pmatrix} = \begin{pmatrix} 0 \\ 0 \\ 0 \end{pmatrix}.$$

Thus, $\mathbf{Av} = \mathbf{0}$ does not imply $\mathbf{A} = \mathbf{0}$ or $\mathbf{v} = \mathbf{0}$.

(b) Let $\mathbf{A} = \begin{pmatrix} 1 & 0 & 0 \\ 0 & 0 & 0 \\ 0 & 0 & 0 \end{pmatrix}$ and $\mathbf{B} = \begin{pmatrix} 0 & 0 & 0 \\ 0 & 0 & 0 \\ 0 & 0 & 1 \end{pmatrix}$. Then

$$\mathbf{AB} = \begin{pmatrix} 1 & 0 & 0 \\ 0 & 0 & 0 \\ 0 & 0 & 0 \end{pmatrix} \begin{pmatrix} 0 & 0 & 0 \\ 0 & 0 & 0 \\ 0 & 0 & 1 \end{pmatrix} = \begin{pmatrix} 0 & 0 & 0 \\ 0 & 0 & 0 \\ 0 & 0 & 0 \end{pmatrix}.$$

Thus, $\mathbf{AB} = \mathbf{0}$ does not imply $\mathbf{A} = \mathbf{0}$ or $\mathbf{B} = \mathbf{0}$.

11.34 $\begin{pmatrix} 4 & -2 & -1 \\ 0 & 5 & -1 \\ -2 & 0 & -1 \end{pmatrix} \begin{pmatrix} -3 & 2 & 6 \\ -1 & 3 & -5 \\ 7 & 0 & 3 \end{pmatrix} = \boxed{\begin{pmatrix} -17 & 2 & 31 \\ -12 & 15 & -28 \\ -1 & -4 & -15 \end{pmatrix}}.$

11.35

(a) Suppose that such a matrix $\mathbf{P} = \begin{pmatrix} p_{11} & p_{12} & p_{13} \\ p_{21} & p_{22} & p_{23} \\ p_{31} & p_{32} & p_{33} \end{pmatrix}$ exists. Then

$$\mathbf{P}\begin{pmatrix} 1 & 0 & 0 \\ 0 & 0 & 0 \\ 0 & 0 & 0 \end{pmatrix} = \begin{pmatrix} 0 & 1 & 0 \\ 0 & 0 & 0 \\ 0 & 0 & 0 \end{pmatrix}.$$

But

$$\mathbf{P}\begin{pmatrix} 1 & 0 & 0 \\ 0 & 0 & 0 \\ 0 & 0 & 0 \end{pmatrix} = \begin{pmatrix} p_{11} & p_{12} & p_{13} \\ p_{21} & p_{22} & p_{23} \\ p_{31} & p_{32} & p_{33} \end{pmatrix}\begin{pmatrix} 1 & 0 & 0 \\ 0 & 0 & 0 \\ 0 & 0 & 0 \end{pmatrix} = \begin{pmatrix} p_{11} & 0 & 0 \\ p_{21} & 0 & 0 \\ p_{31} & 0 & 0 \end{pmatrix}.$$

These two matrices cannot be equal, so no such matrix $\mathbf{P}$ exists.

(b) Let the columns of $\mathbf{P}$ be $\mathbf{p}_1$, $\mathbf{p}_2$, and $\mathbf{p}_3$, in that order. So, the first column of $\mathbf{AP}$ is $\mathbf{Ap}_1$. In order for this to equal the second column of $\mathbf{A}$ for every $\mathbf{A}$, we must have $\mathbf{p}_1 = \begin{pmatrix} 0 \\ 1 \\ 0 \end{pmatrix}$, since this will cause each entry in $\mathbf{Ap}_1$ to be the corresponding entry in the second column of $\mathbf{A}$. Similarly, the product $\mathbf{Ap}_2$ must be the first column of $\mathbf{A}$, so $\mathbf{p}_2 = \begin{pmatrix} 1 \\ 0 \\ 0 \end{pmatrix}$, and $\mathbf{Ap}_3$ is the third column of $\mathbf{A}$, so $\mathbf{p}_3 = \begin{pmatrix} 0 \\ 0 \\ 1 \end{pmatrix}$.

So, we let $\mathbf{P} = \begin{pmatrix} 0 & 1 & 0 \\ 1 & 0 & 0 \\ 0 & 0 & 1 \end{pmatrix}$. Then

$$\begin{pmatrix} a_{11} & a_{12} & a_{13} \\ a_{21} & a_{22} & a_{23} \\ a_{31} & a_{32} & a_{33} \end{pmatrix}\mathbf{P} = \begin{pmatrix} a_{11} & a_{12} & a_{13} \\ a_{21} & a_{22} & a_{23} \\ a_{31} & a_{32} & a_{33} \end{pmatrix}\begin{pmatrix} 0 & 1 & 0 \\ 1 & 0 & 0 \\ 0 & 0 & 1 \end{pmatrix} = \begin{pmatrix} a_{12} & a_{11} & a_{13} \\ a_{22} & a_{21} & a_{23} \\ a_{32} & a_{31} & a_{33} \end{pmatrix}.$$

Hence, for any matrix $\mathbf{A}$, $\mathbf{AP}$ is the matrix that results when we swap the first two columns of $\mathbf{A}$.

11.36 We have $(\mathbf{A} + \mathbf{B})\mathbf{v} = \mathbf{0}$ for all vectors $\mathbf{v}$. Since this holds for all vectors $\mathbf{v}$, we can take $\mathbf{v} = \mathbf{i}$, $\mathbf{j}$, and $\mathbf{k}$, to get $(\mathbf{A} + \mathbf{B})\mathbf{i} = \mathbf{0}$, $(\mathbf{A} + \mathbf{B})\mathbf{j} = \mathbf{0}$, and $(\mathbf{A} + \mathbf{B})\mathbf{k} = \mathbf{0}$, respectively. Therefore, we have $\mathbf{A} + \mathbf{B} = \mathbf{0}$, so $\mathbf{A} = -\mathbf{B}$.

11.37

(a)
$$\begin{vmatrix} -2 & -4 & 1 \\ -5 & 3 & 2 \\ -3 & 1 & 8 \end{vmatrix} = -2\begin{vmatrix} 3 & 2 \\ 1 & 8 \end{vmatrix} - (-4)\begin{vmatrix} -5 & 2 \\ -3 & 8 \end{vmatrix} + \begin{vmatrix} -5 & 3 \\ -3 & 1 \end{vmatrix}$$
$$= -2(3 \cdot 8 - 2 \cdot 1) + 4[(-5) \cdot 8 - 2 \cdot (-3)] + [(-5) \cdot 1 - 3 \cdot (-3)]$$
$$= -2(22) + 4(-34) + 4 = \boxed{-176}.$$

(b)
$$\begin{vmatrix} 1/3 & 1/5 & 6 \\ 4 & -3 & 5 \\ -1 & 6 & -10 \end{vmatrix} = \frac{1}{3}\begin{vmatrix} -3 & 5 \\ 6 & -10 \end{vmatrix} - \frac{1}{5}\begin{vmatrix} 4 & 5 \\ -1 & -10 \end{vmatrix} + 6\begin{vmatrix} 4 & -3 \\ -1 & 6 \end{vmatrix}$$
$$= \frac{1}{3}[(-3) \cdot (-10) - 5 \cdot 6] - \frac{1}{5}[4 \cdot (-10) - 5 \cdot (-1)] + 6[4 \cdot 6 - (-3) \cdot (-1)]$$
$$= \frac{1}{3} \cdot 0 - \frac{1}{5} \cdot (-35) + 6 \cdot 21 = \boxed{133}.$$

11.38 By Cramer's Rule,

$$x = \frac{\begin{vmatrix} -3 & -2 & 5 \\ 4 & -5 & 3 \\ -1 & 2 & 3 \end{vmatrix}}{\begin{vmatrix} 1 & -2 & 5 \\ 2 & -5 & 3 \\ -5 & 2 & 3 \end{vmatrix}}, \quad y = \frac{\begin{vmatrix} 1 & -3 & 5 \\ 2 & 4 & 3 \\ -5 & -1 & 3 \end{vmatrix}}{\begin{vmatrix} 1 & -2 & 5 \\ 2 & -5 & 3 \\ -5 & 2 & 3 \end{vmatrix}}, \quad z = \frac{\begin{vmatrix} 1 & -2 & -3 \\ 2 & -5 & 4 \\ -5 & 2 & -1 \end{vmatrix}}{\begin{vmatrix} 1 & -2 & 5 \\ 2 & -5 & 3 \\ -5 & 2 & 3 \end{vmatrix}}.$$

These determinants are

$$\begin{vmatrix} -3 & -2 & 5 \\ 4 & -5 & 3 \\ -1 & 2 & 3 \end{vmatrix} = -3 \begin{vmatrix} -5 & 3 \\ 2 & 3 \end{vmatrix} - (-2) \begin{vmatrix} 4 & 3 \\ -1 & 3 \end{vmatrix} + 5 \begin{vmatrix} 4 & -5 \\ -1 & 2 \end{vmatrix} = -3(-21) + 2(15) + 5(3) = 108,$$

$$\begin{vmatrix} 1 & -3 & 5 \\ 2 & 4 & 3 \\ -5 & -1 & 3 \end{vmatrix} = \begin{vmatrix} 4 & 3 \\ -1 & 3 \end{vmatrix} - (-3) \begin{vmatrix} 2 & 3 \\ -5 & 3 \end{vmatrix} + 5 \begin{vmatrix} 2 & 4 \\ -5 & -1 \end{vmatrix} = 15 + 3(21) + 5(18) = 168,$$

$$\begin{vmatrix} 1 & -2 & -3 \\ 2 & -5 & 4 \\ -5 & 2 & -1 \end{vmatrix} = \begin{vmatrix} -5 & 4 \\ 2 & -1 \end{vmatrix} - (-2) \begin{vmatrix} 2 & 4 \\ -5 & -1 \end{vmatrix} + (-3) \begin{vmatrix} 2 & -5 \\ -5 & 2 \end{vmatrix} = -3 + 2(18) - 3(-21) = 96,$$

$$\begin{vmatrix} 1 & -2 & 5 \\ 2 & -5 & 3 \\ -5 & 2 & 3 \end{vmatrix} = \begin{vmatrix} -5 & 3 \\ 2 & 3 \end{vmatrix} - (-2) \begin{vmatrix} 2 & 3 \\ -5 & 3 \end{vmatrix} + 5 \begin{vmatrix} 2 & -5 \\ -5 & 2 \end{vmatrix} = -21 + 2(21) + 5(-21) = -84.$$

Hence, $x = 108/(-84) = \boxed{-9/7}$, $y = 168/(-84) = \boxed{-2}$, and $z = 96/(-84) = \boxed{-8/7}$.

11.39 Subtracting the first row from the second row, we get

$$\begin{vmatrix} x-4 & x-3 & x-2 \\ x-1 & x & x+1 \\ x+2 & x+3 & x+4 \end{vmatrix} = \begin{vmatrix} x-4 & x-3 & x-2 \\ (x-1)-(x-4) & x-(x-3) & (x+1)-(x-2) \\ x+2 & x+3 & x+4 \end{vmatrix} = \begin{vmatrix} x-4 & x-3 & x-2 \\ 3 & 3 & 3 \\ x+2 & x+3 & x+4 \end{vmatrix}.$$

Subtracting the first row from the third row, we get

$$\begin{vmatrix} x-4 & x-3 & x-2 \\ 3 & 3 & 3 \\ x+2 & x+3 & x+4 \end{vmatrix} = \begin{vmatrix} x-4 & x-3 & x-2 \\ 3 & 3 & 3 \\ (x+2)-(x-4) & (x+3)-(x-3) & (x+4)-(x-2) \end{vmatrix} = \begin{vmatrix} x-4 & x-3 & x-2 \\ 3 & 3 & 3 \\ 6 & 6 & 6 \end{vmatrix}.$$

Then $\begin{vmatrix} x-4 & x-3 & x-2 \\ 3 & 3 & 3 \\ 6 & 6 & 6 \end{vmatrix} = 2 \begin{vmatrix} x-4 & x-3 & x-2 \\ 3 & 3 & 3 \\ 3 & 3 & 3 \end{vmatrix} = \boxed{0}.$

11.40

(a) We have seen that the entry in row i and column j of the product $\mathbf{AB}$ is the dot product of the ith row of $\mathbf{A}$ and the jth column of $\mathbf{B}$. This is equal to the entry in row j and column i of $(\mathbf{AB})^T$.

By the same token, the entry in row j and column i of the product $\mathbf{B}^T\mathbf{A}^T$ is the dot product of the jth row of $\mathbf{B}^T$ and the ith column of $\mathbf{A}^T$. But the jth row of $\mathbf{B}^T$ is the jth column of $\mathbf{B}$, and the ith column of $\mathbf{A}^T$ is the ith row of $\mathbf{A}$. Hence, the entry in row j and column i of $\mathbf{B}^T\mathbf{A}^T$ is the dot product of the jth column of $\mathbf{B}$ and the ith row of $\mathbf{A}$. This equals the entry in row j and column i of $(\mathbf{AB})^T$, as mentioned above. Since this holds for all possible values of i and j, we have $(\mathbf{AB})^T = \mathbf{B}^T\mathbf{A}^T$.

(b) Suppose $\mathbf{A}$ is an $r \times s$ matrix and $\mathbf{B}$ is an $s \times c$ matrix, so that $\mathbf{AB}$ is defined. Then, we note that $\mathbf{B}^T\mathbf{A}^T$ exists because $\mathbf{B}^T$ is a $c \times s$ matrix while $\mathbf{A}^T$ is an $s \times r$ matrix. Next we note that our solution in part (a) does not ever mention the dimensions of the matrices; every step holds for this part, as well. So, we have $(\mathbf{AB})^T = \mathbf{B}^T\mathbf{A}^T$.

11.41

$$\begin{pmatrix} 4 & 0 & 2 \\ 5 & -1 & -1 \\ 3 & -2 & 7 \\ -1 & 5 & 2 \end{pmatrix} \begin{pmatrix} 7 & -1 \\ 0 & -2 \\ -1 & 2 \end{pmatrix} = \begin{pmatrix} (4)(7) + (0)(0) + (2)(-1) & (4)(-1) + (0)(-2) + (2)(2) \\ (5)(7) + (-1)(0) + (-1)(-1) & (5)(-1) + (-1)(-2) + (-1)(2) \\ (3)(7) + (-2)(0) + (7)(-1) & (3)(-1) + (-2)(-2) + (7)(2) \\ (-1)(7) + (5)(0) + (2)(-1) & (-1)(-1) + (5)(-2) + (2)(2) \end{pmatrix}$$

$$= \boxed{\begin{pmatrix} 26 & 0 \\ 36 & -5 \\ 14 & 15 \\ -9 & -5 \end{pmatrix}}.$$

11.42   In order for $\mathbf{AB}$ to be defined, the number of columns in $\mathbf{A}$ must equal the number of rows in $\mathbf{B}$, so $\mathbf{B}$ has 4 rows. The number of columns in $\mathbf{AB}$ equals the number of columns in $\mathbf{B}$, so because $\mathbf{C} = \mathbf{AB}$ has 6 columns, $\mathbf{B}$ also has 6 columns. Therefore, $\mathbf{B}$ is a $\boxed{4 \times 6}$ matrix.

11.43

(a) Let $\mathbf{A}$ be an $n \times n$ matrix and $\mathbf{B}$ be an $m \times m$ matrix. If $\mathbf{AB}$ is defined, then $n = m$, so $\mathbf{B}$ is an $n \times n$ matrix, and the product $\mathbf{AB}$ is also an $n \times n$ matrix.

(b) $\boxed{\text{No}}$. Suppose $\mathbf{A}$ is an $r \times n$ matrix and $\mathbf{B}$ is an $n \times r$ matrix. Then, $\mathbf{AB}$ is an $r \times r$ matrix. If $r \neq n$, then $\mathbf{A}$ and $\mathbf{B}$ are not square, but $\mathbf{AB}$ is.

Challenge Problems

11.44   From the given information, we have $\|\mathbf{u}\| = \|\mathbf{v}\| = \|\mathbf{w}\| = 1$ and $\mathbf{u} \cdot \mathbf{v} = \mathbf{u} \cdot \mathbf{w} = \mathbf{v} \cdot \mathbf{w} = 0$. Then

$$\mathbf{r} \cdot \mathbf{u} = (a\mathbf{u} + b\mathbf{v} + c\mathbf{w}) \cdot \mathbf{u} = a\mathbf{u} \cdot \mathbf{u} + b\mathbf{u} \cdot \mathbf{v} + c\mathbf{u} \cdot \mathbf{w} = a\|\mathbf{u}\|^2 = a.$$

Similarly,

$$\mathbf{r} \cdot \mathbf{v} = (a\mathbf{u} + b\mathbf{v} + c\mathbf{w}) \cdot \mathbf{v} = a\mathbf{u} \cdot \mathbf{v} + b\mathbf{v} \cdot \mathbf{v} + c\mathbf{v} \cdot \mathbf{w} = b\|\mathbf{v}\|^2 = b,$$

and

$$\mathbf{r} \cdot \mathbf{w} = (a\mathbf{u} + b\mathbf{v} + c\mathbf{w}) \cdot \mathbf{w} = a\mathbf{u} \cdot \mathbf{w} + b\mathbf{v} \cdot \mathbf{w} + c\mathbf{w} \cdot \mathbf{w} = c\|\mathbf{w}\|^2 = c.$$

11.45

(a) Let θ be the angle between $\mathbf{u}$ and $\mathbf{v}$. Then $\mathbf{u} \cdot \mathbf{v} = \|\mathbf{u}\|\|\mathbf{v}\| \cos \theta$. But

$$\mathbf{u} \cdot \mathbf{v} = (\mathbf{i} + \mathbf{j} + z\mathbf{k}) \cdot (2\mathbf{i} - \mathbf{j} + 3\mathbf{k}) = 1 \cdot 2 + 1 \cdot (-1) + z \cdot 3 = 3z + 1,$$

$\|\mathbf{u}\| = \sqrt{1^2 + 1^2 + z^2} = \sqrt{z^2 + 2}$, and $\|\mathbf{v}\| = \sqrt{2^2 + (-1)^2 + 3^2} = \sqrt{14}$. Therefore, we have

$$3z + 1 = \sqrt{14(z^2 + 2)} \cos \theta.$$

For $\theta = \frac{\pi}{2}$, we have $\cos \theta = \cos \frac{\pi}{2} = 0$, so $3z + 1 = 0$, which means $z = \boxed{-1/3}$.

(b) For $\theta = \frac{\pi}{3}$, we have $\cos\theta = \cos\frac{\pi}{3} = \frac{1}{2}$, so $3z + 1 = \sqrt{14(z^2 + 2)} \cdot \frac{1}{2}$. Multiplying both sides by 2, we get $6z + 2 = \sqrt{14(z^2 + 2)}$. Squaring both sides, we get $(6z + 2)^2 = 14(z^2 + 2)$, which expands as $36z^2 + 24z + 4 = 14z^2 + 28$, so $22z^2 + 24z - 24 = 0$, or $11z^2 + 12z - 12 = 0$. By the quadratic formula, we have

$$z = \frac{-12 \pm \sqrt{12^2 - 4 \cdot 11 \cdot (-12)}}{2 \cdot 11} = \frac{-12 \pm \sqrt{672}}{22} = \frac{-12 \pm 4\sqrt{42}}{22} = \frac{-6 \pm 2\sqrt{42}}{11}.$$

The negative root is extraneous, because it makes $3z + 1$ negative, but our initial equation is $3z + 1 = \sqrt{14(z^2 + 2)} \cdot \frac{1}{2}$. Therefore, the only value of z that satisfies the equation is $\boxed{(-6 + 2\sqrt{42})/11}$.

(c) Let $c = \cos\theta$, so $3z + 1 = (\sqrt{14(z^2 + 2)})c$. Our goal is to find the largest value of c for which there is a real value of z that satisfies this equation. Squaring both sides, we get

$$9z^2 + 6z + 1 = 14(z^2 + 2)c^2 = 14c^2z^2 + 28c^2.$$

Rearranging this equation gives $(14c^2 - 9)z^2 - 6z + (28c^2 - 1) = 0$, which is a quadratic equation in z. So, we seek the largest value of c such that this quadratic has real roots. This quadratic equation has a real root in z if and only if its discriminant is nonnegative. The discriminant is

$$(-6)^2 - 4(14c^2 - 9)(28c^2 - 1) = 36 - (1568c^4 - 1064c^2 + 36) = -1568c^4 + 1064c^2 = 56c^2(19 - 28c^2).$$

This quantity is nonnegative if and only if $c^2 \le \frac{19}{28}$. Therefore, we can find a value of z such that the angle between $\mathbf{u}$ and $\mathbf{v}$ is θ if and only if $\cos^2\theta \le \frac{19}{28}$. So, the desired maximum value of $\cos^2\theta$ is $\boxed{\frac{19}{28}}$.

11.46 Subtracting the second row from the first row, we have

$$\begin{vmatrix} 1 & a & a^2 \\ 1 & b & b^2 \\ 1 & c & c^2 \end{vmatrix} = \begin{vmatrix} 0 & a-b & a^2-b^2 \\ 1 & b & b^2 \\ 1 & c & c^2 \end{vmatrix} = -(a-b)\begin{vmatrix} 1 & b^2 \\ 1 & c^2 \end{vmatrix} + (a^2-b^2)\begin{vmatrix} 1 & b \\ 1 & c \end{vmatrix}.$$

We can take out a factor of $a - b$, to get

$$-(a-b)\begin{vmatrix} 1 & b^2 \\ 1 & c^2 \end{vmatrix} + (a^2-b^2)\begin{vmatrix} 1 & b \\ 1 & c \end{vmatrix} = -(a-b)\begin{vmatrix} 1 & b^2 \\ 1 & c^2 \end{vmatrix} + (a-b)(a+b)\begin{vmatrix} 1 & b \\ 1 & c \end{vmatrix} = (a-b)\left[-\begin{vmatrix} 1 & b^2 \\ 1 & c^2 \end{vmatrix} + (a+b)\begin{vmatrix} 1 & b \\ 1 & c \end{vmatrix} \right]$$

$$= (a-b)[-(c^2-b^2) + (a+b)(c-b)] = (a-b)[(b^2-c^2) - (a+b)(b-c)]$$

$$= (a-b)[(b+c)(b-c) - (a+b)(b-c)] = (a-b)(b-c)[(b+c) - (a+b)]$$

$$= \boxed{(a-b)(b-c)(c-a)}.$$

We also could have tackled this problem with a little clever insight into polynomials. Clearly, the determinant, if nonzero, has at most degree 3, since every term in the expansion has degree 3. Moreover, since the determinant has two equal rows if $a = b$, the determinant is 0 if $a = b$. So, we conclude that $a - b$ is a factor of the polynomial that is the determinant. Similarly, $b - c$ and $c - a$ are also factors of the polynomial, so the desired polynomial is $k(a-b)(b-c)(c-a)$ for some constant k. Choosing $a = 0$, $b = 1$, $c = 2$ gives a determinant of 2, so we must have $2 = k(a-b)(b-c)(c-a) = k(0-1)(1-2)(2-0)$, which gives $k = 1$, and the desired polynomial is $\boxed{(a-b)(b-c)(c-a)}$, as before.

11.47 If $\mathbf{M}^T = -\mathbf{M}$, then $\det(\mathbf{M}^T) = \det(-\mathbf{M}) = (-1)^3 \det(\mathbf{M}) = -\det(\mathbf{M})$. But $\det(\mathbf{M}^T) = \det(\mathbf{M})$, so $\det(\mathbf{M}) = -\det(\mathbf{M})$. Therefore, $\det(\mathbf{M}) = 0$.

11.48

(a) If $\mathbf{A}\mathbf{v} = \lambda\mathbf{v}$, then $\mathbf{A}\mathbf{v} - \lambda\mathbf{v} = \mathbf{0}$, so $(\mathbf{A} - \lambda\mathbf{I})\mathbf{v} = \mathbf{0}$. If $\det(\mathbf{A} - \lambda\mathbf{I}) \ne 0$, then the equation $(\mathbf{A} - \lambda\mathbf{I})\mathbf{v} = \mathbf{0}$ has a unique solution, which is clearly $\mathbf{v} = \mathbf{0}$. Therefore, if $(\mathbf{A} - \lambda\mathbf{I})\mathbf{v} = \mathbf{0}$ for some nonzero vector $\mathbf{v}$, then $\det(\mathbf{A} - \lambda\mathbf{I}) = 0$.

(b)   We have

$$\det(\mathbf{A} - \lambda \mathbf{I}) = \begin{vmatrix} -8 - \lambda & 1 & 3 \\ -4 & 1 - \lambda & 7 \\ 8 & 1 & -1 - \lambda \end{vmatrix}$$

$$= (-8 - \lambda)\begin{vmatrix} 1 - \lambda & 7 \\ 1 & -1 - \lambda \end{vmatrix} - \begin{vmatrix} -4 & 7 \\ 8 & -1 - \lambda \end{vmatrix} + 3\begin{vmatrix} -4 & 1 - \lambda \\ 8 & 1 \end{vmatrix}$$

$$= -(\lambda + 8)[(1 - \lambda)(-1 - \lambda) - 7 \cdot 1] - [(-4)(-1 - \lambda) - 7 \cdot 8] + 3[(-4)(1) - (1 - \lambda)(8)]$$

$$= -(\lambda + 8)(\lambda^2 - 8) - (4\lambda - 52) + 3(8\lambda - 12)$$

$$= (-\lambda^3 - 8\lambda^2 + 8\lambda + 64) + (-4\lambda + 52) + (24\lambda - 36)$$

$$= -\lambda^3 - 8\lambda^2 + 28\lambda + 80.$$

This polynomial factors as $-(\lambda - 4)(\lambda + 2)(\lambda + 10) = 0$, so the values of λ such that $\det(\mathbf{A} - \lambda \mathbf{I}) = 0$ are $\lambda = \boxed{4, -2, \text{ and } -10}$.

(c)   Let $\mathbf{v} = \begin{pmatrix} x \\ y \\ z \end{pmatrix}$. Then for each eigenvalue λ from part (a), we want to find a vector $\mathbf{v}$ such that $\mathbf{Av} = \lambda\mathbf{v}$, or

$$-8x + y + 3z = \lambda x,$$
$$-4x + y + 7z = \lambda y,$$
$$8x + y - z = \lambda z.$$

For $\lambda = 4$, this system becomes

$$-12x + y + 3z = 0,$$
$$-4x - 3y + 7z = 0,$$
$$8x + y - 5z = 0.$$

Subtracting the third equation from the first equation to eliminate y, we get $-20x + 8z = 0$, so $20x = 8z$, or $5x = 2z$. Let $x = 2$ and $z = 5$. Then from the first equation, we have $y = 12x - 3z = 9$. These values satisfy all three equations, so for $\lambda = 4$, we can take $\mathbf{v} = \boxed{\begin{pmatrix} 2 \\ 9 \\ 5 \end{pmatrix}}$.

For $\lambda = -2$, this system becomes

$$-6x + y + 3z = 0,$$
$$-4x + 3y + 7z = 0,$$
$$8x + y + z = 0.$$

Subtracting the third equation from the first equation to eliminate y, we get $-14x + 2z = 0$, so $14x = 2z$, or $z = 7x$. Let $x = 1$ and $z = 7$. Then from the first equation, we have $y = 6x - 3z = -15$. These values satisfy all three equations, so for $\lambda = -2$, we can take $\mathbf{v} = \boxed{\begin{pmatrix} 1 \\ -15 \\ 7 \end{pmatrix}}$.

Finally, for $\lambda = -10$, this system becomes

$$2x + y + 3z = 0,$$
$$-4x + 11y + 7z = 0,$$
$$8x + y + 9z = 0.$$

Subtracting the first equation from the third equation to eliminate y, we get $6x + 6z = 0$, so $z = -x$. Let $x = 1$ and $z = -1$. Then from the first equation, we have $y = -2x - 3z = 1$. These values satisfy all three equations, so for $\lambda = -10$, we can take $\mathbf{v} = \boxed{\begin{pmatrix} 1 \\ 1 \\ -1 \end{pmatrix}}$.

11.49 Computing the first few powers of $\mathbf{A}$, we find

$$\mathbf{A} = \begin{pmatrix} 1 & 1 & 0 \\ 0 & 1 & 1 \\ 0 & 0 & 1 \end{pmatrix}, \mathbf{A}^2 = \begin{pmatrix} 1 & 2 & 1 \\ 0 & 1 & 2 \\ 0 & 0 & 1 \end{pmatrix}, \mathbf{A}^3 = \begin{pmatrix} 1 & 3 & 3 \\ 0 & 1 & 3 \\ 0 & 0 & 1 \end{pmatrix}, \mathbf{A}^4 = \mathbf{A} \cdot \mathbf{A}^3 = \begin{pmatrix} 1 & 4 & 6 \\ 0 & 1 & 4 \\ 0 & 0 & 1 \end{pmatrix}, \mathbf{A}^5 = \mathbf{A} \cdot \mathbf{A}^4 = \begin{pmatrix} 1 & 5 & 10 \\ 0 & 1 & 5 \\ 0 & 0 & 1 \end{pmatrix}.$$

It appears that $\mathbf{A}^n = \begin{pmatrix} 1 & n & \frac{n(n-1)}{2} \\ 0 & 1 & n \\ 0 & 0 & 1 \end{pmatrix}$ for all positive integers n. We prove this using induction.

The claim is true for $n = 1$, so assume that it is true for some positive integer k, i.e. $\mathbf{A}^k = \begin{pmatrix} 1 & k & \frac{k(k-1)}{2} \\ 0 & 1 & k \\ 0 & 0 & 1 \end{pmatrix}$. Then

$$\mathbf{A}^{k+1} = \mathbf{A} \cdot \mathbf{A}^k$$

$$= \begin{pmatrix} 1 & 1 & 0 \\ 0 & 1 & 1 \\ 0 & 0 & 1 \end{pmatrix} \begin{pmatrix} 1 & k & \frac{k(k-1)}{2} \\ 0 & 1 & k \\ 0 & 0 & 1 \end{pmatrix} = \begin{pmatrix} 1 & k+1 & \frac{k(k-1)}{2} + k \\ 0 & 1 & k+1 \\ 0 & 0 & 1 \end{pmatrix} = \begin{pmatrix} 1 & k+1 & \frac{k^2-k+2k}{2} \\ 0 & 1 & k+1 \\ 0 & 0 & 1 \end{pmatrix}$$

$$= \begin{pmatrix} 1 & k+1 & \frac{k^2+k}{2} \\ 0 & 1 & k+1 \\ 0 & 0 & 1 \end{pmatrix} = \begin{pmatrix} 1 & k+1 & \frac{k(k+1)}{2} \\ 0 & 1 & k+1 \\ 0 & 0 & 1 \end{pmatrix}.$$

Thus, the claim is true for $n = k + 1$, so by induction, it holds for all positive integers n.

11.50 Since $\mathbf{A}$ has 3 columns and $\mathbf{B}$ has 2 rows, $\mathbf{M}$ must be a 3×2 matrix. So, we must have

$$\begin{pmatrix} 5 & 3 & -2 \\ -1 & 5 & 4 \end{pmatrix} \begin{pmatrix} m_{11} & m_{12} \\ m_{21} & m_{22} \\ m_{31} & m_{32} \end{pmatrix} = \begin{pmatrix} -6 & -20 \\ 8 & 24 \end{pmatrix}.$$

Equating the entries in the first column of the product to the corresponding entries in $\mathbf{B}$ gives us the two equations

$$5m_{11} + 3m_{21} - 2m_{31} = -6,$$
$$-m_{11} + 5m_{21} + 4m_{31} = 8.$$

We can't produce any more equations involving m_{11}, m_{21}, and m_{31}. Since we only have two linear equations for these three variables, we cannot find a unique solution (if there is a solution). But we are able to find solutions. For example, adding 5 times the second equation to the first, we get $28m_{21} + 18m_{31} = 34$. We can't eliminate any more variables, so we can choose any value for m_{21} and then find corresponding m_{11} and m_{31}. For example, if we let $m_{21} = -2$, we have $18m_{31} = 34 - 28m_{21} = 90$, so $m_{31} = 5$. From either of our original equations, we have $m_{11} = 2$.

Turning to the second column of the product and the corresponding entries in $\mathbf{B}$, we have

$$5m_{12} + 3m_{22} - 2m_{32} = -20,$$
$$-m_{12} + 5m_{22} + 4m_{32} = 24.$$

We can again choose any value for one of the variables, and then find values for the other two. For example, if $m_{12} = -3$, we will find $m_{22} = 1$ and $m_{32} = 4$. Checking, we find that we do indeed have

$$\begin{pmatrix} 5 & 3 & -2 \\ -1 & 5 & 4 \end{pmatrix} \begin{pmatrix} 2 & -3 \\ -2 & 1 \\ 5 & 4 \end{pmatrix} = \begin{pmatrix} -6 & -20 \\ 8 & 24 \end{pmatrix},$$

so $\mathbf{M} = \begin{pmatrix} 2 & -3 \\ -2 & 1 \\ 5 & 4 \end{pmatrix}$ satisfies the problem, but it is not the only matrix that does so. For example, the matrix

$\mathbf{M} = \begin{pmatrix} -9 & 8 \\ 7 & -8 \\ -9 & 18 \end{pmatrix}$ also satisfies the problem.

11.51

(a) Let $\mathbf{A} = \begin{pmatrix} 1 & 0 \\ 0 & -1 \end{pmatrix}$ and $\mathbf{B} = \begin{pmatrix} 0 & 1 \\ -1 & 0 \end{pmatrix}$. Then

$$\mathbf{AB} = \begin{pmatrix} 1 & 0 \\ 0 & -1 \end{pmatrix} \begin{pmatrix} 0 & 1 \\ -1 & 0 \end{pmatrix} = \begin{pmatrix} 0 & 1 \\ 1 & 0 \end{pmatrix}$$

and

$$\mathbf{BA} = \begin{pmatrix} 0 & 1 \\ -1 & 0 \end{pmatrix} \begin{pmatrix} 1 & 0 \\ 0 & -1 \end{pmatrix} = \begin{pmatrix} 0 & -1 \\ -1 & 0 \end{pmatrix}.$$

Thus, $\mathbf{AB} = -\mathbf{BA}$. But $\det(\mathbf{A}) = -1$ and $\det(\mathbf{B}) = 1$, so both $\mathbf{A}$ and $\mathbf{B}$ are nonsingular.

(b) If $\mathbf{AB} = -\mathbf{BA}$, then $\det(\mathbf{AB}) = \det(-\mathbf{BA})$. Since $-\mathbf{BA}$ is a 3×3 matrix, $\det(-\mathbf{BA}) = (-1)^3 \det(\mathbf{BA}) = -\det(\mathbf{BA})$, so $\det(\mathbf{AB}) = -\det(\mathbf{BA})$. We also have $\det(\mathbf{BA}) = \det(\mathbf{B})\det(\mathbf{A}) = \det(\mathbf{A})\det(\mathbf{B}) = \det(\mathbf{AB})$, so $\det(\mathbf{AB}) = -\det(\mathbf{BA})$ gives us $\det(\mathbf{AB}) = -\det(\mathbf{AB})$, which means that $\det(\mathbf{AB}) = 0$. Since $\det(\mathbf{AB}) = \det(\mathbf{A})\det(\mathbf{B})$, we must have $\det(\mathbf{A}) = 0$ or $\det(\mathbf{B}) = 0$. In other words, at least one of $\mathbf{A}$ and $\mathbf{B}$ must be singular.

11.52 Let

$$\mathbf{A}^{2006} = \begin{pmatrix} b_{11} & b_{12} & b_{13} \\ b_{21} & b_{22} & b_{23} \\ b_{31} & b_{32} & b_{33} \end{pmatrix}.$$

Since all the entries of $\mathbf{A}$ are even, all the entries of $\mathbf{A}^{2006}$ are even, i.e. b_{ij} is even for all i, j. Then

$$\det(\mathbf{I}-\mathbf{A}^{2006}) = \begin{vmatrix} 1-b_{11} & -b_{12} & -b_{13} \\ -b_{21} & 1-b_{22} & -b_{23} \\ -b_{31} & -b_{32} & 1-b_{33} \end{vmatrix} = (1-b_{11}) \begin{vmatrix} 1-b_{22} & -b_{23} \\ -b_{32} & 1-b_{33} \end{vmatrix} - (-b_{12}) \begin{vmatrix} -b_{21} & -b_{23} \\ -b_{31} & 1-b_{33} \end{vmatrix} + (-b_{13}) \begin{vmatrix} -b_{21} & 1-b_{22} \\ -b_{31} & -b_{32} \end{vmatrix}.$$

The first term expands as

$$(1-b_{11}) \begin{vmatrix} 1-b_{22} & -b_{23} \\ -b_{32} & 1-b_{33} \end{vmatrix} = (1-b_{11})[(1-b_{22})(1-b_{33}) - (-b_{23})(-b_{32})].$$

Since all of the b_{ij} are even, the expressions $1 - b_{11}$, $1 - b_{22}$, and $1 - b_{33}$ are odd and $(-b_{23})(-b_{32})$ is even. Therefore, the expression $(1 - b_{11})[(1 - b_{22})(1 - b_{33}) - (-b_{23})(-b_{32})]$ is odd.

Since b_{12} and b_{13} are even, so are

$$(-b_{12}) \begin{vmatrix} -b_{21} & -b_{23} \\ -b_{31} & 1-b_{33} \end{vmatrix} \quad \text{and} \quad (-b_{13}) \begin{vmatrix} -b_{21} & 1-b_{22} \\ -b_{31} & -b_{32} \end{vmatrix}.$$

Therefore, $\det(\mathbf{I}-\mathbf{A}^{2006})$ is the sum of one odd and two even terms, which means $\det(\mathbf{I}-\mathbf{A}^{2006})$ is odd. In particular, it cannot be equal to 0.

CHAPTER 12

Vectors and Matrices in Three Dimensions, Part 2

Exercises for Section 12.1

12.1.1 The vector from $(-4, 5, 0)$ to a point on the line must be in the same or opposite direction as the vector from $(-4, 5, 0)$ to $(-5, -1, 2)$. Therefore, if we take

$$\mathbf{v} = \boxed{\begin{pmatrix} -4 \\ 5 \\ 0 \end{pmatrix}} \quad \text{and} \quad \mathbf{w} = \begin{pmatrix} -5 \\ -1 \\ 2 \end{pmatrix} - \begin{pmatrix} -4 \\ 5 \\ 0 \end{pmatrix} = \boxed{\begin{pmatrix} -1 \\ -6 \\ 2 \end{pmatrix}},$$

then the line consists of all points (x, y, z) such that $\begin{pmatrix} x \\ y \\ z \end{pmatrix} = \mathbf{v} + \mathbf{w}t$.

These values of $\mathbf{v}$ and $\mathbf{w}$ are not unique; we can take $\mathbf{v}$ to be any vector to a point on the line, and $\mathbf{w}$ to be any nonzero scalar multiple of $\begin{pmatrix} -1 \\ -6 \\ 2 \end{pmatrix}$.

12.1.2 We can write

$$\begin{pmatrix} x \\ y \\ z \end{pmatrix} = \begin{pmatrix} 4 - 2t \\ 2 + t \\ 2 - 3t \end{pmatrix} = \begin{pmatrix} 4 \\ 2 \\ 2 \end{pmatrix} + \begin{pmatrix} -2 \\ 1 \\ -3 \end{pmatrix} t,$$

so a vector that is parallel to the graph is $\boxed{\begin{pmatrix} -2 \\ 1 \\ -3 \end{pmatrix}}$. (Any nonzero scalar multiple of this vector is parallel to the graph as well.)

12.1.3 $\mathcal{P}$ is the graph of

$$\begin{pmatrix} x \\ y \\ z \end{pmatrix} = \begin{pmatrix} -2 \\ -1 \\ 4 \end{pmatrix} + s \begin{pmatrix} 2 \\ -1 \\ 1 \end{pmatrix} + t \begin{pmatrix} -3 \\ 0 \\ 2 \end{pmatrix}.$$

In order for $(10, -4, 3)$ to be on the plane, we must have

$$10 = -2 + 2s - 3t,$$
$$-4 = -1 - s,$$
$$3 = 4 + s + 2t.$$

From the second equation, we have $s = 3$. Substituting this into the third equation gives $t = -2$. The values $(s, t) = (3, -2)$ also satisfy the first equation, so letting $(s, t) = (3, -2)$ does give $(x, y, z) = (10, -4, 3)$ in our equation for $\mathcal{P}$. This means $\mathcal{P}$ passes through $(10, -4, 3)$.

To show that $\mathcal{P}$ does not pass through $(-1, 5, 2)$, we must show the equation for $\mathcal{P}$ cannot give $(x, y, z) = (-1, 5, 2)$ for any values of s and t. If such values of s and t exist, we must have

$$-1 = -2 + 2s - 3t,$$
$$5 = -1 - s,$$
$$2 = 4 + s + 2t.$$

The second equation gives us $s = -6$. Substituting this into the third equation gives $t = 2$. However, the values $(s, t) = (-6, 2)$ do not satisfy the first equation. So, there are no values of s and t for which $(x, y, z) = (-1, 5, 2)$, which means the $\mathcal{P}$ does not pass through this point.

12.1.4   Let A be $(-1, 2, 4)$, B be $(3, -1, 2)$, and C be $(5, 1, 6)$. Then, the plane through A, B, and C is the plane through A generated by $\vec{AB}$ and $\vec{AC}$. We have

$$\vec{AB} = \vec{B} - \vec{A} = 4\mathbf{i} - 3\mathbf{j} - 2\mathbf{k},$$
$$\vec{AC} = \vec{C} - \vec{A} = 6\mathbf{i} - \mathbf{j} + 2\mathbf{k}.$$

If $(x, 7, 10)$ is in the plane through A generated by $\vec{AB}$ and $\vec{AC}$, then we must have

$$x\mathbf{i} + 7\mathbf{j} + 10\mathbf{k} = \vec{A} + s\vec{AB} + t\vec{AC} = (-1 - 14s + 6t)\mathbf{i} + (2 - 3s - t)\mathbf{j} + (4 - 2s + 2t)\mathbf{k}.$$

This gives us the system

$$x = -1 + 4s + 6t,$$
$$7 = 2 - 3s - t,$$
$$10 = 4 - 2s + 2t.$$

The last two equations give us $(s, t) = (-2, 1)$. Substituting these into the first equation gives us $x = \boxed{-3}$.

Exercises for Section 12.2

12.2.1   As described in the text, the vector $\begin{pmatrix} 4 \\ -3 \\ 12 \end{pmatrix}$ is normal to the graph of $4x - 3y + 12z = 5$. (Note that the negative of this vector is also normal to the plane.) The unit vector in the direction of this normal is

$$\frac{\begin{pmatrix} 4 \\ -3 \\ 12 \end{pmatrix}}{\left\| \begin{pmatrix} 4 \\ -3 \\ 12 \end{pmatrix} \right\|} = \frac{\begin{pmatrix} 4 \\ -3 \\ 12 \end{pmatrix}}{\sqrt{4^2 + (-3)^2 + 12^2}} = \frac{\begin{pmatrix} 4 \\ -3 \\ 12 \end{pmatrix}}{\sqrt{169}} = \frac{\begin{pmatrix} 4 \\ -3 \\ 12 \end{pmatrix}}{13} = \boxed{\begin{pmatrix} \frac{4}{13} \\ -\frac{3}{13} \\ \frac{12}{13} \end{pmatrix}}.$$

Multiplying this vector by -1 gives the other unit vector that is normal to the plane, $\boxed{\begin{pmatrix} -\frac{4}{13} \\ \frac{3}{13} \\ -\frac{12}{13} \end{pmatrix}}$

12.2.2   Let $\mathbf{v} = v_1\mathbf{i} + v_2\mathbf{j} + v_3\mathbf{k}$ and let (x_0, y_0, z_0) be a point on the plane. Then, we have $n_1 x_0 + n_2 y_0 + n_3 z_0 = d$ since (x_0, y_0, z_0) is on the plane, and we have $n_1 v_1 + n_2 v_2 + n_3 v_3 = 0$ because $\mathbf{n} \cdot \mathbf{v} = 0$. Adding $n_1 x_0 + n_2 y_0 + n_3 z_0 = d$ and $n_1 v_1 + n_2 v_2 + n_3 v_3 = 0$ gives

$$n_1(x_0 + v_1) + n_2(y_0 + v_2) + n_3(z_0 + v_3) = d,$$

so $(x_0 + v_1, y_0 + v_2, z_0 + v_3)$ is on the plane. The vector between $(x_0 + v_1, y_0 + v_2, z_0 + v_3)$ and (x_0, y_0, z_0) is $\mathbf{v}$, as desired.

12.2.3   Let $\mathbf{v} = 3\mathbf{i} + 2\mathbf{j} - 2\mathbf{k}$ and $\mathbf{w} = 4\mathbf{i} - \mathbf{j} + 8\mathbf{k}$. Then the projection of $\mathbf{v}$ onto $\mathbf{w}$ is

$$\frac{\mathbf{v} \cdot \mathbf{w}}{\|\mathbf{w}\|^2}\mathbf{w} = \frac{3 \cdot 4 + 2 \cdot (-1) + (-2) \cdot 8}{4^2 + (-1)^2 + 8^2}(4\mathbf{i} - \mathbf{j} + 8\mathbf{k}) = \boxed{-\tfrac{8}{27}\mathbf{i} + \tfrac{2}{27}\mathbf{j} - \tfrac{16}{27}\mathbf{k}}.$$

12.2.4   The vector $\begin{pmatrix} 2 \\ -1 \\ 3 \end{pmatrix}$ is normal to the graph of $2x - y + 3z = 5$, as is any scalar multiple of this vector, $\begin{pmatrix} 2k \\ -k \\ 3k \end{pmatrix}$.

Hence, $2k = 3$, $-k = b$, and $3k = c$. Then $k = 3/2$, so $b = -k = \boxed{-3/2}$ and $c = 3k = \boxed{9/2}$.

12.2.5   The distance from $(3, b, -5)$ to the plane $x + 2y - 2z - 8 = 0$ is

$$\frac{|3 + 2b - 2 \cdot (-5) - 8|}{\sqrt{1^2 + 2^2 + (-2)^2}} = \frac{|2b + 5|}{\sqrt{9}} = \frac{|2b + 5|}{3}.$$

If this distance is 8, then $|2b + 5| = 24$, so $2b + 5 = \pm 24$. If $2b + 5 = 24$, then $b = 19/2$, and if $2b + 5 = -24$, then $b = -29/2$. So the solutions are $b = \boxed{19/2}$ and $\boxed{-29/2}$.

12.2.6   Let $\mathbf{v}_1, \mathbf{v}_2, \mathbf{v}_3$ and $\mathbf{v}_4$ be four vectors in three dimensions. We have two cases to consider.

Case 1: $\mathbf{v}_1, \mathbf{v}_2$, and $\mathbf{v}_3$ are linearly dependent. Then, there exist constants a_1, a_2, a_3, not all 0, such that $a_1\mathbf{v}_1 + a_2\mathbf{v}_2 + a_3\mathbf{v}_3 = \mathbf{0}$. Therefore, we have $a_1\mathbf{v}_1 + a_2\mathbf{v}_2 + a_3\mathbf{v}_3 + 0\mathbf{v}_4 = \mathbf{0}$, so there are constants a_1, a_2, a_3, a_4, not all 0, such that $a_1\mathbf{v}_1 + a_2\mathbf{v}_2 + a_3\mathbf{v}_3 + a_4\mathbf{v}_4 = \mathbf{0}$, which means that $\mathbf{v}_1, \mathbf{v}_2, \mathbf{v}_3$ and $\mathbf{v}_4$ are linearly dependent.

Case 2: $\mathbf{v}_1, \mathbf{v}_2$, and $\mathbf{v}_3$ are linearly independent. Let $\mathbf{v}_4 = x_4\mathbf{i} + y_4\mathbf{j} + z_4\mathbf{k}$ As shown in the book, if $\mathbf{v}_1, \mathbf{v}_2$, and $\mathbf{v}_3$ are linearly independent, then the graph of all (x, y, z) such that $x\mathbf{i} + y\mathbf{j} + z\mathbf{k}$ is a linear combination of $\mathbf{v}_1, \mathbf{v}_2$, and $\mathbf{v}_3$ is all of $\mathbb{R}^3$. Specifically, (x_4, y_4, z_4) is in the graph, so $\mathbf{v}_4$ is a linear combination of $\mathbf{v}_1, \mathbf{v}_2$, and $\mathbf{v}_3$. Therefore, we have $\mathbf{v}_4 = a_1\mathbf{v}_1 + a_2\mathbf{v}_2 + a_3\mathbf{v}_3$ for some constants a_1, a_2, a_3. This means we have $a_1\mathbf{v}_1 + a_2\mathbf{v}_2 + a_3\mathbf{v}_3 + (-1)\mathbf{v}_4 = \mathbf{0}$, so $\mathbf{v}_1, \mathbf{v}_2, \mathbf{v}_3$ and $\mathbf{v}_4$ are linearly dependent.

For any set of vectors $\mathbf{v}_1, \mathbf{v}_2, \ldots, \mathbf{v}_n$ with $n > 4$, we use the fact that any set of four vectors is linearly dependent. Because $\mathbf{v}_1, \mathbf{v}_2, \mathbf{v}_3$ and $\mathbf{v}_4$ are linearly dependent, there exist constants a_1, a_2, a_3, a_4, not all zero, such that $a_1\mathbf{v}_1 + a_2\mathbf{v}_2 + a_3\mathbf{v}_3 + a_4\mathbf{v}_4 = \mathbf{0}$. So, we have $a_1\mathbf{v}_1 + a_2\mathbf{v}_2 + a_3\mathbf{v}_3 + a_4\mathbf{v}_4 + 0\mathbf{v}_5 + \cdots + 0\mathbf{v}_n = \mathbf{0}$ and not all of a_1, a_2, a_3, a_4 are 0, so $\mathbf{v}_1, \mathbf{v}_2, \ldots, \mathbf{v}_n$ are linearly dependent.

Exercises for Section 12.3

12.3.1

(a) $\mathbf{a} \times \mathbf{b} = \begin{vmatrix} \mathbf{i} & \mathbf{j} & \mathbf{k} \\ 3 & 4 & -1 \\ -1 & 7 & 0 \end{vmatrix} = \begin{vmatrix} 4 & -1 \\ 7 & 0 \end{vmatrix}\mathbf{i} - \begin{vmatrix} 3 & -1 \\ -1 & 0 \end{vmatrix}\mathbf{j} + \begin{vmatrix} 3 & 4 \\ -1 & 7 \end{vmatrix}\mathbf{k} = 7\mathbf{i} + \mathbf{j} + 25\mathbf{k} = \boxed{\begin{pmatrix} 7 \\ 1 \\ 25 \end{pmatrix}}.$

(b) $\mathbf{b} \times \mathbf{c} = \begin{vmatrix} \mathbf{i} & \mathbf{j} & \mathbf{k} \\ -1 & 7 & 0 \\ -4 & -1 & 3 \end{vmatrix} = \begin{vmatrix} 7 & 0 \\ -1 & 3 \end{vmatrix}\mathbf{i} - \begin{vmatrix} -1 & 0 \\ -4 & 3 \end{vmatrix}\mathbf{j} + \begin{vmatrix} -1 & 7 \\ -4 & -1 \end{vmatrix}\mathbf{k} = 21\mathbf{i} + 3\mathbf{j} + 29\mathbf{k} = \boxed{\begin{pmatrix} 21 \\ 3 \\ 29 \end{pmatrix}}.$

(c) $\quad \mathbf{c} \times \mathbf{a} = \begin{vmatrix} \mathbf{i} & \mathbf{j} & \mathbf{k} \\ -4 & -1 & 3 \\ 3 & 4 & -1 \end{vmatrix} = \begin{vmatrix} -1 & 3 \\ 4 & -1 \end{vmatrix} \mathbf{i} - \begin{vmatrix} -4 & 3 \\ 3 & -1 \end{vmatrix} \mathbf{j} + \begin{vmatrix} -4 & -1 \\ 3 & 4 \end{vmatrix} \mathbf{k} = -11\mathbf{i} + 5\mathbf{j} - 13\mathbf{k} = \boxed{\begin{pmatrix} -11 \\ 5 \\ -13 \end{pmatrix}}.$

12.3.2 Let $\vec{A} = -\mathbf{i} + 3\mathbf{j} + 4\mathbf{k}$, $\vec{B} = 2\mathbf{i} + \mathbf{j} - 5\mathbf{k}$, and $\vec{C} = -7\mathbf{i} - \mathbf{j}$. Then

$$\vec{AB} = \vec{B} - \vec{A} = (2\mathbf{i} + \mathbf{j} - 5\mathbf{k}) - (-\mathbf{i} + 3\mathbf{j} + 4\mathbf{k}) = 3\mathbf{i} - 2\mathbf{j} - 9\mathbf{k},$$
$$\vec{AC} = \vec{C} - \vec{A} = (-7\mathbf{i} - \mathbf{j}) - (-\mathbf{i} + 3\mathbf{j} + 4\mathbf{k}) = -6\mathbf{i} - 4\mathbf{j} - 4\mathbf{k}.$$

A vector that is orthogonal to both these vectors is

$$\vec{AB} \times \vec{AC} = (3\mathbf{i} - 2\mathbf{j} - 9\mathbf{k}) \times (-6\mathbf{i} - 4\mathbf{j} - 4\mathbf{k}) = \begin{vmatrix} \mathbf{i} & \mathbf{j} & \mathbf{k} \\ 3 & -2 & -9 \\ -6 & -4 & -4 \end{vmatrix}$$

$$= \begin{vmatrix} -2 & -9 \\ -4 & -4 \end{vmatrix} \mathbf{i} - \begin{vmatrix} 3 & -9 \\ -6 & -4 \end{vmatrix} \mathbf{j} + \begin{vmatrix} 3 & -2 \\ -6 & -4 \end{vmatrix} \mathbf{k} = -28\mathbf{i} + 66\mathbf{j} - 24\mathbf{k}.$$

Hence, the equation of the plane is of the form $-28x + 66y - 24z = d$ for some constant d. Substituting the point $(-1, 3, 4)$, we get $d = -28(-1) + 66(3) - 24(4) = 130$. Therefore, the equation of the plane is $-28x + 66y - 24z = 130$, or dividing by -2, we get $\boxed{14x - 33y + 12z = -65}$.

12.3.3 We have $\mathbf{u} \times \mathbf{v} = \mathbf{u} \times \mathbf{w}$ if and only if $\mathbf{u} \times \mathbf{v} - \mathbf{u} \times \mathbf{w} = \mathbf{0}$, so $\mathbf{u} \times (\mathbf{v} - \mathbf{w}) = \mathbf{0}$. Thus, we want to take $\mathbf{u}$, $\mathbf{v}$, and $\mathbf{w}$ such that $\mathbf{u}$ is parallel to $\mathbf{v} - \mathbf{w}$. So for example, we can take $\mathbf{u}$ and $\mathbf{v}$ to be any nonzero vectors such that $\mathbf{u} + \mathbf{v} \neq \mathbf{0}$, and let $\mathbf{w} = \mathbf{u} + \mathbf{v}$. Then

$$\mathbf{u} \times \mathbf{w} = \mathbf{u} \times (\mathbf{u} + \mathbf{v}) = \mathbf{u} \times \mathbf{u} + \mathbf{u} \times \mathbf{v} = \mathbf{u} \times \mathbf{v},$$

as desired. For example, let $\mathbf{u} = \mathbf{i}$, $\mathbf{v} = \mathbf{j}$, and $\mathbf{w} = \mathbf{i} + \mathbf{j}$. Then, we have $\mathbf{u} \times \mathbf{v} = \mathbf{i} \times \mathbf{j} = \mathbf{k}$ and $\mathbf{u} \times \mathbf{w} = \mathbf{i} \times (\mathbf{i} + \mathbf{j}) = \mathbf{i} \times \mathbf{i} + \mathbf{i} \times \mathbf{j} = \mathbf{0} + \mathbf{k} = \mathbf{k}$, so $\mathbf{u} \times \mathbf{v} = \mathbf{u} \times \mathbf{w}$ but $\mathbf{v} \neq \mathbf{w}$.

12.3.4 Let $\vec{A} = -\mathbf{i} + 3\mathbf{j}$, $\vec{B} = -2\mathbf{i} + 5\mathbf{j} + 6\mathbf{k}$, and $\vec{C} = 6\mathbf{i} - 4\mathbf{k}$. Then

$$\vec{AB} = \vec{B} - \vec{A} = (-2\mathbf{i} + 5\mathbf{j} + 6\mathbf{k}) - (-\mathbf{i} + 3\mathbf{j}) = -\mathbf{i} + 2\mathbf{j} + 6\mathbf{k},$$
$$\vec{AC} = \vec{C} - \vec{A} = (6\mathbf{i} - 4\mathbf{k}) - (-\mathbf{i} + 3\mathbf{j}) = 7\mathbf{i} - 3\mathbf{j} - 4\mathbf{k}.$$

The area of $\triangle ABC$ is $\frac{1}{2}(AB)(AC) \sin \theta$. We also have $\|\vec{AB} \times \vec{AC}\| = (AB)(AC) \sin \theta$, so the desired area is $\frac{1}{2}\|\vec{AB} \times \vec{AC}\|$. We have

$$\vec{AB} \times \vec{AC} = (-\mathbf{i} + 2\mathbf{j} + 6\mathbf{k}) \times (7\mathbf{i} - 3\mathbf{j} - 4\mathbf{k}) = \begin{vmatrix} \mathbf{i} & \mathbf{j} & \mathbf{k} \\ -1 & 2 & 6 \\ 7 & -3 & -4 \end{vmatrix} = \begin{vmatrix} 2 & 6 \\ -3 & -4 \end{vmatrix} \mathbf{i} - \begin{vmatrix} -1 & 6 \\ 7 & -4 \end{vmatrix} \mathbf{j} + \begin{vmatrix} -1 & 2 \\ 7 & -3 \end{vmatrix} \mathbf{k} = 10\mathbf{i} + 38\mathbf{j} - 11\mathbf{k}.$$

Therefore, the desired area is $\frac{1}{2}\|\vec{AB} \times \vec{AC}\| = \frac{1}{2}\|10\mathbf{i} + 38\mathbf{j} - 11\mathbf{k}\| = \frac{1}{2}\sqrt{10^2 + 38^2 + (-11)^2} = \frac{1}{2}\sqrt{1665} = \boxed{\frac{3\sqrt{185}}{2}}.$

12.3.5 Let θ be the angle between $\mathbf{u}$ and $\mathbf{v}$. Then $\mathbf{u} \cdot \mathbf{v} = \|\mathbf{u}\|\|\mathbf{v}\| \cos \theta$ and $\|\mathbf{u} \times \mathbf{v}\| = \|\mathbf{u}\|\|\mathbf{v}\| \sin \theta$, so

$$(\mathbf{u} \cdot \mathbf{v})^2 + \|\mathbf{u} \times \mathbf{v}\|^2 = (\|\mathbf{u}\|\|\mathbf{v}\| \cos \theta)^2 + (\|\mathbf{u}\|\|\mathbf{v}\| \sin \theta)^2 = \|\mathbf{u}\|^2\|\mathbf{v}\|^2 \cos^2 \theta + \|\mathbf{u}\|^2\|\mathbf{v}\|^2 \sin^2 \theta$$
$$= \|\mathbf{u}\|^2\|\mathbf{v}\|^2(\cos^2 \theta + \sin^2 \theta) = \|\mathbf{u}\|^2\|\mathbf{v}\|^2.$$

Therefore, $\|\mathbf{u} \times \mathbf{v}\|^2 = \|\mathbf{u}\|^2\|\mathbf{v}\|^2 - (\mathbf{u} \cdot \mathbf{v})^2$.

12.3.6

(a) Let $\mathbf{a} = \begin{pmatrix} a_1 \\ a_2 \\ a_3 \end{pmatrix}$, $\mathbf{b} = \begin{pmatrix} b_1 \\ b_2 \\ b_3 \end{pmatrix}$, and $\mathbf{c} = \begin{pmatrix} c_1 \\ c_2 \\ c_3 \end{pmatrix}$. Then

$$\mathbf{b} \times \mathbf{c} = \begin{vmatrix} \mathbf{i} & \mathbf{j} & \mathbf{k} \\ b_1 & b_2 & b_3 \\ c_1 & c_2 & c_3 \end{vmatrix} = \begin{vmatrix} b_2 & b_3 \\ c_2 & c_3 \end{vmatrix} \mathbf{i} - \begin{vmatrix} b_1 & b_3 \\ c_1 & c_3 \end{vmatrix} \mathbf{j} + \begin{vmatrix} b_1 & b_2 \\ c_1 & c_2 \end{vmatrix} \mathbf{k} = (b_2 c_3 - b_3 c_2)\mathbf{i} + (b_3 c_1 - b_1 c_3)\mathbf{j} + (b_1 c_2 - b_2 c_1)\mathbf{k},$$

and

$$\mathbf{a} \times (\mathbf{b} \times \mathbf{c}) = \begin{vmatrix} \mathbf{i} & \mathbf{j} & \mathbf{k} \\ a_1 & a_2 & a_3 \\ b_2 c_3 - b_3 c_2 & b_3 c_1 - b_1 c_3 & b_1 c_2 - b_2 c_1 \end{vmatrix}$$

$$= \begin{vmatrix} a_2 & a_3 \\ b_3 c_1 - b_1 c_3 & b_1 c_2 - b_2 c_1 \end{vmatrix} \mathbf{i} - \begin{vmatrix} a_1 & a_3 \\ b_2 c_3 - b_3 c_2 & b_1 c_2 - b_2 c_1 \end{vmatrix} \mathbf{j} + \begin{vmatrix} a_1 & a_2 \\ b_2 c_3 - b_3 c_2 & b_3 c_1 - b_1 c_3 \end{vmatrix} \mathbf{k}$$

$$= (a_2 b_1 c_2 - a_2 b_2 c_1 - a_3 b_3 c_1 + a_3 b_1 c_3)\mathbf{i}$$
$$\quad - (a_1 b_1 c_2 - a_1 b_2 c_1 - a_3 b_2 c_3 + a_3 b_3 c_2)\mathbf{j}$$
$$\quad + (a_1 b_3 c_1 - a_1 b_1 c_3 - a_2 b_2 c_3 + a_2 b_3 c_2)\mathbf{k}$$

$$= \begin{pmatrix} a_2 b_1 c_2 + a_3 b_1 c_3 - a_2 b_2 c_1 - a_3 b_3 c_1 \\ a_1 b_2 c_1 + a_3 b_2 c_3 - a_1 b_1 c_2 - a_3 b_3 c_2 \\ a_1 b_3 c_1 + a_2 b_3 c_2 - a_1 b_1 c_3 - a_2 b_2 c_3 \end{pmatrix}.$$

Also,

$$(\mathbf{a} \cdot \mathbf{c})\mathbf{b} - (\mathbf{a} \cdot \mathbf{b})\mathbf{c} = (a_1 c_1 + a_2 c_2 + a_3 c_3) \begin{pmatrix} b_1 \\ b_2 \\ b_3 \end{pmatrix} - (a_1 b_1 + a_2 b_2 + a_3 b_3) \begin{pmatrix} c_1 \\ c_2 \\ c_3 \end{pmatrix}$$

$$= \begin{pmatrix} a_1 b_1 c_1 + a_2 b_1 c_2 + a_3 b_1 c_3 \\ a_1 b_2 c_1 + a_2 b_2 c_2 + a_3 b_2 c_3 \\ a_1 b_3 c_1 + a_2 b_3 c_2 + a_3 b_3 c_3 \end{pmatrix} - \begin{pmatrix} a_1 b_1 c_1 + a_2 b_2 c_1 + a_3 b_3 c_1 \\ a_1 b_1 c_2 + a_2 b_2 c_2 + a_3 b_3 c_2 \\ a_1 b_1 c_3 + a_2 b_2 c_3 + a_3 b_3 c_3 \end{pmatrix}$$

$$= \begin{pmatrix} a_2 b_1 c_2 + a_3 b_1 c_3 - a_2 b_2 c_1 - a_3 b_3 c_1 \\ a_1 b_2 c_1 + a_3 b_2 c_3 - a_1 b_1 c_2 - a_3 b_3 c_2 \\ a_1 b_3 c_1 + a_2 b_3 c_2 - a_1 b_1 c_3 - a_2 b_2 c_3 \end{pmatrix}.$$

Hence, $\mathbf{a} \times (\mathbf{b} \times \mathbf{c}) = (\mathbf{a} \cdot \mathbf{c})\mathbf{b} - (\mathbf{a} \cdot \mathbf{b})\mathbf{c}$.

(b) Applying the result we proved in part (a), we have both $\mathbf{b} \times (\mathbf{c} \times \mathbf{a}) = (\mathbf{b} \cdot \mathbf{a})\mathbf{c} - (\mathbf{b} \cdot \mathbf{c})\mathbf{a}$ and $\mathbf{c} \times (\mathbf{a} \times \mathbf{b}) = (\mathbf{c} \cdot \mathbf{b})\mathbf{a} - (\mathbf{c} \cdot \mathbf{a})\mathbf{b}$ in addition to $\mathbf{a} \times (\mathbf{b} \times \mathbf{c}) = (\mathbf{a} \cdot \mathbf{c})\mathbf{b} - (\mathbf{a} \cdot \mathbf{b})\mathbf{c}$. Adding all three equations, we get

$$\mathbf{a} \times (\mathbf{b} \times \mathbf{c}) + \mathbf{b} \times (\mathbf{c} \times \mathbf{a}) + \mathbf{c} \times (\mathbf{a} \times \mathbf{b}) = [(\mathbf{a} \cdot \mathbf{c})\mathbf{b} - (\mathbf{a} \cdot \mathbf{b})\mathbf{c}] + [(\mathbf{b} \cdot \mathbf{a})\mathbf{c} - (\mathbf{b} \cdot \mathbf{c})\mathbf{a}] + [(\mathbf{c} \cdot \mathbf{b})\mathbf{a} - (\mathbf{c} \cdot \mathbf{a})\mathbf{b}] = 0.$$

Exercises for Section 12.4

12.4.1   The vectors emanating from the vertex $(-2, 1, 5)$ that correspond to the edges of the parallelepiped are

$$\begin{pmatrix} 4 \\ -1 \\ -2 \end{pmatrix} - \begin{pmatrix} -2 \\ 1 \\ 5 \end{pmatrix} = \begin{pmatrix} 6 \\ -2 \\ -7 \end{pmatrix}, \quad \begin{pmatrix} -1 \\ 2 \\ -1 \end{pmatrix} - \begin{pmatrix} -2 \\ 1 \\ 5 \end{pmatrix} = \begin{pmatrix} 1 \\ 1 \\ -6 \end{pmatrix}, \quad \begin{pmatrix} 4 \\ -3 \\ 0 \end{pmatrix} - \begin{pmatrix} -2 \\ 1 \\ 5 \end{pmatrix} = \begin{pmatrix} 6 \\ -4 \\ -5 \end{pmatrix}.$$

Therefore, the volume of the parallelepiped is the absolute value of

$$\begin{vmatrix} 6 & -2 & -7 \\ 1 & 1 & -6 \\ 6 & -4 & -5 \end{vmatrix} = -42,$$

or $\boxed{42}$.

12.4.2 We saw that if $\mathbf{a} = \begin{pmatrix} a_1 \\ a_2 \\ a_3 \end{pmatrix}$, $\mathbf{b} = \begin{pmatrix} b_1 \\ b_2 \\ b_3 \end{pmatrix}$, and $\mathbf{c} = \begin{pmatrix} c_1 \\ c_2 \\ c_3 \end{pmatrix}$, then $\mathbf{a} \cdot (\mathbf{b} \times \mathbf{c}) = \begin{vmatrix} a_1 & a_2 & a_3 \\ b_1 & b_2 & b_3 \\ c_1 & c_2 & c_3 \end{vmatrix}$. Hence,

$$\mathbf{c} \cdot (\mathbf{a} \times \mathbf{b}) = \begin{vmatrix} c_1 & c_2 & c_3 \\ a_1 & a_2 & a_3 \\ b_1 & b_2 & b_3 \end{vmatrix} = - \begin{vmatrix} a_1 & a_2 & a_3 \\ c_1 & c_2 & c_3 \\ b_1 & b_2 & b_3 \end{vmatrix} = \begin{vmatrix} a_1 & a_2 & a_3 \\ b_1 & b_2 & b_3 \\ c_1 & c_2 & c_3 \end{vmatrix} = \mathbf{a} \cdot (\mathbf{b} \times \mathbf{c}),$$

so $(\mathbf{a} \times \mathbf{b}) \cdot \mathbf{c} = \mathbf{c} \cdot (\mathbf{a} \times \mathbf{b}) = \mathbf{a} \cdot (\mathbf{b} \times \mathbf{c})$.

Geometrically speaking, the absolute value of each expression equals the volume of the parallelepiped with edges $\mathbf{a}$, $\mathbf{b}$, and $\mathbf{c}$. If you picture yourself standing at the origin and looking along the interior diagonal of the parallelepiped from the origin, you'll see that in both expressions, the vectors appear in the same order, counterclockwise or clockwise, around the diagonal. Therefore, the two given expressions equal the same signed volume of the parallelepiped.

12.4.3 Since each two rows are orthogonal, the parallelepiped with the rows as edges is a rectangular prism. The volume of a rectangular prism equals the product of its dimensions, so the volume of the parallelepiped with the rows as edges is the product of the magnitudes of the rows. This parallelepiped has volume $|\det(\mathbf{M})|$, so $|\det(\mathbf{M})|$ equals the product of the magnitudes of the rows.

Exercises for Section 12.5

12.5.1 We have $\mathbf{A}\mathbf{A}^{-1} = \mathbf{I}$, so $\det(\mathbf{A}\mathbf{A}^{-1}) = \det(\mathbf{I}) = 1$. We also have $\det(\mathbf{A}\mathbf{A}^{-1}) = \det(\mathbf{A})\det(\mathbf{A}^{-1})$, so $\det(\mathbf{A})\det(\mathbf{A}^{-1}) = 1$, which means $\det(\mathbf{A})$ and $\det(\mathbf{A}^{-1})$ are reciprocals.

12.5.2 A matrix does not have an inverse if and only if its determinant is zero. We have

$$\det \begin{pmatrix} 1 & 5 & 5 \\ -2 & 6 & x \\ x & -2 & 1 \end{pmatrix} = 6 + 5x^2 + 20 - (-2x) - (-10) - 30x = 5x^2 - 28x + 36.$$

Factoring and setting this equal to 0 gives $(5x - 18)(x - 2) = 0$. Therefore, the values of x for which the matrix does not have an inverse are $\boxed{\frac{18}{5} \text{ and } 2}$.

12.5.3 $\boxed{\text{No}}$. Note that

$$\begin{pmatrix} 0 & 0 & 0 \\ 0 & 1 & 0 \\ 0 & 0 & 0 \end{pmatrix} + \begin{pmatrix} 1 & 0 & 0 \\ 0 & 0 & 0 \\ 0 & 0 & 1 \end{pmatrix} = \begin{pmatrix} 1 & 0 & 0 \\ 0 & 1 & 0 \\ 0 & 0 & 1 \end{pmatrix}.$$

The addends on the left are singular, but their sum is not.

12.5.4 Let $\mathbf{A} = \begin{pmatrix} a_{11} & a_{12} & a_{13} \\ 0 & a_{22} & a_{23} \\ 0 & 0 & a_{33} \end{pmatrix}$.

(a) We have $\det(\mathbf{A}) = a_{11}a_{22}a_{33}$, so $\det(\mathbf{A}) \neq 0$ if and only if a_{11}, a_{22}, and a_{33} are nonzero. A matrix is invertible if and only if its determinant is nonzero, so an upper triangular matrix is invertible if and only if the entries on its main diagonal are nonzero.

(b) Let $\mathbf{M}$ be the inverse of upper triangular matrix, so we must have

$$\begin{pmatrix} m_{11} & m_{12} & m_{13} \\ m_{21} & m_{22} & m_{23} \\ m_{31} & m_{32} & m_{33} \end{pmatrix} \begin{pmatrix} a_{11} & a_{12} & a_{13} \\ 0 & a_{22} & a_{23} \\ 0 & 0 & a_{33} \end{pmatrix} = \begin{pmatrix} 1 & 0 & 0 \\ 0 & 1 & 0 \\ 0 & 0 & 1 \end{pmatrix}.$$

The entry in the second row of the first column of $\mathbf{MA}$ is $m_{21}a_{11}$. Since $\mathbf{MA} = \mathbf{I}$, this entry must equal 0. But a_{11} is nonzero since $\mathbf{A}$ is an invertible upper triangular matrix. Therefore, $m_{21} = 0$. Similarly, the entry in the third row of the first column of $\mathbf{MA}$ must be 0. This entry equals $m_{31}a_{11}$ and $a_{11} \neq 0$, so $m_{31} = 0$. Finally, the entry in the third row of the second column of $\mathbf{MA}$ equals 0. This entry equals $m_{31}a_{12} + m_{32}a_{22}$. Since $m_{31} = 0$ and a_{22} is nonzero, we must have $m_{32} = 0$. Combining this with $m_{21} = m_{31} = 0$, we see that $\mathbf{M}$ is an upper triangular matrix.

Review Problems

12.35

(a) The equation $\mathbf{p} \cdot \mathbf{v} = 4$ becomes $x + 2y - 5z = 4$, which is the equation of a $\boxed{\text{plane}}$.

(b) The equation $\mathbf{p} \cdot \mathbf{v} = \mathbf{p} \cdot \mathbf{w}$ becomes

$$\begin{pmatrix} x \\ y \\ z \end{pmatrix} \cdot \begin{pmatrix} 1 \\ 2 \\ -5 \end{pmatrix} = \begin{pmatrix} x \\ y \\ z \end{pmatrix} \cdot \begin{pmatrix} -3 \\ 4 \\ -2 \end{pmatrix},$$

or $x + 2y - 5z = -3x + 4y - 2z$, which simplifies as $4x - 2y - 3z = 0$. This is the equation of a $\boxed{\text{plane}}$.

We also could have written $\mathbf{p} \cdot \mathbf{v} = \mathbf{p} \cdot \mathbf{w}$ as $\mathbf{p} \cdot (\mathbf{v} - \mathbf{w}) = 0$, so the graph consists of all (x, y, z) such that $\begin{pmatrix} x \\ y \\ z \end{pmatrix}$ is orthogonal to the constant vector given by $\mathbf{v} - \mathbf{w}$. Therefore, the graph is a plane through the origin with $\mathbf{v} - \mathbf{w}$ as a normal vector.

(c) If $\mathbf{p} \cdot \mathbf{v} = 4$ and $\mathbf{p} \cdot \mathbf{v} = \mathbf{p} \cdot \mathbf{w}$, then it follows that $\mathbf{p} \cdot \mathbf{v} = 4$ and $\mathbf{p} \cdot \mathbf{w} = 4$. Hence, the graph in this part is the intersection of the graphs in parts (a) and (b). The graphs in parts (a) and (b) are different planes, so the graph must be a $\boxed{\text{line}}$. (The two planes must intersect, because the vector $\begin{pmatrix} 1 \\ 2 \\ -5 \end{pmatrix}$ is normal to the first plane, and the vector $\begin{pmatrix} 4 \\ -2 \\ -3 \end{pmatrix}$ is normal to the second plane, and these vectors are not linearly dependent.)

12.36 The equation of the line passing through $(2, 4, 2)$ and $(1, 6, -1)$ is given by

$$\begin{pmatrix} x \\ y \\ z \end{pmatrix} = \begin{pmatrix} 2 \\ 4 \\ 2 \end{pmatrix} + \begin{pmatrix} 1 - 2 \\ 6 - 4 \\ -1 - 2 \end{pmatrix} t = \begin{pmatrix} 2 \\ 4 \\ 2 \end{pmatrix} + \begin{pmatrix} -1 \\ 2 \\ -3 \end{pmatrix} t = \begin{pmatrix} 2 - t \\ 4 + 2t \\ 2 - 3t \end{pmatrix}.$$

Hence, if $(a, 2, b)$ lies on this line, then $a = 2 - t$, $4 + 2t = 2$, and $b = 2 - 3t$. From the equation $4 + 2t = 2$, we have $t = -1$. Then $a = 2 - t = \boxed{3}$ and $b = 2 - 3t = \boxed{5}$.

12.37   The vector $\mathbf{n}_1 = \begin{pmatrix} 3 \\ -2 \\ 8 \end{pmatrix}$ is normal to the plane $3x - 2y + 8z = 5$, and the vector $\mathbf{n}_2 = \begin{pmatrix} a \\ b \\ -4 \end{pmatrix}$ is normal to

the plane $ax + by - 4z = 7$. These planes are parallel if and only if one of the vectors $\mathbf{n}_1$ and $\mathbf{n}_2$ is a scalar multiple of the other. The z-coordinate of $\mathbf{n}_1$ is 8 and the z-coordinate of $\mathbf{n}_2$ is -4, so we must have $\mathbf{n}_1 = -2\mathbf{n}_2$.

Comparing the x- and y-coordinates, we get $3 = -2a$ and $-2 = -2b$, so $a = \boxed{-3/2}$ and $b = \boxed{1}$.

12.38

(a)   The equation of the line passing through $(5, 4, 1)$ and $(11, 1, -8)$ is given by

$$\begin{pmatrix} x \\ y \\ z \end{pmatrix} = \begin{pmatrix} 5 \\ 4 \\ 1 \end{pmatrix} + \begin{pmatrix} 11 - 5 \\ 1 - 4 \\ -8 - 1 \end{pmatrix} t = \begin{pmatrix} 5 \\ 4 \\ 1 \end{pmatrix} + \begin{pmatrix} 6 \\ -3 \\ -9 \end{pmatrix} t = \begin{pmatrix} 5 + 6t \\ 4 - 3t \\ 1 - 9t \end{pmatrix}.$$

The equation of the line passing through $(5, -1, 2)$ and $(0, 14, 7)$ is given by

$$\begin{pmatrix} x \\ y \\ z \end{pmatrix} = \begin{pmatrix} 5 \\ -1 \\ 2 \end{pmatrix} + \begin{pmatrix} 0 - 5 \\ 14 - (-1) \\ 7 - 2 \end{pmatrix} s = \begin{pmatrix} 5 \\ -1 \\ 2 \end{pmatrix} + \begin{pmatrix} -5 \\ 15 \\ 5 \end{pmatrix} s = \begin{pmatrix} 5 - 5s \\ -1 + 15s \\ 2 + 5s \end{pmatrix}.$$

Hence, if (x, y, z) lies on both lines, then s and t must satisfy the equations $5 + 6t = 5 - 5s$, $4 - 3t = -1 + 15s$, and $1 - 9t = 2 + 5s$. Multiplying the second equation by 2, we get $8 - 6t = -2 + 30s$. Adding this to the equation $5 + 6t = 5 - 5s$, we get $13 = 3 + 25s$. Solving for s, we find $s = 2/5$. Substituting into the equation $1 - 9t = 2 + 5s$, we get $1 - 9t = 4$. Solving for t, we find $t = -1/3$. Then

$$\begin{pmatrix} 5 + 6t \\ 4 - 3t \\ 1 - 9t \end{pmatrix} = \begin{pmatrix} 3 \\ 5 \\ 4 \end{pmatrix} \quad \text{and} \quad \begin{pmatrix} 5 - 5s \\ -1 + 15s \\ 2 + 5s \end{pmatrix} = \begin{pmatrix} 3 \\ 5 \\ 4 \end{pmatrix},$$

so the two lines intersect at $\boxed{(3, 5, 4)}$.

(b)   The line ℓ is parallel to the vector $\begin{pmatrix} 6 \\ -3 \\ -9 \end{pmatrix}$, and the line m is parallel to the vector $\begin{pmatrix} -5 \\ 15 \\ 5 \end{pmatrix}$. Then a vector that is normal to the plane containing both ℓ and m is given by the cross product of these two vectors, which is

$$\begin{vmatrix} \mathbf{i} & \mathbf{j} & \mathbf{k} \\ 6 & -3 & -9 \\ -5 & 15 & 5 \end{vmatrix} = \begin{vmatrix} -3 & -9 \\ 15 & 5 \end{vmatrix} \mathbf{i} - \begin{vmatrix} 6 & -9 \\ -5 & 5 \end{vmatrix} \mathbf{j} + \begin{vmatrix} 6 & -3 \\ -5 & 15 \end{vmatrix} \mathbf{k}$$

$$= 120\mathbf{i} + 15\mathbf{j} + 75\mathbf{k}$$

$$= \begin{pmatrix} 120 \\ 15 \\ 75 \end{pmatrix}.$$

Hence, the equation of the plane is of the form $120x + 15y + 75z = d$ for some constant d. Substituting the point $(3, 5, 4)$, we get $d = 120(3) + 15(5) + 75(4) = 735$, so the equation of the plane is given by $120x + 15y + 75z = 735$, or dividing by 15, we get $\boxed{8x + y + 5z = 49}$.

12.39   *Solution 1:* Let $P = (x, y, z)$, $P_1 = (x_1, y_1, z_1)$, $P_2 = (x_2, y_2, z_2)$, and $P_3 = (x_3, y_3, z_3)$. Then

$$\vec{P_1 P} = \begin{pmatrix} x - x_1 \\ y - y_1 \\ z - z_1 \end{pmatrix}, \quad \vec{P_1 P_2} = \begin{pmatrix} x_2 - x_1 \\ y_2 - y_1 \\ z_2 - z_1 \end{pmatrix}, \quad \vec{P_1 P_3} = \begin{pmatrix} x_3 - x_1 \\ y_3 - y_1 \\ z_3 - z_1 \end{pmatrix},$$

so

$$\begin{vmatrix} x - x_1 & y - y_1 & z - z_1 \\ x_2 - x_1 & y_2 - y_1 & z_2 - z_1 \\ x_3 - x_1 & y_3 - y_1 & z_3 - z_1 \end{vmatrix}$$

is the signed area of the parallelepiped whose edges are $\overrightarrow{P_1P}$, $\overrightarrow{P_1P_2}$, and $\overrightarrow{P_1P_3}$. This volume is 0 if and only if P lies in the same plane as P_1, P_2, and P_3.

Solution 2: The determinant is clearly of the form $ax + by + cz + d$, where a, b, c, and d are constants, so the graph of the equation formed when we set this determinant equal to 0 is a plane. We can show that this is the desired plane if we can show that the determinant is 0 when (x, y, z) equals each of (x_1, y_1, z_1), (x_2, y_2, z_2), and (x_3, y_3, z_3) (and therefore all three points are on the graph of the equation). When $(x, y, z) = (x_1, y_1, z_1)$, the first row of the determinant consists of three 0s, so the determinant is 0. If (x, y, z) equals (x_2, y_2, z_2), then the first and second rows are the same, so the determinant is 0. If (x, y, z) equals (x_3, y_3, z_3), then the first and third rows are the same, so, again, the determinant is 0. Therefore, the graph of the given equation is the desired plane.

12.40

(a) We have

$$a\mathbf{v} + b\mathbf{w} = a \begin{pmatrix} 4 \\ -1 \\ 2 \end{pmatrix} + b \begin{pmatrix} -1 \\ 2 \\ 3 \end{pmatrix} = \begin{pmatrix} 4a - b \\ -a + 2b \\ 2a + 3b \end{pmatrix}.$$

If this is equal to $\mathbf{p} = \begin{pmatrix} 9 \\ -4 \\ 1 \end{pmatrix}$, then $4a - b = 9$, $-a + 2b = -4$, and $2a + 3b = 1$. Solving the system $4a - b = 9$, $-a + 2b = -4$ gives $a = 2$, $b = -1$. Note that these values of a and b satisfy the third equation, $2a + 3b = 1$. Therefore, the only solution is $(a, b) = \boxed{(2, -1)}$.

(b) We will show that it is impossible to have two different pairs (a, b) of scalars that satisfy $\mathbf{t} = a\mathbf{v} + b\mathbf{w}$. We let (a_1, b_1) and (a_2, b_2) be pairs of scalars such that $a_1\mathbf{v} + b_1\mathbf{w} = a_2\mathbf{v} + b_2\mathbf{w}$. We will show that we must have $(a_1, b_1) = (a_2, b_2)$.

Rearranging $a_1\mathbf{v} + b_1\mathbf{w} = a_2\mathbf{v} + b_2\mathbf{w}$ gives $(a_1 - a_2)\mathbf{v} - (b_1 - b_2)\mathbf{w} = \mathbf{0}$. Since $\mathbf{v}$ and $\mathbf{w}$ are linearly independent, we must have $a_1 - a_2 = b_1 - b_2 = 0$, so $a_1 = a_2$ and $b_1 = b_2$. Therefore, the pair of scalars (a, b) such that $\mathbf{r} = a\mathbf{v} + b\mathbf{w}$ is unique.

12.41

(a) If $\mathbf{u} \cdot \mathbf{w} = \mathbf{v} \cdot \mathbf{w}$, then $\mathbf{u} \cdot \mathbf{w} - \mathbf{v} \cdot \mathbf{w} = 0$, so $(\mathbf{u} - \mathbf{v}) \cdot \mathbf{w} = 0$. This holds for all vectors $\mathbf{w}$, so in particular, we can take $\mathbf{w} = \mathbf{u} - \mathbf{v}$ to get $(\mathbf{u} - \mathbf{v}) \cdot (\mathbf{u} - \mathbf{v}) = 0$. But $(\mathbf{u} - \mathbf{v}) \cdot (\mathbf{u} - \mathbf{v}) = \|\mathbf{u} - \mathbf{v}\|^2$, so $\|\mathbf{u} - \mathbf{v}\|^2 = 0$. Hence, $\|\mathbf{u} - \mathbf{v}\| = 0$, which means $\mathbf{u} - \mathbf{v} = \mathbf{0}$, or $\mathbf{u} = \mathbf{v}$.

We could also have noted that $\mathbf{u} \cdot \mathbf{i} = \mathbf{v} \cdot \mathbf{i}$, $\mathbf{u} \cdot \mathbf{j} = \mathbf{v} \cdot \mathbf{j}$, and $\mathbf{u} \cdot \mathbf{k} = \mathbf{v} \cdot \mathbf{k}$, so each component of $\mathbf{u}$ equals the corresponding component of $\mathbf{v}$. Therefore, we have $\mathbf{u} = \mathbf{v}$.

(b) If $\mathbf{u} \times \mathbf{w} = \mathbf{v} \times \mathbf{w}$, then $\mathbf{u} \times \mathbf{w} - \mathbf{v} \times \mathbf{w} = \mathbf{0}$, so $(\mathbf{u} - \mathbf{v}) \times \mathbf{w} = \mathbf{0}$. Let $\mathbf{t} = \mathbf{u} - \mathbf{v} = \begin{pmatrix} t_1 \\ t_2 \\ t_3 \end{pmatrix}$, so $\mathbf{t} \times \mathbf{w} = \mathbf{0}$ for all vectors $\mathbf{w}$.

Taking $\mathbf{w} = \mathbf{i}$, we get

$$\mathbf{t} \times \mathbf{i} = \begin{vmatrix} \mathbf{i} & \mathbf{j} & \mathbf{k} \\ t_1 & t_2 & t_3 \\ 1 & 0 & 0 \end{vmatrix} = \begin{vmatrix} t_2 & t_3 \\ 0 & 0 \end{vmatrix} \mathbf{i} - \begin{vmatrix} t_1 & t_3 \\ 1 & 0 \end{vmatrix} \mathbf{j} + \begin{vmatrix} t_1 & t_2 \\ 1 & 0 \end{vmatrix} \mathbf{k} = t_3\mathbf{j} - t_2\mathbf{k}.$$

This is equal to $\mathbf{0}$, so $t_2 = t_3 = 0$. Similarly, $\mathbf{t} \times \mathbf{j} = -t_3\mathbf{i} + t_1\mathbf{k} = \mathbf{0}$, so $t_1 = 0$. Therefore, $\mathbf{t} = \mathbf{0}$, which means $\mathbf{u} = \mathbf{v}$.

12.42   The cross product is equal to

$$\begin{vmatrix} \mathbf{i} & \mathbf{j} & \mathbf{k} \\ 3 & 2 & -1 \\ -3 & 4 & 2 \end{vmatrix} = \begin{vmatrix} 2 & -1 \\ 4 & 2 \end{vmatrix}\mathbf{i} - \begin{vmatrix} 3 & -1 \\ -3 & 2 \end{vmatrix}\mathbf{j} + \begin{vmatrix} 3 & 2 \\ -3 & 4 \end{vmatrix}\mathbf{k}$$

$$= [2\cdot 2 - (-1)\cdot 4]\mathbf{i} - [3\cdot 2 - (-1)\cdot(-3)]\mathbf{j} + [3\cdot 4 - 2\cdot(-3)]\mathbf{k} = \boxed{8\mathbf{i} - 3\mathbf{j} + 18\mathbf{k}}.$$

12.43   Let $\mathbf{v} = \begin{pmatrix} x \\ y \\ z \end{pmatrix}$. Then

$$\mathbf{i} \times \mathbf{v} = \begin{vmatrix} \mathbf{i} & \mathbf{j} & \mathbf{k} \\ 1 & 0 & 0 \\ x & y & z \end{vmatrix} = \begin{vmatrix} 0 & 0 \\ y & z \end{vmatrix}\mathbf{i} - \begin{vmatrix} 1 & 0 \\ x & z \end{vmatrix}\mathbf{j} + \begin{vmatrix} 1 & 0 \\ x & y \end{vmatrix}\mathbf{k} = -z\mathbf{j} + y\mathbf{k}.$$

If this is equal to $\mathbf{k}$, then $y = 1$ and $z = 0$, but x can be any real number, and we have $\mathbf{v} = x\mathbf{i} + \mathbf{j}$. Thus, $\mathbf{i} \times \mathbf{v} = \mathbf{k}$ does not imply $\mathbf{v} = \mathbf{j}$.

12.44   Let $\mathbf{a} = \begin{pmatrix} 4 \\ -1 \\ 2 \end{pmatrix}$ and $\mathbf{b} = \begin{pmatrix} -1 \\ 0 \\ 4 \end{pmatrix}$. Then $\mathbf{a} \times \mathbf{b}$ is orthogonal to both these vectors, so it is orthogonal to the plane generated by these vectors. This cross product is equal to

$$\mathbf{a} \times \mathbf{b} = \begin{vmatrix} \mathbf{i} & \mathbf{j} & \mathbf{k} \\ 4 & -1 & 2 \\ -1 & 0 & 4 \end{vmatrix} = \begin{vmatrix} -1 & 2 \\ 0 & 4 \end{vmatrix}\mathbf{i} - \begin{vmatrix} 4 & 2 \\ -1 & 4 \end{vmatrix}\mathbf{j} + \begin{vmatrix} 4 & -1 \\ -1 & 0 \end{vmatrix}\mathbf{k} = -4\mathbf{i} - 18\mathbf{j} - \mathbf{k}.$$

Hence, the equation of the plane is of the form $-4x - 18y - z = d$, where d is some constant. Since this plane passes through the origin, we have $d = 0$. Therefore, the equation of the plane is $-4x - 18y - z = 0$, or multiplying by -1, we get $\boxed{4x + 18y + z = 0}$.

12.45   We can express the parametric equations as

$$\begin{pmatrix} x \\ y \\ z \end{pmatrix} = \begin{pmatrix} 3 \\ 4 \\ 0 \end{pmatrix} + \begin{pmatrix} -2 \\ 1 \\ -5 \end{pmatrix} t.$$

As discussed in the text, the graph of this equation is a line that has direction vector $\begin{pmatrix} -2 \\ 1 \\ -5 \end{pmatrix}$. If we take $t = 0$, then

we see that the point $(3, 4, 0)$ lies on the line. The vector joining $(-2, 3, 5)$ to $(3, 4, 0)$ is $\begin{pmatrix} 3 - (-2) \\ 4 - 3 \\ 0 - 5 \end{pmatrix} = \begin{pmatrix} 5 \\ 1 \\ -5 \end{pmatrix}$.

This vector and the direction vector are in the desired plane. The cross product of these vectors is orthogonal to both vectors, so it is normal to the plane. This cross product is

$$\begin{vmatrix} \mathbf{i} & \mathbf{j} & \mathbf{k} \\ -2 & 1 & -5 \\ 5 & 1 & -5 \end{vmatrix} = \begin{vmatrix} 1 & -5 \\ 1 & -5 \end{vmatrix}\mathbf{i} - \begin{vmatrix} -2 & -5 \\ 5 & -5 \end{vmatrix}\mathbf{j} + \begin{vmatrix} -2 & 1 \\ 5 & 1 \end{vmatrix}\mathbf{k}$$

$$= [1\cdot(-5) - 1\cdot(-5)]\mathbf{i} - [(-2)\cdot(-5) - (-5)\cdot 5]\mathbf{j} - [(-2)\cdot 1 - 1\cdot 5]\mathbf{k}$$

$$= -35\mathbf{j} - 7\mathbf{k}$$

$$= \begin{pmatrix} 0 \\ -35 \\ -7 \end{pmatrix},$$

so the equation of the plane is of the form $-35y - 7z = d$, for some constant d. Substituting the point $(-2, 3, 5)$, we get $d = -35(3) - 7(5) = -140$, so the equation of the plane is $-35y - 7z = -140$, or dividing both sides by -7, we get $\boxed{5y + z = 20}$.

12.46 Let $\mathbf{a} = 2\mathbf{i}$, $\mathbf{b} = \mathbf{j}$, and $\mathbf{c} = 2\mathbf{k}$. Then

$$\mathbf{a} \times \mathbf{b} = (2\mathbf{i}) \times \mathbf{j} = 2(\mathbf{i} \times \mathbf{j}) = 2\mathbf{k} = \mathbf{c},$$

and

$$\mathbf{b} \times \mathbf{c} = \mathbf{j} \times (2\mathbf{k}) = 2(\mathbf{j} \times \mathbf{k}) = 2\mathbf{i} = \mathbf{a},$$

but

$$\mathbf{c} \times \mathbf{a} = (2\mathbf{k}) \times (2\mathbf{i}) = 4(\mathbf{k} \times \mathbf{i}) = 4\mathbf{j},$$

which is not equal to $\mathbf{b}$.

12.47

(a) Let $\mathbf{w} = \begin{pmatrix} w_1 \\ w_2 \\ w_3 \end{pmatrix}$. Then

$$\mathbf{v} \times \mathbf{w} = \begin{vmatrix} \mathbf{i} & \mathbf{j} & \mathbf{k} \\ 4 & -1 & 2 \\ w_1 & w_2 & w_3 \end{vmatrix} = \begin{vmatrix} -1 & 2 \\ w_2 & w_3 \end{vmatrix} \mathbf{i} - \begin{vmatrix} 4 & 2 \\ w_1 & w_3 \end{vmatrix} \mathbf{j} + \begin{vmatrix} 4 & -1 \\ w_1 & w_2 \end{vmatrix} \mathbf{k} = (-2w_2 - w_3)\mathbf{i} + (2w_1 - 4w_3)\mathbf{j} + (w_1 + 4w_2)\mathbf{k}$$

$$= \begin{pmatrix} -2w_2 - w_3 \\ 2w_1 - 4w_3 \\ w_1 + 4w_2 \end{pmatrix}.$$

Also, $\begin{pmatrix} 0 & -2 & -1 \\ 2 & 0 & -4 \\ 1 & 4 & 0 \end{pmatrix} \begin{pmatrix} w_1 \\ w_2 \\ w_3 \end{pmatrix} = \begin{pmatrix} -2w_2 - w_3 \\ 2w_1 - 4w_3 \\ w_1 + 4w_2 \end{pmatrix}$, so we can take $\mathbf{M} = \boxed{\begin{pmatrix} 0 & -2 & -1 \\ 2 & 0 & -4 \\ 1 & 4 & 0 \end{pmatrix}}$.

(b) The function $f(\mathbf{w}) = \mathbf{u} \times \mathbf{w}$ satisfies $f(k\mathbf{w}) = kf(\mathbf{w})$ and $f(\mathbf{v} + \mathbf{w}) = f(\mathbf{v}) + f(\mathbf{w})$ for any scalar k and any vectors $\mathbf{v}$ and $\mathbf{w}$. (We proved these fundamental properties of the cross product in the text.) Therefore, f is a linear function that maps vectors to vectors, so there is a unique matrix $\mathbf{N}$ such that $f(\mathbf{w}) = \mathbf{N}\mathbf{w}$ for all $\mathbf{w}$. Let's see if we can find the matrix in terms of the components of $\mathbf{u}$!

Let $\mathbf{u} = \begin{pmatrix} u_1 \\ u_2 \\ u_3 \end{pmatrix}$. Then

$$\mathbf{u} \times \mathbf{w} = \begin{vmatrix} \mathbf{i} & \mathbf{j} & \mathbf{k} \\ u_1 & u_2 & u_3 \\ w_1 & w_2 & w_3 \end{vmatrix} = \begin{vmatrix} u_2 & u_3 \\ w_2 & w_3 \end{vmatrix} \mathbf{i} - \begin{vmatrix} u_1 & u_3 \\ w_1 & w_3 \end{vmatrix} \mathbf{j} + \begin{vmatrix} u_1 & u_2 \\ w_1 & w_2 \end{vmatrix} \mathbf{k}$$

$$= (-u_3 w_2 + u_2 w_3)\mathbf{i} + (u_3 w_1 - u_1 w_3)\mathbf{j} + (-u_2 w_1 + u_1 w_2)\mathbf{k} = \begin{pmatrix} -u_3 w_2 + u_2 w_3 \\ u_3 w_1 - u_1 w_3 \\ -u_2 w_1 + u_1 w_2 \end{pmatrix}.$$

Also, $\begin{pmatrix} 0 & -u_3 & u_2 \\ u_3 & 0 & -u_1 \\ -u_2 & u_1 & 0 \end{pmatrix} \begin{pmatrix} w_1 \\ w_2 \\ w_3 \end{pmatrix} = \begin{pmatrix} -u_3 w_2 + u_2 w_3 \\ u_3 w_1 - u_1 w_3 \\ -u_2 w_1 + u_1 w_2 \end{pmatrix}$, so we can take $\mathbf{N} = \boxed{\begin{pmatrix} 0 & -u_3 & u_2 \\ u_3 & 0 & -u_1 \\ -u_2 & u_1 & 0 \end{pmatrix}}$.

12.48 In Exercise 12.3.5, we showed that $\|\mathbf{u} \times \mathbf{v}\|^2 = \|\mathbf{u}\|^2\|\mathbf{v}\|^2 - (\mathbf{u} \cdot \mathbf{v})^2$. So if $\mathbf{u} \cdot \mathbf{v} = 0$ and $\mathbf{u} \times \mathbf{v} = \mathbf{0}$, then $\|\mathbf{u}\|^2\|\mathbf{v}\|^2 = 0$, so $\|\mathbf{u}\| = 0$ or $\|\mathbf{v}\| = 0$. Therefore, $\mathbf{u} = \mathbf{0}$ or $\mathbf{v} = \mathbf{0}$.

12.49 Let O be the origin, and let U, V, and W be points such that
$\vec{U} = \begin{pmatrix} u_1 \\ u_2 \\ u_3 \end{pmatrix}$, $\vec{V} = \begin{pmatrix} v_1 \\ v_2 \\ v_3 \end{pmatrix}$, and $\vec{W} = \begin{pmatrix} w_1 \\ w_2 \\ w_3 \end{pmatrix}$. The first determinant is the

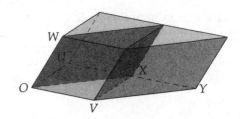

volume of the shown parallelepiped with edges $\overline{OU}$, $\overline{OV}$, and $\overline{OW}$. Let
this be parallelepiped $\mathcal{P}$. Let X be the point such that $\vec{X} = \vec{V} + \vec{U}$, so X is a
vertex of $\mathcal{P}$. The second determinant in the problem is the volume of the
parallelepiped with edges $\overline{OX}$, $\overline{OV}$, and $\overline{OW}$. Let this be parallelepiped
Q, and let Y be the vertex of Q such that $OXYV$ is a face of Q.

Both $OVYX$ and $OVXU$ are parallelograms with $\overline{OV}$ as a base, and with height to that base equal to the distance
between parallel lines $\overleftrightarrow{UY}$ and $\overleftrightarrow{OV}$. Therefore, these parallelograms have equal area. Parallelograms $OVYX$ and
$OVXU$ are faces of Q and $\mathcal{P}$, respectively. They have equal area, and they are in the same plane. Point W is a
vertex on the face opposite the respective parallelogram in both Q and $\mathcal{P}$, so the altitudes to these equal area faces
in the two parallelepipeds are equal. Since the two parallelepipeds have faces of equal area, and equal altitudes
to these faces, the parallelepipeds have equal volume, which means the given determinants are equal.

12.50 In each case, we find the inverse of the matrix by first figuring out the effect when a matrix $\mathbf{M}$ is multiplied
by the given matrix.

(a) Let the matrix be $\mathbf{A}$. We have
$$\mathbf{AM} = \begin{pmatrix} 0 & 1 & 0 \\ 1 & 0 & 0 \\ 0 & 0 & 1 \end{pmatrix} \begin{pmatrix} m_{11} & m_{12} & m_{13} \\ m_{21} & m_{22} & m_{23} \\ m_{31} & m_{32} & m_{33} \end{pmatrix} = \begin{pmatrix} m_{21} & m_{22} & m_{23} \\ m_{11} & m_{12} & m_{13} \\ m_{31} & m_{32} & m_{33} \end{pmatrix}.$$

Multiplying $\mathbf{M}$ by $\mathbf{A}$ swaps the first two rows of $\mathbf{M}$. Multiplying $\mathbf{AM}$ on the left again by $\mathbf{A}$ swaps the rows
back, so $\mathbf{A}^2\mathbf{M} = \mathbf{M}$. Therefore, $\mathbf{A}^2 = \mathbf{I}$, so $\mathbf{A}^{-1} = \mathbf{A}$. Checking, we have
$$\begin{pmatrix} 0 & 1 & 0 \\ 1 & 0 & 0 \\ 0 & 0 & 1 \end{pmatrix} \begin{pmatrix} 0 & 1 & 0 \\ 1 & 0 & 0 \\ 0 & 0 & 1 \end{pmatrix} = \begin{pmatrix} 1 & 0 & 0 \\ 0 & 1 & 0 \\ 0 & 0 & 1 \end{pmatrix}.$$

(b) Let the matrix be $\mathbf{B}$. We have
$$\mathbf{BM} = \begin{pmatrix} 0 & 0 & 1 \\ 1 & 0 & 0 \\ 0 & 1 & 0 \end{pmatrix} \begin{pmatrix} m_{11} & m_{12} & m_{13} \\ m_{21} & m_{22} & m_{23} \\ m_{31} & m_{32} & m_{33} \end{pmatrix} = \begin{pmatrix} m_{31} & m_{32} & m_{33} \\ m_{11} & m_{12} & m_{13} \\ m_{21} & m_{22} & m_{23} \end{pmatrix}.$$

Multiplying $\mathbf{M}$ by $\mathbf{B}$ "cycles" the rows of $\mathbf{M}$, moving the third row to the first row, and moving the other
two rows down one row. So, multiplying by $\mathbf{B}$ twice more gives us $\mathbf{M}$ back again. Therefore, we have
$\mathbf{B}^3\mathbf{M} = \mathbf{M}$, which means $\mathbf{B}^3 = \mathbf{I}$. This gives us $\mathbf{B}^2\mathbf{B} = \mathbf{I}$, so $\mathbf{B}^{-1} = \mathbf{B}^2$. Testing, we find that
$$\mathbf{B}^2 = \begin{pmatrix} 0 & 0 & 1 \\ 1 & 0 & 0 \\ 0 & 1 & 0 \end{pmatrix} \begin{pmatrix} 0 & 0 & 1 \\ 1 & 0 & 0 \\ 0 & 1 & 0 \end{pmatrix} = \begin{pmatrix} 0 & 1 & 0 \\ 0 & 0 & 1 \\ 1 & 0 & 0 \end{pmatrix},$$

and
$$\mathbf{B}^2\mathbf{B} = \begin{pmatrix} 0 & 1 & 0 \\ 0 & 0 & 1 \\ 1 & 0 & 0 \end{pmatrix} \begin{pmatrix} 0 & 0 & 1 \\ 1 & 0 & 0 \\ 0 & 1 & 0 \end{pmatrix} = \begin{pmatrix} 1 & 0 & 0 \\ 0 & 1 & 0 \\ 0 & 0 & 1 \end{pmatrix}.$$

We also can note that if $\mathbf{M}$ is the inverse of $\mathbf{B}$, then $\mathbf{BM} = \mathbf{I}$. Setting the result of the product $\mathbf{BM}$ equal
to $\mathbf{I}$, we can read off the entries of $\mathbf{M}$ and find that
$$\mathbf{B}^{-1} = \mathbf{M} = \boxed{\begin{pmatrix} 0 & 1 & 0 \\ 0 & 0 & 1 \\ 1 & 0 & 0 \end{pmatrix}}.$$

(c) Let the matrix be $\mathbf{C}$. We have

$$\mathbf{CM} = \begin{pmatrix} 0 & 0 & -1 \\ 1 & 0 & 0 \\ 0 & 1 & 0 \end{pmatrix} \begin{pmatrix} m_{11} & m_{12} & m_{13} \\ m_{21} & m_{22} & m_{23} \\ m_{31} & m_{32} & m_{33} \end{pmatrix} = \begin{pmatrix} -m_{31} & -m_{32} & -m_{33} \\ m_{11} & m_{12} & m_{13} \\ m_{21} & m_{22} & m_{23} \end{pmatrix}.$$

So, multiplying $\mathbf{M}$ by $\mathbf{C}$ "cycles" the rows just like $\mathbf{B}$ in part (b) and also multiplies the new first row by -1. Therefore, multiplying $\mathbf{M}$ by $\mathbf{C}$ three times gives $-\mathbf{M}$, which means $\mathbf{C}^3\mathbf{M} = -\mathbf{M}$. So, multiplying by $\mathbf{C}^3$ three more times gives us $\mathbf{M}$ back again: $\mathbf{C}^6\mathbf{M} = \mathbf{C}^3(-\mathbf{M}) = -(-\mathbf{M}) = \mathbf{M}$, which means $\mathbf{C}^6 = \mathbf{I}$. Since $\mathbf{C}^5\mathbf{C} = \mathbf{I}$, we have $\mathbf{C}^{-1} = \mathbf{C}^5$.

We also could have equated our result of $\mathbf{CM}$ to $\mathbf{I}$ to read off the entries of $\mathbf{M}$, and we find that

$$\mathbf{C}^{-1} = \boxed{\begin{pmatrix} 0 & 1 & 0 \\ 0 & 0 & 1 \\ -1 & 0 & 0 \end{pmatrix}}.$$

12.51 We seek a matrix $\mathbf{M}$ such that $\mathbf{MAB} = \mathbf{I}$. Multiplying both sides on the right by $\mathbf{B}^{-1}$ gives $\mathbf{MABB}^{-1} = \mathbf{IB}^{-1}$, so $\mathbf{MA} = \mathbf{B}^{-1}$. Multiplying both sides on the right by $\mathbf{A}^{-1}$ gives $\mathbf{MAA}^{-1} = \mathbf{B}^{-1}\mathbf{A}^{-1}$, so $\mathbf{M} = \boxed{\mathbf{B}^{-1}\mathbf{A}^{-1}}$. Note that this is *not* necessarily the same as $\mathbf{A}^{-1}\mathbf{B}^{-1}$, because matrix multiplication is not commutative.

12.52 Let $\mathbf{A}$ be the invertible matrix and $\mathbf{B}$ be the non-invertible matrix. Since $\mathbf{B}$ does not have an inverse, we have $\det(\mathbf{B}) = 0$. Therefore, we have $\det(\mathbf{P}) = \det(\mathbf{AB}) = \det(\mathbf{A})\det(\mathbf{B}) = 0$, which means $\mathbf{P}$ $\boxed{\text{does not have an inverse}}$.

12.53 We showed in a Review Problem in the previous chapter that $(\mathbf{AB})^T = \mathbf{B}^T\mathbf{A}^T$. Letting $\mathbf{B} = \mathbf{A}^{-1}$, we have $(\mathbf{AA}^{-1})^T = (\mathbf{A}^{-1})^T\mathbf{A}^T$, so $\mathbf{I}^T = (\mathbf{A}^{-1})^T\mathbf{A}^T$. Since $\mathbf{I}^T = \mathbf{I}$, we have $(\mathbf{A}^{-1})^T\mathbf{A}^T = \mathbf{I}$, which means the inverse of $\mathbf{A}^T$ is $(\mathbf{A}^{-1})^T$.

Challenge Problems

12.54

(a) Let $\mathbf{v} = \begin{pmatrix} a \\ b \\ c \end{pmatrix}$ and $\mathbf{f} = \begin{pmatrix} a \\ b \\ 0 \end{pmatrix}$. The point $(a, b, 0)$ lies on the graph of $z = 0$, and the vector $\mathbf{v} - \mathbf{f} = \begin{pmatrix} 0 \\ 0 \\ c \end{pmatrix}$ is normal to the graph of $z = 0$, so $\mathbf{f}$ is the projection of $\mathbf{v}$ onto the xy-plane. Furthermore,

$$\begin{pmatrix} 1 & 0 & 0 \\ 0 & 1 & 0 \\ 0 & 0 & 0 \end{pmatrix} \begin{pmatrix} a \\ b \\ c \end{pmatrix} = \begin{pmatrix} a \\ b \\ 0 \end{pmatrix},$$

so we can take

$$\mathbf{M} = \boxed{\begin{pmatrix} 1 & 0 & 0 \\ 0 & 1 & 0 \\ 0 & 0 & 0 \end{pmatrix}}.$$

(b) Let $\mathbf{v} = \begin{pmatrix} a \\ b \\ c \end{pmatrix}$. We seek a vector $\mathbf{f} = \begin{pmatrix} s \\ t \\ u \end{pmatrix}$ such that (s, t, u) lies on the graph of $x + y + z = 0$, so $s + t + u = 0$.

Also, we want $\mathbf{v} - \mathbf{f} = \begin{pmatrix} a - s \\ b - t \\ c - u \end{pmatrix}$ to be normal to the graph of $x + y + z = 0$. In other words,

$$\begin{pmatrix} a - s \\ b - t \\ c - u \end{pmatrix} = \begin{pmatrix} k \\ k \\ k \end{pmatrix}$$

for some scalar k, so $a - s = k$, $b - t = k$, and $c - u = k$. Adding all these equations, we get $(a+b+c) - (s+t+u) = 3k$. But $s + t + u = 0$, so $a + b + c = 3k$, which means $k = (a + b + c)/3$.

Then

$$s = a - k = a - \frac{a + b + c}{3} = \frac{2a - b - c}{3},$$
$$t = b - k = b - \frac{a + b + c}{3} = \frac{-a + 2b - c}{3},$$
$$u = c - k = c - \frac{a + b + c}{3} = \frac{-a - b + 2c}{3},$$

so

$$\mathbf{f} = \begin{pmatrix} s \\ t \\ u \end{pmatrix} = \begin{pmatrix} \frac{2a-b-c}{3} \\ \frac{-a+2b-c}{3} \\ \frac{-a-b+2c}{3} \end{pmatrix}.$$

Furthermore,

$$\begin{pmatrix} \frac{2}{3} & -\frac{1}{3} & -\frac{1}{3} \\ -\frac{1}{3} & \frac{2}{3} & -\frac{1}{3} \\ -\frac{1}{3} & -\frac{1}{3} & \frac{2}{3} \end{pmatrix} \begin{pmatrix} a \\ b \\ c \end{pmatrix} = \begin{pmatrix} \frac{2a-b-c}{3} \\ \frac{-a+2b-c}{3} \\ \frac{-a-b+2c}{3} \end{pmatrix},$$

so we can take

$$\mathbf{N} = \boxed{\begin{pmatrix} \frac{2}{3} & -\frac{1}{3} & -\frac{1}{3} \\ -\frac{1}{3} & \frac{2}{3} & -\frac{1}{3} \\ -\frac{1}{3} & -\frac{1}{3} & \frac{2}{3} \end{pmatrix}}.$$

12.55 There are many possible answers. We'll show how to find one pair of vectors $\mathbf{u}$ and $\mathbf{w}$ that satisfy the problem. First, we can take

$$\mathbf{u} = \mathbf{v} \times \mathbf{i} = \begin{vmatrix} \mathbf{i} & \mathbf{j} & \mathbf{k} \\ 3 & -2 & 1 \\ 1 & 0 & 0 \end{vmatrix} = \begin{vmatrix} -2 & 1 \\ 0 & 0 \end{vmatrix} \mathbf{i} - \begin{vmatrix} 3 & 1 \\ 1 & 0 \end{vmatrix} \mathbf{j} + \begin{vmatrix} 3 & -2 \\ 1 & 0 \end{vmatrix} \mathbf{k} = \mathbf{j} + 2\mathbf{k},$$

which is orthogonal to $\mathbf{v}$.

Now, we want $\mathbf{w}$ to be orthogonal to both $\mathbf{u}$ and $\mathbf{v}$. We can simply take

$$\mathbf{w} = \mathbf{u} \times \mathbf{v} = \begin{vmatrix} \mathbf{i} & \mathbf{j} & \mathbf{k} \\ 0 & 1 & 2 \\ 3 & -2 & 1 \end{vmatrix} = \begin{vmatrix} 1 & 2 \\ -2 & 1 \end{vmatrix} \mathbf{i} - \begin{vmatrix} 0 & 2 \\ 3 & 1 \end{vmatrix} \mathbf{j} + \begin{vmatrix} 0 & 1 \\ 3 & -2 \end{vmatrix} \mathbf{k} = 5\mathbf{i} + 6\mathbf{j} - 3\mathbf{k},$$

which is orthogonal to both $\mathbf{u}$ and $\mathbf{v}$.

These aren't the only possible answers. We can let $\mathbf{u}$ be the cross product of $\mathbf{v}$ and any vector that is not a scalar multiple of $\mathbf{v}$, and then let $\mathbf{w} = \mathbf{u} \times \mathbf{v}$. By this process of constructing $\mathbf{u}$ and $\mathbf{w}$, we guarantee that they are orthogonal to each other, and orthogonal to $\mathbf{v}$.

12.56 Let

$$\mathbf{P} = \frac{1}{\det(\mathbf{A})} \begin{pmatrix} A_{11} & -A_{21} & A_{31} \\ -A_{12} & A_{22} & -A_{32} \\ A_{13} & -A_{23} & A_{33} \end{pmatrix} \begin{pmatrix} a_{11} & a_{12} & a_{13} \\ a_{21} & a_{22} & a_{23} \\ a_{31} & a_{32} & a_{33} \end{pmatrix}.$$

We wish to show that $\mathbf{P} = \mathbf{I}$. First, we confirm that $p_{11} = p_{22} = p_{33} = 1$. We have

$$p_{11} = \frac{1}{\det(\mathbf{A})}(a_{11}A_{11} - a_{21}A_{21} + a_{31}A_{31}),$$

$$p_{22} = \frac{1}{\det(\mathbf{A})}(-a_{12}A_{12} + a_{22}A_{22} - a_{32}A_{32}),$$

$$p_{33} = \frac{1}{\det(\mathbf{A})}(a_{13}A_{13} - a_{23}A_{23} + a_{33}A_{33}).$$

Each of the expressions in parentheses above is an expansion by minors expression for $\det(\mathbf{A})$, where we are expanding along the first, second, and third columns of $\mathbf{A}$ in the expressions for p_{11}, p_{22}, and p_{33}, respectively. Therefore, we have $p_{11} = p_{22} = p_{33} = 1$.

We also have to show that each of the other entries in $\mathbf{P}$ is 0. We'll only show one here; showing that each of the others is 0 can be done in the same way. We have

$$p_{12} = \frac{1}{\det(\mathbf{A})}(a_{12}A_{11} - a_{22}A_{21} + a_{32}A_{31}) = \frac{1}{\det(\mathbf{A})}\left(a_{12}\begin{vmatrix} a_{22} & a_{23} \\ a_{32} & a_{33} \end{vmatrix} - a_{22}\begin{vmatrix} a_{12} & a_{13} \\ a_{32} & a_{33} \end{vmatrix} + a_{32}\begin{vmatrix} a_{12} & a_{13} \\ a_{22} & a_{23} \end{vmatrix}\right)$$

$$= \frac{1}{\det(\mathbf{A})}(a_{12}(a_{22}a_{33} - a_{23}a_{32}) - a_{22}(a_{12}a_{33} - a_{32}a_{13}) + a_{32}(a_{12}a_{23} - a_{22}a_{13}))$$

$$= 0.$$

Multiplying out each of the other non-diagonal entries produces similar convenient cancellation that shows that each of these equals 0, so $\mathbf{P} = \mathbf{I}$.

12.57 Let the vector be $\mathbf{v} = x\mathbf{i} + y\mathbf{j} + z\mathbf{k}$. The angle between $\mathbf{v}$ and $\mathbf{i}$ is α, so $\mathbf{v} \cdot \mathbf{i} = \|\mathbf{v}\|\|\mathbf{i}\| \cos\alpha = \|\mathbf{v}\| \cos\alpha$. But $\mathbf{v} \cdot \mathbf{i} = x$, so $x = \|\mathbf{v}\| \cos\alpha$. Similarly, we find that $y = \|\mathbf{v}\| \cos\beta$ and $z = \|\mathbf{v}\| \cos\gamma$. We then have

$$\|\mathbf{v}\|^2 = x^2 + y^2 + z^2 = \|\mathbf{v}\|^2 \cos^2\alpha + \|\mathbf{v}\|^2 \cos^2\beta + \|\mathbf{v}\|^2 \cos^2\gamma = \|\mathbf{v}\|^2(\cos^2\alpha + \cos^2\beta + \cos^2\gamma).$$

Since $\|\mathbf{v}\|^2 \neq 0$, we can divide by $\|\mathbf{v}\|^2$ to give $\cos^2\alpha + \cos^2\beta + \cos^2\gamma = 1$.

12.58 Let $\mathbf{w} = \|\mathbf{v}\|\mathbf{u} + \|\mathbf{u}\|\mathbf{v}$. Let α be the angle between $\mathbf{u}$ and $\mathbf{w}$, and let β be the angle between $\mathbf{v}$ and $\mathbf{w}$, where $0 \le \alpha, \beta \le \pi$. Then

$$\cos\alpha = \frac{\mathbf{u} \cdot \mathbf{w}}{\|\mathbf{u}\|\|\mathbf{w}\|} = \frac{\mathbf{u} \cdot (\|\mathbf{v}\|\mathbf{u} + \|\mathbf{u}\|\mathbf{v})}{\|\mathbf{u}\|\|\mathbf{w}\|} = \frac{\|\mathbf{v}\|\mathbf{u} \cdot \mathbf{u} + \|\mathbf{u}\|\mathbf{u} \cdot \mathbf{v}}{\|\mathbf{u}\|\|\mathbf{w}\|} = \frac{\|\mathbf{v}\|\|\mathbf{u}\|^2 + \|\mathbf{u}\|\mathbf{u} \cdot \mathbf{v}}{\|\mathbf{u}\|\|\mathbf{w}\|} = \frac{\|\mathbf{u}\|\|\mathbf{v}\| + \mathbf{u} \cdot \mathbf{v}}{\|\mathbf{w}\|}.$$

Similarly,

$$\cos\beta = \frac{\mathbf{v} \cdot \mathbf{w}}{\|\mathbf{v}\|\|\mathbf{w}\|} = \frac{\mathbf{v} \cdot (\|\mathbf{v}\|\mathbf{u} + \|\mathbf{u}\|\mathbf{v})}{\|\mathbf{v}\|\|\mathbf{w}\|} = \frac{\|\mathbf{v}\|\mathbf{u} \cdot \mathbf{v} + \|\mathbf{u}\|\mathbf{v} \cdot \mathbf{v}}{\|\mathbf{v}\|\|\mathbf{w}\|} = \frac{\|\mathbf{v}\|\mathbf{u} \cdot \mathbf{v} + \|\mathbf{u}\|\|\mathbf{v}\|^2}{\|\mathbf{v}\|\|\mathbf{w}\|} = \frac{\mathbf{u} \cdot \mathbf{v} + \|\mathbf{u}\|\|\mathbf{v}\|}{\|\mathbf{w}\|}.$$

Thus, we have $\cos\alpha = \cos\beta$. Since $0 \le \alpha, \beta \le \pi$, we must have $\alpha = \beta$, so $\mathbf{w}$ bisects the angle between $\mathbf{u}$ and $\mathbf{v}$.

12.59

(a) Taking $\mathbf{a} = \mathbf{b} = \mathbf{v}$ in (ii), we get $\mathbf{v} \times \mathbf{v} = -\mathbf{v} \times \mathbf{v}$, so $\mathbf{v} \times \mathbf{v} = \mathbf{0}$.

(b) If $\mathbf{v} = v_1\mathbf{i} + v_2\mathbf{j} + v_3\mathbf{k}$ and $\mathbf{w} = w_1\mathbf{i} + w_2\mathbf{j} + w_3\mathbf{k}$, then repeatedly applying (iv) gives

$$\mathbf{v} \times \mathbf{w} = (v_1\mathbf{i} + v_2\mathbf{j} + v_3\mathbf{k}) \times (w_1\mathbf{i} + w_2\mathbf{j} + w_3\mathbf{k})$$

$$= v_1w_1(\mathbf{i} \times \mathbf{i}) + v_1w_2(\mathbf{i} \times \mathbf{j}) + v_1w_3(\mathbf{i} \times \mathbf{k})$$

$$+ v_2w_1(\mathbf{j} \times \mathbf{i}) + v_2w_2(\mathbf{j} \times \mathbf{j}) + v_2w_3(\mathbf{j} \times \mathbf{k})$$

$$+ v_3w_1(\mathbf{k} \times \mathbf{i}) + v_3w_2(\mathbf{k} \times \mathbf{j}) + v_3w_3(\mathbf{k} \times \mathbf{k}).$$

From part (a), we have $\mathbf{i} \times \mathbf{i} = \mathbf{j} \times \mathbf{j} = \mathbf{k} \times \mathbf{k} = 0$. From (i) and (ii), we have $\mathbf{j} \times \mathbf{i} = -\mathbf{i} \times \mathbf{j} = -\mathbf{k}$, $\mathbf{k} \times \mathbf{j} = -\mathbf{j} \times \mathbf{k} = -\mathbf{i}$, and $\mathbf{i} \times \mathbf{k} = -\mathbf{k} \times \mathbf{i} = -\mathbf{j}$. Hence,

$$
\begin{aligned}
\mathbf{v} \times \mathbf{w} &= v_1 w_1 (\mathbf{i} \times \mathbf{i}) + v_1 w_2 (\mathbf{i} \times \mathbf{j}) + v_1 w_3 (\mathbf{i} \times \mathbf{k}) \\
&\quad + v_2 w_1 (\mathbf{j} \times \mathbf{i}) + v_2 w_2 (\mathbf{j} \times \mathbf{j}) + v_2 w_3 (\mathbf{j} \times \mathbf{k}) \\
&\quad + v_3 w_1 (\mathbf{k} \times \mathbf{i}) + v_3 w_2 (\mathbf{k} \times \mathbf{j}) + v_3 w_3 (\mathbf{k} \times \mathbf{k}) \\
&= v_1 w_2 \mathbf{k} - v_1 w_3 \mathbf{j} - v_2 w_1 \mathbf{k} + v_2 w_3 \mathbf{i} + v_3 w_1 \mathbf{j} - v_3 w_2 \mathbf{i} \\
&= (v_2 w_3 - v_3 w_2)\mathbf{i} - (v_1 w_3 - v_3 w_1)\mathbf{j} + (v_1 w_2 - v_2 w_1)\mathbf{k} \\
&= \begin{vmatrix} v_2 & v_3 \\ w_2 & w_3 \end{vmatrix} \mathbf{i} - \begin{vmatrix} v_1 & v_3 \\ w_1 & w_3 \end{vmatrix} \mathbf{j} + \begin{vmatrix} v_1 & v_2 \\ w_1 & w_2 \end{vmatrix} \mathbf{k}.
\end{aligned}
$$

(c)   Using the same notation as in part (b), since $\mathbf{v} \times \mathbf{w} = (v_2 w_3 - v_3 w_2)\mathbf{i} - (v_1 w_3 - v_3 w_1)\mathbf{j} + (v_1 w_2 - v_2 w_1)\mathbf{k}$,

$$
\begin{aligned}
(\mathbf{v} \times \mathbf{w}) \cdot \mathbf{v} &= v_1(v_2 w_3 - v_3 w_2) - v_2(v_1 w_3 - v_3 w_1) + v_3(v_1 w_2 - v_2 w_1) \\
&= v_1 v_2 w_3 - v_1 v_3 w_2 - v_1 v_2 w_3 + v_2 v_3 w_1 + v_1 v_3 w_2 - v_2 v_3 w_1 \\
&= 0.
\end{aligned}
$$

We then have

$$
(\mathbf{v} \times \mathbf{w}) \cdot \mathbf{w} = (-\mathbf{w} \times \mathbf{v}) \cdot \mathbf{w} = -(\mathbf{w} \times \mathbf{v}) \cdot \mathbf{w} = 0,
$$

where we use the result we proved at the start of this part in the final step. Hence, $\mathbf{v} \times \mathbf{w}$ is orthogonal to both $\mathbf{v}$ and $\mathbf{w}$.

12.60   From the identity $\mathbf{a} \times (\mathbf{b} \times \mathbf{c}) = (\mathbf{a} \cdot \mathbf{c})\mathbf{b} - (\mathbf{a} \cdot \mathbf{b})\mathbf{c}$ (proved as an Exercise in Section 12.3), we get

$$
(\mathbf{a} \times \mathbf{b}) \times (\mathbf{a} \times \mathbf{c}) = [(\mathbf{a} \times \mathbf{b}) \cdot \mathbf{c}]\mathbf{a} - [(\mathbf{a} \times \mathbf{b}) \cdot \mathbf{a}]\mathbf{c}.
$$

But $\mathbf{a} \times \mathbf{b}$ is orthogonal to $\mathbf{a}$, so $(\mathbf{a} \times \mathbf{b}) \cdot \mathbf{a} = 0$. Furthermore, as proved in an Exercise in Section 12.4, we have $(\mathbf{a} \times \mathbf{b}) \cdot \mathbf{c} = \mathbf{a} \cdot (\mathbf{b} \times \mathbf{c})$, so

$$
(\mathbf{a} \times \mathbf{b}) \times (\mathbf{a} \times \mathbf{c}) = [(\mathbf{a} \times \mathbf{b}) \cdot \mathbf{c}]\mathbf{a} - [(\mathbf{a} \times \mathbf{b}) \cdot \mathbf{a}]\mathbf{c} = [(\mathbf{a} \times \mathbf{b}) \cdot \mathbf{c}]\mathbf{a} - (0)\mathbf{c} = [(\mathbf{a} \times \mathbf{b}) \cdot \mathbf{c}]\mathbf{a} = [\mathbf{a} \cdot (\mathbf{b} \times \mathbf{c})]\mathbf{a}.
$$

12.61   Let $\mathbf{e} = \mathbf{a} \times \mathbf{b}$, so $(\mathbf{a} \times \mathbf{b}) \cdot (\mathbf{c} \times \mathbf{d}) = \mathbf{e} \cdot (\mathbf{c} \times \mathbf{d})$. From the identity $\mathbf{a} \cdot (\mathbf{b} \times \mathbf{c}) = (\mathbf{a} \times \mathbf{b}) \cdot \mathbf{c}$,

$$
\mathbf{e} \cdot (\mathbf{c} \times \mathbf{d}) = (\mathbf{e} \times \mathbf{c}) \cdot \mathbf{d} = [(\mathbf{a} \times \mathbf{b}) \times \mathbf{c}] \cdot \mathbf{d}.
$$

But $(\mathbf{a} \times \mathbf{b}) \times \mathbf{c} = -\mathbf{c} \times (\mathbf{a} \times \mathbf{b}) = (\mathbf{a} \cdot \mathbf{c})\mathbf{b} - (\mathbf{b} \cdot \mathbf{c})\mathbf{a}$, so

$$
[(\mathbf{a} \times \mathbf{b}) \times \mathbf{c}] \cdot \mathbf{d} = [(\mathbf{a} \cdot \mathbf{c})\mathbf{b} - (\mathbf{b} \cdot \mathbf{c})\mathbf{a}] \cdot \mathbf{d} = (\mathbf{a} \cdot \mathbf{c})(\mathbf{b} \cdot \mathbf{d}) - (\mathbf{a} \cdot \mathbf{d})(\mathbf{b} \cdot \mathbf{c}).
$$

12.62   *Solution 1: Geometry.* The expressions $\frac{\mathbf{v} \cdot \mathbf{a}}{\|\mathbf{a}\|^2}\mathbf{a}$, $\frac{\mathbf{v} \cdot \mathbf{b}}{\|\mathbf{b}\|^2}\mathbf{b}$, and $\frac{\mathbf{v} \cdot \mathbf{c}}{\|\mathbf{c}\|^2}\mathbf{c}$ are the projections of $\mathbf{v}$ onto $\mathbf{a}$, $\mathbf{b}$, and $\mathbf{c}$. Consider the parallelepiped with these three vectors as edges, as shown at right. Because each two of $\mathbf{a}$, $\mathbf{b}$, and $\mathbf{c}$ are orthogonal, this parallelepiped is a right rectangular prism (or box), and $\mathbf{v}$ is a space diagonal of the box. From face $OADB$ of the box, we have $\vec{D} = \operatorname{proj}_{\mathbf{a}} \mathbf{v} + \operatorname{proj}_{\mathbf{b}} \mathbf{v}$, and from rectangle $OCVD$, we have $\mathbf{v} = \vec{D} + \operatorname{proj}_{\mathbf{c}} \mathbf{v} = \operatorname{proj}_{\mathbf{a}} \mathbf{v} + \operatorname{proj}_{\mathbf{b}} \mathbf{v} + \operatorname{proj}_{\mathbf{c}} \mathbf{v}$, as desired.

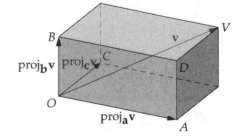

Solution 2: Algebra. Since $\mathbf{a}$, $\mathbf{b}$, $\mathbf{c}$ are linearly independent, there is a unique triplet of constants (α, β, γ) such that

$$
\mathbf{v} = \alpha \mathbf{a} + \beta \mathbf{b} + \gamma \mathbf{c}.
$$

Since $\mathbf{b} \cdot \mathbf{a} = \mathbf{c} \cdot \mathbf{a} = 0$ (we are given that $\mathbf{a}$ is orthogonal to $\mathbf{b}$ and $\mathbf{c}$), we have

$$\mathbf{v} \cdot \mathbf{a} = (\alpha\mathbf{a} + \beta\mathbf{b} + \gamma\mathbf{c}) \cdot \mathbf{a} = \alpha\mathbf{a} \cdot \mathbf{a} + \beta\mathbf{b} \cdot \mathbf{a} + \gamma\mathbf{c} \cdot \mathbf{a} = \alpha \|\mathbf{a}\|^2.$$

Therefore, we have $\alpha = \frac{\mathbf{v} \cdot \mathbf{a}}{\|\mathbf{a}\|^2}$. Similarly, we have $\beta = \frac{\mathbf{v} \cdot \mathbf{b}}{\|\mathbf{b}\|^2}$ and $\gamma = \frac{\mathbf{v} \cdot \mathbf{c}}{\|\mathbf{c}\|^2}$, so

$$\mathbf{v} = \frac{\mathbf{v} \cdot \mathbf{a}}{\|\mathbf{a}\|^2}\mathbf{a} + \frac{\mathbf{v} \cdot \mathbf{b}}{\|\mathbf{b}\|^2}\mathbf{b} + \frac{\mathbf{v} \cdot \mathbf{c}}{\|\mathbf{c}\|^2}\mathbf{c}.$$

12.63 We prove that no such matrix $\mathbf{M}$ exists by contradiction. Let $\mathbf{v}_1$, $\mathbf{v}_2$, and $\mathbf{v}_3$ be linearly independent vectors in three dimensions. If we can find a vector $\mathbf{u}$ for each vector $\mathbf{v}$ such that $\mathbf{Mu} = \mathbf{v}$, then there exist vectors $\mathbf{u}_1$, $\mathbf{u}_2$, and $\mathbf{u}_3$ such that $\mathbf{Mu}_1 = \mathbf{v}_1$, $\mathbf{Mu}_2 = \mathbf{v}_2$, and $\mathbf{Mu}_3 = \mathbf{v}_3$. Since $\mathbf{u}_1$, $\mathbf{u}_2$, and $\mathbf{u}_3$ are in two dimensions, they are linearly dependent. Therefore, there are some constants c_1, c_2, c_3, not all 0, such that $c_1\mathbf{u}_1 + c_2\mathbf{u}_2 + c_3\mathbf{u}_3 = \mathbf{0}$. Multiplying each of our equations with $\mathbf{M}$ by the respective c_i, and then adding all three equations, we have

$$c_1\mathbf{Mu}_1 + c_2\mathbf{Mu}_2 + c_3\mathbf{Mu}_3 = c_1\mathbf{v}_1 + c_2\mathbf{v}_2 + c_3\mathbf{v}_3.$$

Therefore, we have

$$c_1\mathbf{v}_1 + c_2\mathbf{v}_2 + c_3\mathbf{v}_3 = \mathbf{M}(c_1\mathbf{u}_1 + c_2\mathbf{u}_2 + c_3\mathbf{u}_3) = \mathbf{M0} = \mathbf{0},$$

which contradicts the assumption that $\mathbf{v}_1$, $\mathbf{v}_2$, and $\mathbf{v}_3$ are linearly independent. Therefore, no such matrix $\mathbf{M}$ exists.

12.64 We know how to find the area of the the triangle with the three points as vertices, and how to find the distance between each pair of the points. We can combine these to find any altitude of the triangle. Let $\mathbf{v}$ be the vector pointing from $(0,1,2)$ to $(4,1,-3)$, and let $\mathbf{w}$ be the vector pointing from $(0,1,2)$ to $(-2,5,1)$, so $\mathbf{v} = 4\mathbf{i} - 5\mathbf{k}$, and $\mathbf{w} = -2\mathbf{i} + 4\mathbf{j} - \mathbf{k}$. The area of the triangle with the three points as vertices is $\frac{1}{2}\|\mathbf{v} \times \mathbf{w}\|$. We have

$$\mathbf{v} \times \mathbf{w} = \begin{vmatrix} \mathbf{i} & \mathbf{j} & \mathbf{k} \\ 4 & 0 & -5 \\ -2 & 4 & -1 \end{vmatrix} = \begin{vmatrix} 0 & -5 \\ 4 & -1 \end{vmatrix}\mathbf{i} - \begin{vmatrix} 4 & -5 \\ -2 & -1 \end{vmatrix}\mathbf{j} + \begin{vmatrix} 4 & 0 \\ -2 & 4 \end{vmatrix}\mathbf{k} = 20\mathbf{i} + 14\mathbf{j} + 16\mathbf{k}.$$

Therefore, the area of the triangle is $\frac{1}{2}\|\mathbf{v} \times \mathbf{w}\| = \frac{1}{2}\sqrt{20^2 + 14^2 + 16^2} = \frac{1}{2}\sqrt{852} = \sqrt{213}$.

But the area, A, of this triangle is also given by $A = \frac{1}{2}bh$, where b is the distance between the points $(4,1,-3)$ and $(-2,5,1)$, and h is the distance between the point $(0,1,2)$ and the line through $(4,1,-3)$ and $(-2,5,1)$. The distance between the points $(4,1,-3)$ and $(-2,5,1)$ is $b = \sqrt{[4-(-2)]^2 + (1-5)^2 + (-3-1)^2} = \sqrt{68} = 2\sqrt{17}$. Hence,

$$h = \frac{2A}{b} = \frac{2\sqrt{213}}{b} = \frac{2\sqrt{213}}{2\sqrt{17}} = \boxed{\sqrt{\frac{213}{17}}}.$$

As an alternative solution, we also could have found a vector from a point on the line to $(0,1,2)$ such that the vector is normal to the line. The length of this vector is the desired distance. We let $\mathbf{r}$ be the vector from $(4,1,-3)$ to $(0,1,2)$, and let $\mathbf{d}$ be the direction vector of the line. Then, the difference $\mathbf{r} - \text{proj}_{\mathbf{d}}\mathbf{r}$ is normal to the line, and goes from the line to $(0,1,2)$. The magnitude of this difference is the desired distance.

12.65 Let $\mathbf{b} = \vec{AB}$, $\mathbf{c} = \vec{AC}$, and $\mathbf{d} = \vec{AD}$. From the given information about the volume of the parallelepiped with edges $\overline{AB}$, $\overline{AC}$, and $\overline{AD}$, we have

$$\left| \det \begin{pmatrix} b_1 & b_2 & b_3 \\ c_1 & c_2 & c_3 \\ d_1 & d_2 & d_3 \end{pmatrix} \right| = 80.$$

We then relate each of the desired volumes to this volume.

(a) Since $\overline{AD}$ is the interior diagonal of $\mathcal{P}$, $\vec{AD} = \mathbf{d}$ is the sum of the vectors corresponding to the three edges of $\mathcal{P}$ emanating from A. Two of these vectors are $\vec{AB} = \mathbf{b}$ and $\vec{AC} = \mathbf{c}$, so the third vector is $\mathbf{d} - \mathbf{b} - \mathbf{c}$. Hence, the volume of $\mathcal{P}$ is the absolute value of

$$\begin{vmatrix} b_1 & b_2 & b_3 \\ c_1 & c_2 & c_3 \\ d_1 - b_1 - c_1 & d_2 - b_2 - c_2 & d_3 - b_3 - c_3 \end{vmatrix}.$$

Adding the first row to the third row, we get

$$\begin{vmatrix} b_1 & b_2 & b_3 \\ c_1 & c_2 & c_3 \\ d_1 - b_1 - c_1 & d_2 - b_2 - c_2 & d_3 - b_3 - c_3 \end{vmatrix} = \begin{vmatrix} b_1 & b_2 & b_3 \\ c_1 & c_2 & c_3 \\ d_1 - c_1 & d_2 - c_2 & d_3 - c_3 \end{vmatrix}.$$

Adding the second row to the third row, we get

$$\begin{vmatrix} b_1 & b_2 & b_3 \\ c_1 & c_2 & c_3 \\ d_1 - c_1 & d_2 - c_2 & d_3 - c_3 \end{vmatrix} = \begin{vmatrix} b_1 & b_2 & b_3 \\ c_1 & c_2 & c_3 \\ d_1 & d_2 & d_3 \end{vmatrix}.$$

We know that the absolute value of this determinant is 80, so the volume of $\mathcal{P}$ is $\boxed{80}$.

(b) Let $\mathbf{u}$, $\mathbf{v}$, and $\mathbf{w}$ be the vectors corresponding to the edges of Q emanating from A. If $\mathbf{b}$ is the face diagonal of the face formed from $\mathbf{v}$ and $\mathbf{w}$, then $\mathbf{b} = \mathbf{v} + \mathbf{w}$. Similarly, we can set $\mathbf{c} = \mathbf{u} + \mathbf{w}$ and $\mathbf{d} = \mathbf{u} + \mathbf{v}$. Adding all three equations, we get $\mathbf{b} + \mathbf{c} + \mathbf{d} = 2(\mathbf{u} + \mathbf{v} + \mathbf{w})$. Then $\mathbf{u} + \mathbf{v} + \mathbf{w} = \frac{\mathbf{b}+\mathbf{c}+\mathbf{d}}{2}$, which makes

$$\mathbf{u} = (\mathbf{u} + \mathbf{v} + \mathbf{w}) - (\mathbf{v} + \mathbf{w}) = \frac{\mathbf{b} + \mathbf{c} + \mathbf{d}}{2} - \mathbf{b} = \frac{-\mathbf{b} + \mathbf{c} + \mathbf{d}}{2}.$$

Similarly, $\mathbf{v} = \frac{\mathbf{b}-\mathbf{c}+\mathbf{d}}{2}$ and $\mathbf{w} = \frac{\mathbf{b}+\mathbf{c}-\mathbf{d}}{2}$. Therefore, the volume of Q is the absolute value of

$$\begin{vmatrix} \frac{-b_1+c_1+d_1}{2} & \frac{-b_2+c_2+d_2}{2} & \frac{-b_3+c_3+d_3}{2} \\ \frac{b_1-c_1+d_1}{2} & \frac{b_2-c_2+d_2}{2} & \frac{b_3-c_3+d_3}{2} \\ \frac{b_1+c_1-d_1}{2} & \frac{b_2+c_2-d_2}{2} & \frac{b_2+c_2-d_2}{2} \end{vmatrix}. \tag{12.1}$$

Each entry in the first column is a linear combination of b_1, c_1, and d_1. Similarly, each entry in the second column is a linear combination of b_2, c_2, and d_2, and each entry in the third column is a linear combination of b_3, c_3, and d_3. So, we can express this matrix as the product of two matrices whose determinants we can find easily:

$$\begin{pmatrix} \frac{-b_1+c_1+d_1}{2} & \frac{-b_2+c_2+d_2}{2} & \frac{-b_3+c_3+d_3}{2} \\ \frac{b_1-c_1+d_1}{2} & \frac{b_2-c_2+d_2}{2} & \frac{b_3-c_3+d_3}{2} \\ \frac{b_1+c_1-d_1}{2} & \frac{b_2+c_2-d_2}{2} & \frac{b_2+c_2-d_2}{2} \end{pmatrix} = \begin{pmatrix} -\frac{1}{2} & \frac{1}{2} & \frac{1}{2} \\ \frac{1}{2} & -\frac{1}{2} & \frac{1}{2} \\ \frac{1}{2} & \frac{1}{2} & -\frac{1}{2} \end{pmatrix} \begin{pmatrix} b_1 & b_2 & b_3 \\ c_1 & c_2 & c_3 \\ d_1 & d_2 & d_3 \end{pmatrix},$$

we have

$$\begin{vmatrix} \frac{-b_1+c_1+d_1}{2} & \frac{-b_2+c_2+d_2}{2} & \frac{-b_3+c_3+d_3}{2} \\ \frac{b_1-c_1+d_1}{2} & \frac{b_2-c_2+d_2}{2} & \frac{b_3-c_3+d_3}{2} \\ \frac{b_1+c_1-d_1}{2} & \frac{b_2+c_2-d_2}{2} & \frac{b_2+c_2-d_2}{2} \end{vmatrix} = \begin{vmatrix} -\frac{1}{2} & \frac{1}{2} & \frac{1}{2} \\ \frac{1}{2} & -\frac{1}{2} & \frac{1}{2} \\ \frac{1}{2} & \frac{1}{2} & -\frac{1}{2} \end{vmatrix} \begin{vmatrix} b_1 & b_2 & b_3 \\ c_1 & c_2 & c_3 \\ d_1 & d_2 & d_3 \end{vmatrix} = \frac{1}{8} \begin{vmatrix} -1 & 1 & 1 \\ 1 & -1 & 1 \\ 1 & 1 & -1 \end{vmatrix} \cdot \begin{vmatrix} b_1 & b_2 & b_3 \\ c_1 & c_2 & c_3 \\ d_1 & d_2 & d_3 \end{vmatrix}.$$

The determinant of the first matrix is 4, and the absolute value of the second is 80, so the volume of Q is $\boxed{40}$.

We also could have manipulated the determinant on line (12.1), like we manipulated the determinant in part (a). We add the first row to the second, and then to the third, to find

$$\begin{vmatrix} \frac{-b_1+c_1+d_1}{2} & \frac{-b_2+c_2+d_2}{2} & \frac{-b_3+c_3+d_3}{2} \\ \frac{b_1-c_1+d_1}{2} & \frac{b_2-c_2+d_2}{2} & \frac{b_3-c_3+d_3}{2} \\ \frac{b_1+c_1-d_1}{2} & \frac{b_2+c_2-d_2}{2} & \frac{b_2+c_2-d_2}{2} \end{vmatrix} = \begin{vmatrix} \frac{-b_1+c_1+d_1}{2} & \frac{-b_2+c_2+d_2}{2} & \frac{-b_3+c_3+d_3}{2} \\ d_1 & d_2 & d_3 \\ \frac{b_1+c_1-d_1}{2} & \frac{b_2+c_2-d_2}{2} & \frac{b_2+c_2-d_2}{2} \end{vmatrix} = \begin{vmatrix} \frac{-b_1+c_1+d_1}{2} & \frac{-b_2+c_2+d_2}{2} & \frac{-b_3+c_3+d_3}{2} \\ d_1 & d_2 & d_3 \\ c_1 & c_2 & c_3 \end{vmatrix}.$$

Factoring $-\frac{1}{2}$ out of the first row, then adding the second and third rows to the first gives:

$$\begin{vmatrix} \frac{-b_1+c_1+d_1}{2} & \frac{-b_2+c_2+d_2}{2} & \frac{-b_3+c_3+d_3}{2} \\ d_1 & d_2 & d_3 \\ c_1 & c_2 & c_3 \end{vmatrix} = -\frac{1}{2}\begin{vmatrix} b_1-c_1-d_1 & b_2-c_2-d_2 & b_3-c_3-d_3 \\ d_1 & d_2 & d_3 \\ c_1 & c_2 & c_3 \end{vmatrix} = -\frac{1}{2}\begin{vmatrix} b_1 & b_2 & b_3 \\ d_1 & d_2 & d_3 \\ c_1 & c_2 & c_3 \end{vmatrix}.$$

Finally, exchanging the second and third rows gives

$$-\frac{1}{2}\begin{vmatrix} b_1 & b_2 & b_3 \\ d_1 & d_2 & d_3 \\ c_1 & c_2 & c_3 \end{vmatrix} = -\frac{1}{2}(-1)\begin{vmatrix} b_1 & b_2 & b_3 \\ c_1 & c_2 & c_3 \\ d_1 & d_2 & d_3 \end{vmatrix},$$

and again we have a volume of 40.

12.66

(a) As explained in the text, such a matrix $\mathbf{M}$ exists if the function $f(\mathbf{w}) = \text{proj}_\mathbf{v}\mathbf{w}$ is linear. We have

$$f(a\mathbf{w}) = \text{proj}_\mathbf{v}(a\mathbf{w}) = \frac{(a\mathbf{w})\cdot\mathbf{v}}{\|\mathbf{v}\|^2}\mathbf{v} = a\frac{\mathbf{w}\cdot\mathbf{v}}{\|\mathbf{v}\|^2}\mathbf{v} = a(\text{proj}_\mathbf{v}\mathbf{w}) = af(\mathbf{w})$$

and

$$f(\mathbf{u}+\mathbf{w}) = \text{proj}_\mathbf{v}(\mathbf{u}+\mathbf{w}) = \frac{(\mathbf{u}+\mathbf{w})\cdot\mathbf{v}}{\|\mathbf{v}\|^2}\mathbf{v} = \frac{\mathbf{u}\cdot\mathbf{v}+\mathbf{w}\cdot\mathbf{v}}{\|\mathbf{v}\|^2}\mathbf{v}$$
$$= \frac{\mathbf{u}\cdot\mathbf{v}}{\|\mathbf{v}\|^2}\mathbf{v} + \frac{\mathbf{w}\cdot\mathbf{v}}{\|\mathbf{v}\|^2}\mathbf{v} = \text{proj}_\mathbf{v}\mathbf{u} + \text{proj}_\mathbf{v}\mathbf{w} = f(\mathbf{u}) + f(\mathbf{w}).$$

Since f is linear, there exists a matrix $\mathbf{M}$ such that $f(\mathbf{w}) = \mathbf{M}\mathbf{w}$. Therefore, there exists a matrix $\mathbf{M}$ such that $\text{proj}_\mathbf{v}\mathbf{w} = \mathbf{M}\mathbf{w}$ for all $\mathbf{w}$.

(b) The matrix $\mathbf{M}$ is not invertible, so $\det(\mathbf{M}) = \boxed{0}$. To see why this is the case, note that if $\mathbf{M}$ had an inverse $\mathbf{M}^{-1}$, then multiplying both sides of the equation $\mathbf{M}\mathbf{w} = \mathbf{0}$ by $\mathbf{M}^{-1}$ would give us $\mathbf{w} = \mathbf{0}$ as the only vector whose projection onto $\mathbf{v}$ is $\mathbf{0}$. However, the projection of any vector orthogonal to $\mathbf{v}$ onto $\mathbf{v}$ is $\mathbf{0}$, so the equation $\mathbf{M}\mathbf{w} = \mathbf{0}$ does not have a unique solution. Therefore, $\mathbf{M}$ is not invertible, which means its determinant is 0.

(c) Let $\mathbf{w} = \begin{pmatrix} x \\ y \\ z \end{pmatrix}$. Then the projection of $\mathbf{w}$ onto $\mathbf{v}$ is

$$\frac{\mathbf{w}\cdot\mathbf{v}}{\|\mathbf{v}\|^2}\mathbf{v} = \frac{3x-2y+5z}{3^2+(-2)^2+5^2}\begin{pmatrix} 3 \\ -2 \\ 5 \end{pmatrix} = \frac{3x-2y+5z}{38}\begin{pmatrix} 3 \\ -2 \\ 5 \end{pmatrix} = \begin{pmatrix} \frac{3(3x-2y+5z)}{38} \\ \frac{2(3x-2y+5z)}{38} \\ \frac{5(3x-2y+5z)}{38} \end{pmatrix}.$$

Also,

$$\begin{pmatrix} \frac{9}{38} & -\frac{6}{38} & \frac{15}{38} \\ -\frac{6}{38} & \frac{4}{38} & -\frac{10}{38} \\ \frac{15}{38} & -\frac{10}{38} & \frac{25}{38} \end{pmatrix}\begin{pmatrix} x \\ y \\ z \end{pmatrix} = \begin{pmatrix} \frac{9}{38}x - \frac{6}{38}y + \frac{15}{38}z \\ -\frac{6}{38}x + \frac{4}{38}y - \frac{10}{38}z \\ \frac{15}{38}x - \frac{10}{38}y + \frac{25}{38}z \end{pmatrix} = \begin{pmatrix} \frac{3(3x-2y+5z)}{38} \\ \frac{2(3x-2y+5z)}{38} \\ \frac{5(3x-2y+5z)}{38} \end{pmatrix},$$

so we can take

$$\mathbf{M} = \begin{pmatrix} \frac{9}{38} & -\frac{6}{38} & \frac{15}{38} \\ -\frac{6}{38} & \frac{4}{38} & -\frac{10}{38} \\ \frac{15}{38} & -\frac{10}{38} & \frac{25}{38} \end{pmatrix}.$$

As an additional challenge, see if you can use the expression for $\mathbf{M}$ above to generalize this problem. That is, find a matrix $\mathbf{P}$ in terms of a, b, and c such that for any vector $\mathbf{v}$, the product $\mathbf{Pv}$ is the projection of $\mathbf{v}$ onto $a\mathbf{i} + b\mathbf{j} + c\mathbf{k}$. Here's a hint: we also could have solved this part by projecting $\mathbf{i}$, $\mathbf{j}$, and $\mathbf{k}$ onto $\mathbf{v}$.

(d) A scalar multiple of $\mathbf{v}$ is its own projection onto $\mathbf{v}$. Since $\mathbf{Mw}$ is the projection of $\mathbf{w}$ onto $\mathbf{v}$, it's already a scalar multiple of $\mathbf{v}$. So, the projection of $\mathbf{Mw}$ onto $\mathbf{v}$ is simply $\mathbf{Mw}$. Just for fun, let's check:

$$\mathbf{M}^2 = \begin{pmatrix} \frac{9}{38} & -\frac{6}{38} & \frac{15}{38} \\ -\frac{6}{38} & \frac{4}{38} & -\frac{10}{38} \\ \frac{15}{38} & -\frac{10}{38} & \frac{25}{38} \end{pmatrix} \begin{pmatrix} \frac{9}{38} & -\frac{6}{38} & \frac{15}{38} \\ -\frac{6}{38} & \frac{4}{38} & -\frac{10}{38} \\ \frac{15}{38} & -\frac{10}{38} & \frac{25}{38} \end{pmatrix} = \begin{pmatrix} \frac{342}{38^2} & -\frac{228}{38^2} & \frac{570}{38^2} \\ -\frac{228}{38^2} & \frac{152}{38^2} & -\frac{380}{38^2} \\ \frac{570}{38^2} & -\frac{380}{38^2} & \frac{950}{38^2} \end{pmatrix} = \begin{pmatrix} \frac{9}{38} & -\frac{6}{38} & \frac{15}{38} \\ -\frac{6}{38} & \frac{4}{38} & -\frac{10}{38} \\ \frac{15}{38} & -\frac{10}{38} & \frac{25}{38} \end{pmatrix} = \mathbf{M}.$$

12.67 Let B_1 and B_2 be the points on lines k_1 and k_2, respectively, such that B_1B_2 is the shortest distance between the two lines. $\overline{B_1B_2}$ must be perpendicular to both lines. To see why, suppose $\overline{B_1B_2}$ is not perpendicular to k_1. Then, let the plane through B_2 perpendicular to k_1 meet k_1 at C, so $\triangle B_1CB_2$ is a right triangle with right angle at C. Therefore, we have $B_2C < B_1B_2$, so B_1B_2 is not the shortest distance between the two lines.

Since $\overline{B_1B_2}$ is perpendicular to both lines, it must be parallel to $\mathbf{v}_1 \times \mathbf{v}_2$. To relate $\vec{A_1A_2}$ to $\overline{B_1B_2}$, $\vec{v}_1$, and $\vec{v}_2$, we note that

$$\vec{A_1A_2} = \vec{A_1B_1} + \vec{B_1B_2} + \vec{B_2A_2}.$$

Since A_1 and B_1 are on k_1, and A_2 and B_2 are on k_2, we have $\vec{A_1B_1} = c_1\mathbf{v}_1$ and $\vec{A_2B_2} = c_2\mathbf{v}_2$. So, we have

$$\vec{A_1A_2} \cdot (\mathbf{v}_1 \times \mathbf{v}_2) = (\vec{A_1B_1} + \vec{B_1B_2} + \vec{B_2A_2}) \cdot (\mathbf{v}_1 \times \mathbf{v}_2)$$
$$= (c_1\mathbf{v}_1 + \vec{B_1B_2} + c_2\mathbf{v}_2) \cdot (\mathbf{v}_1 \times \mathbf{v}_2)$$
$$= c_1\mathbf{v}_1 \cdot (\mathbf{v}_1 \times \mathbf{v}_2) + \vec{B_1B_2} \cdot (\mathbf{v}_1 \times \mathbf{v}_2) + c_2\mathbf{v}_2 \cdot (\mathbf{v}_1 \times \mathbf{v}_2)$$
$$= 0 + \vec{B_1B_2} \cdot (\mathbf{v}_1 \times \mathbf{v}_2) + 0.$$

Since $\overline{B_1B_2}$ is parallel to $\mathbf{v}_1 \times \mathbf{v}_2$, we have

$$|\vec{A_1A_2} \cdot (\mathbf{v}_1 \times \mathbf{v}_2)| = \left\| \vec{B_1B_2} \cdot (\mathbf{v}_1 \times \mathbf{v}_2) \right\| = \left\| \vec{B_1B_2} \right\| \left\| \mathbf{v}_1 \times \mathbf{v}_2 \right\|.$$

(We introduce the absolute value on the left to account for the fact that $\vec{B_1B_2}$ can be in the same or opposite direction as $\mathbf{v}_1 \times \mathbf{v}_2$.) Since $\left\| \vec{B_1B_2} \right\| = B_1B_2$, we have

$$B_1B_2 = \frac{|\vec{A_1A_2} \cdot (\mathbf{v}_1 \times \mathbf{v}_2)|}{\left\| \mathbf{v}_1 \times \mathbf{v}_2 \right\|}.$$

13

Exercises for Section 13.1

13.1.1

(a) Let A be the origin. Since the midpoint of $\overline{AC}$ is the same as the midpoint of $\overline{BD}$, we have $(\vec{A}+\vec{C})/2 = (\vec{B}+\vec{D})/2$. Since $\vec{A} = \mathbf{0}$, we have $\vec{C} = \vec{B} + \vec{D}$. But the head of $\vec{B} + \vec{D}$ is the fourth vertex of the parallelogram with consecutive vertices at B, A (the origin), and D, so $ABCD$ is a parallelogram.

(b) Let $ABCD$ be the quadrilateral. Let E be the midpoint of $\overline{AB}$, F be the midpoint of $\overline{BC}$, G be the midpoint of $\overline{CD}$, and H be the midpoint of $\overline{DA}$. We will show that $EFGH$ is a parallelogram by showing that the midpoints of diagonals $\overline{EG}$ and $\overline{FH}$ coincide. If M is the midpoint of $\overline{EG}$, then

$$\vec{M} = \frac{\vec{E} + \vec{G}}{2} = \frac{\frac{\vec{A}+\vec{B}}{2} + \frac{\vec{C}+\vec{D}}{2}}{2} = \frac{\vec{A} + \vec{B} + \vec{C} + \vec{D}}{4}.$$

If N is the midpoint of $\overline{FH}$, then

$$\vec{N} = \frac{\vec{F} + \vec{H}}{2} = \frac{\frac{\vec{B}+\vec{C}}{2} + \frac{\vec{D}+\vec{A}}{2}}{2} = \frac{\vec{A} + \vec{B} + \vec{C} + \vec{D}}{4}.$$

Since $\vec{M} = \vec{N}$, points M and N are the same. Because the midpoints of $\overline{EG}$ and $\overline{FH}$ are the same, the diagonals of $EFGH$ bisect each other, which means that $EFGH$ is a parallelogram.

13.1.2 Suppose $ABCD$ is a kite with $AB = BC$ and $CD = DA$. To show that $\overline{BD} \perp \overline{AC}$, we can show that $\vec{BD} \cdot \vec{CA} = 0$. The equal side lengths give $\vec{AB} \cdot \vec{AB} = \vec{BC} \cdot \vec{BC}$ and $\vec{CD} \cdot \vec{CD} = \vec{DA} \cdot \vec{DA}$. Letting B be the origin simplifies the first equation to $\vec{A} \cdot \vec{A} = \vec{C} \cdot \vec{C}$. We can write $\vec{CD} \cdot \vec{CD} = \vec{DA} \cdot \vec{DA}$ as

$$(\vec{D} - \vec{C}) \cdot (\vec{D} - \vec{C}) = (\vec{D} - \vec{A}) \cdot (\vec{D} - \vec{A}).$$

Expanding both sides gives

$$\vec{D} \cdot \vec{D} - 2\vec{C} \cdot \vec{D} + \vec{C} \cdot \vec{C} = \vec{D} \cdot \vec{D} - 2\vec{A} \cdot \vec{D} + \vec{A} \cdot \vec{A}.$$

The $\vec{D} \cdot \vec{D}$ terms cancel and $\vec{A} \cdot \vec{A} = \vec{C} \cdot \vec{C}$ eliminates two more terms and leaves $-2\vec{C} \cdot \vec{D} = -2\vec{A} \cdot \vec{D}$, which rearranges as $\vec{D} \cdot (\vec{A} - \vec{C}) = 0$.

Looking back at the equation we wish to prove, we see that when B is the origin, the equation $\vec{BD} \cdot \vec{CA} = 0$ is $\vec{D} \cdot (\vec{A} - \vec{C}) = 0$, which we have just proved above. So, from $\vec{D} \cdot (\vec{A} - \vec{C}) = 0$ (and the fact that B is the origin), we have $(\vec{D} - \vec{B}) \cdot (\vec{A} - \vec{C}) = 0$, which means $\vec{BD} \cdot \vec{CA} = 0$. Therefore, the diagonals of kite $ABCD$ are perpendicular.

13.1.3   We have

$$4\vec{MN} = 4\left(\vec{N} - \vec{M}\right) = 4\left(\frac{\vec{B}+\vec{D}}{2} - \frac{\vec{A}+\vec{C}}{2}\right) = 2\vec{B} + 2\vec{D} - 2\vec{A} - 2\vec{C},$$

and

$$\vec{AB} + \vec{AD} + \vec{CB} + \vec{CD} = \vec{B} - \vec{A} + \vec{D} - \vec{A} + \vec{B} - \vec{C} + \vec{D} - \vec{C} = 2\vec{B} + 2\vec{D} - 2\vec{A} - 2\vec{C},$$

so $\vec{AB} + \vec{AD} + \vec{CB} + \vec{CD} = 4\vec{MN}$.

13.1.4   Let $ABCD$ be the quadrilateral. The square of the length of the median from the midpoint of $\overline{AB}$ to the midpoint of $\overline{CD}$ is

$$\left(\frac{\vec{A}+\vec{B}}{2} - \frac{\vec{C}+\vec{D}}{2}\right) \cdot \left(\frac{\vec{A}+\vec{B}}{2} - \frac{\vec{C}+\vec{D}}{2}\right).$$

Setting this equal to the square of the length of the other median gives

$$\left(\frac{\vec{A}+\vec{B}}{2} - \frac{\vec{C}+\vec{D}}{2}\right) \cdot \left(\frac{\vec{A}+\vec{B}}{2} - \frac{\vec{C}+\vec{D}}{2}\right) = \left(\frac{\vec{B}+\vec{C}}{2} - \frac{\vec{A}+\vec{D}}{2}\right) \cdot \left(\frac{\vec{B}+\vec{C}}{2} - \frac{\vec{A}+\vec{D}}{2}\right).$$

We multiply both sides by 4 to get rid of the fractions:

$$(\vec{A} + \vec{B} - \vec{C} - \vec{D}) \cdot (\vec{A} + \vec{B} - \vec{C} - \vec{D}) = (\vec{B} + \vec{C} - \vec{A} - \vec{D}) \cdot (\vec{B} + \vec{C} - \vec{A} - \vec{D}).$$

We'll clearly have a lot of terms in common when we expand both sides. Rather than expanding right away, we start with a little clever regrouping, rewriting the equation as

$$((\vec{B} - \vec{D}) - (\vec{C} - \vec{A})) \cdot ((\vec{B} - \vec{D}) - (\vec{C} - \vec{A})) = ((\vec{B} - \vec{D}) + (\vec{C} - \vec{A})) \cdot ((\vec{B} - \vec{D}) + (\vec{C} - \vec{A})).$$

Now, the cancellation when we expand is clear. Expanding gives

$$(\vec{B} - \vec{D}) \cdot (\vec{B} - \vec{D}) - 2(\vec{B} - \vec{D}) \cdot (\vec{C} - \vec{A}) + (\vec{C} - \vec{A}) \cdot (\vec{C} - \vec{A}) = (\vec{B} - \vec{D}) \cdot (\vec{B} - \vec{D}) + 2(\vec{B} - \vec{D}) \cdot (\vec{C} - \vec{A}) + (\vec{C} - \vec{A}) \cdot (\vec{C} - \vec{A}).$$

Simplifying this equation gives $4(\vec{B} - \vec{D}) \cdot (\vec{C} - \vec{A}) = 0$, from which we deduce that $\vec{DB} \cdot \vec{AC} = 0$, which means that $\overline{BD} \perp \overline{AC}$. All of these steps are reversible, so we have $\overline{BD} \perp \overline{AC}$ if and only if the medians of $ABCD$ are equal in length.

13.1.5   Letting C be the origin, we have $CA^2 + CB^2 = \vec{A} \cdot \vec{A} + \vec{B} \cdot \vec{B}$. We also have $\vec{M} = \frac{\vec{A}+\vec{B}}{2}$, so

$$2CM^2 + \frac{AB^2}{2} = 2\left(\frac{\vec{A}+\vec{B}}{2}\right) \cdot \left(\frac{\vec{A}+\vec{B}}{2}\right) + \frac{1}{2}(\vec{B} - \vec{A}) \cdot (\vec{B} - \vec{A})$$
$$= \frac{1}{2}(\vec{A} \cdot \vec{A} + 2\vec{A} \cdot \vec{B} + \vec{B} \cdot \vec{B}) + \frac{1}{2}(\vec{B} \cdot \vec{B} - 2\vec{B} \cdot \vec{A} + \vec{A} \cdot \vec{A})$$
$$= \vec{A} \cdot \vec{A} + \vec{B} \cdot \vec{B}.$$

Therefore, we have $CA^2 + CB^2 = 2CM^2 + \frac{AB^2}{2}$.

13.1.6   We are given that $\vec{AB} \cdot \vec{CD} = \vec{AC} \cdot \vec{BD} = 0$, so

$$(\vec{B} - \vec{A}) \cdot (\vec{D} - \vec{C}) = (\vec{C} - \vec{A}) \cdot (\vec{D} - \vec{B}).$$

Expanding both sides gives

$$\vec{B} \cdot \vec{D} - \vec{B} \cdot \vec{C} - \vec{A} \cdot \vec{D} + \vec{A} \cdot \vec{C} = \vec{C} \cdot \vec{D} - \vec{C} \cdot \vec{B} - \vec{A} \cdot \vec{D} + \vec{A} \cdot \vec{B}.$$

Cancelling the common terms (and noting that $\vec{B} \cdot \vec{C} = \vec{C} \cdot \vec{B}$), we have

$$\vec{B} \cdot \vec{D} + \vec{A} \cdot \vec{C} = \vec{C} \cdot \vec{D} + \vec{A} \cdot \vec{B}.$$

Rearranging this gives $\vec{B} \cdot \vec{D} - \vec{B} \cdot \vec{A} - \vec{C} \cdot \vec{D} + \vec{C} \cdot \vec{A} = 0$, from which we have $(\vec{B} - \vec{C}) \cdot (\vec{D} - \vec{A}) = 0$, so $\vec{CB} \cdot \vec{AD} = 0$. Therefore, $\vec{CB}$ and $\vec{AD}$ are orthogonal, so $\overline{BC}$ and $\overline{AD}$ are orthogonal.

Exercises for Section 13.2

13.2.1 Let the vertices be A, B, and C, and the centroid be G. We have

$$\vec{G} = \frac{\vec{A} + \vec{B} + \vec{C}}{3} = \frac{1}{3}\left(\begin{pmatrix} 1 \\ 4 \\ -1 \end{pmatrix} + \begin{pmatrix} -4 \\ 5 \\ 2 \end{pmatrix} + \begin{pmatrix} 6 \\ -1 \\ 7 \end{pmatrix}\right) = \begin{pmatrix} 1 \\ 8/3 \\ 8/3 \end{pmatrix}.$$

Therefore, the centroid of the triangle is $\boxed{\left(1, \frac{8}{3}, \frac{8}{3}\right)}$.

13.2.2

(a) The Angle Bisector Theorem tells us that if D is on $\overline{BC}$ such that $\overline{AD}$ bisects $\angle BAC$, then $BD/CD = AB/AC = c/b$. We showed in the text that if point D is on $\overline{BC}$ such that $BD/CD = k$, then $\vec{D} = \frac{\vec{B}+k\vec{C}}{1+k}$. So, we have

$$\vec{D} = \frac{\vec{B} + \frac{c}{b}\vec{C}}{1 + \frac{c}{b}} = \boxed{\frac{b\vec{B} + c\vec{C}}{b + c}}.$$

(b) Continuing the reasoning from part (a), we see that if E and F are on $\overline{AC}$ and $\overline{AB}$, respectively, such that $\overline{BE}$ and $\overline{CF}$ are angle bisectors of $\triangle ABC$, then $\vec{E} = \frac{a\vec{A}+c\vec{C}}{a+c}$ and $\vec{F} = \frac{a\vec{A}+b\vec{B}}{a+b}$. Looking at the forms of $\vec{D}$, $\vec{E}$, and $\vec{F}$, we have a natural guess for $\vec{I}$. Suppose we let

$$\vec{I} = \boxed{\frac{a\vec{A} + b\vec{B} + c\vec{C}}{a + b + c}}.$$

Then, we have

$$\vec{AD} = \vec{D} - \vec{A} = \frac{b\vec{B} + c\vec{C}}{b+c} - \vec{A} = \frac{b\vec{B} + c\vec{C} - b\vec{A} - c\vec{A}}{b+c},$$

$$\vec{AI} = \vec{I} - \vec{A} = \frac{a\vec{A} + b\vec{B} + c\vec{C}}{a+b+c} - \vec{A} = \frac{a\vec{A} + b\vec{B} + c\vec{C} - a\vec{A} - b\vec{A} - c\vec{A}}{a+b+c} = \frac{b\vec{B} + c\vec{C} - b\vec{A} - c\vec{A}}{a+b+c}.$$

Therefore, we have $\vec{AI} = \frac{b+c}{a+b+c}\vec{AD}$, which means I is on $\overline{AD}$. Similarly, I is on each of the other angle bisectors of $\triangle ABC$, and we have found the desired expression for $\vec{I}$.

13.2.3 We must show that

$$3(PA^2 + PB^2 + PC^2) - AB^2 - BC^2 - CA^2 \geq 0.$$

Applying vectors to the problem, we have

$$3(PA^2 + PB^2 + PC^2) - AB^2 - BC^2 - CA^2 = 3((\vec{P} - \vec{A}) \cdot (\vec{P} - \vec{A}) + (\vec{P} - \vec{B}) \cdot (\vec{P} - \vec{B}) + (\vec{P} - \vec{C}) \cdot (\vec{P} - \vec{C}))$$
$$- (\vec{B} - \vec{A}) \cdot (\vec{B} - \vec{A}) - (\vec{C} - \vec{B}) \cdot (\vec{C} - \vec{B}) - (\vec{A} - \vec{C}) \cdot (\vec{A} - \vec{C})$$
$$= 3(\vec{P} \cdot \vec{P} - 2\vec{P} \cdot \vec{A} + \vec{A} \cdot \vec{A} + \vec{P} \cdot \vec{P} - 2\vec{P} \cdot \vec{B} + \vec{B} \cdot \vec{B} + \vec{P} \cdot \vec{P} - 2\vec{P} \cdot \vec{C} + \vec{C} \cdot \vec{C})$$
$$- \vec{B} \cdot \vec{B} + 2\vec{A} \cdot \vec{B} - \vec{A} \cdot \vec{A} - \vec{C} \cdot \vec{C} + 2\vec{B} \cdot \vec{C} - \vec{B} \cdot \vec{B} - \vec{A} \cdot \vec{A} + 2\vec{C} \cdot \vec{A} - \vec{C} \cdot \vec{C}$$
$$= 9\vec{P} \cdot \vec{P} - 6\vec{P} \cdot (\vec{A} + \vec{B} + \vec{C}) + \vec{A} \cdot \vec{A} + \vec{B} \cdot \vec{B} + \vec{C} \cdot \vec{C} + 2(\vec{A} \cdot \vec{B} + \vec{B} \cdot \vec{C} + \vec{C} \cdot \vec{A})$$
$$= 9\vec{P} \cdot \vec{P} - 6\vec{P} \cdot (\vec{A} + \vec{B} + \vec{C}) + (\vec{A} + \vec{B} + \vec{C}) \cdot (\vec{A} + \vec{B} + \vec{C})$$
$$= \left(3\vec{P} - (\vec{A} + \vec{B} + \vec{C})\right) \cdot \left(3\vec{P} - (\vec{A} + \vec{B} + \vec{C})\right)$$
$$= \left\|3\vec{P} - (\vec{A} + \vec{B} + \vec{C})\right\|^2.$$

This final expression is clearly nonnegative, so $3(PA^2 + PB^2 + PC^2) - AB^2 - BC^2 - CA^2 \geq 0$, as desired.

We have equality when $3\vec{P} - (\vec{A} + \vec{B} + \vec{C}) = 0$, which occurs when $\vec{P} = (\vec{A} + \vec{B} + \vec{C})/3$. Therefore, equality holds if and only if P is the $\boxed{\text{centroid of } \triangle ABC}$.

Note: We could have simplified the above calculations by letting P be the origin. We then find that

$$3(PA^2 + PB^2 + PC^2) - AB^2 - BC^2 - CA^2 = \left\| \vec{A} + \vec{B} + \vec{C} \right\|^2,$$

and equality holds in the desired inequality if and only if $\vec{A} + \vec{B} + \vec{C} = 0$. Therefore, $(\vec{A} + \vec{B} + \vec{C})/3 = 0$, so the origin (which is point P) is the centroid, as before. (Note that we cannot deduce that the orthocenter is the origin from $\vec{A} + \vec{B} + \vec{C} = 0$. When H is the orthocenter of $\triangle ABC$, we only have $\vec{H} = \vec{A} + \vec{B} + \vec{C} = 0$ if the circumcenter is the origin.)

13.2.4   We showed in the text that $\vec{OG} = \vec{OH}/3$, where G is the centroid of $\triangle ABC$. Therefore, we have $3\vec{OG} = \vec{OH}$, so $3(\vec{G} - \vec{O}) = \vec{H} - \vec{O}$, which means $\vec{H} = 3\vec{G} - 2\vec{O}$. Since $\vec{G} = (\vec{A} + \vec{B} + \vec{C})/3$ for any $\triangle ABC$, we have $\vec{H} = 3\vec{G} - 2\vec{O} = \vec{A} + \vec{B} + \vec{C} - 2\vec{O}$, as desired.

13.2.5   Let the tetrahedron be $ABCD$ such that we have

$$\frac{\vec{A} + \vec{B} + \vec{C}}{3} = \begin{pmatrix} 17 \\ 0 \\ 3 \end{pmatrix}, \quad \frac{\vec{B} + \vec{C} + \vec{D}}{3} = \begin{pmatrix} -2 \\ 13 \\ 3 \end{pmatrix}, \quad \frac{\vec{C} + \vec{D} + \vec{A}}{3} = \begin{pmatrix} 1 \\ 2 \\ 15 \end{pmatrix}, \quad \frac{\vec{D} + \vec{A} + \vec{B}}{3} = \begin{pmatrix} 1 \\ -1 \\ 2 \end{pmatrix}.$$

Adding these four equations gives

$$\vec{A} + \vec{B} + \vec{C} + \vec{D} = \begin{pmatrix} 17 \\ 14 \\ 23 \end{pmatrix}.$$

Subtracting 3 times each of our first four equations from the equation above gives us the four vertices of the tetrahedron:

$$\boxed{A = (23, -25, 14), \qquad B = (14, 8, -22), \qquad C = (14, 17, 17), \qquad D = (-34, 14, 14).}$$

13.2.6   We let the circumcenter of $\triangle ABC$ be the origin, O. Since U is the reflection of O over $\overline{BC}$ and $OB = OC$, the quadrilateral $OBUC$ is a rhombus. Every rhombus is also a parallelogram, so $\vec{U} = \vec{B} + \vec{C}$. Similarly, we have $\vec{V} = \vec{A} + \vec{C}$ and $\vec{W} = \vec{A} + \vec{B}$. From here we offer a couple solutions.

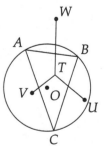

Solution 1: Guess! We have $\vec{UA} = \vec{A} - \vec{U} = \vec{A} - \vec{B} - \vec{C}$. Next, we check convenient points, looking for one such that the vector from U to the convenient point is in the same (or opposite) direction as $\vec{UA}$.

First, we try the circumcenter. We have $\vec{UO} = -\vec{B} - \vec{C}$. No good; that's not a constant times $\vec{UA}$.

Next, we try the orthocenter of $\triangle ABC$, point H. We have $\vec{UH} = \vec{A} + \vec{B} + \vec{C} - \vec{B} - \vec{C} = \vec{A}$. Another loser.

Next, the center of the nine-point circle of $\triangle ABC$, point T. We have $\vec{UT} = \frac{\vec{A} + \vec{B} + \vec{C}}{2} - \vec{B} - \vec{C} = \frac{\vec{A} - \vec{B} - \vec{C}}{2} = \frac{\vec{UA}}{2}$. Woohoo! We found it! Similarly, T is on $\overleftrightarrow{VB}$ and $\overleftrightarrow{WC}$, so the lines are concurrent at the center of the nine-point circle of $\triangle ABC$.

Solution 2: Symmetry. If there is a point Z at which the three lines meet, the symmetry of the problem suggests that there is some constant k such that $\vec{Z} = k(\vec{A} + \vec{B} + \vec{C})$. As before, we have $\vec{UA} = \vec{A} - \vec{B} - \vec{C}$. We hope to find k

such that $\vec{UZ}$ is some multiple of $\vec{UA}$. We have

$$\vec{UZ} = \vec{Z} - \vec{U} = k(\vec{A} + \vec{B} + \vec{C}) - (\vec{B} + \vec{C}) = k\vec{A} + (k-1)\vec{B} + (k-1)\vec{C}.$$

If this is a multiple of $\vec{A} - \vec{B} - \vec{C}$, then we must have $k = -(k-1)$, from which we have $k = \frac{1}{2}$, so point Z is the center of the nine-point circle of $\triangle ABC$. Similarly, the center of the nine-point circle of $\triangle ABC$ is on all three lines.

Challenge Problems

13.19 We have

$$
\begin{aligned}
WX^2 + YZ^2 - XY^2 - ZW^2 &= (\vec{X} - \vec{W}) \cdot (\vec{X} - \vec{W}) + (\vec{Z} - \vec{Y}) \cdot (\vec{Z} - \vec{Y}) - (\vec{Y} - \vec{X}) \cdot (\vec{Y} - \vec{X}) - (\vec{W} - \vec{Z}) \cdot (\vec{W} - \vec{Z}) \\
&= \vec{X} \cdot \vec{X} - 2\vec{W} \cdot \vec{X} + \vec{W} \cdot \vec{W} + \vec{Z} \cdot \vec{Z} - 2\vec{Y} \cdot \vec{Z} + \vec{Y} \cdot \vec{Y} \\
&\quad - \vec{Y} \cdot \vec{Y} + 2\vec{X} \cdot \vec{Y} - \vec{X} \cdot \vec{X} - \vec{W} \cdot \vec{W} + 2\vec{Z} \cdot \vec{W} - \vec{Z} \cdot \vec{Z} \\
&= 2(\vec{X} \cdot \vec{Y} - \vec{Y} \cdot \vec{Z} + \vec{Z} \cdot \vec{W} - \vec{W} \cdot \vec{X}) \\
&= 2(\vec{Y} \cdot (\vec{X} - \vec{Z}) - \vec{W} \cdot (\vec{X} - \vec{Z})) \\
&= 2((\vec{Y} - \vec{W}) \cdot (\vec{X} - \vec{Z})) = 2\vec{WY} \cdot \vec{ZX}.
\end{aligned}
$$

13.20 Let $ABCD$ be the quadrilateral. We have

$$
\begin{aligned}
AB^2 + BC^2 + CD^2 + DA^2 - AC^2 - BD^2 &= (\vec{B} - \vec{A}) \cdot (\vec{B} - \vec{A}) + (\vec{C} - \vec{B}) \cdot (\vec{C} - \vec{B}) + (\vec{D} - \vec{C}) \cdot (\vec{D} - \vec{C}) + (\vec{A} - \vec{D}) \cdot (\vec{A} - \vec{D}) \\
&\quad - (\vec{C} - \vec{A}) \cdot (\vec{C} - \vec{A}) - (\vec{D} - \vec{B}) \cdot (\vec{D} - \vec{B}) \\
&= \vec{A} \cdot \vec{A} + \vec{B} \cdot \vec{B} + \vec{C} \cdot \vec{C} + \vec{D} \cdot \vec{D} + 2\vec{A} \cdot \vec{C} + 2\vec{B} \cdot \vec{D} \\
&\quad - 2\vec{A} \cdot \vec{B} - 2\vec{B} \cdot \vec{C} - 2\vec{C} \cdot \vec{D} - 2\vec{D} \cdot \vec{A}.
\end{aligned}
$$

We'd like to show that this expression equals 0. Here, we might seem stuck, but thinking about what we'd like to prove, that $ABCD$ is a parallelogram, helps us get unstuck. $ABCD$ is a parallelogram if and only if $\vec{A} + \vec{C} = \vec{B} + \vec{D}$. We could prove that $\vec{A} + \vec{C} - \vec{B} - \vec{D} = 0$ if we can show that $\|\vec{A} + \vec{C} - \vec{B} - \vec{D}\| = 0$. Now that we know what to look for, we see look back at that huge expression above and see that

$$AB^2 + BC^2 + CD^2 + DA^2 - AC^2 - BD^2 = (\vec{A} + \vec{C} - \vec{B} - \vec{D}) \cdot (\vec{A} + \vec{C} - \vec{B} - \vec{D}).$$

Therefore, we have $AB^2 + BC^2 + CD^2 + DA^2 = AC^2 + BD^2$ if and only if $\|\vec{A} + \vec{C} - \vec{B} - \vec{D}\| = 0$, which is true if and only if $ABCD$ is a parallelogram, and our proof is complete.

13.21 Let T be the midpoint of $\overline{MN}$. We have $\vec{M} = (\vec{A} + \vec{B})/2$ and $\vec{N} = (\vec{C} + \vec{D})/2$, so

$$\vec{T} = \frac{\vec{M} + \vec{N}}{2} = \frac{\vec{A} + \vec{B} + \vec{C} + \vec{D}}{4}.$$

We have $\vec{P} = (\vec{A} + \vec{C})/2$ and $\vec{Q} = (\vec{B} + \vec{D})/2$, so $(\vec{P} + \vec{Q})/2 = (\vec{A} + \vec{B} + \vec{C} + \vec{D})/4 = \vec{T}$, which means T is the midpoint of $\overline{PQ}$. Similarly, we have $\vec{R} = (\vec{A} + \vec{D})/2$ and $\vec{S} = (\vec{B} + \vec{C})/2$, so $(\vec{R} + \vec{S})/2 = (\vec{A} + \vec{B} + \vec{C} + \vec{D})/4 = \vec{T}$ and T is the midpoint of $\overline{RS}$, as well. Therefore, $\overleftrightarrow{MN}$, $\overleftrightarrow{PQ}$, and $\overleftrightarrow{RS}$ are concurrent at T.

13.22 Letting X be the origin, we have

$$
\begin{aligned}
2(AX^2 + BX^2 + CX^2 + DX^2) &- (AB^2 + BC^2 + CD^2 + DA^2) \\
&= 2(\vec{A} \cdot \vec{A} + \vec{B} \cdot \vec{B} + \vec{C} \cdot \vec{C} + \vec{D} \cdot \vec{D}) - ((\vec{B} - \vec{A}) \cdot (\vec{B} - \vec{A}) \\
&\quad + (\vec{C} - \vec{B}) \cdot (\vec{C} - \vec{B}) + (\vec{D} - \vec{C}) \cdot (\vec{D} - \vec{C}) + (\vec{A} - \vec{D}) \cdot (\vec{A} - \vec{D}))
\end{aligned}
$$

When we expand the dot products, all the terms of the form $\vec{A} \cdot \vec{A}$ cancel, and we are left with

$$-2(\vec{A} \cdot \vec{B} + \vec{B} \cdot \vec{C} + \vec{C} \cdot \vec{D} + \vec{D} \cdot \vec{A}) = -2(\vec{B} \cdot (\vec{A} + \vec{C}) + \vec{D} \cdot (\vec{A} + \vec{C}))$$
$$= -2(\vec{B} + \vec{D}) \cdot (\vec{A} + \vec{C}).$$

Therefore, the original expression equals 0 if and only if one of the following conditions is met:

- $\vec{B} + \vec{D} = \vec{0}$. In this case, the origin is the midpoint of $\overline{BD}$. The origin is the intersection of the diagonals, so this means that $\overline{AC}$ bisects diagonal $\overline{BD}$.

- $\vec{A} + \vec{C} = \vec{0}$. For the same reasons as in the previous case, this means that diagonal $\overline{BD}$ bisects diagonal $\overline{AC}$.

- $\vec{B} + \vec{D}$ and $\vec{A} + \vec{C}$ are nonzero and orthogonal to each other. Since A and C are on the same line through the origin, the vector $\vec{A} + \vec{C}$ is parallel to this line. Similarly, the vector $\vec{B} + \vec{D}$ is parallel to diagonal $\overline{BD}$. Therefore, if $(\vec{A} + \vec{C}) \perp (\vec{B} + \vec{D})$, then $\overline{AC} \perp \overline{BD}$.

13.23 We showed in the text that if H is the orthocenter of $\triangle ABC$ *and the origin is the circumcenter of $\triangle ABC$*, then we have $\vec{H} = \vec{A} + \vec{B} + \vec{C}$. But in this problem, the origin is not the circumcenter of $\triangle ABC$, because A, B, and C are not equidistant from the origin. Therefore, we cannot find the orthocenter in the way Mr. Assumption tried.

13.24 Let O be the origin, so $\vec{A} \cdot \vec{A} = \vec{B} \cdot \vec{B} = \vec{C} \cdot \vec{C}$. We have $\vec{M} = (\vec{A} + \vec{B})/2$ and

$$\vec{X} = \frac{\vec{A} + \vec{C} + \vec{M}}{3} = \frac{\vec{A} + \vec{C} + \frac{\vec{A}+\vec{B}}{2}}{3} = \frac{\vec{A}}{2} + \frac{\vec{B}}{6} + \frac{\vec{C}}{3} = \frac{1}{6}(3\vec{A} + \vec{B} + 2\vec{C}).$$

Therefore, we have

$$\vec{CM} \cdot \vec{OX} = (\vec{M} - \vec{C}) \cdot \vec{X} = \frac{1}{2} \cdot \frac{1}{6}(\vec{A} + \vec{B} - 2\vec{C}) \cdot (3\vec{A} + \vec{B} + 2\vec{C})$$
$$= \frac{1}{12}\left(3\vec{A} \cdot \vec{A} + \vec{B} \cdot \vec{B} - 4\vec{C} \cdot \vec{C} + 4\vec{A} \cdot \vec{B} - 4\vec{A} \cdot \vec{C}\right).$$

Since $\vec{A} \cdot \vec{A} = \vec{B} \cdot \vec{B} = \vec{C} \cdot \vec{C}$, we have $3\vec{A} \cdot \vec{A} + \vec{B} \cdot \vec{B} - 4\vec{C} \cdot \vec{C} = 0$, so

$$\vec{CM} \cdot \vec{OX} = \frac{1}{12}\left(4\vec{A} \cdot \vec{B} - 4\vec{A} \cdot \vec{C}\right) = \frac{1}{3}\vec{A} \cdot (\vec{B} - \vec{C}).$$

This quantity equals 0 if and only if $\overleftrightarrow{AO} \perp \overleftrightarrow{BC}$. The line through O that is perpendicular to $\overleftrightarrow{BC}$ is the perpendicular bisector of $\overline{BC}$, because O is the circumcenter of $\triangle ABC$. Therefore, the condition that $\overleftrightarrow{AO} \perp \overleftrightarrow{BC}$ means that A is on the perpendicular bisector of $\overline{BC}$, which means that $AB = AC$.

We still have to show that if $AB = AC$, then we have $\overleftrightarrow{CM} \perp \overleftrightarrow{OX}$. If $AB = AC$, then $\vec{AB} \cdot \vec{AB} = \vec{AC} \cdot \vec{AC}$, which means $(\vec{B} - \vec{A}) \cdot (\vec{B} - \vec{A}) = (\vec{C} - \vec{A}) \cdot (\vec{C} - \vec{A})$. Expanding and rearranging, and using the fact that $\vec{B} \cdot \vec{B} = \vec{C} \cdot \vec{C}$, we have $2\vec{A} \cdot \vec{B} - 2\vec{A} \cdot \vec{C} = 0$, from which we have $\vec{A} \cdot (\vec{B} - \vec{C}) = 0$. (We also could have noted that $AB = AC$ and $OB = OC$ mean that both A and O are on the perpendicular bisector of $\overline{BC}$, so $\overleftrightarrow{AO} \perp \overleftrightarrow{BC}$, which means $\vec{A} \cdot (\vec{B} - \vec{C}) = 0$.) We can then reverse our steps above to find that $\vec{CM} \cdot \vec{X} = 0$.

Therefore, we have $\overleftrightarrow{CM} \perp \overleftrightarrow{OX}$ if and only if $AB = AC$.

13.25 We let the center of the circle be the origin. This point is also the circumcenter of each of the four triangles, so the vectors to the vertices of the new quadrilateral are in question are

$$\vec{A} + \vec{B} + \vec{C}, \quad \vec{B} + \vec{C} + \vec{D}, \quad \vec{C} + \vec{D} + \vec{A}, \quad \vec{D} + \vec{A} + \vec{B}.$$

Therefore the vectors along the four sides of the quadrilateral are

$$\vec{A} + \vec{B} + \vec{C} - (\vec{B} + \vec{C} + \vec{D}) = \vec{A} - \vec{D} = \vec{DA},$$

$$\vec{B} + \vec{C} + \vec{D} - (\vec{C} + \vec{D} + \vec{A}) = \vec{B} - \vec{A} = \vec{AB},$$

$$\vec{C} + \vec{D} + \vec{A} - (\vec{D} + \vec{A} + \vec{B}) = \vec{C} - \vec{B} = \vec{BC},$$

$$\vec{D} + \vec{A} + \vec{B} - (\vec{A} + \vec{B} + \vec{C}) = \vec{D} - \vec{C} = \vec{CD}.$$

Each vector representing a side of the new quadrilateral is the same as a vector representing the corresponding side of $ABCD$, so the each side of the new quadrilateral has the same length and is parallel to the corresponding side of $ABCD$. Therefore, the quadrilaterals are congruent.

13.26

(a) Let O be the origin, so $\|\vec{A}\| = \|\vec{B}\| = \|\vec{C}\| = R$, which means

$$OH^2 = (\vec{A} + \vec{B} + \vec{C}) \cdot (\vec{A} + \vec{B} + \vec{C}) = \vec{A} \cdot \vec{A} + \vec{B} \cdot \vec{B} + \vec{C} \cdot \vec{C} + 2(\vec{A} \cdot \vec{B} + \vec{B} \cdot \vec{C} + \vec{C} \cdot \vec{A}) = 3R^2 + 2(\vec{A} \cdot \vec{B} + \vec{B} \cdot \vec{C} + \vec{C} \cdot \vec{A}).$$

We also have $\vec{A} \cdot \vec{B} = \|\vec{A}\|\|\vec{B}\| \cos \angle AOB = R^2 \cos \angle AOB$. The desired equation has cosines of the angles of $\triangle ABC$, so we must relate $\angle AOB$ to the angles of $\triangle ABC$.

If $\triangle ABC$ is acute, then $\angle AOB$ is a central angle that subtends the same arc in which $\angle ACB$ is inscribed. So we have $\angle AOB = 2\angle ACB$, which means that $\cos \angle AOB = \cos 2C$. If $\angle ACB$ is right, then $\overline{AB}$ is a diameter of the circumcircle, and $\angle AOB = 180°$. This gives us $\cos \angle AOB = \cos 180° = \cos 2C$ once again. Finally, if $\angle ACB$ is obtuse, then $\angle AOB$ subtends an arc of the circumcircle on which point C lies. In this case, the measure of $\angle C$ is one-half the major arc $\overset{\frown}{AB}$, which has measure $360° - \overset{\frown}{ACB} = 360° - \angle AOB$. So, we have $\cos 2C = \cos(360° - \angle AOB) = \cos \angle AOB$, since $\cos(360° - \theta) = \cos \theta$ for any angle θ.

In all of these cases, we have $\cos \angle AOB = \cos 2C$, so

$$\vec{A} \cdot \vec{B} = \|\vec{A}\|\|\vec{B}\| \cos \angle AOB = R^2 \cos 2C.$$

Similarly, we have $\vec{B} \cdot \vec{C} = R^2 \cos 2A$ and $\vec{C} \cdot \vec{A} = R^2 \cos 2B$. Substituting into our expression for OH^2, we have
$$OH^2 = 3R^2 + 2(R^2 \cos 2C + R^2 \cos 2A + R^2 \cos 2B) = R^2(3 + 2(\cos 2A + \cos 2B + \cos 2C)).$$

(b) Applying the cosine double angle identity, we have

$$OH^2 = R^2(3 + 2(2\cos^2 A - 1 + 2\cos^2 B - 1 + 2\cos^2 C - 1)).$$

Dividing by R^2, we have $3 + 2(2\cos^2 A + 2\cos^2 B + 2\cos^2 C - 3) = \frac{OH^2}{R^2}$. Since $\frac{OH^2}{R^2}$ must be nonnegative, we have
$$3 + 2(2\cos^2 A + 2\cos^2 B + 2\cos^2 C - 3) \geq 0.$$

Isolating $\cos^2 A + \cos^2 B + \cos^2 C$ gives the desired $\cos^2 A + \cos^2 B + \cos^2 C \geq \frac{3}{4}$.

13.27 Suppose there is a point V through which all three lines pass. Let M be the midpoint of $\overline{AB}$. Since $\overleftrightarrow{MV}$ must be parallel to $\overline{CX}$, the vector from M to V must be in the same or opposite direction as the vector from C to X. Therefore, we must have $\vec{MV} = k_1 \vec{CX}$ for some constant k_1. Since $\vec{MV} = \vec{V} - \vec{M}$, $\vec{CX} = \vec{X} - \vec{C}$, and $\vec{M} = (\vec{A} + \vec{B})/2$, we have

$$\vec{V} - \frac{\vec{A}}{2} - \frac{\vec{B}}{2} = k_1(\vec{X} - \vec{C}). \qquad (13.1)$$

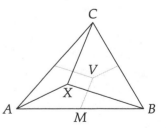

Similarly, we must also have

$$\vec{V} - \frac{\vec{B}}{2} - \frac{\vec{C}}{2} = k_2(\vec{X} - \vec{A}),$$

$$\vec{V} - \frac{\vec{C}}{2} - \frac{\vec{A}}{2} = k_3(\vec{X} - \vec{B}),$$

for some constants k_2 and k_3. From here, we present two solutions.

Solution 1: Solve for $\vec{V}$. We can solve for $\vec{V}$ by adding all three equations to get

$$3\vec{V} - \vec{A} - \vec{B} - \vec{C} = (k_1 + k_2 + k_3)\vec{X} - k_1\vec{C} - k_2\vec{A} - k_3\vec{B}.$$

Solving for $\vec{V}$ gives

$$\vec{V} = \frac{(k_1 + k_2 + k_3)\vec{X} + (1 - k_1)\vec{C} + (1 - k_2)\vec{A} + (1 - k_3)\vec{B}}{3}.$$

But what are k_1, k_2, and k_3?

Substituting this expression for $\vec{V}$ back into Equation (13.1), we have

$$\frac{(k_1 + k_2 + k_3)\vec{X} + (1 - k_1)\vec{C} + (1 - k_2)\vec{A} + (1 - k_3)\vec{B}}{3} - \frac{\vec{A}}{2} - \frac{\vec{B}}{2} = k_1(\vec{X} - \vec{C}). \qquad (13.2)$$

This equation will hold for all $\vec{X}$, $\vec{A}$, $\vec{B}$, and $\vec{C}$ if the coefficient of each vector on the left matches the corresponding coefficient on the right. This is the case if

$$\frac{k_1 + k_2 + k_3}{3} = k_1,$$

$$\frac{1 - k_2}{3} - \frac{1}{2} = 0,$$

$$\frac{1 - k_3}{3} - \frac{1}{2} = 0,$$

$$\frac{1 - k_2}{3} = -k_1.$$

The second and third equations give us $k_2 = k_3 = -\frac{1}{2}$. The fourth equation then gives us $k_1 = -\frac{1}{2}$. These values also satisfy the first equation. So, we have found constants k_1, k_2, and k_3 such that Equation (13.2) holds for all $\vec{X}$, $\vec{A}$, $\vec{B}$, and $\vec{C}$. Putting these values for the constants back into our original equations for $\vec{V}$ gives us

$$\vec{V} - \frac{\vec{A}}{2} - \frac{\vec{B}}{2} = -\frac{1}{2}(\vec{X} - \vec{C}),$$

$$\vec{V} - \frac{\vec{B}}{2} - \frac{\vec{C}}{2} = -\frac{1}{2}(\vec{X} - \vec{A}),$$

$$\vec{V} - \frac{\vec{C}}{2} - \frac{\vec{A}}{2} = -\frac{1}{2}(\vec{X} - \vec{B}).$$

We see that all three equations are satisfied if $\vec{V} = (-\vec{X} + \vec{A} + \vec{B} + \vec{C})/2$. Therefore, there is a point V such that all three equations are satisfied, which tells us that the three lines described in the problem are concurrent.

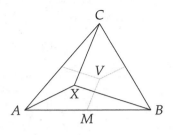

Solution 2: Educated Guessing. We must prove that there is some vector $\vec{V}$ and some constants k_1, k_2, and k_3 such that we have

$$\vec{V} - \frac{\vec{A}}{2} - \frac{\vec{B}}{2} = k_1(\vec{X} - \vec{C}),$$

$$\vec{V} - \frac{\vec{B}}{2} - \frac{\vec{C}}{2} = k_2(\vec{X} - \vec{A}),$$

$$\vec{V} - \frac{\vec{C}}{2} - \frac{\vec{A}}{2} = k_3(\vec{X} - \vec{B}).$$

From the diagram at right, it looks like $VM = \frac{1}{2}CX$, and that $\vec{MV}$ is in the opposite direction of $\vec{CX}$. So, it looks like we'll have $\vec{VM} = -\frac{1}{2}\vec{CX}$. It looks like V has a similar relationship to each of the other midpoints we care about in the problem, so we guess that we have $k_1 = k_2 = k_3 = -\frac{1}{2}$. We substitute these values into the equations and then proceed as above to show that there is a $\vec{V}$ that satisfies all three equations for all $\vec{X}$, $\vec{A}$, $\vec{B}$, and $\vec{C}$.

13.28 We have

$$\vec{uV} = \vec{V} - \vec{U} = \frac{\vec{B} + \vec{C} + \vec{D}}{3} - \frac{\vec{A} + \vec{B} + \vec{C}}{3} = \frac{1}{3}(\vec{D} - \vec{A}),$$

$$\vec{YX} = \vec{X} - \vec{Y} = \frac{\vec{D} + \vec{E} + \vec{F}}{3} - \frac{\vec{A} + \vec{E} + \vec{F}}{3} = \frac{1}{3}(\vec{D} - \vec{A}).$$

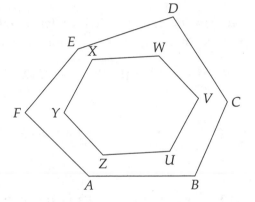

Since $\vec{UV} = \vec{YX}$, we have $\overline{UV} \parallel \overline{XY}$ and $UV = XY$. Similarly, we have $\vec{VW} = \vec{ZY}$ and $\vec{WX} = \vec{UZ}$. Therefore, each two opposite sides of $UVWXYZ$ are parallel and equal in length.

13.29 We let the foot of the altitude from D to face ABC be the origin, both because this point is a special point of $\triangle ABC$ (the orthocenter), and because this gives us $\vec{A} \cdot \vec{D} = \vec{B} \cdot \vec{D} = \vec{C} \cdot \vec{D} = 0$. From the given $\angle BDC = 90°$, we have $\vec{BD} \cdot \vec{CD} = 0$, which gives us $(\vec{D} - \vec{B}) \cdot (\vec{D} - \vec{C}) = 0$. Expanding the product on the left, and using the fact that $\vec{B} \cdot \vec{D} = \vec{C} \cdot \vec{D} = 0$, we have $\vec{D} \cdot \vec{D} + \vec{B} \cdot \vec{C} = 0$, so $\vec{D} \cdot \vec{D} = -\vec{B} \cdot \vec{C}$.

We can show that $\angle ADB = 90°$ by showing that $\vec{AD} \cdot \vec{BD} = 0$. We have

$$\vec{AD} \cdot \vec{BD} = (\vec{D} - \vec{A}) \cdot (\vec{D} - \vec{B}) = \vec{D} \cdot \vec{D} - \vec{D} \cdot \vec{B} - \vec{A} \cdot \vec{D} + \vec{A} \cdot \vec{B}.$$

Since $\vec{D} \cdot \vec{B} = \vec{A} \cdot \vec{D} = 0$, we have $\vec{AD} \cdot \vec{BD} = \vec{D} \cdot \vec{D} + \vec{A} \cdot \vec{B}$. Substituting $\vec{D} \cdot \vec{D} = -\vec{B} \cdot \vec{C}$ from above, we have $\vec{AD} \cdot \vec{BD} = -\vec{B} \cdot \vec{C} + \vec{A} \cdot \vec{B} = \vec{B} \cdot (\vec{A} - \vec{C})$. Since the origin is the orthocenter, $\vec{B}$ is parallel to the altitude from B to $\overline{AC}$. Therefore, $\vec{B}$ is orthogonal to $\vec{AC}$, which means $\vec{B} \cdot (\vec{A} - \vec{C}) = 0$. This gives us $\vec{AD} \cdot \vec{BD} = 0$, so $\angle ADB = 90°$. We can go through essentially the same steps to show that $\angle ADC = 90°$ as well.

13.30 If it is possible for Harrison to construct such a hexagon, then let $ABCDEF$ be the hexagon. Then, the vectors to the six points given are $\frac{\vec{A}+\vec{B}}{2}$, $\frac{\vec{B}+\vec{C}}{2}$, $\frac{\vec{C}+\vec{D}}{2}$, $\frac{\vec{D}+\vec{E}}{2}$, $\frac{\vec{E}+\vec{F}}{2}$, $\frac{\vec{F}+\vec{A}}{2}$, in some order. The sum of the first, third, and fifth of these equals the sum of the second, fourth, and sixth. Another way of saying this is that the centroid of the triangle with vertices at the first, third, and fifth midpoints is the same as the centroid of the triangle with vertices at the second, fourth, and sixth midpoints.

Looking at the six given points, we see that the x-coordinate of one of the points is odd, while the x-coordinates of the other five points are even. So, no matter how we split the six points into two groups of three, the sum of the x-coordinates will be odd and even in the other. Therefore, it is impossible to divide the six points into two groups of three points such that sum of the x-coordinates in one group is the same as the sum of the x-coordinates in the other.

13.31 We have

$$\vec{AX} \cdot \vec{CX} - \vec{CB} \cdot \vec{AX} = (\vec{X} - \vec{A}) \cdot (\vec{X} - \vec{C}) - (\vec{B} - \vec{C}) \cdot (\vec{X} - \vec{A})$$
$$= \vec{X} \cdot \vec{X} - \vec{X} \cdot \vec{C} - \vec{A} \cdot \vec{X} + \vec{A} \cdot \vec{C} - \vec{B} \cdot \vec{X} + \vec{B} \cdot \vec{A} + \vec{C} \cdot \vec{X} - \vec{C} \cdot \vec{A}$$
$$= \vec{X} \cdot \vec{X} - \vec{B} \cdot \vec{X} - \vec{A} \cdot \vec{X} + \vec{B} \cdot \vec{A}$$
$$= (\vec{X} - \vec{A}) \cdot (\vec{X} - \vec{B}) = \vec{AX} \cdot \vec{BX}.$$

We have $\vec{AX} \cdot \vec{BX} = 0$ if and only if $\overline{AX} \perp \overline{BX}$ or if X is A or B. Therefore, X is on the circle with diameter $\overline{AB}$. Conversely, if X is on the circle with diameter $\overline{AB}$, then $\vec{AX} \cdot \vec{BX} = 0$, and we can reverse the steps above to show that $\vec{AX} \cdot \vec{CX} = \vec{CB} \cdot \vec{AX}$.

13.32 Let the center of the circle be the origin. For all i with $1 \le i \le n$, we have

$$\vec{P_iX} \cdot \vec{P_iX} = (\vec{X} - \vec{P_i}) \cdot (\vec{X} - \vec{P_i}) = \vec{X} \cdot \vec{X} + \vec{P_i} \cdot \vec{P_i} - 2\vec{X} \cdot \vec{P_i}.$$

Since the P_i are on the circle centered at the origin with radius r, we have $\vec{P_i} \cdot \vec{P_i} = \|P_i\|^2 = r^2$. Since the polygon is regular, the sum of the $\vec{P_i}$ is $\mathbf{0}$ (for the same reason that the sum of the nth roots of unity is 0). Therefore, we have

$$(P_1X)^2 + (P_2X)^2 + (P_3X)^2 + \cdots + (P_nX)^2 = n\vec{X} \cdot \vec{X} + (\vec{P_1} \cdot \vec{P_1} + \vec{P_2} \cdot \vec{P_2} + \cdots + \cdots + \vec{P_n} \cdot \vec{P_n}) - 2\vec{X} \cdot (\vec{P_1} + \vec{P_2} + \cdots + \vec{P_n})$$
$$= n\vec{X} \cdot \vec{X} + nr^2 - 2\vec{X} \cdot \mathbf{0}.$$

Since $\vec{X} \cdot \vec{X}$ is the square of the distance between X and the origin, and the origin is the center of the circle, C, we have the desired

$$(P_1X)^2 + (P_2X)^2 + (P_3X)^2 + \cdots + (P_nX)^2 = n(r^2 + CX^2).$$

13.33 Let the circumcenter be the origin. Therefore, we must have $\|\vec{A}\| = \|\vec{H} - \vec{A}\|$. Since $\vec{H} = \vec{A} + \vec{B} + \vec{C}$ and $\|\vec{A}\| = \|\vec{B}\| = \|\vec{C}\| = R$, where R is the circumradius, we have $R = \|(\vec{A} + \vec{B} + \vec{C}) - \vec{A}\| = \|\vec{B} + \vec{C}\|$. Squaring both sides gives

$$R^2 = \|\vec{B} + \vec{C}\|^2 = (\vec{B} + \vec{C}) \cdot (\vec{B} + \vec{C}).$$

Expanding the dot product, and noting that $\vec{B} \cdot \vec{B} = \vec{C} \cdot \vec{C} = R^2$, we have

$$R^2 = \vec{B} \cdot \vec{B} + 2\vec{B} \cdot \vec{C} + \vec{C} \cdot \vec{C} = 2R^2 + 2\|\vec{B}\|\|\vec{C}\| \cos \angle BOC = 2R^2 + 2R^2 \cos \angle BOC.$$

We therefore have $\cos \angle BOC = -\frac{1}{2}$. Since $0 < \angle BOC < \pi$, we have $\angle BOC = \frac{2\pi}{3}$. We have $\angle BAC = \frac{1}{2}\angle BOC$ because $\angle BAC$ is inscribed in the same arc that central angle $\angle BOC$ subtends. Therefore, the only possible value of $\angle A$ is $\boxed{\frac{\pi}{3}}$.

13.34 Let the given points be A, B, and C, respectively. Any segment connecting two vertices of an octahedron is either an edge or one of the three space diagonals connecting opposite vertices. We note that $AB = BC = 6$ and $AC = 6\sqrt{2}$, so $\overline{AB}$ and $\overline{BC}$ are edges and $\overline{AC}$ is a space diagonal. We can find the fourth vertex D of the octahedron in the plane of $\triangle ABC$ by noting that $ABCD$ is a square, so $\vec{CD} = \vec{BA}$. We have $\vec{BA} = (2-6)\mathbf{i} + (6-4)\mathbf{j} + (-8-(-4))\mathbf{k} = -4\mathbf{i} + 2\mathbf{j} - 4\mathbf{k}$. Since $\vec{CD} = \vec{BA} = -4\mathbf{i} + 2\mathbf{j} - 4\mathbf{k}$ and point C is $(4, 8, 0)$, we find that D is $(-4+4, 2+8, -4+0) = \boxed{(0, 10, -4)}$.

Let E and F be the other vertices of the octahedron, and let M be the center of square $ABCD$. To find E and F, we note that $\vec{ME}$ and $\vec{MF}$ are normal to plane $ABCD$, and in opposite directions. Therefore, the vector $\vec{AM} \times \vec{BM}$ is in the same direction as one of $\vec{ME}$ and $\vec{MF}$, and in the opposite direction of the other.

The center of square $ABCD$ is the midpoint of $\overline{AC}$, which is $(3,7,-4)$. We then have $\vec{AM} = \mathbf{i} + \mathbf{j} + 4\mathbf{k}$ and $\vec{BM} = -3\mathbf{i} + 3\mathbf{j}$. We then find

$$\vec{AM} \times \vec{BM} = \begin{vmatrix} \mathbf{i} & \mathbf{j} & \mathbf{k} \\ 1 & 1 & 4 \\ -3 & 3 & 0 \end{vmatrix} = -12\mathbf{i} - 12\mathbf{j} + 6\mathbf{k}.$$

Therefore, one of $\vec{ME}$ and $\vec{MF}$ is in the direction of $-12\mathbf{i} - 12\mathbf{j} + 6\mathbf{k}$ and the other is in the opposite direction. Because the octahedron is regular, each of these vectors has length equal to half the length of a space diagonal of the octahedron. Therefore, we have $\|\vec{ME}\| = \|\vec{MF}\| = AC/2 = 3\sqrt{2}$. The vector $-12\mathbf{i} - 12\mathbf{j} + 6\mathbf{k}$ has direction

$$\frac{-12\mathbf{i} - 12\mathbf{j} + 6\mathbf{k}}{\|-12\mathbf{i} - 12\mathbf{j} + 6\mathbf{k}\|} = \frac{-12\mathbf{i} - 12\mathbf{j} + 6\mathbf{k}}{\sqrt{(-12)^2 + (-12)^2 + 6^2}} = \frac{-12\mathbf{i} - 12\mathbf{j} + 6\mathbf{k}}{18} = -\frac{2}{3}\mathbf{i} - \frac{2}{3}\mathbf{j} + \frac{1}{3}\mathbf{k}.$$

So, we can take

$$\vec{ME} = 3\sqrt{2}\left(-\frac{2}{3}\mathbf{i} - \frac{2}{3}\mathbf{j} + \frac{1}{3}\mathbf{k}\right) = -2\sqrt{2}\mathbf{i} - 2\sqrt{2}\mathbf{j} + \sqrt{2}\mathbf{k},$$

$$\vec{MF} = -3\sqrt{2}\left(-\frac{2}{3}\mathbf{i} - \frac{2}{3}\mathbf{j} + \frac{1}{3}\mathbf{k}\right) = 2\sqrt{2}\mathbf{i} + 2\sqrt{2}\mathbf{j} - \sqrt{2}\mathbf{k},$$

Since M is $(3,7,-4)$, we find that E is $\boxed{(3 - 2\sqrt{2}, 7 - 2\sqrt{2}, -4 + \sqrt{2})}$ and F is $\boxed{(3 + 2\sqrt{2}, 7 + 2\sqrt{2}, -4 - \sqrt{2})}$.

13.35 Let $a = a_1 + a_2 i$ and $b = b_1 + b_2 i$. The desired area equals $\frac{1}{2}|a||b| \sin\theta$, where θ is the angle formed by the segments connecting a and b to the origin. We have seen expressions like this when working with the cross product. So, let's view this problem in terms of vectors. Let A and B be the points $(a_1, a_2, 0)$ and $(b_1, b_2, 0)$ in space, and let O be the origin. The area of the parallelogram with $\overline{OA}$ and $\overline{OB}$ as sides is $\|\vec{OA} \times \vec{OB}\|$. This is the same parallelogram as described in the problem, so its area equals the desired area. We have

$$\|\vec{OA} \times \vec{OB}\| = \|(a_1 b_2 - a_2 b_1)\mathbf{k}\| = |a_1 b_2 - a_2 b_1|.$$

We also have

$$\left|\frac{1}{2}(a\bar{b} - \bar{a}b)\right| = \left|\frac{1}{2}((a_1 + a_2 i)(b_1 - b_2 i) - (a_1 - a_2 i)(b_1 + b_2 i))\right| = \frac{1}{2}|-2a_1 b_2 i + 2a_2 b_1 i| = |a_1 b_2 - a_2 b_1|,$$

which equals the expression we found earlier for the area of the parallelogram. Therefore, the area of the parallelogram equals $\left|\frac{1}{2}(a\bar{b} - \bar{a}b)\right|$, as desired.

13.36 The perpendicularity suggests vectors. We let A be the origin. Since $\overline{AX} \perp \overline{DY}$, we have $\vec{AX} \cdot \vec{DY} = 0$, which becomes $\vec{X} \cdot \vec{DY} = 0$ when we let A be the origin. We have $\vec{Y} = (\vec{A} + \vec{B} + \vec{C})/3 = (\vec{B} + \vec{C})/3$ since Y is the centroid of $\triangle ABC$ and A is the origin. We also have $\vec{X} = (\vec{B} + \vec{C} + \vec{D})/3$ because X is the centroid of $\triangle BCD$. Therefore, $\vec{X} \cdot \vec{DY} = 0$ becomes

$$\left(\frac{\vec{B} + \vec{C} + \vec{D}}{3}\right) \cdot \left(\frac{\vec{B} + \vec{C}}{3} - \vec{D}\right) = 0.$$

We multiply both sides by 9 to get rid of the fractions and have

$$(\vec{B} + \vec{C} + \vec{D}) \cdot (\vec{B} + \vec{C} - 3\vec{D}) = 0.$$

Expanding the dot product on the left gives

$$\vec{B} \cdot \vec{B} + 2\vec{B} \cdot \vec{C} - 2\vec{B} \cdot \vec{D} + \vec{C} \cdot \vec{C} - 2\vec{C} \cdot \vec{D} - 3\vec{D} \cdot \vec{D} = 0. \tag{13.3}$$

From $AB = 6$, we have $\vec{B} \cdot \vec{B} = AB^2 = 36$. Similarly, we have $\vec{C} \cdot \vec{C} = AC^2 = 8^2 = 64$. Letting the desired AD be t, we have $\vec{D} \cdot \vec{D} = t^2$, and Equation (13.3) becomes

$$100 - 3t^2 + 2\vec{B} \cdot \vec{C} - 2\vec{B} \cdot \vec{D} - 2\vec{C} \cdot \vec{D} = 0. \tag{13.4}$$

We have $BC^2 = \left\| \vec{B} - \vec{C} \right\|^2 = (\vec{B} - \vec{C}) \cdot (\vec{B} - \vec{C}) = \vec{B} \cdot \vec{B} - 2\vec{B} \cdot \vec{C} + \vec{C} \cdot \vec{C} = 100 - 2\vec{B} \cdot \vec{C}$. Since $BC^2 = AD^2 = t^2$, we have $t^2 = 100 - 2\vec{B} \cdot \vec{C}$, so $2\vec{B} \cdot \vec{C} = 100 - t^2$. Similarly, from $BD^2 = \left\| \vec{B} - \vec{D} \right\|^2 = (\vec{B} - \vec{D}) \cdot (\vec{B} - \vec{D})$, we find $2\vec{B} \cdot \vec{D} = t^2 - 28$, and from $CD^2 = \left\| \vec{C} - \vec{D} \right\|^2 = (\vec{C} - \vec{D}) \cdot (\vec{C} - \vec{D})$, we find $2\vec{C} \cdot \vec{D} = t^2 + 28$. Substituting these into Equation (13.4) gives

$$100 - 3t^2 + (100 - t^2) - (t^2 - 28) - (t^2 + 28) = 0,$$

from which we find $6t^2 = 200$, so $t = \sqrt{200/6} = \sqrt{100/3} = \boxed{10\sqrt{3}/3}$. Notice that the answer equals $\sqrt{(AB^2 + AC^2)/3}$. Is this a coincidence?

13.37 We let O be the origin, and we let R be the radius of the sphere. Because P and Q are opposite vertices of a parallelepiped with edges $\overline{PU}$, $\overline{PV}$, and $\overline{PW}$, we have

$$\vec{Q} = \vec{P} + (\vec{U} - \vec{P}) + (\vec{V} - \vec{P}) + (\vec{W} - \vec{P}).$$

Therefore, we have

$$\begin{aligned} \left\| \vec{Q} \right\|^2 &= (\vec{P} + (\vec{U} - \vec{P}) + (\vec{V} - \vec{P}) + (\vec{W} - \vec{P})) \cdot (\vec{P} + (\vec{U} - \vec{P}) + (\vec{V} - \vec{P}) + (\vec{W} - \vec{P})) \\ &= \vec{P} \cdot \vec{P} + 2\vec{P} \cdot (\vec{U} - \vec{P}) + 2\vec{P} \cdot (\vec{V} - \vec{P}) + 2\vec{P} \cdot (\vec{W} - \vec{P}) \\ &\quad + (\vec{U} - \vec{P}) \cdot (\vec{U} - \vec{P}) + (\vec{V} - \vec{P}) \cdot (\vec{V} - \vec{P}) + (\vec{W} - \vec{P}) \cdot (\vec{W} - \vec{P}) \\ &\quad + 2(\vec{U} - \vec{P}) \cdot (\vec{V} - \vec{P}) + 2(\vec{V} - \vec{P}) \cdot (\vec{W} - \vec{P}) + 2(\vec{W} - \vec{P}) \cdot (\vec{U} - \vec{P}). \end{aligned}$$

Each of the last three dot products is 0 because any two edges that share a vertex of the parallelepiped are orthogonal. Expanding the remaining terms, we see that all terms of the form $2\vec{P} \cdot \vec{U}$ appear twice, once added and once subtracted, so they all cancel. Since $\left\| \vec{U} \right\| = \left\| \vec{V} \right\| = \left\| \vec{W} \right\| = R$, we are left with

$$\left\| \vec{Q} \right\|^2 = \left\| \vec{U} \right\|^2 + \left\| \vec{V} \right\|^2 + \left\| \vec{W} \right\|^2 - 2\left\| \vec{P} \right\|^2 = 3R^2 - 2\left\| \vec{P} \right\|^2.$$

Therefore, all possible Q are a distance of $3R^2 - 2\left\| \vec{P} \right\|^2$ from the origin. (As an extra challenge, show that every point that is $3R^2 - 2\left\| \vec{P} \right\|^2$ from the origin can possibly be point Q.)

13.38 We use the intersection of the diagonals as the origin, and we use complex numbers. As usual, each lowercase letter is the complex number corresponding to the point denoted by the respective uppercase letter.

$$g_2 - g_1 = \frac{y + z + o}{3} - \frac{w + x + o}{3} = \frac{1}{3}(y + z - w - x).$$

We wish to show that the quotient $(h_2 - h_1)/(g_2 - g_1)$ is imaginary. Unfortunately, h_2 and h_1 are a lot harder to deal with. We know that $\overleftrightarrow{OH_2} \perp \overleftrightarrow{ZW}$, so $h_2/(z - w)$ is imaginary. Can we find what this quotient equals? For that, we turn to trigonometry.

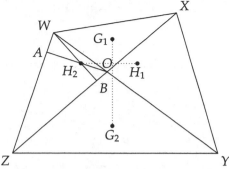

We focus on $\angle WOZ$ because $\angle XOY$ of $\triangle XOY$ has the same measure. So, if we can express h_2 in terms of this angle, then the same angle will appear in our corresponding expression for h_1. We let the feet of the altitudes from O and W be A and B, as shown. From quadrilateral ZAH_2B, we have $\angle AH_2B + \angle AZB = \pi$. We also have $\angle AH_2B + \angle BH_2O = \pi$, so $\angle BH_2O = \angle AZB$, which means $\triangle OH_2B \sim \triangle WZB$ (since $\angle WBZ = \angle OBH_2 = \pi/2$). From this similarity, we have $\frac{OH_2}{OB} = \frac{WZ}{WB}$, so

$$OH_2 = WZ \cdot \frac{OB}{WB} = WZ \cot \angle WOZ.$$

We also have $\overleftrightarrow{OH_2} \perp \overline{WZ}$, so to get h_2, we rotate $w - z$ by $\frac{\pi}{2}$, and scale the result by $\cot \angle WOZ$. Therefore, with the diagram oriented as shown, we have $h_2 = i(w-z) \cot \angle WOZ$. Similarly, we have $h_1 = i(y - x) \cot \angle XOY$. We have $\angle WOZ = \angle XOY$, so we have $h_2 - h_1 = i(w + x - y - z) \cot \angle WOZ$. Combining this with our earlier expression for $g_2 - g_1$, we have

$$\frac{h_2 - h_1}{g_2 - g_1} = \frac{i(w + x - y - z) \cot \angle WOZ}{\frac{1}{3}(y + z - w - x)} = \frac{i(w + x - y - z) \cot \angle WOZ}{-\frac{1}{3}(w + x - y - z)} = -3i \cot \angle WOZ.$$

Since this quotient is imaginary, we have $\overleftrightarrow{G_1 G_2} \perp \overleftrightarrow{H_1 H_2}$.

13.39 We let the centroid of $ABCD$ be the origin. We therefore have $\vec{A} + \vec{B} + \vec{C} + \vec{D} = \mathbf{0}$. We wish to show that $\vec{E} + \vec{F} + \vec{G} + \vec{H} = \mathbf{0}$.

In order to use the information relating the altitudes of $ABCD$ to $EFGH$, we note that the altitude from A to face BCD is parallel to $\overline{AE}$. Since $\vec{DC} \times \vec{DB}$ is normal to face BCD, this cross product is in the same or opposite direction as $\vec{AE}$. We choose the orientation of $ABCD$ such that these vectors have the same direction. (If the orientation is the opposite, the rest of the solution is essentially the same.)

We have

$$\vec{DC} \times \vec{DB} = (\vec{C} - \vec{D}) \times (\vec{B} - \vec{D}) = \vec{C} \times \vec{B} - \vec{C} \times \vec{D} - \vec{D} \times \vec{B} + \vec{D} \times \vec{D} = \vec{C} \times \vec{B} - \vec{C} \times \vec{D} - \vec{D} \times \vec{B} + \mathbf{0}.$$

We also have $\|\vec{DC} \times \vec{DB}\| = 2[BCD]$. Letting V be the volume of $ABCD$, we have $h_a[BCD]/3 = V$, so $h_a = 3V/[BCD]$. Since $\|\vec{AE}\| = k/h_a$, we have

$$\|\vec{AE}\| = \frac{k}{h_a} = \frac{k[BCD]}{3V} = \frac{k\|\vec{DC} \times \vec{DB}\|}{6V}.$$

Since $\vec{AE}$ and $\vec{DC} \times \vec{DB}$ are in the same direction, we have

$$\vec{AE} = \frac{k}{6V}(\vec{C} \times \vec{B} - \vec{C} \times \vec{D} - \vec{D} \times \vec{B}).$$

Similarly, we can show that

$$\vec{BF} = \frac{k}{6V}(\vec{AC} \times \vec{AD}) = \frac{k}{6V}(\vec{C} \times \vec{D} - \vec{A} \times \vec{D} - \vec{C} \times \vec{A}),$$

$$\vec{CG} = \frac{k}{6V}(\vec{BA} \times \vec{BD}) = \frac{k}{6V}(\vec{A} \times \vec{D} - \vec{A} \times \vec{B} - \vec{B} \times \vec{D}),$$

$$\vec{DH} = \frac{k}{6V}(\vec{CA} \times \vec{CB}) = \frac{k}{6V}(\vec{A} \times \vec{B} - \vec{A} \times \vec{C} - \vec{C} \times \vec{B}).$$

Taking the sum of these four equations gives us a lot of cancellation, and leaves

$$\vec{AE} + \vec{BF} + \vec{CG} + \vec{DH} = \frac{k}{6V}(-\vec{D} \times \vec{B} - \vec{C} \times \vec{A} - \vec{B} \times \vec{D} - \vec{A} \times \vec{C}).$$

Since $\vec{B} \times \vec{D} = -\vec{D} \times \vec{B}$ and $\vec{A} \times \vec{C} = -\vec{D} \times \vec{A}$, the four terms on the right above sum to $\mathbf{0}$ as well, and we have

$$\vec{AE} + \vec{BF} + \vec{CG} + \vec{DH} = \mathbf{0}.$$

Writing each vector as a difference of vectors gives

$$\vec{E} - \vec{A} + \vec{F} - \vec{B} + \vec{G} - \vec{C} + \vec{H} - \vec{D} = \mathbf{0}.$$

Since $\vec{A} + \vec{B} + \vec{C} + \vec{D} = \mathbf{0}$, we have $\vec{E} + \vec{F} + \vec{G} + \vec{H} = \mathbf{0}$, as desired. Therefore, the centroid of $EFGH$ is the same as the centroid of $ABCD$.

13.40 The interior angles of a hexagon sum to 4π, so if all the angles of hexagon $ABCDEF$ are equal, then each is equal to $\frac{4\pi}{6} = \frac{2\pi}{3}$.

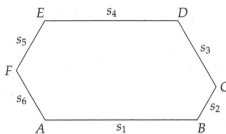

Let $s_1 = AB$, $s_2 = BC$, $s_3 = CD$, $s_4 = DE$, $s_5 = EF$, and $s_6 = FA$. Let a be the complex number corresponding to A, and similarly for the other points. Without loss of generality, assume that hexagon $ABCDEF$ is oriented so that $\overline{AB}$ is parallel to the real axis, i.e. $b - a$ is real.

Then we can let $b - a = s_1$, for some real number s_1. Since $\angle ABC = \frac{2\pi}{3}$, the argument of $c - b$, which corresponds to side $\overline{BC}$, is $\frac{\pi}{3}$ greater than the argument of $b - a$, which corresponds to $\overline{AB}$. Therefore, letting $\zeta = e^{\pi i/3}$, we have $c - b = \zeta s_2$ for some real number s_2. Similarly, we have $d - c = \zeta^2 s_3$, $e - d = \zeta^3 s_4$, $f - e = \zeta^4 s_5$, and $a - f = \zeta^5 s_6$ for some real numbers s_3, s_4, s_5, and s_6. Adding all of the equations corresponding to the sides of the hexagon, we find

$$s_1 + \zeta s_2 + \zeta^2 s_3 + \zeta^3 s_4 + \zeta^4 s_5 + \zeta^5 s_6 = 0.$$

Since $\zeta = e^{\pi i/3}$, we have $\zeta^3 = e^{\pi i} = -1$, which makes the above equation

$$s_1 + \zeta s_2 + \zeta^2 s_3 - s_4 - \zeta s_5 - \zeta^2 s_6 = 0,$$

which we can rearrange as

$$(s_1 - s_4) + \zeta(s_2 - s_5) + \zeta^2(s_3 - s_6) = 0.$$

We could now substitute $e^{\pi i/3} = (1 + i\sqrt{3})/2$ and do some algebra to finish, but that's not nearly as illuminating (or as fun!) as reasoning geometrically.

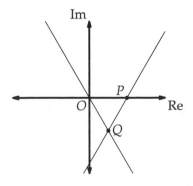

Let $p = s_1 - s_4$, $q = s_2 - s_5$, and $r = s_3 - s_6$, so that our equation above is $p + \zeta q + \zeta^2 r = 0$. Let P correspond to p in the complex plane, and suppose $s_1 - s_4 > 0$, so P is on the positive real axis, as shown. (The proof is essentially the same if $s_1 - s_4 < 0$.) Let Q correspond to $p + \zeta q$. Writing the equation $p + \zeta q + \zeta^2 r = 0$ as $p + \zeta q = -\zeta^2 r$, we see that Q also corresponds to $-\zeta^2 r$. Since r is real, the point corresponding to $-\zeta^2 r$ is on the line through the origin and the second and fourth quadrants that makes and angle of $\pi/3$ with the positive real axis. Likewise, since q is real, the point corresponding to $p + \zeta q$ must be on the line shown through P that makes an angle of $\pi/3$ with the positive real axis. (There are two lines through P that make an angle of $\pi/3$ with the positive real axis; we take the one that does not pass through the second quadrant, as shown.)

Point Q must be on both of these lines, so it is the shown intersection point. Letting O be the origin, we therefore have $\angle POQ = \angle OPQ = \pi/3$, so $\angle OQP = \pi/3$ and the triangle is equilateral. This gives us $OP = PQ = QO$. Since $OP = |p|$, $PQ = |\zeta q| = |q|$, and $OQ = |-r\zeta^2| = |r||-\zeta^2| = |r|$, we have $|p| = |q| = |r|$, which gives us $|s_1 - s_4| = |s_2 - s_5| = |s_3 - s_6|$, so $|AB - DE| = |BC - EF| = |CD - FA|$, as desired.

See if you can find another solution by proving that the equation $p + \zeta q + \zeta^2 r = 0$ implies that the triangle with vertices p, ζq, and $\zeta^2 r$ is equilateral.

13.41 Let $a = BC$, $c = AB$, and $b = AC$. From Exercise 13.2.2, we have $\vec{D} = \frac{b\vec{B}+c\vec{C}}{b+c}$, $\vec{E} = \frac{a\vec{A}+c\vec{C}}{a+c}$, and $\vec{F} = \frac{a\vec{A}+b\vec{B}}{a+b}$. We simplify the problem a bit by letting A be the origin. We choose A rather than one of the other vertices because this choice gives $\vec{ED}$ and $\vec{FD}$ symmetric forms. We therefore have

$$\vec{ED} = \vec{D} - \vec{E} = \frac{b\vec{B}+c\vec{C}}{b+c} - \frac{c\vec{C}}{a+c} = \frac{1}{(a+c)(b+c)}((a+c)(b\vec{B}+c\vec{C}) - (b+c)c\vec{C})$$

$$= \frac{1}{(a+c)(b+c)}(ab\vec{B} + ac\vec{C} + bc\vec{B} + c^2\vec{C} - bc\vec{C} - c^2\vec{C})$$

$$= \frac{1}{(a+c)(b+c)}(ab\vec{B} + ac\vec{C} + bc(\vec{B} - \vec{C})).$$

Similarly, we have

$$\vec{FD} = \frac{1}{(a+b)(b+c)}(ab\vec{B} + ac\vec{C} + bc(\vec{C} - \vec{B})).$$

We are given that $\vec{ED} \cdot \vec{FD} = 0$, so

$$(ab\vec{B} + ac\vec{C} + bc(\vec{B} - \vec{C})) \cdot (ab\vec{B} + ac\vec{C} + bc(\vec{C} - \vec{B})) = 0.$$

Since the first two terms in each of the vectors in the dot product above are the same, but the last terms of each are opposites, we have

$$(ab\vec{B} + ac\vec{C} + bc(\vec{B} - \vec{C})) \cdot (ab\vec{B} + ac\vec{C} + bc(\vec{C} - \vec{B})) = (ab\vec{B} + ac\vec{C}) \cdot (ab\vec{B} + ac\vec{C}) - b^2c^2(\vec{C} - \vec{B}) \cdot (\vec{C} - \vec{B}).$$

Expanding this expression, and noting that $\vec{B} \cdot \vec{B} = AB^2 = c^2$, $\vec{C} \cdot \vec{C} = AC^2 = b^2$, and $(\vec{C} - \vec{B}) \cdot (\vec{C} - \vec{B}) = BC^2 = a^2$, we have

$$(ab\vec{B} + ac\vec{C}) \cdot (ab\vec{B} + ac\vec{C}) - b^2c^2(\vec{C} - \vec{B}) \cdot (\vec{C} - \vec{B}) = a^2b^2\vec{B} \cdot \vec{B} + 2a^2bc\vec{B} \cdot \vec{C} + a^2c^2\vec{C} \cdot \vec{C} - b^2c^2a^2$$

$$= a^2b^2c^2 + 2a^2bc\vec{B} \cdot \vec{C} + a^2b^2c^2 - b^2c^2a^2 = 2a^2bc\vec{B} \cdot \vec{C} + a^2b^2c^2.$$

Setting this equal to 0 gives $\vec{B} \cdot \vec{C} = -\frac{1}{2}bc$. Since $\vec{B} \cdot \vec{C} = bc \cos \angle BAC$, we have $\cos \angle BAC = -\frac{1}{2}$, from which we see that the only possible value of $\angle BAC$ is $\boxed{120°}$.

To see that such a triangle is possible, consider an isosceles triangle ABC with $\angle BAC = 120°$, so $\angle ABC = \angle ACB = 30°$. There are many ways to show that $\angle EDF = 90°$ in such a triangle. For example, by symmetry, we have $\overline{EF} \parallel \overline{BC}$. Then, since $\overline{BE}$ bisects $\angle B$, we have $\angle FEB = \angle EBC = \angle FBE$, which means $\triangle BEF$ is isosceles with $EF = BF$. Similarly, $\triangle CEF$ is isosceles with $EF = EC$. Letting X be the foot of the altitude from E to $\overline{CD}$, triangle EXC is a 30-60-90 triangle. Therefore, $EX = EC/2 = EF/2$, which means altitude $\overline{DZ}$ from D to $\overline{EF}$ also has length $EF/2$. Since $EZ = FZ = EF/2 = DZ$ and $\overline{DZ} \perp \overline{EF}$, triangle DZE is a 45-45-90 triangle, so $\angle EDF = 2\angle EDZ = 90°$.

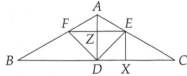

13.42 Let $\omega = e^{\pi i/7}$, a primitive 14th root of unity, so $\omega^{14} = 1$. We place the diagram in the complex plane so that A, B, C, and D correspond to the complex numbers $a = 0$, $b = k$, $c = k\omega$ (so that $AB = AC$ and $\angle BAC = \frac{\pi}{7}$), and $d = 1$, where we take k to be a positive real number such that $|b - c| = 1$.

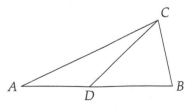

Since $\angle BAC = \frac{\pi}{7}$ and $AC = AB$, we have $\angle ABC = \angle ACB = \frac{3\pi}{7}$. Since $\angle ABC = \frac{3\pi}{7}$, we see that the angle $\overline{BC}$ makes with the x-axis on the right is $\pi - \frac{3\pi}{7} = \frac{4\pi}{7}$. Therefore, the argument of $c - b$ is $\frac{4\pi}{7}$. Since the magnitude of $c - b$ is 1, we have $c - b = \omega^4$. We also have $c - b = k\omega - k = k(\omega - 1)$, so $k(\omega - 1) = \omega^4$, which means

$$k = \frac{\omega^4}{\omega - 1}.$$

Then $CD = |c - d| = |k\omega - 1| = \left|\frac{\omega^5}{\omega-1} - 1\right| = \left|\frac{\omega^5-\omega+1}{\omega-1}\right| = \frac{|\omega^5-\omega+1|}{|\omega-1|}$. As usual, we get rid of the magnitudes by considering the square of the desired length:

$$CD^2 = \frac{|\omega^5 - \omega + 1|^2}{|\omega - 1|^2} = \frac{(\omega^5 - \omega + 1)\overline{(\omega^5 - \omega + 1)}}{(\omega - 1)\overline{(\omega - 1)}}.$$

We'll handle the numerator and denominator separately. But first, we note that $\omega^7 = e^{\pi i} = -1$ and $\omega^{14} = e^{2\pi i} = 1$, and that $\overline{\omega^k} = (\overline{\omega})^k = (1/\omega)^k = 1/\omega^k = \omega^{14}/\omega^k = \omega^{14-k}$. Using these relationships, we find

$$(\omega^5 - \omega + 1)\overline{(\omega^5 - \omega + 1)} = (\omega^5 - \omega + 1)(\overline{\omega^5} - \overline{\omega} + \overline{1}) = (\omega^5 - \omega + 1)(\omega^9 - \omega^{13} + 1) = (\omega^5 - \omega + 1)(-\omega^2 + \omega^6 + 1)$$

$$= \omega^5(-\omega^2 + \omega^6 + 1) - \omega(-\omega^2 + \omega^6 + 1) + 1(-\omega^2 + \omega^6 + 1)$$

$$= -\omega^7 + \omega^{11} + \omega^5 + \omega^3 - \omega^7 - \omega - \omega^2 + \omega^6 + 1$$

$$= 1 - \omega^4 + \omega^5 + \omega^3 + 1 - \omega - \omega^2 + \omega^6 + 1$$

$$= \omega^6 + \omega^5 - \omega^4 + \omega^3 - \omega^2 - \omega + 3.$$

Similarly, we have

$$(\omega - 1)\overline{(\omega - 1)} = (\omega - 1)(\overline{\omega} - 1) = (\omega - 1)(\omega^{13} - 1)$$

$$= \omega^{14} - \omega^{13} - \omega + 1 = 1 + \omega^6 - \omega + 1 = \omega^6 - \omega + 2.$$

At first, it doesn't look like we can do anything with the ratio of our expressions for $|\omega^5 - \omega + 1|^2$ and $|\omega - 1|^2$. In the past, we have simplified expressions involving roots of unity by deriving equations that are satisfied by those roots of unity. We try the same here, starting with our earlier observation that $\omega^7 = -1$. Rearranging this gives $\omega^7 + 1 = 0$, and factoring gives

$$\omega^7 + 1 = (\omega + 1)(\omega^6 - \omega^5 + \omega^4 - \omega^3 + \omega^2 - \omega + 1)$$

Since $\omega \neq -1$, we have $\omega^6 - \omega^5 + \omega^4 - \omega^3 + \omega^2 - \omega + 1 = 0$, which means $\omega^6 = \omega^5 - \omega^4 + \omega^3 - \omega^2 + \omega - 1$. Substituting this into our expressions for $|\omega^5 - \omega + 1|^2$ and $|\omega - 1|^2$ gives

$$(\omega^5 - \omega + 1)\overline{(\omega^5 - \omega + 1)} = \omega^6 + \omega^5 - \omega^4 + \omega^3 - \omega^2 - \omega + 3$$

$$= (\omega^5 - \omega^4 + \omega^3 - \omega^2 + \omega - 1) + \omega^5 - \omega^4 + \omega^3 - \omega^2 - \omega + 3$$

$$= 2\omega^5 - 2\omega^4 + 2\omega^3 - 2\omega^2 + 2 = 2(\omega^5 - \omega^4 + \omega^3 - \omega^2 + 1).$$

and

$$(\omega - 1)\overline{(\omega - 1)} = \omega^6 - \omega + 2 = (\omega^5 - \omega^4 + \omega^3 - \omega^2 + \omega - 1) - \omega + 2 = \omega^5 - \omega^4 + \omega^3 - \omega^2 + 1.$$

Therefore, we have $CD^2 = \dfrac{2(\omega^5 - \omega^4 + \omega^3 - \omega^2 + 1)}{\omega^5 - \omega^4 + \omega^3 - \omega^2 + 1} = 2$, which means $CD = \boxed{\sqrt{2}}$.